Selective Oxidation by Heterogeneous Catalysis

FUNDAMENTAL AND APPLIED CATALYSIS

Series Editors: M. V. Twigg

Johnson Matthey
Catalytic Systems Division
Royston, Hertfordshire, United Kingdom

M. S. Spencer

Department of Chemistry
Cardiff University
Cardiff, United Kingdom

CATALYST CHARACTERIZATION: Physical Techniques for Solid Materials
Edited by Boris Imelik and Jacques C. Vedrine

CATALYTIC AMMONIA SYNTHESIS: Fundamentals and Practice
Edited by J. R. Jennings

CHEMICAL KINETICS AND CATALYSIS
R. A. van Santen and J. W. Niemantsverdriet

DYNAMIC PROCESSES ON SOLID SURFACES
Edited by Kenzi Tamaru

ELEMENTARY PHYSICOCHEMICAL PROCESSES ON SOLID SURFACES
V. P. Zhdanov

PRINCIPLES OF CATALYST DEVELOPMENT
James T. Richardson

SELECTIVE OXIDATION BY HETEROGENEOUS CATALYSIS
Gabriele Centi, Fabrizio Cavani, and Ferruccio Trifirò

A Continuation Order Plan is available for this series. A continuation order will bring delivery of each new volume immediately upon publication. Volumes are billed only upon actual shipment. For further information please contact the publisher.

Selective Oxidation by Heterogeneous Catalysis

Gabriele Centi

University of Messina
Messina, Italy

Fabrizio Cavani

and

Ferruccio Trifirò

University of Bologna
Bologna, Italy

Kluwer Academic / Plenum Publishers
New York, Boston, Dordrecht, London, Moscow

Chemistry Library

Library of Congress Cataloging-in-Publication Data

Centi, G. (Gabriele), 1955–
    Selective oxidation by heterogeneous catalysis/Gabriele Centi, Fabrizio Cavani,
Ferruccio Trifirò.
        p.   cm. — (Fundamental and applied catalysis)
    Includes bibliographic references and index.
    ISBN 0-306-46265-6
    1. Oxidation. 2. Heterogeneous catalysis. I. Cavani, Fabrizio. II. Trifirò, Ferruccio. III.
Title. IV. Series.

TP156.O9 C46 2000
660'.28443—dc21

                                                    99-055323

ISBN 0-306-46265-6

©2001 Kluwer Academic/Plenum Publishers, New York
233 Spring Street, New York, New York 10013

http://www.wkap.nl

10 9 8 7 6 5 4 3 2 1

A C.I.P. record for this book is available from the Library of Congress

Printed in the United States of America ℒℒ ✓

FOREWORD to the
Fundamental and Applied Catalysis Series

Catalysis is important academically and industrially. It plays an essential role in the manufacture of a wide range of products, from gasoline and plastics to fertilizers and herbicides, which would otherwise be unobtainable or prohibitively expensive. There are few chemical- or oil-based material items in modern society that do not depend in some way on a catalytic stage in their manufacture. Apart from manufacturing processes, catalysis is finding other important and ever-increasing uses; for example, successful applications of catalysis in the control of pollution and its use in environmental control are certain to increase in the future.

The commercial importance of catalysis and the diverse intellectual challenges of catalytic phenomena have stimulated study by a broad spectrum of scientists, including chemists, physicists, chemical engineers, and material scientists. Increasing research activity over the years has brought deeper levels of understanding, and these have been associated with a continually growing amount of published material. As recently as sixty years ago, Rideal and Taylor could still treat the subject comprehensively in a single volume, but by the 1950s Emmett required six volumes, and no conventional multivolume text could now cover the whole of catalysis in any depth. In view of this situation, we felt there was a need for a collection of monographs, each one of which would deal at an advanced level with a selected topic, so as to build a catalysis reference library. This is the aim of the present series, *Fundamental and Applied Catalysis*. Some books in the series deal with particular techniques used in the study of catalysts and catalysis: these cover the scientific basis of the technique, details of its practical applications, and examples of its usefulness. An industrial process or a class of catalysts forms the basis of other books, with information on the fundamental science of the topic, the use of the process or catalysts, and engineering aspects. Single topics in catalysis are also treated in the series, with books giving the theory of the underlying science

v

and relating it to catalytic practice. We believe that this approach provides a collection that is of value to both academic and industrial workers. The series editors welcome comments on the series and suggestions of topics for future volumes.

Martyn Twigg
Michael Spencer

Royston and Cardiff

PREFACE

Selective catalytic oxidation is one of the major areas of industrial petrochemical production. Roughly one-quarter of the principal organic chemicals are synthesized using catalytic oxidation processes in either the gas or the liquid phase. The worldwide market corresponds to about $50 billion, but it assumes even greater importance if one takes into account the fact that the products made through oxidation mechanisms are usually key intermediates for subsequent processes (e.g., monomers, comonomers, or modifiers for polymers).

This book surveys recent developments in the field of selective oxidation. These developments, driven by the need for alternative, less expensive reagents and the mandate to reduce the environmental impact of chemical production processes, now necessitate a complete reanalysis of this area of research. Although providing an updated discussion of the state of the art especially in the principal new areas of selective oxidation (e.g., alkane selective transformation) is one of our objectives, it is not the central focus of the volume. We have concentrated instead on outlining open questions and new opportunities, as well as offering some guidelines for future research and useful tools for the interpretation and analysis of data and the design of new catalysts and reactions.

There are two key aspects to our treatment of the subject: *concept by example* and *outlook for the future*. The first is the approach used throughout the book, which is not to provide a more or less critical overview of the literature, but rather to select some representative examples, for which in-depth analyses make it possible to clarify the basic but new concepts required for a better understanding of the new opportunities in the field and the design of new catalysts or catalytic reactions. Attention is focused not only on the catalyst itself but also on the use of the catalyst within the process, thus illustrating the relationship between catalyst design and process engineering. The second aspect refers to identifying new directions of both basic and industrial research that can lead to new breakthroughs in the field and taking a fresh look at long-established ideas and concepts to ascertain their limitations, and by so doing to delineate guidelines for future research.

While this book is not just for specialists, it does assume a basic knowledge of homogeneous and heterogeneous catalysis. The discussion is aimed at both graduate students and industrial scientists already working in catalysis who would like to have a concise overview of the new possibilities in this field. Specialists in selective oxidation will also find suggestions for novel research projects and ways to reconsider their own work from a different perspective.

Chapter 1 includes an introduction to the field of selective oxidation and the major basic and technological directions of research, and discusses the forces that drive innovation. Chapter 2 deals with developing technological and industrial opportunities in the field with emphasis on new options in terms of alternative raw materials, reactor technologies, and the use of oxygen instead of air. Chapter 3 continues the discussion, analyzing problematic aspects using more specific examples, and reviews the most recent developments.

Chapter 4 provides a detailed discussion of two specific examples of industrial alkane oxidation processes: (i) the synthesis of maleic anhydride from *n*-butane, and (ii) the ammoxidation of propane to acrylonitrile. The first is the only existing commercial process, and the second is at the pilot plant stage, but its outlook for industrialization is good. The emphasis in this case is on an integrated view of how to control the surface reactivity. The various factors that determine the surface selectivity and the specificity of the catalyst for this reaction are discussed as are alternative process options and patented catalytic systems.

Chapter 5 discusses the problem of controlling the reactivity of solid catalysts in the new alkane oxidation reactions and what still remains to be done before these catalysts can be used in an industrial setting. In particular, the oxidative dehydrogenation of alkanes and new reactions of (amm)oxidation of C2–C6 alkanes are analyzed, with discussion focused especially on the question of how to address the problems of catalyst design in the context of the new research directions and trends.

Chapter 6 deals with applications for solid catalysts, which can be thought of as being on the borderline between homogeneous and heterogeneous catalysis. Three cases are discussed: (i) selective oxidation in the liquid phase with solid micro- or mesoporous material, (ii) heteropoly compounds as molecular-type catalysts, and (iii) solid Wacker-type catalysts. Chapter 7 emphasizes some new concepts and possible strategies in selective oxidation in order to offer suggestions on new and unconventional directions for research.

Chapter 8 is devoted to a discussion of reaction mechanisms in selective oxidation and structure–activity relationships. Rather than being an analysis of the state of the art in this field, the presentation focuses on developing a general "philosophy" in the approach. "Established" published reaction models are discussed to determine whether they can really be considered fundamental for selective oxidation or should instead be thought of as special cases of more general phenomena. The discussion of this mode of approach and its limitations allows some general guidelines to be delineated for future research on the basic aspects of selective transformations at oxide surfaces.

The general scope of the book is thus to offer suggestions for new and innovative directions of research as well as indications on how to reconsider selective oxidation from different perspectives. We believe, in fact, that selective oxidation is not a mature field of research and that important breakthroughs can be derived through basic and applied research, but that these advances will only come about through less conventional approaches in terms of both catalyst design and analysis of the data.

Gabriele Centi
Fabrizio Cavani
Ferruccio Trifirò

Messina and Bologna

CONTENTS

CHAPTER 3. NEW TECHNOLOGICAL AND INDUSTRIAL OPPORTUNITIES: EXAMPLES

CHAPTER 4. CONTROL OF THE SURFACE REACTIVITY OF SOLID CATALYSTS: INDUSTRIAL PROCESSES OF ALKANE OXIDATION

CHAPTER 7.   NEW CONCEPTS AND NEW STRATEGIES IN SELECTIVE OXIDATION

CHAPTER 8. NEW ASPECTS OF THE MECHANISMS OF SELECTIVE OXIDATION AND STRUCTURE/ACTIVITY RELATIONSHIPS

TRENDS AND OUTLOOK IN SELECTIVE OXIDATION

An Introduction

1.1. INTRODUCTION

A large segment of the modern chemical industry is based on catalytic selective oxidation processes.[1-10] Indeed, more than 60% of the chemicals and intermediates synthesized via catalytic processes are products of oxidation. Rough estimates place the worth of world products that have undergone a catalytic oxidation step at $20 to $40 billion.[4] Figure 1.1 shows the global share of selective oxidation processes with respect to the total organic chemical production in 1991. If one considers catalysts alone, the value of oxidation catalysts produced commercially in the United States was $105 million (in 1990), second only to polymers. Total catalytic oxidation is also becoming increasingly important as a method for destroying trace pollutants and contaminants in gaseous streams.[11,12]

One of the most important applications of selective oxidation catalysis is the functionalization of hydrocarbons. Today, catalytic oxidation is the basis for the synthesis of a large percentage of the monomers or modifiers used for the production of fibers and plastic. The major industrial applications of catalytic oxidation for the synthesis of monomers are listed in Table 1.1.[13]

The data in Table 1.1 indicate some important limitations in catalytic oxidation, which can be summarized as follows:

1. Because of the formation of undesired by-products, none of the reactions runs at maximum selectivity, and few reactions attain total or close-to-total conversion.
2. Processes can generate coproducts that are not always of economic interest.
3. Some raw materials and products are suspected or proven carcinogens.
4. Some processes require expensive oxidizing agents.

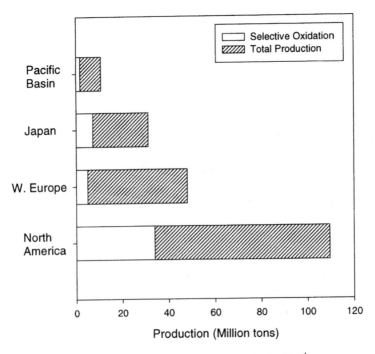

Figure 1.1. World organic chemical production in 1991.[4]

The use of stoichiometric oxidants such as permanganates or chromium salts is a common practice in fine chemicals production, and catalytic oxidation is always an environmentally more acceptable alternative to these stoichiometric oxidations. Molecular oxygen is, of course, the most desirable oxidant. However, hydrogen peroxide is also relatively inexpensive, has a high active oxygen content, and is environmentally friendly because the coproduct is water. Table 1.2 characterizes the commonly used oxidants.[14]

In today's petrochemical industry, most of the available technologies are considered "mature," and even considerable research effort is not expected to yield significant improvements. However, as profit margins can sometimes be increased by improving selectivities by one percentage point or less, it is still desirable to invest in improving catalyst performance. Moreover, the polymer industry continues to demand monomers that are not only less expensive but also increasingly pure.

Technological innovations in catalytic oxidation processes are thus expected from:

1. Introducing substantial modifications in technology: New raw materials and catalysts are being used, and new process technologies, new machinery, and new equipment are being developed.

TABLE 1.1. Major Catalytic Oxidation Process for the Synthesis of Monomers[13]

Reactants	Products/coproducts	Heterogeneous/homogeneous	Conversion, %[a]	Selectivity, %
Methanol/air	Formaldehyde	Heterogeneous	99	94
n-Butane/air	Maleic anhydride	Heterogeneous	70–80	65–70
Benzene/air	Maleic anhydride	Heterogeneous	98	75
o-Xylene/air	Phthalic anhydride	Heterogeneous	100	79
Naphthalene/air	Phthalic anhydride	Heterogeneous	100	84
Propene/air/ammonia	Acrylonitrile	Heterogeneous	97–99	73–82
Propene/air	Acrolein	Heterogeneous	>90	80–85
Acrolein/air	Acrylic acid	Heterogeneous	>95	90–95
Isobutene/air	Methacrolein	Heterogeneous	>97	85–90
Methacrolein/air	Methacrylic acid	Heterogeneous	70–75	80–90
Ethene/O_2	Ethene oxide	Heterogeneous	90	80
Ethene/air/HCl	1,2-Dichloroethane	Heterogeneous	>95	>95
Ethene/O_2/HCl	1,2-Dichloroethane	Heterogeneous	>95[b]	>97
Propene/hydroperoxide	Epoxide/Alcohol	Homogeneous	10	90
p-Xylene/air	Terephthalic acid	Homogeneous	90	85–95
Cumene/air	Phenol/acetone	Homogeneous	35–40	90
Cyclohexane/air	Cyclohexanone	Homogeneous	5–15	85–90
Cyclohexanone/HNO_3	Adipic acid	Homogeneous	>95	92–96
Cyclohexanone/NH_2OH/H_2SO_4	Cyclohexanone oxime, $(NH_4)_2SO_4$	Homogeneous	100	>98
Ethene/acetic acid/air	Vinyl acetate	Homogeneous	20–35	98–99

[a]Conversion per pass on a hydrocarbon basis.
[b]When the ethene/O_2/HCl ratio is close to stoichiometry; otherwise, the conversion per pass depends on the ethene excess.

2. Fine-tuning existing processes: Work continues on optimizing all the process components, eliminating bottlenecks, and revamping existing plants.

3. Complying with new requirements and regulations: Processes are being developed that have a less severe environmental impact, which translates into lower pollutant emissions, less waste, and greater safety. There is also an increasing demand for purer products for the production of high-quality polymers.

The release of carbon dioxide is an environmental issue that is expected to have a considerable effect on the development of future catalytic oxidation processes. In some cases, combustion of a hydrocarbon, which is more exothermic than selective conversion, helps the overall process economics in that it generates high-value steam. Nevertheless, limitations on carbon dioxide emissions imposed by future regulations could provide additional impetus for improving process selectivity.

Generally speaking, the evolution of almost all technological fields follows a particular pattern. After the publication of a breakthrough result, there is an exponential increase in research and knowledge, which then results in the technological application of the discovery followed by gradually increasing research effort aimed at consolidation of the results and analysis of further possible fields of application. Most of the selective oxidation industrial processes are in a late stage of this evolution.

A superficial analysis might suggest that there are no major driving industrial forces stimulating research in this area and that activity is thus focused mainly on improving basic knowledge or tuning existing catalytic oxidation processes. This, however, is not true because several new challenges—broadly distinguished as problem-driven and curiosity-driven challenges—are stimulating research in this field. To the first group belongs research focused on:

- Using new, less costly or more environmentally acceptable, raw materials.
- Reducing waste or eliminating cogeneration of by-products.
- Improving product quality.
- Reducing complexity and number of stages of the processes.
- Reducing energy use or losses.
- Reducing manpower requirements or required handling of hazardous products.
- Creating multifunctional processes (to produce different kinds of products in the same continuous plant, but for only a fraction of the year for each product).
- Reducing risk and environmental impact.

TABLE 1.2. Oxygen Donor Oxidants[14]

Donor oxidant	Active oxygen, %	By-product	Comments
O_2	100	None	No by-product is generated if reducing agent-free, nonradical chain aerobic oxygenation can be achieved
H_2O_2	47	H_2O	Environmentally attractive
O_3	33	O_2	Potentially environmentally attractive
NaClO	21.6	NaCl	Inorganic salt by-products are to be avoided in general; in some cases, ClO^- can produce toxic and carcinogenic chlorocarbons by-products
tert-BuOOH	17.8	tert-BuOH	Commercially important in catalyzed oxygenations
$C_5H_{11}NO_2$[a]	13.7	$C_5H_{11}NO$	
$KHSO_5$	10.5	$KHSO_4$	Water-compatible, but generates a marginally toxic salt
C_6H_5IO	7.3	C_6H_5I	Metal-catalyzed oxidations are often quite selective, but the cost is prohibitive

[a]N-methylmorpholine N-oxide (MMNO).

To the second group of challenges belongs research aimed at developing:

- New ways to generate active oxygen species.
- New low-temperature processes to oxidize complex molecules.
- New kinds of superselective catalytic materials.
- New types of reactor options that allow the use of new classes of catalytic materials.
- Combined homogeneous and heterogeneous oxidation catalysts.

There are thus several challenges stimulating an intense research effort in the field of selective oxidation, a very dynamic area for future activity. The general trends and outlook in selective oxidation are dealt with in this chapter as an introduction to the more specific discussions in the rest of the book.

1.2. TECHNOLOGICAL AND INDUSTRIAL DEVELOPMENTS

The recent trends and developments in industrial selective oxidation technology can be summarized as follows:

1. The use of new raw materials and alternative oxidizing agents: Alkanes are increasingly replacing aromatics and alkenes as raw materials. In liquid phase oxidations, hydrogen peroxide is being used more frequently as an oxidizing agent in place of traditional oxygen transfer agents. In some cases, the hydrogen peroxide is generated *in situ*.
2. The development of new catalytic systems and processes: Heterogeneous rather than homogeneous catalysts are being used, and oxidative dehydrogenation is gradually replacing simple dehydrogenation. New processes are being developed that generate fewer or no undesired coproducts.
3. The conversion of air-based to oxygen-based processes in gas phase oxidation to reduce polluting emissions.
4. The fine-tuning of existing processes to improve each stage of the overall process: Profit margins are being increased through changes in process engineering rather than through economies of scale.

As most innovations in catalytic oxidation either result from the use of new raw materials or are environmentally driven, the first two points noted above can be considered primarily technological developments, whereas the last two are primarily modifications or improvements of existing processes.

1.2.1. New Raw Materials

The petrochemical industry is based in large part on the conversion of alkenes because of the ease and economy with which they can be obtained from petroleum, and as they are easily functionalized, they are versatile raw materials. However, the industry is moving toward the direct use of alkanes, which can be obtained from both petroleum and natural gas and are even more economical.

The primary existing and under-development alkane selective oxidation processes are: (i) methane to formaldehyde or ethene, (ii) methane to vinyl chloride in the presence of HCl, (iii) ethane to 1,2-dichloroethane in the presence of HCl, (iv) ethane to acetic acid, (v) propane to acrylic acid, (vi) propane to acrylonitrile in the presence of ammonia, (vii) *n*-butane to maleic anhydride, (viii) isobutane to methacrylic acid, (ix) *n*-pentane to phthalic anhydride, (x) isobutane to *tert*-butyl alcohol, and (xi) cyclohexane to cyclohexanone and cyclohexanol. Some of these alkanes—e.g., *n*-butane and *n*-pentane—will be more available in the near future, as environmental regulations impose increasingly stringent limits on the light-alkane content of gasoline. In addition, the increasing demand for chemicals such as methyl *tert*-butyl ether (MTBE) favors the development of new synthetic routes. For example, the conversion of *n*-butane to isobutane and the conversion of isobutane to *tert*-butyl alcohol by direct monoxygenation might be of considerable industrial interest (*tert*-butyl alcohol is an intermediate in MTBE synthesis).

Of the processes listed above, only the oxidation of *n*-butane to maleic anhydride has been commercialized. Butane is a particularly attractive material for this process because of its low cost, ready availability, and low toxicity. The fact of this commercialization refutes the commonly held notion that alkanes cannot be used as raw materials because of their low reactivity, which generally translates into low selectivity. The use of butenes as the raw material in maleic anhydride synthesis was abandoned because of the many by-products, whereas *n*-butane oxidation is a very clean selective oxidation (only traces of acetic acid as a by-product, apart from carbon oxides). In Chapter 4 *n*-butane oxidation to maleic anhydride will be discussed in more detail, while the mechanistic aspects of this reaction and the reasons for the differences in the catalytic behavior of butenes and *n*-butane will be dealt with in Chapter 8.

The current situation in regard to the other processes is as follows:

1. Enormous efforts have been devoted to the study of the oxydimerization of methane to ethene (oxidative coupling), as indicated by the more than 500 publications and patents issued in this area in the last few years. However, the goal of a 30% C2 (ethane and ethene) yield, which makes the process economically advantageous compared to steam cracking, has not yet been reached. The best reported yields are close to 24–25%.

2. The oxychlorination of ethane to 1,2-dichloroethane has been studied for many years, but commercialization has probably been hindered by the low purity of the product obtained.

3. The ammoxidation of propane to acrylonitrile is very close to commercialization and a number of different processes have been proposed. British Petroleum (BP) has announced the imminent commercialization of a process that uses V/Sb/Al/O catalysts. Although the yield is lower than that achieved with propene (40 vs. 75%), process economics are driven primarily by the difference in price between propane and propene. This process will be discussed in more detail in Chapter 4.

4. In the selective oxidation of *n*-pentane to phthalic anhydride, the continuing challenge is high selectivity. The reaction is discussed in more detail in Chapter 5.

5. The gas phase oxidation of isobutane over heteropolycompound catalysts gives methacrolein and methacrylic acid. The major obstacle to the commercialization of this process is the structural stability of the catalyst and the low yield achieved. Under the reaction conditions, heteropoly compound catalysts tend to progressively decompose, which leads to performance degradation. The conventional process for methyl methacrylate synthesis uses hazardous raw materials, such as hydrogen cyanide, and produces large quantities of ammonium sulfate. This coproduct must be pyrolized or disposed of as waste material. Thus there is a strong incentive for the development of an alternative process for methyl methacrylate synthesis. This reaction will also be discussed in more detail in Chapter 5.

Any discussion of the raw materials in catalytic oxidation must relate to alternative oxidizing agents as well as to alternative hydrocarbons. In gas phase oxidation, the oxidant is always molecular oxygen, although recent developments also indicate that in special cases it may be possible to use other oxidants such as nitrous oxide (for synthesis of phenol from benzene, see Chapter 7). In liquid phase oxidation, the problem is more complex. Molecular oxygen is the oxygen transfer agent in very few liquid phase oxidations; e.g., in alkylaromatic oxidation, as in many other cases, the reaction mechanism is a free-radical autoxidation that is initiated by metal-catalyzed oxidation of the hydrocarbon to a hydroperoxide. However, this reaction using molecular oxygen requires severe reaction conditions (high temperature, special solvents, corrosive cocatalysts) and can only be applied successfully for certain kinds of side-chain oxidations of alkylaromatics, usually to obtain acids that can be easily recovered from the solution, e.g., by crystallization. In liquid phase oxidation, and especially in fine chemicals production, catalytic oxidation is being explored as an alternative to the widely used stoichiometric oxidation.

The nature of the oxidizing agent must also be considered. Of the agents listed in Table 1.2, the most desirable is hydrogen peroxide. In liquid phase oxidation, hydrogen peroxide can be handled more easily and safely than molecular oxygen. The economic advantages of hydrogen peroxide versus molecular oxygen are well demonstrated in the industrial production of hydroxybenzenes. Since the worldwide production of the three dihydroxybenzene isomers is only about 100,000 tons/year, these isomers are classified as fine chemicals. At present, dihydroxybenzenes are usually obtained by the autoxidation of p-diisopropylbenzene and m-diisopropylbenzene using molecular oxygen as the initial oxidizing agent. However, the direct hydroxylation of phenol with hydrogen peroxide using a titanium silicalite (TS-1) catalyst is a successful new option. The same catalyst can be used for a variety of other selective oxidations of industrial interest such as propene to propene oxide, but in this case the process cannot be commercialized at present because of the high cost of hydrogen peroxide with respect to the final product value.

Challenges in this area are thus to develop processes for hydrogen peroxide production that are less expensive, *in situ* or *ex situ*, but integrated in the selective oxidation process. *In situ* generation of hydrogen peroxide in the final oxidation medium, as in alkene epoxidation and alkane hydroxylation, has certain advantages, not the least of which is the elimination of expensive purification units for the hydrogen peroxide.[15,16] *In situ* generation can be achieved through the use of appropriate redox couples (such as alkyl anthraquinone/alkyl anthrahydroquinone) and a catalyst (such as titanium silicalite) for substrate oxidation.

Even more interesting is the *in situ* generation of hydrogen peroxide from mixtures of hydrogen and oxygen. In this case, the catalyst has a metal component that can generate hydrogen peroxide and a component that can cause substrate oxidation with the hydrogen peroxide generated. Tatsumi *et al.*[17] reported a palladium-containing titanium silicalite catalyst for the hydroxylation of benzene and hexane, and more recently Hölderich *et al.*[18] used a similar catalyst and system for propene oxide synthesis from propene. Kuznetsova *et al.*[19,20] used supported platinum catalysts, and Tabushi *et al.*[21] used platinum with metalloporphyrins for alkene epoxidation with a hydrogen/oxygen mixture. In the latter case, molecular hydrogen acts as the required reductant when molecular oxygen reacts with alkenes in metalloporphyrin-based systems (typical coreductants are $NaBH_4$ and Zn/CH_3COOH).[22] More recently, Neumann and Levin[23] used a metal-substituted heteropoly compound as the catalyst for cyclohexene epoxidation with molecular oxygen. The reductant was Pt/H_2.

The *in situ* generation of hydrogen peroxide from hydrogen and oxygen may well represent an inexpensive and versatile method for liquid phase oxidation of organic substrates, but several scientific and technical problems must be solved before it can actually be applied. *In situ* generation of hydrogen peroxide is discussed in more detail in Chapter 7.

1.2.2. Conversion of Air-Based to Oxygen-Based Processes

The conversion of air-based to oxygen-based processes is an interesting example of how it is possible not only to improve performance and process economics, but also to reduce the environmental impact of the production. Three examples of this conversion can be considered: (i) synthesis of formaldehyde from methanol, (ii) ethene epoxidation to ethene oxide, and (iii) oxychlorination of ethene to 1,2-dichloroethane.

Any evaluation of the advantages and disadvantages of the use of oxygen versus air must take into account the need for additional capital investment, the relative operating costs, safety issues, and, above all, the need to comply with existing legislation regarding gaseous emissions.

In the past, the decision to convert air-based processes was dictated primarily by economic considerations, such as a general improvement in productivity and/or selectivity for the desired product. In recent years, however, restrictions on pollutant emissions have become a driving factor. Indeed, in some cases, the extremely low levels of allowable pollutants make it necessary to recycle reactor outlet streams. When such streams are recycled, it is important that no inert materials, especially nitrogen, are present so that the amount of vent gas is minimized and the pollutant is concentrated as much as possible for combustion.

An important example of a process that can benefit from using oxygen in place of air is oxychlorination of ethene in the production of 1,2-dichloroethane, an intermediate in the synthesis of vinyl chloride (monomer for PVC plastic). Pollutant concentrations in the air-based oxychlorination process for a plant with a production capacity of 300,000 tons/year of vinyl chloride and a vent stream flow of 18,000–28,000 m^3/h are summarized in Table 1.3.[24]

Thermal combustion of the dilute stream from this air-based process would require a significant amount of energy and a substantial investment in equipment. Although catalytic combustion of chlorinated hydrocarbons is possible,[24,25] the cost of the technology to achieve complete conversion, stable behavior, and the elimination of even more toxic components (such as dioxins and PCB) at sub-ppb levels is not competitive with off-gas recycling in the oxygen-based process. Introducing

TABLE 1.3. Pollutant Concentration in
Oxychlorination Off-Gas[24]

Component	Concentration[a]
Oxygen	4–7 vol %
Carbon monoxide	0.4–0.7 vol %
Ethene	0.2–0.9 vol %
Ethane	0.03–0.15 vol %
Methyl chloride	2–30 vpm
Vinyl chloride	15–60 vpm
Ethyl chloride	30–400 vpm
1,2-dichloroethane	5–100 vpm
Hydrocarbon absorbant	50–250 vpm

[a]Depends on the quality of the oxychlorination catalyst.

recycling would reduce the gaseous emissions by more than 95% by volume and the resulting emissions would be much more concentrated in combustible compounds, thus significantly reducing the cost of combustion. A look at emissions combustion costs shows that the operating and utilities costs for the oxygen-based process are 10% less than the equivalent costs for the air-based process.[26] The difference constitutes a significant portion of the overall production cost of vinyl chloride and more than compensates for the higher cost of oxygen. Further, as shown in Table 1.1, it is possible to achieve better performance with the oxygen-based process.

Further aspects of the advantages of using pure oxygen instead of air in heterogeneous selective oxidation processes will be discussed in Chapter 2.

1.2.3. Fine-Tuning Existing Oxidation Processes

In many cases, considerable effort has been directed toward improving performance in all stages of a system, from pretreatment of the reagents to storage of the product. Such research is generally directed toward maximizing operating flexibility, increasing efficiency in the use of raw materials and utilities, and minimizing waste. Catalysts have been improved, computer simulations have been developed to optimize parameters, and catalyst beds have been redesigned using a single bed with a catalyst of varying activity (structured reactors). Almost all oxidation plants have been renovated along these lines, and it is not unusual for actual productivities to be two to three times or more higher than the plant's original productivity. Thus, although innovation requires stepwise improvements (in the sense of a radical change in production), much industrial research effort is still being devoted to incremental improvements, which requires thorough knowledge of the basics of the catalytic reaction and catalyst/reactor system.

An interesting illustration of this point is found in the case of methanol oxidation to formaldehyde. Until a few years ago, methanol was oxidized through an air-based process that used methanol in concentrations of less than 6% because of the flammability of the air/methanol mixture. Any increase in the productivity of existing plants would require an increase in the concentration of methanol in the feedstock. Figure 1.2 shows the various possible operating conditions. Zone 1 corresponds to operation with less than stoichiometric amounts of oxygen, where complete conversion of methanol is not achieved. Zone 2 is the flammability zone for the methanol/oxygen mixture. Zone 3 corresponds to operation with oxygen-enriched air. Zone 4 represents the optimum operating conditions, where total conversion is achieved without problems of flammability, as well as the lowest operating costs.

Under the conditions of Zone 4, the maximum possible concentration of methanol in the feedstock is 10–12%, which doubles productivity. This concentration requires an oxygen content no higher than 10%. Both of these conditions can

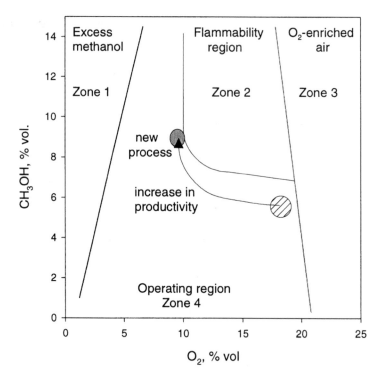

Figure 1.2. Operative regions for methanol oxidation to formaldehyde and change of process conditions from old to new options.[13]

be achieved by recycling the reactor outlet stream, which is rich in nitrogen. Recycling also considerably reduces problems related to the emission of toxic substances, such as formaldehyde. However, the increased concentration of methanol leads to an increase in the heat developed in the first part of the catalytic bed, which can cause local catalyst overheating and rapid catalyst deactivation. One possible solution is to charge the first part of the catalytic bed with a less active catalyst. As shown in Figure 1.3, the longitudinal temperature profile of the reactor used in this latter approach is characterized by two maxima corresponding to the two zones containing catalysts of different reactivity, and the heat of reaction is more uniformly distributed along the catalyst bed.

In general, the amount of unconverted methanol is 1–2% of the methanol in the feedstock, and because it is generally not economically feasible to separate this fraction, methanol is allowed to accumulate in the aqueous formaldehyde products. However, as many polymerization processes require high-purity formaldehyde, adding a small adiabatic step to the oxidation reactor to increase the methanol

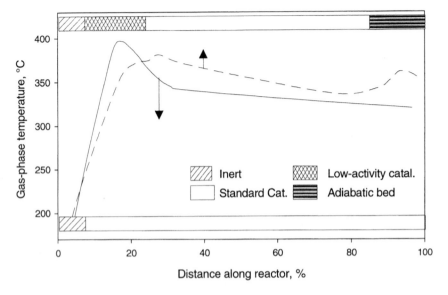

Figure 1.3. Axial temperature profile in reactors during the synthesis of formaldehyde at low (solid line) and high (broken line) methanol concentration. The bars at the top and bottom indicate the relative structure of the reactor catalytic bed.[13]

conversion and thereby increase the purity of the formaldehyde might be worth-while.

In recent years, changes of the type discussed above have led to major redesigning of the methanol oxidation process, even though the conversion and selectivity of the original process were sufficiently high that significant improvements in performance did not seem possible (Table 1.1).

1.2.4. Reducing the Number of Process Steps

Reducing the number of steps in a chemical transformation is advantageous not only in terms of reduced capital cost, but especially to: (i) avoid storage of possibly dangerous products, (ii) reduce the risks and the cost of environmental control in complex plants, and (iii) increase process flexibility. Sometimes, simplification of a process just involves eliminating a purification step, so that the product of one reactor goes straight into another without isolation of an intermediate. This option is discussed in more detail in Chapter 5 in regard to alternative possibilities in light alkane selective oxidation.

Table 1.4 lists potential targets for new selective oxidation processes[27] for which there are fewer steps and reduced process complexity compared to existing commercial technology. In some cases, the number of steps is reduced by eliminat-

TABLE 1.4. Potential Targets for New Selective Oxidation Processes

Target product	Possible reactant(s)	Competing technology
Methanol	Methane	Two-step, via CH_4 steam-reforming
Formaldehyde	Methane	Three-step, via methanol
Ethene	Methane, ethane	Cracking of naphtha
Propene oxide	Propene	Coproduct with *tert*-butyl alcohol or styrene
Phenol	Benzene	Coproduct with acetone
Acetic acid	Ethane	Multistep, from methanol
1,4-Butanediol, Furan	*n*-Butane, butadiene	Reppe acetylene chemistry
Tetrahydrofuran	*n*-Butane, butadiene	Multistep[a]
Ethylene glycol	Ethene	Hydration of ethene oxide
Adipic acid	Cyclohexane	Multistep
Styrene	Ethylbenzene	Coproduct with propene oxide
Methyl methacrylate	Isobutane	Two-step, via methacrolein
Methyl formate	Methane	Three-step, via methanol
Acrylic acid	Propane	Two-step, via acrolein

[a]A recent development is the synthesis of tetrahydrofuran from *n*-butane via maleic anhydride intermediate synthesis.

ing coproducts. Benzene hydroxylation to phenol using nitrous oxide and propene epoxidation to propene oxide using hydrogen peroxide are already at a quasi-commercialization level, as discussed in Chapter 7. Other examples are still being researched and some of them will be discussed in more detail later, particularly in Chapter 5.

1.3. NEW OPPORTUNITIES DERIVED FROM BASIC RESEARCH

Catalytic science has been a driving force for innovation in selective oxidation, both in terms of providing a better understanding of surface catalytic phenomena (a key aspect for the design of new, improved catalysts) and of developing new reactor and engineering options for catalytic transformations.[28] In addition, the development of new catalytic materials (see Chapter 2) and new ways to activate oxygen or hydrocarbons (see Chapters 7 and 8) are continuing challenges which provide the basis for industrial innovation.

The main contribution of basic research to industrial application of selective oxidation reactions derives from the change in catalyst preparation from an "art" to a "science." An interesting example regarding this point is the modification of base catalysts with small amounts of doping agents, a basic step in the preparation of almost all industrial catalysts, but one which has often been based mainly on empirical considerations.

Doping has been used as a way to: (i) add or tune catalyst functionalities (e.g., enhance the rate of catalyst reoxidation), (ii) control the valence state of the active element, and (iii) moderate catalyst acido-base properties. However, there is often very little information on the details of the effective action(s) of the doping agents and thus reliable data on how to control the doping action and how to use the same effect with different catalyst substrates are lacking. Understanding the role of dopants in selective oxidation catalysts is thus a primary research objective and will lead to substantial future progress.

Supports are also commonly used in industrial practice to impart the necessary physicomechanical properties to the active phase, but often are considered only in the role of inerts or as dispersants of the active components. However, evidence not only of the reactivity of the supports with the active component(s), but also of the possibility of tuning their characteristics by epitaxial or constraint effects is increasing. One area of possible future development spurred by basic research is the search for ways to modify structural and reactivity characteristics of active oxide phases by adding specific components that (i) stabilize or induce distortions by epitaxial effects in the active phase, or (ii) enhance the formation of intergrown phases when the active centers are located at the boundary between the intergrown phases. Similarly, it is possible to extend the concept by creating the active oxide phase within ordered microporous matrices such as zeolites, which orient the growth of the oxide, thus creating, e.g., tunable distortions, defects, or chain-type species, not achievable (or not stable) in the bulk oxide, even when synthesized at the level of nanoparticles.

Another basic area of investigation is the identification of the *in operandi* catalyst state, i.e., the state the catalyst is in during the catalytic cycle. Oxidation catalysts generally operate dynamically through redox cycles, which often involve not only the active centers but extend over the whole catalyst. The catalyst thus undergoes extensive surface changes in a cyclical fashion, and its surface is a site of continuous movement and transformation of adsorbed species. This suggests that understanding and control of these factors is fundamental for the development of new selective oxidation catalysts. These aspects will be discussed in detail in Chapter 8.

To date little attention has been directed toward identification of the role of the gas phase composition and how this influences catalyst selection and optimization. For example, in several processes steam is injected together with the hydrocarbon and oxygen, but there are very few published studies concerning its effect. Water can have multiple effects: it can influence catalyst characteristics, inhibit homogeneous gas phase reactions, and enhance the rate of product desorption from the catalyst surface. The feed composition does not concern only steam. Industrially, additives are often continuously added to the feed to improve performance or reduce deactivation. However, specific gas phase chemicals may also be added to tune surface properties, e.g., to kill some unselective sites or even to add catalyst functionalities (e.g., aldehydes on the surface of oxides may have a role in generating surface peroxo species). This area of research, discussed in more detail in Chapter 8, might be a very fruitful one for basic studies with practical implications for the control of catalytic reactivity.

The recent advances in selective oxidation have not come about only through the development of new concepts or materials in catalytic science, but also derive from a stricter integration between reactor engineering and catalytic science. Some of the most interesting recent ones, for instance, in the field of selective functionalization of alkanes, have resulted from the use of new reactors, not traditionally employed in catalytic oxidation. This topic will be discussed in detail in Chapter 3.

It is also worth noting that these new reactor options imply the necessity to completely revise catalytic oxidation science. For example, it is possible to separate the reduction and oxidation steps (DuPont riser reactor, see Chapters 2 and 4), which means that new classes of catalysts (nonselective under the usual conditions) may become interesting. New reactor technologies also have a potential for unusual reaction conditions. With the use of a monolith-type reactor, noble metal–based catalysts, high reaction temperatures, and ultrafast contact times (5 ms),[29] it is possible to considerably improve alkene selectivity in propane oxidative dehydrogenation (65% selectivity at 100% conversion). It should be noted that the same catalyst is a complete oxidation catalyst in ordinary contact time regimes.

Another example of ultrafast oxidation is the conversion of isoprenol (2 methyl-but-1-ene-4-ol) to isoprenal for the synthesis of citral,[30] using a continuous flow reactor operating at 500°C with a residence time of the order of 1 ms. The yield in the extremely sensitive isoprenol product is 95%.

Catalytic oxidation is thus breaking new ground in the field of catalysis with the development of novel concepts and new industrial achievements. Basic research plays a fundamental role in this process when it is directed toward more innovative areas of research and not limited to consolidating more "mature" lines of activity.

1.4. THE ECOLOGICAL ISSUE AS A DRIVING FORCE

1.4.1. Reduction or Elimination of Coproducts

Coproduct inorganic salts have for the most part been disposed of as waste materials in aqueous effluents, but increasingly stringent environmental regulations governing the disposal of these effluents have stimulated the development of processes that do not coproduce inorganic salts. Fine chemical processes that coproduce stoichiometric amounts of inorganic salts include sulfonation, nitration, halogenation, and oxidation. Some selective oxidation processes for the production of bulk chemicals also generate useless inorganic coproducts. Three such processes are the syntheses of propene oxide, cyclohexanone oxides, and methyl methacrylate.

1.4.1.1. Synthesis of Propene Oxide

Increasingly stringent environmental requirements are driving the search for new epoxidation routes. The several approaches currently under study differ primarily in the nature of the oxidizing agent, e.g., hydrogen peroxide or *tert*-butyl hydroperoxide, and in the type of catalyst used.

The epoxidation of propene is an example of a process originally carried out by homogeneous catalysis (ARCO) but for which alternative heterogeneous routes (SHELL) have been successfully developed. The homogeneous ARCO process is catalyzed by molybdenum salts, whereas the heterogeneous SHELL process is catalyzed by titania on silica. In both cases, the oxidant can be either ethylbenzene peroxide (coproduct styrene) or *tert*-butyl hydroperoxide (coproduct isobutene). Both processes are alternatives to the classical route in which propene oxide is formed from the chlorohydrin produced by the reaction of propene and hypochlorous acid. The classical route suffers from the formation of highly toxic chlorinated hydrocarbon by-products, the production of large amounts of chlorine-containing water, and the corrosivity of the reagents.

EniChem/EniTecnologie (the latter formerly Eniricerche) developed a heterogeneous process catalyzed by TS-1, which uses hydrogen peroxide as the oxidizing agent, and its only coproduct is water. Recently, the Olin Corporation piloted a propene epoxidation process based on a molten salt catalyst.[31,32] The role of the molten salt is to control the heat of reaction rather than to catalyze the reaction. Some further aspects of these two new processes will be discussed in Chapter 3.

1.4.1.2. Synthesis of Cyclohexanone Oxime

The synthesis of cyclohexanone oxime and its transformation to caprolactam is presently carried out via the synthetic pathway shown in Scheme 1.1. Major problems are the coproduction of large amounts of ammonium sulfate, the use of toxic hydroxylamine, and the many steps required. When the classic Rashig process is used for the synthesis of hydroxylamine (Scheme 1.1), the amount of coproduced ammonium sulfate is 4.4 tons per ton of caprolactame. However, because ammonium sulfate is no longer used for fertilizer, the only alternative to waste disposal is pyrolysis, which produces sulfur dioxide, water, and nitrogen. In addition to the negative environmental impact of this process, the energy costs are high.

Montedipe (now EniChem) developed an alternative cyclohexanone oxime synthesis process, presently at the demonstration plant stage, in which the ammoximation of cyclohexanone is carried out in the liquid phase by reaction with ammonia and hydrogen peroxide and TS-1 is the catalyst. The process is competitive with processes that use hydroxylamine, such as the BASF HSO and the DSM HPO processes, both of which produce ammonium sulfate in small amounts. Cyclohexanone ammoximation will also be discussed in Chapter 3. Zeolite materials are being developed as catalysts for the Beckmann rearrangement step in the synthesis

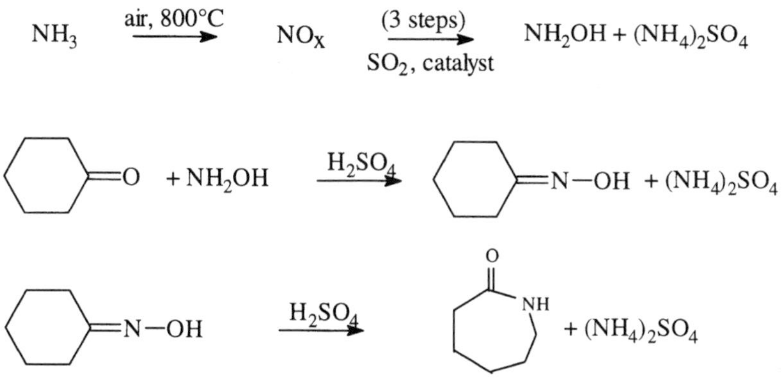

Scheme 1.1. Commercial synthetic pathway for caprolactam.

of caprolactam. In this way, the entire caprolactam synthesis can be achieved without coproduction of ammonium sulfate. Another pathway to cyclohexanone oxime is the gas phase ammoximation of cyclohexanone by reaction with ammonia and molecular oxygen, using a silica-type catalyst.[33,34] However, the selectivity is relatively low (around 50%) and the catalyst quickly deactivates.

1.4.1.3. Synthesis of Methyl Methacrylate

Methyl methacrylate is presently synthesized via the pathway shown in Scheme 1.2. However, this process has serious problems related to the toxicity of HCN and the coproduction of large amounts of ammonium sulfate (2 tons per ton of methyl methacrylate). For these reasons, various alternative processes based on the oxidation of alkene have been proposed (Scheme 1.3).[35]

One of the most attractive of these alternative processes from an economic and environmental point of view is the direct oxidation of isobutane to methacrylic acid. Patents issued to Japanese companies describe the use of heteropoly compounds for the selective oxidation of isobutane (see Chapter 4), although low selectivity and catalyst stability are the main problems in process development.

1.4.2. Use of Alternative Catalysts

In some cases the use of new alternative catalysts considerably reduces the environmental impact of a process and its by-products with minimal modifications required to the process itself. An interesting example in the field of liquid phase alkene oxidation with molecular oxygen is the synthesis of acetaldehyde from ethene.

In the well-known Wacker process, a combination of $CuCl_2$ and $PdCl_2$ is used as the catalyst. $PdCl_2$ is reduced to palladium with the oxidation of ethene to

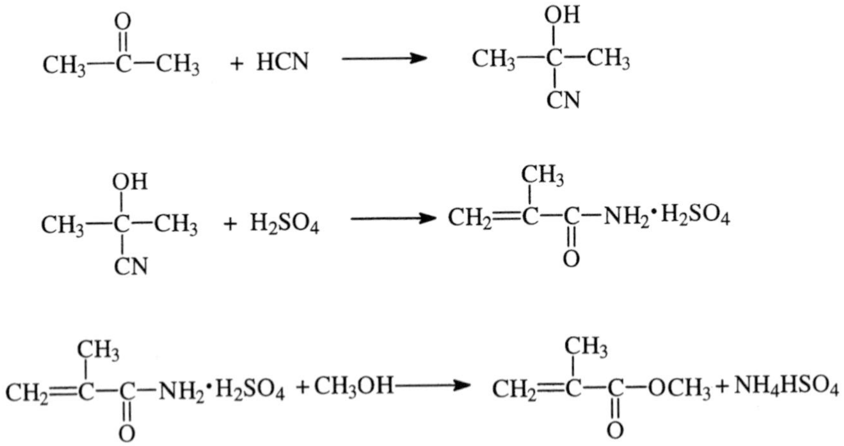

$$CH_3-\overset{\overset{\displaystyle O}{\|}}{C}-CH_3 \ + \ HCN \ \longrightarrow \ CH_3-\overset{\overset{\displaystyle OH}{|}}{\underset{\underset{\displaystyle CN}{|}}{C}}-CH_3$$

$$CH_3-\overset{\overset{\displaystyle OH}{|}}{\underset{\underset{\displaystyle CN}{|}}{C}}-CH_3 \ + \ H_2SO_4 \ \longrightarrow \ CH_2{=}\overset{\overset{\displaystyle CH_3}{|}}{C}-\overset{\overset{\displaystyle}{C}}{\underset{\underset{\displaystyle O}{\|}}{}}-NH_2{\cdot}H_2SO_4$$

$$CH_2{=}\overset{\overset{\displaystyle CH_3}{|}}{C}-\overset{C}{\underset{\underset{\displaystyle O}{\|}}{}}-NH_2{\cdot}H_2SO_4 \ +CH_3OH \longrightarrow \ CH_2{=}\overset{\overset{\displaystyle CH_3}{|}}{C}-\overset{C}{\underset{\underset{\displaystyle O}{\|}}{}}-OCH_3+NH_4HSO_4$$

Scheme 1.2. Commercial synthesis of methyl methacrylate.

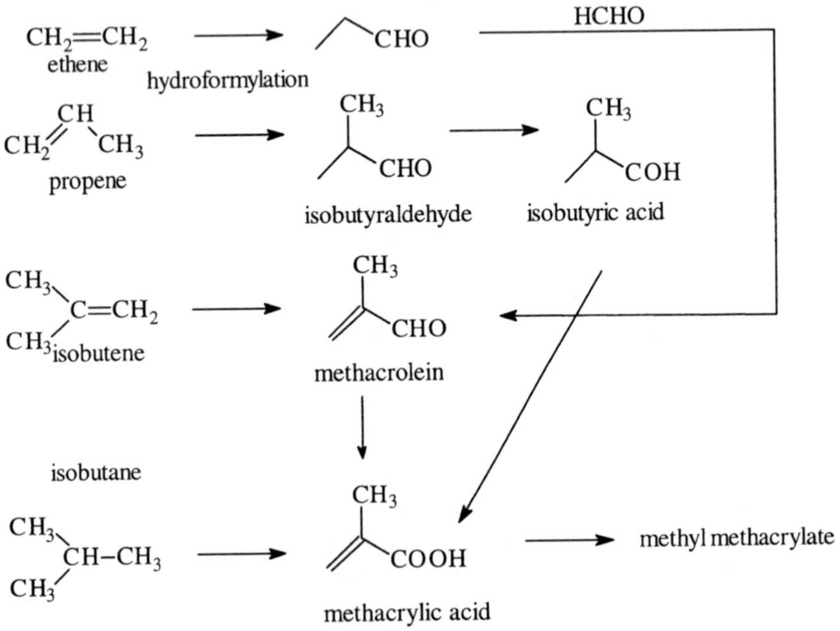

Scheme 1.3. Alternative processes for methyl methacrylate synthesis.

acetaldehyde, and the palladium is then reoxidized by $CuCl_2$. The resulting $CuCl$ is reoxidized by molecular oxygen. The main disadvantage of this process is the production of chlorinated organic waste by-products owing to the high chloride concentration. Disposal of these by-products poses severe environmental problems.

Catalytica[36] developed the catalyst $PdCl_2-Na_yH_{3+x-y}PMo_{12-x}V_xO_{40}$ (x is usually 6) where the V-heteropoly acid acts as the reoxidant for the reduced palladium and is then reoxidized by molecular oxygen. Using V-heteropoly acid the rate of ethene oxidation is about two orders of magnitude higher and thus the quantity of chloride ions necessary to maintain the catalytic activity of Pd^{2+} is about two orders lower, considerably decreasing the formation of waste chlorinated by-products. Another alternative is to carry out the reaction in the gas phase, using either Pd-acetate doped supported V-oxide or solid $PdSO_4$/V-heteropolyacids (see Chapter 7), but fast catalyst deactivation is the main critical problem. On the other hand, when heterogeneous Wacker-type catalysts are used the problem of chlorinated by-products is completely eliminated.

Another example can be found in the side-chain oxidation of substituted alkylaromatics. This reaction is currently carried out using a Co^{2+} homogeneous catalyst, gaseous oxygen as the oxidant, and Mn^{2+} and Br^- ions as cocatalysts. Typical reaction conditions are a temperature of around 160°C (i.e., in an autoclave reactor and under pressure for reasonable rates) and glacial acetic acid as the solvent, the latter selected mainly because of flammability problems. The products are often recovered by crystallization from the mother solution and by-products are low. However, with some types of substituents recovery yield by crystallization is low, and on the aromatic ring the product formation of by-products (some of them toxic) is greater. In this case, alternative catalysts (in the gas phase) and catalyst/oxidant (in the liquid phase) must be used, but no method has yet been developed that reaches the necessary activity/selectivity requirements. On the other hand, with these new catalysts, it is possible to eliminate the formation of toxic by-products and the corrosion problems related to the use of CH_3COOH/Br^- components at high temperatures as well as reduce the risks associated with the possible formation of flammable mixtures.

There are thus several new fields of research where further development is essentially driven by the necessity to develop cleaner processes of selective oxidation that reduce by-products, process complexity, and risks of the process.

1.5. HETEROGENEOUS VERSUS HOMOGENEOUS CATALYSIS IN SELECTIVE OXIDATION

In general, heterogeneous systems offer easier separation of catalysts from the product mixture and the possibility of continuous processing, whereas homogeneous oxidation usually allows more efficient heat transfer and milder operating conditions, as well as higher selectivities. Both heterogeneous gas phase oxidation and homogeneous liquid phase oxidation are used in the petrochemical industry (Table 1.1). Indeed, liquid phase oxidation is often used for the synthesis of fine chemicals, which are often complex molecules with limited thermal stability.

Heterogeneous catalysts are being increasingly studied for liquid phase oxidation as alternatives to the more typically used dissolved metal salts. Important examples of these heterogeneous catalysts are TS-1 and other microporous materials containing transition metals. These samples, including Ti-containing materials, are often erroneously classified as "redox zeolites." In fact, in the case of titanium and H_2O_2 as the oxidizing agent, there is no "redox" chemistry in the sense of change of the valence state of the metal during the catalytic reaction, but rather only the formation of a Ti-hydroperoxo active complex (see Chapter 7). Similar situations are found in several other cases of transition metal–containing microporous materials (zeolite, zeolite-like materials such as the MeAPO family, clays, mesoporous materials), although the evidence is often less clear about whether a "redox" chemistry actually exists. The general problem of these catalysts is the easy leaching of the transition metal, apart from the specific case of TS-1, which makes their industrial application problematic.

Many recent patents and scientific publications have addressed the application of special classes of catalysts to selective oxidation reactions, and much attention has been focused on systems that simulate enzymatic action in the monoxygenation of organic substrates and on molecular-type solids, such as polyoxometallates. The high structural flexibility of the molecular-type solids, combined with the possibility of modulating their composition while maintaining their structural integrity, allows the catalysts to be tailored for specific applications.

Impressive results have been reported when molecular-type solids are used as homogeneous catalysts for liquid phase oxidation, although data on the stability are not available. In the gas phase oxidation these catalysts usually deactivate rapidly because of the rather limited structural stability of the molecular unit (usually the molecular unit in polyoxometallates is the Keggin anion, such as 12-phospho-molybdates, silicomolybdates, or silicotungstates). Metallo-containing zeolite and polyoxometallates thus may be considered bridging catalysts between homogeneous and heterogeneous selective oxidation catalysis, but in general the factors determining the behavior (and therefore catalyst characteristics) are different.

In heterogeneous gas phase oxidation, the reaction performance is dictated by both the bulk and surface properties of the metal oxide catalyst (see Chapter 8). The mechanism generally, but not always, proceeds via activation of molecular oxygen with stepwise formation of electrophilic-type intermediate species, dissociation into atoms, and then incorporation of O^{2-} nucleophilic species into anionic vacancies in the solid. Anionic vacancies are formed by interaction of the solid with the organic substrate.[37] In this mechanism, the reducing agent is the hydrocarbon itself (Mars–van Krevelen-type mechanism).

In gas phase oxidation, the only feasible oxidizing agent is molecular oxygen, although some interesting alternative oxidants are possible in some specific cases (Chapter 7). At the molecular level, all oxygen atoms are potentially useful for formation of the active oxidizing species. The nature of the oxidizing species in terms of reactivity toward the substrate and, hence, stability and lifetime depends only on the physicochemical properties of the oxides. Thus the reaction rate and the distribution of products depend to a significant degree on the nature and physicochemical properties of the catalyst; a catalyst that is active and selective for a given hydrocarbon oxidation will seldom be active and selective for the oxidation of another substrate. Some reactions are only catalyzed by a very specific composition or structure, whereas others, usually less demanding in terms of complexity of the sequence of stages and multifunctional properties of the catalyst, are catalyzed by many different types of catalysts. An example of the first type of reaction is n-butane oxidation to maleic anhydride on a vanadyl pyrophosphate catalyst (Chapter 4) while examples of the second type are methanol oxidation to formaldehyde (Chapter 8), alkane oxidehydrogenation to alkenes (Chapter 5), and o-xylene oxidation to phthalic anhydride.

In attempts to explain selective oxidation reactions catalyzed by metal oxides, researchers have developed concepts that correlate catalytic performance with some property of the material used. Among these are structure sensitivity, synergy between mixed oxides, remote control effects, nature of the metal oxide monolayer, and catalytic polyfunctionality. The development of these concepts has further stimulated the study of solid-state chemistry and has underscored the importance of surface and interface phenomena in catalytic processes. However, as discussed in more detail in Chapter 8 none of these theories appears to be a general explanation of surface catalysis of selective oxidation catalysts.

In homogeneous oxidation, the search for a suitable catalytic system has been focused on metals that can catalyze the transfer of oxygen from the oxygen donor to the organic substrate. In the absence of autoxidation, there are essentially two mechanisms by which this transfer may occur[14,38,39]:

1. Heterolytic activation of an oxidant such as an alkyl hydroperoxide by a metal center and transfer of the activated oxygen to the substrate: The metal center binds the substrate and acts as an electrophilic center. This type of mechanism is the basis of the Sharpless homogeneous asymmetric epoxidation technology[40] and the ARCO homogeneous epoxidation of propene to alkyl hydroperoxide. The latter is catalyzed by Mo(VI) compounds, and the reaction proceeds via a peroxometal mechanism in which oxygen transfer to the alkene is rate-limiting. The same mechanism also operates in the TS-1–catalyzed alkene epoxidation by hydrogen peroxide.

2. Oxidation of a substrate by a metal ion that is reduced in the process and subsequently regenerated by reaction with molecular oxygen: This is the mechanism of the Pd(II)-catalyzed Wacker oxidation of alkenes in which

the metal is in the form of a soluble salt. It is also the mechanism in the metal porphyrin-catalyzed alkene epoxidation and alkane oxygenation.

Systems such as halogenated metalloporphyrin complexes and transition metal–substituted polyoxometallates have been recently reported to catalyze the liquid phase oxidation of organic substrates. The reactions use molecular oxygen and require no coreductants. While this approach looks interesting, the mechanism by which these systems operate needs to be elucidated. Indeed, it is often not clear whether the mechanism is truly catalytic or whether the catalyst merely initiates the reaction, which then proceeds via the classical free-radical autoxidation route.

REFERENCES

1. F. Trifirò and F. Cavani, *Selective Partial Oxidation of Hydrocarbons and Related Oxidations*, Catalytica Study No. 4193 SO, Catalytica Studies Div.: Montain View, CA (1994).
2. D.J. Hucknall, *Selective Oxidation of Hydrocarbons*, Academic Press: London (1974).
3. K. Weissermel and H.J. Arpe, *Industrial Organic Chemistry, 2nd Ed.*, Verlag Chemie, Weinheim, Germany (1993).
4. S.T. Oyama, A.N. Desikan, and J.W. Hightower, in: *Catalytic Selective Oxidation* (S.T. Oyama and J.W. Hightower, eds.), ACS Symposium Series 523, American Chemical Society: Washington DC (1993), Ch. 1, p. 1.
5. J. Haber, in: *3rd World Congress on Oxidation Catalysis* (R.K. Grasselli, S.T. Oyama, A.M. Gaffney, and J.E. Lyons, eds.), Studies in Surface Science and Catalysis Vol. 110, Elsevier Science: Amsterdam (1997), p. 1.
6. R.K. Grasselli, in: *Surface Properties and Catalysis by Non-Metals* (J.P. Bonelle, B. Delmon, and E. Derouane, eds.), Reidel: Dordrecht (1983), p. 273, 289.
7. R.K. Grasselli, *J. Chem. Educ.* **63**, 216 (1986).
8. K. Bielanski and J. Haber, *Oxygen in Catalysis*, Marcel Dekker: New York (1991).
9. H.H. Kung, *Transition Metal Oxides: Surface Chemistry and Catalysis*, Studies in Surface Chemistry and Catalysis Vol. 45, Elsevier Science, Amsterdam (1990).
10. G. Chinchen, P. Davies, and R.J. Sampson, in: *Catalysis, Science and Development*, Vol. 8 (J.R. Anderson and M. Boudart, eds.), Springer-Verlag: Berlin (1987), p. 1.
11. Catalytica, *Catalysts for the Elimination of Volatile Organic Compounds: Halogenated Compounds*, Environmental Report E4, Catalytica Studies Division: Mountain View, CA (1992).
12. Catalytica, *Catalysts for the Elimination of Volatile Organic Compounds: Nonhalogenated Compounds*, Environmental Report E4, Catalytica Studies Division: Mountain View, CA (1993).
13. F. Cavani and F. Trifirò, *Appl. Catal. A* **88**, 115 (1992).
14. C.L. Hill, A.M. Khenkin, M.S. Weeks, and Y. Hou, in: *Catalytic Selective Oxidation* (S.T. Oyama and J.W. Hightower, eds.), ACS Symposium Series 523, American Chemical Society: Washington DC (1993), p. 67.

15. M.G. Clerici and P. Ingallina, *EP Patent* 526, 945 (1993).
16. C. Ferrini and H.W. Kouvenhove, in: *Proceedings DGMK Conference on Selective Oxidation in Petrochemistry*, Goslar (Germany), DGMK (German Society for Petroleum and Coal Science and Technology): Hamburg (1992), p. 205.
17. T. Tasumi, K. Yuasa, and H. Tominaga, *J. Chem. Soc. Chem. Comm.*, 1446 (1992).
18. W. Laufer, R. Meiers, and W.F. Hölderich, *Proceedings 12th International Zeolite Conference*, Baltimore, July 1998, B-8.
19. N.I. Kuznetsova, A.S. Lisitsyn, A.I. Boronin, and V.A. Likholobov, in: *New Developments in Selective Oxidation* (G. Centi and F. Trifirò, eds.), Studies in Surface Science and Catalysis, Vol. 55, Elsevier Science: Amsterdam (1990), p. 89.
20. N.I. Kuznetsova, A.S. Lisitsyn, and V.A. Likholobov, *React. Kinet. Katal. Lett.* **38**, 205 (1989).
21. I. Tabushi, M. Kodera, and M. Yokoyama, *J. Am. Chem. Soc.* **107**, 4466 (1985).
22. P. Battioni, J.F. Bartoli, P. Ledue, M. Fontecave, and D. Mansuy, *J. Chem. Soc. Chem. Comm.*, 791 (1987).
23. R. Neumann and M. Levin, in: *Dioxygen Activation and Homogeneous Catalytic Oxidation* (L.I. Simandi, ed.), Studies in Surface Science and Catalysis, Vol. 66, Elsevier Science: Amsterdam (1991), p. 121.
24. H. Muller, K. Deller, B. Despeyroux, E. Peldszus, P. Kammerhofer, W. Kuhn, R. Spielmannleitner, and M. Stoger, *Catal. Today* **17**, 383 (1993).
25. I.M. Freidel, A.C. Frost, K. Herbert, F.J. Meyer, and J.C. Summers, *Catal. Today* **17**, 367 (1993).
26. R.G. Markeloff, *Hydrocarbon Process.* **63**, 91 (1984).
27. I. Pasquon, *Catal. Today* **1**, 297 (1987).
28. D. Delmon, in: *3rd World Congress on Oxidation Catalysis* (R.K. Grasselli, S.T. Oyama, A.M. Gaffney, and J.E. Lyons, eds.), Elsevier Science: Amsterdam (1997), Studies in Surface Science and Catalysis Vol. 110, p. 43.
29. M. Huff and L.D. Schmidt, *J. Catal.* **149**, 127 (1994).
30. W.F. Hölderich, in: *New Frontiers in Catalysis* (L. Guczi, F. Solymosi, and P. Tétényi, eds.), Studies in Surface Science and Catalysis Vol. 75, Elsevier Science: Amsterdam (1993), p. 127.
31. B.T. Pennington, *US Patent* 4,785,123 (1988).
32. B.T. Pennington, M.C. Fullington, *US Patent*, 4,943,643 (1990).
33. J.N. Armor, *J. Catal.* **70**, 72 (1981).
34. D. Dreoni, D. Pinelli, and F. Trifirò, in: *New Developments in Selective Oxidation by Heterogeneous Catalysts* (P. Ruiz and B. Delmon, eds.), Studies in Surface Science and Catalysis Vol. 72, Elsevier Science: Amsterdam (1992), p. 109.
35. R.V. Porcelli and B. Juran, *Hydrocarbon Process.* **65**, 37 (1986).
36. J.A. Cusumano, *CHEMTECH* **22**, 482 (1992).
37. J. Haber, in: *New Developments in Selective Oxidation by Heterogeneous Catalysts* (P. Ruiz and B. Delmon, eds.), Studies in Surface Science and Catalysis Vol. 72, Elsevier Science: Amsterdam (1992), p. 279.
38. R.A. Sheldon, *CHEMTECH* **21**, 566 (1991).

39. R.A. Sheldon, in: *Heterogeneous Catalysis and Fine Chemicals II* (M. Guisnet, J. Barrault, C. Bouchoule, D. Duprez, G. Perot, R. Maurel, and C. Montassier, eds.), Studies in Surface Science and Catalysis Vol. 59, Elsevier Science: Amsterdam (1991), p. 33.
40. T. Katsuki and K.B. Sharpless, *J. Am. Chem. Soc.* **102**, 5976 (1980).

2

NEW TECHNOLOGICAL AND INDUSTRIAL OPPORTUNITIES
Options

2.1. USE OF ALTERNATIVE RAW MATERIALS

2.1.1. Alkanes as Raw Materials for Selective Oxidation Reactions

The possibility of developing new lower-environmental-impact and lower-cost processes has recently generated interest in the transformation of light alkanes to valuable oxygenated compounds and alkenes by means of oxidation. As shown in Table 2.1, natural gas contains considerable amounts of heavier alkanes in addition to methane.

The availability and the low cost of these hydrocarbons accounts for the surge in interest in the development of new processes based on these compounds as alternatives to those using traditional raw materials such as alkenes and, sometimes, aromatics. In other cases traditional raw materials might still be used, but the reaction to transform it might be completely different, leading to considerable savings in both investment and operating costs, with particular gains in terms of

TABLE 2.1. Composition of Natural Gas from Various Locations

	C_1	C_2	C_3, C_4	$C_{\geq 5}$	N_2	CO_2	H_2S	He
Groningen, The Netherlands	81	2.8	0.6	0.1	14.3	0.9	—	0.03
Ekofisk, Norway	85	8.4	3.9	0.15	0.4	2	0.001	—
Söhlingen, Germany	85	1.5	0.1	—	12.5	0.5	—	0.03
Süd-Oldenburg, Germany	77	0.1	—	—	7	8	8	—
Tenguiz, Russia	42	8.5	8.5	22	0.8	2.6	16	—
Panhandle, Texas	73.2	6.1	4.8	0.6	14.3	0.3	—	0.7
Arun, Indonesia	75	5.5	3.4	0.8	0.3	15	0.01	—
Bearberry, Canada	4	—	—	—	1	5	90	—

conserving energy. The transformation of methane to syngas by direct selective oxidation is one example of an alternative to the conventional process via steam-reforming.

2.1.1.1. Advantages and Targets in Using an Alkane Feedstock

Advantages to the use of alkanes as raw materials for some specific applications are shown in Figure 2.1, which compares the heats of reaction for some current technologies versus the corresponding alternative oxidation processes for the synthesis of specific intermediates for the petrochemical industry, including:

1. The synthesis of methanol through the conventional technologies of methane steam-reforming, followed by syngas conversion to methanol versus the alternative direct oxidation of methane to methanol.
2. The steam-cracking of natural gas to yield ethene, converted to acetaldehyde through the Wacker process and then oxidized to acetic acid versus the alternative one-step oxidation of ethane to acetic acid.
3. The synthesis of acrylonitrile by propane dehydrogenation, followed by ammoxidation of the alkene versus the direct ammoxidation of propane.

It is evident that the paradox of using endothermal processes (steam-reforming, steam-cracking, and dehydrogenation) to produce chemical building blocks, followed by the exothermal transformation to other chemicals is by far less logical than a direct single exothermal step of transformation. Of course, this is a simplistic approach to the problem, as there are other factors involved, and as the alkene intermediates are employed for a variety of applications, there have to be processes to produce them. However, there are some further considerations that also should be taken into account:

1. The endothermal, energy-intensive processes for the production of building blocks for the petrochemical industry (syngas and alkenes) require very large plants, with high investment costs. If there is a market need for a particular compound, it can be more economical to set up a specific, mid-sized plant for the production of that chemical starting from cheap natural gas components.
2. Refineries are now being altered dramatically, as a result of changes in the demand for reformulated gasolines and diesel fuels. Modifications in the FCC and catalytic-reforming processes have been directed toward responding to the changing requirements of the petrochemical industry, where demand for isobutene, high-purity alkenes, and hydrogen has increased while that for aromatics has gone down. These modifications have a considerable effect on the market value of alkenes, with variations that are hard to predict even for the near future. The possibility of developing a process that does not depend on the alkene supply, i.e.,

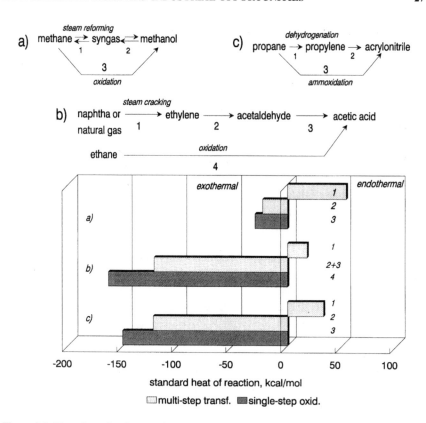

Figure 2.1. Heat of reaction for some industrial multistep transformations of alkanes to intermediates for the petrochemical industry and for alternative oxidative single-step transformations.

one that starts from the natural gas components, would avoid problems associated with these variations in price.

3. Interest is now growing in processes for the production of base chemicals that might overcome the thermodynamic barriers associated with strongly endothermal processes. For instance, the exothermal partial oxidation of CH_4 to CO/H_2, as an alternative to methane steam-reforming, and oxidehydrogenation of alkanes, as an alternative to dehydrogenation, are transformations that draw advantage from more favorable equilibria and can take place at much lower temperatures.

In addition to energy savings and greater process simplicity, in some cases new processes might represent a considerable advantage from the environmental point of view with respect to traditional multistep industrial processes starting from conventional raw materials. This is the case in the synthesis of methylmethacrylate by the one-step oxidation of isobutane to methacrylic acid, followed by esterification of the latter with methanol. Indeed, currently methylmethacrylate is produced by the acetone–cyanohydrin route, a multistep process with several drawbacks that make it environmentally nonsustainable.

The targets that have to be reached to enable industrial exploitation of the reactions now being researched are the following:

1. Achievement of high productivity and/or high selectivity for the desired product: One of the most striking examples in this direction is the synthesis of ethene by methane oxidative coupling, where a yield of 30 mol%, with 70–80 mol% selectivity, has been claimed as the minimum to make the process economically feasible with respect to steam-cracking of naphtha. At the moment, only yields up to 22–24 mol% have been reported.

2. Development of stable catalysts that exhibit an overall industrially acceptable lifetime and can be easily regenerated.

3. Development of oxidation catalysts able to quickly furnish large amounts of oxygen to the activated alkane substrates without undergoing structural collapse and able to quickly recover the original oxidation state by contact with oxygen. This would allow two-stage oxidation of the organic substrate, thus achieving better control of the selectivity.

4. Absence of carbon monoxide among the by-products: In the oxidehydrogenation of light alkanes to alkenes, the by-products carbon dioxide and carbon monoxide can easily be separated from the alkane/alkene mixture by absorption in basic water and in a demethanizer column under pressure, respectively. However, in the case of methane oxidative coupling or ethane oxidative dehydrogenation, the separation of carbon monoxide from the unreacted alkane and from the alkene produced is rather expensive. Further when it is not immediately separated, e.g., in an integrated process of production of the alkene followed by its transformation to some other chemical, its production has to be minimized. In fact the formation of a minimum quantity of inert gases makes it possible to decrease the purge stream when the unconverted alkane is recycled. Therefore, the elimination of carbon monoxide, through the development of a suitable catalyst and the optimization of reaction conditions, would lead to a considerable decrease in investment and operating costs.

5. Absence of traces of corrosive by-products (e.g., organic acids) when the reaction is aimed at the synthesis of products which are themselves noncorrosive, e.g., alkene. This might allow considerable savings in the cost of materials for construction of reactors and downstream apparatus.

6. Use of the same reactor and separation technology as for the conventional process, with only minor expenses for revamping. This is the case in the ammoxidation of propane to acrylonitrile, where comparable reaction conditions might

allow the use of the same reactor as for propene ammoxidation, and a similar distribution of products might lead to only minor modifications in the separation section.

7. Integration of the oxidation plant with a process that employs the product of partial oxidation itself and can operate without purification of the effluent stream of the oxidation reactor. This might facilitate the process flow sheet, with smaller investments and lower operating costs. Some specific cases will be discussed in the following sections.

2.1.1.2. Key Questions in Alkane Functionalization

Many reviews and monographs have been published in recent years that analyze the fundamental aspects of the oxidative activation and transformation of light alkanes over heterogeneous catalysts.[1-25] The general picture that emerges on the basis of the most important factors examined in these reviews clearly shows that the problems of alkane conversion and selectivity for the desired product have to be solved within a complex framework of interrelated aspects. In particular, the interaction between the catalyst surface and the gas phase and the way in which surface phenomena are affected by homogeneous phenomena play important roles in conditioning the final performance in terms of product distribution.

Some of the points that have been considered key factors in determining the pathway of alkane transformation to the product of selective oxidation are as follows:

1. Activation of oxygen and of the alkane, specifically: (i) the role of adsorbed oxygen species,[1-4,6,7,11] and (ii) the importance of the mode of alkane adsorption.[1-3,5]
2. Reactivity of the reactant and of the product(s): (i) the mechanism of activation of the C–H bond (heterolytic vs. homolytic activation),[11,13] and (ii) the stability of the products.[9,10,12]
3. Mechanism of transformation of the reactant: (i) the importance of non-desorption of reaction intermediates,[9,10] (ii) the importance of the relative ratio between intermediate alkene (oxi)dehydrogenation and oxygen insertion in affecting the selectivity for oxygenated products,[9,10] (iii) the role of the nature of the reaction intermediate (hydrocarbon fragment) in determining the direction of oxidative transformation,[11] (iv) the contribution of homogeneous reactions, especially for applications that require temperatures higher than 400–450°C,[12] and (v) the effect of coadsorbates in facilitating the dissociative adsorption of saturated hydrocarbons.[12]

A better understanding of each of these factors means a better comprehension of the overall process of alkane transformation and, hence, a greater possibility of increasing the catalytic performance to make these processes more attractive to industry. The following features are fundamental for the control of catalytic performance:

- Surface of the catalyst, i.e., the nature of the active sites and the way in which the surface is affected by the bulk features, specifically: (i) the density of the active sites,[3,9,10] which can be explained by the "site isolation" theory[26]; (ii) the role of surface acidity[8,11]; and (iii) the need for intrinsic surface polyfunctionality.[9,10]
- Structure of the catalyst: (i) the redox properties of the metal in transition metal oxide-based catalysts, in terms of reducibility and reoxidizability of the active sites, and the metal–oxygen bond strength[3,5]; (ii) the reactivity of specific crystal faces in the different transformations that constitute the reaction network[2,3,6]; (iii) the role of structural defects in favoring the mobility of ionic species in the bulk[1–3]; (iv) the importance of cooperative effects of different phases in obtaining catalysts with improved performances[12,22]; and (v) the importance of the interaction between the support and the active phase in modifying the latter's catalytic properties.[12]

Most of these aspects may be applied in general to hydrocarbon oxidation, while some of them are specific for alkane transformations.

2.1.1.3. Processes of Industrial Interest Using an Alkane Feedstock

Table 2.2 lists alkane selective oxidation reactions of industrial interest and indicates their stage of development. After the oil crisis, which led to a steady increase in interest in the use of natural gas components as the preferred raw materials, research in this area began moving from n-butane exploitation, with the successful development of the process for maleic anhydride synthesis,[27,28] toward the employment of lighter and less reactive alkanes and the development of methods that make use of these components directly through exothermal processes.

The challenge of transforming methane to liquid fuel was first approached by studying the oxidative coupling to ethane and ethene (with an interest in the pyrolytic dimerization of methyl chloride, which was being produced by methane chlorination or oxihydrochlorination) and the oxidation to methanol. More recently the oxidation to syngas appears to be one of the most promising routes for methane valorization.[29–31]

The possibility of a single-step transformation of ethane to acetic acid is based upon high chemical stability of ethene, one of the most favored products in ethane oxidative conversion.[32–34] For the same reason, one possible commercial exploitation of ethane depends on a reaction in which the alkene itself becomes the raw material for the transformation to a valuable chemical, either in the same reactor

TABLE 2.2. Industrial Processes and Processes under Study or Development for the Oxidative Transformation of Light Alkanes (C1–C6)

Raw material	Product	Stage of development
Methane	Methanol	Pilot plant
Methane	Syngas	Pilot plant
Methane	Ethylene	Pilot plant
Ethane	1,2 Dichloroethane, vinyl chloride	Pilot plant
Ethane	Acetaldehyde	Research
Ethane	Acetic acid	Research
Ethane	Ethylene	Research
Propane	Acrolein, acrylic acid	Research
Propane	Propyl alcohol	Research
Propane	Acrylonitrile	Demonstration plant
Propane	Propylene	Research
n-Butane	Acetic acid	Industrial
n-Butane	Maleic anhydride	Industrial
n-Butane	Butadiene	Industrial, abandoned
Isobutane	Methacrylic acid	Pilot plant
Isobutane	Isobutene	Research
Isobutane	tert-Butyl alcohol	Research
n-Pentane	Phthalic anhydride	Research
Cyclohexane	Cyclohexanol, cyclohexanone	Industrial
Cyclohexane	Cyclohexanone	Research

(the second reaction must be carried out under conditions that are close to those necessary for ethane activation), or in an *integrated process*, where the ethene-containing stream exiting from the first reactor is the feedstock (after necessary makeup) for the second reactor. One example of this is the single-step or two-step ethane oxychlorination to 1,2-dichloroethane and to vinyl chloride, where the ethene produced can be easily transformed into the chlorinated products. While this process was studied in the 1970s,[35,36] it did not reach the stage of industrial application, but interest is now growing again, as documented by recent patents registered by companies that supply PVC.[37]

Apart from the oxidation of *n*-butane (Chapter 4), a second heterogeneous gas phase process that is expected to be commercialized successfully is the ammoxidation of propane to acrylonitrile, developed independently by both BP America and Mitsubishi, using fluidized-bed technology.[38–44] Here as well the stability of the product, which saves it from nonselective consecutive reactions, has been a key factor in process development.

Work is now being published that describes attempts to transform isobutane into methacrolein and methacrylic acid in a single step (a process that might replace acetone–cyanohydrin technology) and propane to acrylic acid.[45–53] Interest on the part of industry for this kind of reaction goes back to the 1980s.[54] In these reactions a further problem arises owing to the high reactivity of the desired products, which are not stable under the conditions necessary for alkane activation and undergo nonselective oxidative transformations.

The availability of C5 alkanes, as a consequence of environmental regulations that require a lower gasoline vapor pressure, and the discovery that n-pentane can be transformed with unexpected selectivity to valuable compounds, i.e., maleic anhydride and phthalic anhydride, might create new prospects in the exploitation of saturated raw materials.[55–66] In this case the key factor for commercialization is the achievement of high selectivity for phthalic anhydride, possibly close to that obtained from o-xylene.

The enormous number of papers devoted to the reaction of oxidative dehydrogenation of alkanes is an indication of the scientific and industrial interest in alternatives to catalytic and thermal dehydrogenation/cracking reactions, which suffer from energy drawbacks.[5,9,17–20] In these cases the advantage gained from the use of a cheaper raw material and an exothermal process must be weighed against drawbacks such as the loss of valuable hydrogen (coproduced in dehydrogenation and in steam-cracking), the difficulty in separating carbon monoxide from the alkane (in the case of ethane oxidehydrogenation, but also of methane oxidative coupling to ethene), and the formation of traces of corrosive by-products. While in the case of ethane, the problem concerns mainly the low reactivity of the molecule (the selectivity for the alkene is usually high, owing to the low ethene reactivity and the nature of the mechanism involved), in the case of oxidehydrogenation of propane and n-butane to the corresponding alkenes the selectivity problem is the primary concern. With the latter molecules, low selectivity is due to the formation of allylic oxygenated compounds, which are readily transformed to carbon oxides.

2.1.2. New Oxidants

Several different nonconventional gas phase oxidizing agents have been proposed to activate alkanes as well as other hydrocarbons such as SO_2, CO_2,[67] N_2O,[68–90] and O_3[91–94] (see also Chapters 7 and 8). These oxidants obviously do not offer substantial advantages with respect to the traditional ones unless they exhibit very specific oxidizing properties, being able to perform oxidative attacks that cannot otherwise be carried out. In the case of alkanes and less reactive hydrocarbons, the advantage would lie in the possibility of developing very active oxidizing species, able to activate the alkane under relatively mild conditions, under which the desired product (usually much more reactive than the reactant itself) or some reaction intermediate can be saved from undesired consecutive reactions. However, these very active oxidizing species should also not be indiscriminate in their attack.

Another factor influencing the choice unconventional oxidizing agents is low cost coupled with wide availability. This may be the case for CO_2 but not for O_3 and N_2O, unless it is possible to obtain the latter two as by-products of some other on-site chemical process. Indeed N_2O is a coproduct in the synthesis of adipic acid by cyclohexanone/cyclohexanol oxidation with HNO_3; in fact approximately 1 mole of N_2O is produced for every mole of adipic acid, in addition to the NO_x formed by HNO_3 reduction. An adipic acid production plant that generates 600 million lb/year cogenerates about 200 million lb of N_2O.[89] The reaction might also be integrated in an overall adipic acid process (Chapter 7).[89] The phenol produced by benzene oxidation with N_2O might be reduced to cyclohexanone with conventional technologies, the latter then being converted to adipic acid by oxidation with HNO_3.

Also of considerable importance is the possibility of using greenhouse gases for chemical applications, thus decreasing their negative impact on the environment. N_2O is a powerful greenhouse gas and it is known that it contributes to atmospheric ozone depletion. Its destruction with contemporaneous valorization for the production of chemicals obviously represents the best solution from both an economic and an environmental perspective.

2.1.2.1. Nitrous Oxide

Nitrous oxide has been studied as the oxidizing agent for the oxidation of ethane to ethene and acetaldehyde. On MoO_3, which is a nonselective catalyst for the oxidehydrogenation of ethane with O_2, N_2O decomposes at mild temperatures generating the very active radical species O^-, which oxidizes ethane.[68,69] Approximately 35% selectivity for acetaldehyde and 35% for ethene can be reached at about 5% ethane conversion, using a silica-supported MoO_3 catalyst.[68,69] The active site is a $Mo^{6+}-O^-$ species, developed by interaction of a reduced Mo^{5+} center with N_2O. The active species abstracts an H atom from ethane, thus generating the ethyl radical. The latter further reacts with an $Mo=O$ surface site, forming an ethoxy species, which can either decompose to ethene or react with a surface $Mo-OH$ species yielding acetaldehyde (with the possible formation of intermediate ethanol). Another advantage related to the absence of O_2 is the fact that the ethyl radical cannot form the ethylperoxy radical, precursor of carbon oxides formation.

Also in the case of silica-supported vanadium oxide the use of N_2O for ethane oxidation leads to considerable improvement in selectivity with respect to the use of molecular oxygen.[70,71] Here, too, it was suggested that reduced V^{4+} sites are responsible for N_2O decomposition and generation of the active oxygen species. Addition of alkali metal dopants to V_2O_5 and to MoO_3, with formation of mixed metal vanadates and metal molybdates, considerably modifies the surface properties of the transition metals (i.e., reducibility and acid properties) with a positive effect on both activity and selectivity.

Anshits et al.[72,73] postulated that uncharged oxygen species may play a role in the oxidative conversion of ethane. In the decomposition of N_2O on metal-exchanged ZSM5 zeolites, it was proposed that uncharged oxygen atoms $O^{\bullet}$

form rather than O^-, on the basis of the observed formation of cyclopropane from $CH_4/C_2H_4/N_2O$ mixtures. It was postulated that this species forms owing to the presence of strong acid sites (typical of zeolites) and of transition metal ions. It also was suggested that the same oxygen species are formed by N_2O decomposition over $Co^{2+}-MgO$.[74]

Another reaction that has been widely studied using N_2O as the oxidizing agent is the hydroxylation of benzene to phenol (see also Chapter 7).[75–90] Phenol is currently produced primarily through cumene oxidation to the hydroperoxide, followed by its decomposition to phenol with the coproduction of acetone. The technology is well established industrially, but the formation of equimolar amounts of acetone constitutes a major drawback. The alternative route of direct transformation of benzene therefore is of great industrial interest. The insertion of O into the C–H bond of benzene needs an oxidizing agent with electrophilic characteristics.

Electrophilic oxygen species can be generated starting from O_2 and letting it interact with some metal oxides; however, this leads to reactive but nonselective species.[1] On the other hand, the liquid phase hydroxylation of phenol to diphenols with H_2O_2 catalyzed by TS-1 (see Chapter 3) occurs through the formation of a cyclic electrophilic active species including a Ti hydroperoxo species. However, the presence of the OH group on phenol is essential to make the hydroxylation reaction feasible, since the same reaction is not carried out efficiently on benzene. Hence, researchers have studied the possibility of carrying out the synthesis of phenol using an oxidizing agent such as N_2O, able to develop very specific oxygen species *in situ*.

The reaction was first studied using a V_2O_5-based catalyst,[79] but it was soon found that the best performance among a number of zeolites and metal oxides was obtained with a H-ZSM5 zeolite in protonic form,[80–82] or where Al^{3+} cations were in part replaced by Fe^{3+}[77] or Ga^{3+} ions.[83,84] The more widely accepted mechanism is an acid-catalyzed one, even though the role of a redox mechanism catalyzed by metal impurities or framework metals has not been excluded.[77] By-products of the reaction are benzoquinone, catechol, resorcinol, and CO_2. In practice all the by-products form by consecutive reactions on phenol. Therefore an increase in residence time leads to a decrease in selectivity; also an increase in temperature has a negative effect on the selectivity for phenol, mainly because of the increase in the contribution of combustion reactions.

In the absence of benzene, decomposition of N_2O in a closed system at temperatures lower than 300°C leads to the evolution of only N_2, while all the released O is left on the catalyst. If this O-loaded zeolite is made to react with benzene at room temperature, only phenol is obtained.[85] Table 2.3 summarizes the most relevant results reported in the literature for benzene hydroxylation with N_2O.

TABLE 2.3. Key Results Reported in the Literature for
Benzene Oxidation to Phenol with N_2O

Catalyst	SiO_2/Al_2O_3	Selectivity to phenol, %	Production of phenol	Reference
H-ZSM5	> 90	95	3.2	83
H-ZSM5	33	98	1.8	81
H-Fe-ZSM5	100	99	3.0	86
H-Ga-ZSM5		98	10	78

Product.: mmole phenol/(g_{cat} h)

The biggest problem of the catalyst involves rapid deactivation owing to coke formation.[87,88] Consecutive reactions on phenol lead to the formation of coke. The more active the catalyst, the more rapid the deactivation rate and the lower the selectivity for phenol. The activity was found to increase in the order H-Al-ZSM5 < H-Ga-ZSM5 < H-Fe-ZSM5.

Apart from the zeolite composition (framework metal ion), the strength of the acid sites, the molar ratio Si/Al, and the pretreatment conditions also have a significant effect on catalytic performance.[77] As it is believed that the strongest acid sites are those responsible for coke formation (and hence for deactivation) selective poisoning of these sites by silylation was found to improve the long-term stability of the catalyst performance, with only a small loss in the initial activity.[78]

Further aspects of the behavior and use of N_2O in selective oxidation are discussed in Chapter 7.

2.1.2.2. Ozone

Ozone is another nonconventional oxidizing agent of interest (see also Chapter 8). In particular, it has attracted increasing attention recently as an alternative oxidant in the combustion of volatile organic compounds (VOCs) owing to its strong oxidizing ability, which allows for mild reaction conditions.[91–94] It has been proposed that in the oxidation of aromatics and lower aliphatic alcohols, catalyzed by MnO_2, the rate-determining step in oxidation by O_3 is its own decomposition. In ethanol oxidation the mechanism proposed involves the decomposition of O_3 to an adsorbed peroxide species, which reacts with an adsorbed ethoxide species to form CO_2.[94] Molecular oxygen and water are the coproducts obtained by ozone decomposition.

2.1.2.3. In Situ Generated Oxidants in the Liquid Phase

Processes for liquid phase oxidation of organic substrates involving the *in situ* generation of the oxidizing species (see also Chapters 3 and 7) have been proposed as alternatives to those that make use of *ex situ* prepared oxidants. Advantages to the *in situ* generation essentially relate to the possibility of overcoming typical drawbacks of monoxygen donors, and in particular[95]: (i) The low content of active O_2; e.g., in H_2O_2/H_2O solutions the content of H_2O_2 is always less than 47%. (ii) The considerable quantities of by-products and coproducts that are associated with the use of classical oxidants. For instance, in the case of *tert*-butylhydroperoxide and cumylhydroperoxide (used as oxidants in the reaction of propene epoxidation), *tert*-butyl alcohol and styrene which are valuable products, are coproduced but the market for them is rarely in balance with the demand for the main product. (iii) Their relatively high cost, which for bulk chemicals may be a serious economic consideration.

These drawbacks can be partially circumvented when the monoxygen donors can be generated *in situ* starting from a cheap and widely available oxidant, such as O_2, and a coreductant. Of course, the choice for an *in situ* generation process presupposes the total compatibility of the generation reaction with the oxidation reaction and with all the reactants and products involved.

The following oxidants can be generated *in situ*:

1. Metallorganic peroxy species and metal–oxenoid species can be obtained *in situ* by the reaction between the metal and O_2, in the presence of a reductant species (metal hydrides, metallic Zn or Fe, H_2), or by reaction between the metal and H_2O_2. The drawbacks are: (i) the large quantity of reductant that is consumed in the reaction (more than the stoichiometric amount), (ii) the slow reaction kinetics, and (iii) the by-products formed by overoxidation.[96–99]

2. Organic peracids, by reaction between the corresponding carboxylic acid and O_2 in the presence of H_2, or directly by reaction with H_2O_2. Alternatively, the cooxidation route involves oxidation of unsaturated organic substrates (i.e., epoxidation of alkenes) with the combined use of O_2 and of an aldehyde (i.e., acetaldehyde).[100–102] Aldehydes thus effectively act as reductant species, generating either the corresponding peracid or the acylperoxy species *in situ*, which finally epoxides the alkene and is itself converted to the carboxylic acid. The reaction is carried out in the presence of catalysts (metal complexes, heteropolycompounds, redox zeolites) to increase the selectivity of the process. A large excess of aldehyde is usually needed. Applications other than epoxidation have been studied, such as the oxyfunctionalization of alkanes[103,104] and the hydroxylation of benzene.

3. Most examples reported in the literature concern the *in situ* generation of H_2O_2 for the various advantages in using this environmentally friendly reactant.[105] In addition, the very peculiar properties of TS-1 in catalyzing monoxygenation reactions of organic substrates with H_2O_2 make it the most widely studied and characterized heterogeneous catalyst for this class of reactions. An example is the oxidation of alkanes with H_2O_2 generated *in situ*[106–109] using a bifunctional catalyst made of Pd (or Pt) supported over TS-1. Reducing agents for O_2 other than H_2 also

have been reported. In particular, CO can reduce O_2 to hydrogen peroxide in the presence of a Pd-based catalyst in an acidic aqueous medium. This system has been used for the oxidation of methane to formic acid and of ethane to acetic acid,[110] and for the hydroxylation of benzene to phenol.[111]

Other aspects of the *in situ* generation of H_2O_2 are discussed in Chapters 3 and 7.

2.2. NEW REACTOR TECHNOLOGY OPTIONS

Optimization of the catalytic performance in terms of reactant conversion, yield, productivity, and selectivity to the desired product is related not only to a thorough knowledge of the nature of the catalyst and interactions between reacting components and surface-active phases, reaction mechanism, thermodynamics, and kinetics but also to the development and use of a suitable reactor configuration, where all these features can be successfully exploited.

Industrial reactors used in the petrochemical industry for exothermic reactions, apart from a few exceptions, are either fixed-bed (adiabatic or nonadiabatic) or fluidized-bed reactors when the heat developed is too high to be managed in a fixed-bed apparatus. In the last few decades, interest has been mainly directed toward control of these reactors, strictly in terms of understanding the complex phenomena that occur at the interfaces between the different phases that are present in the reaction environment and heat and mass transfer effects on the reaction kinetics.

However, in recent years people working in the field of oxidation have realized that in order to achieve substantial improvement in the performance of several processes of hydrocarbon selective oxidation it is necessary to develop new reactor configurations, in order to overcome the fundamental drawbacks of conventional technologies. An important example in this direction is seen in the circulating fluidized bed reactor, which has been proposed over the years for a number of selective oxidation reactions, and has finally found commercial application in *n*-butane selective oxidation to maleic anhydride. This type of reactor may soon be developed for other applications as well. The principle exploited in this kind of configuration is the possibility of decoupling the classical redox mechanism (which operates in the selective oxidation of most hydrocarbons) into two separate steps, each of which can be optimized, thus improving the overall performance. The same principle has been applied for many years in other kinds of reactions, such as dehydrogenation and fluid catalytic cracking (FCC), but in these cases use of the circulating fluidized-bed reactor allows better management of the catalyst rejuvenation stage and better integration of the heat balance.

Another example is found in the monolithic-type reactors, which have their main application in the field of combustion. These reactor configurations, along with other types of "structured reactors," have recently been used to study the oxidation of hydrocarbons under reaction conditions that are completely different from the conventional ones. In particular, a combination of very high temperature and very low residence time seems to be one way to achieve the oxyfunctionalization of less reactive hydrocarbons with unexpected selectivity. It is likely in these cases that a combination of heterogeneous and homogeneous processes (probably the operating mechanism under these "unusual" conditions) is the best compromise between the limited capability of a catalyst to drive intermediates (once the hydrocarbon has been activated on the surface) along a specific pathway toward the final product and the well-known indiscriminacy of homogeneous radical-chain processes. Finally, membrane technology offers interesting potential advantages in allowing better control of the reaction kinetics.

In this section emerging new reactor technologies for the gas phase oxidation of hydrocarbons are reviewed and discussed, and compared with conventional technologies for the selective oxidation of hydrocarbons. Problems that are usually met in carrying out strongly exothermal reactions are analyzed and solutions aimed at overcoming them are examined.

2.2.1. Fixed-Bed Reactors

Fixed-bed reactors for gas–solid systems are normally used as continuous tubular or chamber reactors. In petrochemistry they are used for several oxidation reactions, including ethene oxidation to ethene oxide, methanol oxidation to formaldehyde, o-xylene or naphthalene oxidation to phthalic anhydride, n-butane or benzene oxidation to maleic anhydride, ethanol oxidation to acetic acid, ammoxidation of propene to acrylonitrile, and oxychlorination of ethene to 1,2-dichloroethane. They can be broadly classified into adiabatic and nonadiabatic reactors.

Disadvantages of conventional multitubular nonadiabatic (and nonisothermal) packed-bed arrangements, which become serious drawbacks for strongly exothermal reactions, are discussed in the following sections.

2.2.1.1. The Problem of Hot Spots in Both the Axial and Radial Directions

Hot zones derive from the poor heat-transfer properties of catalytic particles and from the more rapid reaction rate (corresponding to more heat released per unit time and unit volume) at the beginning of the tubular reactor. These zones, where the temperature of the reaction medium can be much more than 50°C higher than that of the inlet and outlet fluid, are highly undesirable, since they lead to several drawbacks, such as: (i) a faster reaction rate for those reactions that are characterized by a higher apparent activation energy, such as combustion of the hydrocarbon and of products; (ii) lower catalyst effectiveness, owing to the higher value of the Thiele modulus as compared to colder parts in the catalytic bed; and (iii) a faster catalyst deactivation rate, since all those phenomena that typically lead to the deactivation of the active component (e.g., fouling, sublimation of the active phase, phenomena of sintering and recrystallization, with a reduction in the surface area and redistribution of the components) are kinetically favored. It is worth mentioning that the possibility of improving the productivity by building multitubular reactors with larger tubes is often limited by the development of radial temperature gradients and thus by the thermal conduction properties of the catalytic bed.

Temperature gradients also lead to concentration gradients, since the reaction rate is higher in the axial position of the tube. Concentration gradients may lead to a different product distribution in the axial zone and in proximity of the reactor wall, especially when the reaction rates have different orders of reaction with respect to the reactants.

2.2.1.2. Heat and Mass Gradients

For strongly exothermic reactions thermal gradients can develop either inside the catalytic particle or, more frequently, at the interface between the fluid phase and the catalyst surface, thus in the stagnant film that develops over the latter. These gradients lead to the possibility of *multiple steady states* relative to the catalyst particle itself, in the same way as occurs, at macroscopic levels, in homogeneous reactors where there are exothermal reactions.

In regard to the analysis of intraparticle gradients, it is useful to define the following dimensionless number (Prater number for a catalyst surface):

$$\beta = \Delta T_{max}/T_S = -\Delta H C_{A_S} D_{eff_A}/(T_S \lambda_{eff}) \qquad (2.1)$$

where ΔT_{max} is the maximum difference in temperature in the particle with respect to the catalyst surface. The value of β ranges from 0.8 (for exothermic reactions) to -0.8 (for endothermic reactions). T_S and C_{A_S} are the temperature and the concentration of component A at the external catalyst surface. D_{eff_A} is the effective diffusivity of component A, and λ_{eff} is the effective thermal conductivity of the catalyst particle. This relationship represents the maximum increase (or decrease) in temperature in the grain relative to the external catalyst surface. For values of β less than 0, the catalyst efficiency is less than 1 (i.e., for endothermic reactions), while for exothermic reactions β is greater than 0. For exothermic reactions, the mean temperature of the particle may be substantially higher than that of the fluid phase, but, on the other hand, the concentration of the A reactant at the catalyst surface may be substantially lower than that in the gas phase, owing to the fact that concentration gradients may develop through the extragranular layer. These two effects counteract one another, and as a result the catalyst efficiency can assume different values, depending on the value of: (i) the Prater number, (ii) the activation energy relative to the reaction, and (iii) the ratio between the rate constant and the diffusivity, i.e, the Thiele modulus, which is defined, for a first-order irreversible reaction, by the relationship

$$\Phi = L'(k/D_{eff_A})^{0.5} \tag{2.2}$$

This relationship is illustrated in Figure 2.2. In the Φ vs. η (effectiveness) plot for exothermic reactions ($\beta > 0$), the efficiency can be less than, equal to, or greater than 1.[112] In addition, for the highest value of β (> 0.4–0.5, for strongly exothermic reactions) different solutions for the same value of the Thiele modulus can be obtained. Thus there are two different states of stable operation, which correspond to two different values of internal gradients in the particle: (i) a hotter state, with a very high reaction rate and hence very high efficiency; and (ii) a colder state, with a lower temperature and lower concentration gradients. Moreover, transition from one state to the other is possible, owing to events such as runaway or estinguishing of the reaction.[113] When the maximum difference in temperature is lower than 5 K, the effect of temperature on the grain efficiency can be neglected. Criteria of uniqueness of state have been derived by many authors; for a first-order reaction

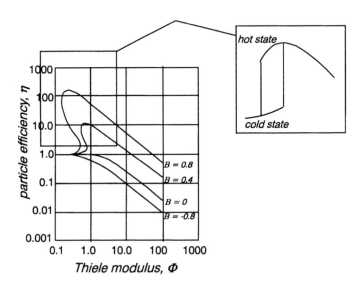

Figure 2.2. Particle efficiency as a function of the Thiele modulus for different values of β, calculated neglecting the external heat and mass-transfer resistance. The curves are calculated for spherical particles, irreversible first-order reactions, and for $E/(R \cdot T_s) = 20$ ($R = 8.31$ J·mol^{-1}·K^{-1}).

in a spherical pellet it was pointed out[114] that if

$$\beta\gamma < 4\,(1 + \beta) \qquad (2.3)$$

where

$$\gamma = E_a/(RT) \qquad (2.4)$$

a single steady state is observed. Values for $\beta\gamma$ and for β are given in Table 2.4.[115]

Moreover, interparticle temperature gradients may lead to multiple steady states. The temperature gradient in the film may be high, thus leading to considerable differences in temperature between the gas phase and the catalyst surface. In practice, if the reaction is very fast and thus is controlled by mass transfer, the temperature difference between the catalyst surface and the bulk corresponds to the adiabatic temperature rise. Multiple steady states arising from extragranular gradients are not very frequent in mild oxidation reactions, and are more likely to occur only for very fast, exothermic reactions such as combustion.

Multiple steady states also exist for single-phase systems, but a high number of steady states is typical of multiphase systems, especially in the presence of complex reaction networks (i.e., with consecutive reactions).

An overall fixed-bed model that takes into account all phenomena occurring in the reactor (included axial mixing), as well as those occurring in the catalyst particles (i.e., a heterogeneous model, with interfacial and intraparticle concentration gradients) was developed first by Hlavacek and Votruba.[116] This type of model predicts the existence of multiple steady states. For a first-order reaction, multiplicity of steady states is possible if the following relationship is satisfied[117]:

$$S > 4\gamma/(\gamma - 4) \qquad (2.5)$$

where

$$S = \gamma(-\Delta H)C_{A_0}/(\rho_F C_{P_M} T_0) = \gamma \Delta T_{ad}/T_0 = T_{ad}E_a/(RT_0^2) \qquad (2.6)$$

is the dimensionless adiabatic temperature rise referred to the temperature at the reactor inlet, T_0, and γ is the dimensionless apparent activation energy; $\Delta T_{ad} = T_{ad} - T_0$; ΔH is the heat of reaction developed per mole of the reference compound A,

TABLE 2.4. Dimensionless Parameters ($\beta\gamma$) and β for Some Exothermal Reactions[115]

Reaction	$\beta\gamma$	β
Ethylene oxidation	1.76	13.4
Methanol oxidation	0.175	16.0
SO_2 oxidation	0.175	14.8
H_2 oxidation	0.21–2.3	6.7–7.5

C_{A_0} is its initial concentration, C_{p_M} is the thermal capacity of the fluid phase, and ρ_F is the density of the fluid phase.

The heat removal rate curve (from the heat balance applied to the particle, in the absence of radiation losses) indicates the corresponding temperature of the catalyst surface for a defined temperature of the gas phase (inlet reactor temperature) (Figure 2.3). The point is given by the intersection between the rates of heat removal and heat generation. At the highest temperature the reaction rate is significantly higher than at the lowest temperature, and under the former conditions the overall rate is mass-transfer controlled. Indeed, in this high-temperature range the reaction rate itself shows little variation for relatively large temperature variations, as expected from the very low activation energy of diffusion processes. It is shown that for gas phase temperatures below point A there is only one point of intersection between the two rates, which corresponds to steady state operation, while for gas phase temperatures above point B there is also only a single point of intersection, but in the temperature range where the reaction rate is mass-transfer limited.

Point B' is referred to as the "catalyst ignition temperature," while point A' is referred to as the "catalyst extinction temperature." Intersections in the intermediate

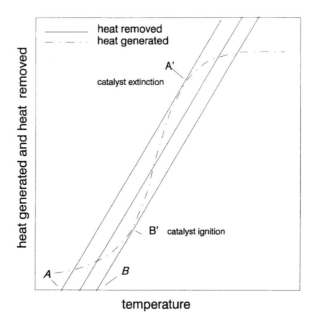

Figure 2.3. Energy balance, applied to the particle, between heat generated and heat removed for an exothermic reaction under autothermal operation.

range of the *S*-shaped curve (i.e., at catalyst temperatures between the ignition and the extinction temperatures) are not stable steady states, and small perturbations lead to a jump to either the low-temperature or the high-temperature region. For instance, if operation is carried out under mass-transfer-limited conditions, a variation in the flow velocity may lead to a decrease in the reaction rate, with a shift to a kinetically controlled regime.

The difference in temperature between the gas phase and the catalyst surface with a mass-transfer-limited steady state may be considerable—more than 100°C. However, when the reaction is kinetically controlled, the temperature of the catalyst surface is essentially the same as that of the gas phase, since the temperature of intersection between the rate of heat removal and the rate of heat generation is very close to the gas phase temperature.

Multiple steady states should be avoided because of the possibility of catalyst ignition or extinction phenomena, but, it has been shown that they do not often occur under ordinary industrial conditions.[118,119]

2.2.1.3. Presence of Multiple Steady States and Runaway Phenomena

The presence of multiple steady states, some of which are stable, while others are unstable and sensitive to small variations, is related to the concept of parametric sensitivity,[120–126] and the latter is very important in defining the conditions under which thermal runaway can occur. These conditions become very stringent in the presence of hot spots in the catalytic bed. Even though these phenomena can also occur in homogeneous systems, temperature gradients that might develop in packed beds are much more important, owing to the poor thermal conductivity of solid particles.

This is one of the most important aspects to be taken into account when analyzing thermal effects in selective oxidation reactions, since operation under conditions of parametric sensitivity may lead to ignition of very exothermic reactions (combustion) and hence to runaway. Thus this situation is typical of highly exothermic reactions characterized by high activation energy.

Parametric sensitivity analysis indicates that for defined reaction conditions (cooling medium temperature, reactor inlet temperature, concentration of the reactant for a certain tube diameter), the tubular reactor may be extremely sensitive to small changes in one of the parameters and suddenly reach much higher temperatures. Particularly critical is the situation of the hot spot, which is very sensitive to relatively small changes in the process variables. Of course, since the hot spot is the hottest part of the reactor, evaluation of the boundary conditions for parametric sensitivity has to be done in reference to it. In other words, a reactor is considered to be unstable when small perturbations from the steady state cause an uncontrolled rise in the hot-spot temperature.

The best way to evaluate the region of parametric sensitivity for a multitubular reactor is to carry out experimental tests in a single-tube pilot plant of the same size and length as the industrial model. However, several criteria have been developed that allow an approximate identification of the critical region of parametric sensitivity. In general, one generally accepted criterion for the assessment of the conditions for thermal runaway consists of the following calculation:

$$N = 4U/(d_T \rho_F C_{p_F} k) \tag{2.7}$$

where k is the kinetic constant, C_{p_F} is the specific heat of the fluid phase, ρ_F is the density of the fluid, d_T is the tube diameter, and N is the dimensionless rate of heat transfer per unit volume. The overall heat-transfer coefficient U accounts for resistance to convective heat transfer at the inner and outer walls of the reactor and to conduction through the wall. By plotting the N/S ratio as a function of S (as defined previously), a curve is obtained that separates the parametric-insensitive region from the parametric-sensitive one. The gas inlet temperature is taken as the reference, since it is assumed to be equal to the wall temperature, which is the situation usually encountered in practical applications. In addition, the wall and coolant temperatures are often assumed to be equal. Generalized charts plotting N/S vs. S for reactions of different orders, for different sets of conditions, have been given by Agnew and Potter.[124]

Figure 2.4 shows a general trend for the runaway diagram. The calculation of S and N allows the determination of the value of d_T below which, for given reaction conditions and for a definite value of the overall heat-transfer coefficient, the

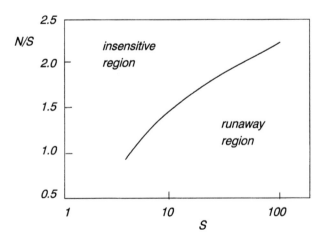

Figure 2.4. Runaway diagram.

operation is carried out outside of the parametric-sensitivity region. One safety measure is to use a design diameter lower than $d_{T_{max}}$, typically $0.67 d_{T_{max}}$.

2.2.1.4. Wide-Residence Time Distribution

Owing to the presence of back-mixing phenomena a wide residence time distribution is found in fixed-bed reactors. This leads to a distribution of residence times inside the reactor, with a worsening of selectivity for the desired product owing to the increased contribution of consecutive reactions (i.e., the combustion of the desired product). When the dimensionless group $D_{ax}/(uL) = 1/Bo$ (where D_{ax} is axial dispersion, u the linear velocity, L the reactor length, Bo the Bodenstein number) is higher than 0.01, considerable deviations from plug flow are expected, which lead to a significant decrease in conversion and worsening of selectivity with respect to the plug flow reactor.

Suitable geometric reactor parameters must be adopted to minimize back-mixing and by-pass phenomena, such as: (i) $Re_p = (du\rho_F/\mu_F)$ higher than 10; (ii) $L/d > 50$; and (iii) $n/m > 0.5$ and $m > 10$, where $L/d_R = n$ and $d_R/d = m$, with d the catalyst average diameter and d_R the reactor diameter. In this way it is also possible to reach a pressure drop along the bed at least equal to the 2500 Pa/m necessary to minimize bypass phenomena.

2.2.1.5. High Pressure Drop

Relatively high pressure drops are also a problem in fixed-bed reactor operations. As noted above, a certain pressure drop is necessary to optimize the radial distribution of fluid components. However, pressure drops exhibited by packed beds can also constitute a limit to the productivity. Moreover, unacceptable levels of pressure drop are often the reason for shutdown of packed-bed reactors.

Pressure drop ΔP per unit length H is usually calculated through the well-known Ergun empirical relationship:

$$\Delta P/H = A[\varepsilon_c^2/(1-\varepsilon_c)^3(\mu_F u/d^2) + B[\varepsilon_c/(1-\varepsilon_c)^3(\rho_F u^2/d) \tag{2.8}$$

where ε_c is the fraction of reactor volume occupied by the catalyst, μ_F is the fluid viscosity, u is the linear velocity, d is the average catalyst particle size, and ρ_F is the fluid density (all calculated at reaction conditions). A and B are two coefficients, the values of which are, respectively, 150 and 1.75 for homogeneous beds of spherical particles all equal in size. $1 - \varepsilon_c$ ranges from 0.42 (for a recently loaded packed bed) to 0.38.

2.2.2. Technologies Designed to Overcome Drawbacks of Fixed-Bed Reactors

Different technological solutions in fixed-bed reactors designed specifically to overcome the problems outlined above are discussed in the following sections.

2.2.2.1. Dual-Bed Reactors

The use of graded catalysts (dual-bed reactors)[127,128] is one of the technical solutions adopted to mitigate the problems of fixed-bed reactor operations discussed above. For instance, catalytic beds are arranged in two or three parts, each part containing a catalyst for which the formulation has been optimized, i.e., less active for the first part of the reactor (close to the reactor inlet), where the reaction rate is the highest, and more active for the final part of the reactor and for finishing the reaction. Control of catalyst activity can be achieved either by a proper formulation of the active phase (i.e., addition of dopants or controlled amounts of the active phase in a supported system), or by dilution of the catalyst with inert particles, the latter being characterized by optimal thermal conductivity properties. This arrangement is usually achieved in reactors for o-xylene oxidation to phthalic anhydride, where the catalytic bed is made up of two sections. In each section the catalyst consists essentially of α-alumina or steatite pellets, impregnated with V_2O_5/TiO_2-based catalysts. The activity of the catalyst in each section is optimized by controlling the vanadia content, as well as by the addition of dopants. In this way the hot-spot temperature may become considerably lower, and the hotter region is spread over a longer reactor length, with a considerable improvement in selectivity for the product of partial oxidation.

2.2.2.2. Distributed Inlet of One Reactant

Differentiation of one reactant (typically oxygen) along the reactor or staged at different reactors in a battery of reactors in series is also sometimes used to improve reactor performance. In this way the reaction is distributed more homogeneously along the entire catalytic bed, resulting in a more uniform temperature profile. This is the case in the reaction of ethene oxychlorination to 1,2-dichloroethane, where three in-series multitubular reactors are used, and where one-third of the overall oxygen requirement is fed into each reactor (while ethene is fed entirely into the first reactor). This results in an improvement in product purity, since with lower average temperatures in the beds, smaller amounts of undesired by-products (not only carbon oxides, but also ethylchloride and polychlorinated compounds) are formed. Oxygen differentiation also makes it possible to maintain the gas phase composition outside the flammable region.

Another example has been proposed by Papageorgiu and Froment,[129] for the selective oxidation of o-xylene to phthalic anhydride. These authors compared the distribution of products in a fixed-bed multitubular reactor and a three-bed reactor in which the catalyst is divided into three chambers and pure oxygen is fed in partly at the top of the reactor and partly between the catalyst beds (Figure 2.5). Feedstock differentiation allowed better distribution of temperatures in the reactor and a final improvement in the selectivity to phthalic anhydride at the expense of carbon oxides.

2.2.2.3. Periodic Flow Reversal

Periodic flow reversal inducing forced unsteady state conditions has been adopted in the control of autothermal adiabatic reactions with direct regenerative heat exchange.[130,131] In practice, part of the catalytic packed bed also acts as a heat exchanger. The flow reversal allows the reaction front (which develops during unsteady state operation) to move alternately in opposite directions along the bed. In this way, the reaction front heats the cooler part of the bed, thus permitting an autothermal reaction (the heat of reaction leaves the reactor only with the outgoing flow) with a relatively small adiabatic temperature increase. Applications have been developed for catalytic combustion of VOCs in air, and for equilibrium-limited exothermic reactions (SO_2 oxidation, synthesis of methanol and ammonia, NO_x reduction by ammonia).

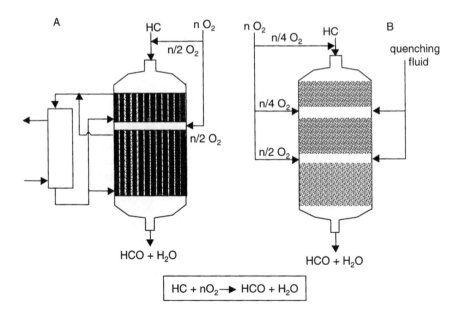

Figure 2.5. Example of oxygen differentiation along the catalytic bed in a multistage multitubular reactor (a), and in a multistage adiabatic reactor with intercooling (b).

This technical solution is also being studied for selective oxidation of hydrocarbons. During reactor start-up an initial unsteady state situation develops, during which the temperature in the catalytic bed progressively increases, until the fully developed temperature profile (including the hotter region) and thus steady state performance is established. Since the hot-spot region is close to the reactor inlet, if the flow to the reactor is continuously reversed before the steady state is attained, a stable situation finally develops with a dual hot-spot temperature profile, characterized by a considerably lower temperature than in the single hot spot that would develop in the traditional flow configuration. The penalty in conversion resulting from the reversal flow configuration is limited, and the gain in selectivity can be considerable.

2.2.2.4. Decoupling of the Exothermal Reaction into Two Steps

In a two-step reactor configuration, the overall heat of reaction is subdivided into two less exothermic steps. In the first step the hydrocarbon is put into contact with the catalyst (which furnishes its surface oxygen to the organic substrate being reduced at the same time), while in the second step the reduced catalyst is restored to its original oxidation state by contact with gaseous oxygen. The two semireactions can be indicated as follows:

$$HC + cat_{ox} \rightarrow HCO + cat_{red} \tag{2.9}$$

$$cat_{red} + O_2 \rightarrow cat_{ox} \tag{2.10}$$

Apart from better control of the overall reaction exothermicity, further advantages of this operation are: (i) the higher selectivity that is usually achieved because the hydrocarbon is never in contact with molecular oxygen; and (ii) the elimination of hazards associated with the possible formation of flammable gas mixtures.

This configuration can be carried out either: (i) in two (or more) parallel reactors, where one (or more) reactor is in the reaction stage and the other is in the catalyst reoxidation stage; or (ii) in a circulating-bed reactor, where the catalyst is continuously transported from the reaction vessel to the regeneration vessel and vice versa. This solution has been developed for the oxidation of n-butane to maleic anhydride (DuPont process)[132-136] in a mobile-bed reactor, for the coupling of methane to ethene by ARCO,[137] and for the ammoxidation of propane to acrylonitrile by Monsanto.[138] The technology of circulating fluidized-bed reactors will be discussed in more detail later.

2.2.2.5. Integration of Exothermal and Endothermal Processes

Another solution aimed at overcoming problems associated with the large amount of heat released consists of heat integration of the exothermal process with an endothermal one.[139,140] In the Oxco process the oxidative coupling of methane occurs in the first zone at the inlet of a fluidized-bed reactor, where there is also complete consumption of oxygen. In the remaining section of the reactor the ethane formed (as well as other alkanes contained in natural gas) is pyrolyzed to ethene (endothermal reaction). Of course, an overall isothermal operation is obtained when the two reactions are properly balanced energetically. This approach might represent a solution to problems associated with hot spots in packed beds if the two reactions can be carried out in the same catalytic bed, with a multifunctional catalyst able to perform in both reactions.

A solution has been proposed by some companies for the dehydrogenation of alkanes and ethylbenzene coupled with the oxidation of part of the hydrogen developed to supply the heat internally for the dehydrogenation, thus avoiding costly reheating.[141–143] A further advantage is that the consumption of the hydrogen produced allows the equilibrium to be shifted in favor of the product. Either dual catalyst systems or a single multifunctional catalyst can be adopted. In the former case the oxidation catalyst can be based on supported palladium or platinum, as in the UOP process for the dehydrogenation of ethylbenzene. However, a single multifunctional catalyst has been proposed for dehydrogenation of alkanes.

In the SMART (styrene monomer advanced reheat technology) process for the dehydrogenation of ethylbenzene, the oxidation catalyst is inserted into the second and third of three in-series reactors. The gas flows radially outward from the center through two catalyst beds separated by screens: first the oxidation catalyst bed and then the conventional dehydrogenation catalyst bed. It has been claimed that this technology yields more than 80% conversion per pass at the same selectivity for styrene as for conventional processes.

2.2.2.6. Radial Flow Reactors

In the radial flow configuration the catalyst is placed between two coaxial cylinders, and the gas flows from either the center (outward flow) or the external part (inward flow). With such a configuration, residence times comparable to those obtained in conventional multitubular reactors are achieved, but with a much smaller pressure drop. This reactor type is used commercially in the synthesis of ammonia.

2.2.3. Fluidized-Bed Reactors

In a fluidized-bed reactor the catalyst particles are suspended by the upward gaseous flow, developing a so-called fluidized state. This type of reactor is employed industrially in several hydrocarbon oxidations, such as: (i) oxychlorination of ethene to 1,2-dichloroethane, (ii) ammoxidation of propene to acrylonitrile, (iii) oxidation of n-butane to maleic anhydride, and (iv) oxidation of o-xylene to phthalic anhydride. In all these cases except for the synthesis of acrylonitrile, there are also alternative processes that depend on the use of fixed-bed, multitubular reactors.

The greatest advantage of fluidized-bed reactors is the possibility of achieving an isothermal catalytic bed, thus avoiding the hot spots typical of multitubular fixed beds, which can interfere with the performance and damage the catalyst itself. This possibility arises from the fluidlike behavior of the solid particles suspended in the fluid stream, where the particle movement disperses the heat generated, and from the large heat exchange surface that can be achieved by immersion of the heat exchanger coils into the fluidized bed. The solid particles must be of suitable dimensions, as well as have the required morphological and mechanical properties.

The gaseous feedstock can be fed in either below a grid placed at the bottom of the reactor or directly inside the catalytic bed, above the grid, which is sometimes done to avoid the formation of flammable compositions outside the catalytic bed.

Disadvantages of fluidized-bed technology are essentially related to:

1. Difficulties in making a reliable scale-up of the reactor (due to the presence of complex fluidodynamic behavior, especially in multiphasic systems).
2. Problems in developing a catalytic material characterized by optimal mechanical and fluidization properties.
3. The intrinsic limitation in productivity due to a maximum feed rate that cannot be increased beyond the elutriation limit for catalyst particles.

The need to balance advantages and drawbacks may justify the choice for a fluidized-bed rather than for a fixed-bed reactor. For instance, in the case of ethene oxychlorination, the almost isothermal profile of the bed leads to higher purity of the 1,2-dichloroethane. This quality is important in affecting the quality of vinyl-chloride (obtained by pyrolysis of 1,2-dichloroethane in a cracker unit downstream from the oxychlorination plant), and therefore the choice of the fluidized bed can be the best one. On the other hand, when improvements in catalyst thermal conductivity and the choice of operating conditions (i.e., differentiation of oxygen feedstock) make it possible to lower the temperature of the hot spot in multitubular reactors to a significant degree, fixed beds enable higher productivity than fluidized beds.

In some cases the choice of the fluidized-bed technology is practically compulsory, e.g., in alkane oxidation, where the heat of reaction developed is generally so high (except in the case of alkane oxidehydrogenation) that it can be removed properly only in a fluidized-bed system. A fluidized bed has been developed by BP America for the ammoxidation of propane to acrylonitrile ($\Delta H^0 = -151$ kcal·mol^{-1} at 25°C) and a pilot plant fluidized bed has also been announced by Mitsubishi. It is worth noting that the heat of reaction developed is much higher, owing to the presence of parallel and consecutive combustion reactions, which are strongly exothermic. Therefore the selectivity of the process determines the heat developed.[144]

The same considerations have been reported for methane oxidative coupling.[145] By assuming a selectivity of 84% for C2 products (including 46% for ethene), the remaining product being CO_2, the overall reaction heat per mole of C2 produced is around 123 kcal·mol^{-1}, which is at the limit for the use of multitubular reactors. However, in this reactor type, excessively long tubes would be required to control the heat transfer and avoid runaway. Alternatively, a dilution of reactants might be accomplished by decreasing the conversion per pass (and increasing the recycle ratio), which would lead to large separation systems, or by adding steam, which would lead to an increase in cost. The fluidized-bed reactor would enable better management of the heat transfer problem.

The different hydrodynamic regimes (Figure 2.6) that are encountered in gas–solid systems are:

1. A fixed-bed (or delayed-bed) regime, for gas velocities below the minimum fluidization velocity u_{mf}, which corresponds to the gas velocity sufficient to support the hydrostatic weight of the catalyst, and hence to suspend it. In the fixed-bed regime an increase in gas velocity does not cause a variation in the bed height but does lead to an increase in pressure drop.
2. A homogeneous fluidization regime, for velocities higher than u_{mf}, in which the pressure drop is constant while the bed expands uniformly with increasing gas velocity until a maximum bed height is reached. Typical values of u_{mf} range from less than 1 cm·s^{-1} to a few cm·s^{-1}.
3. As the velocity continues to increase, a point is reached at which bubbles are formed (heterogeneous or bubbling or aggregative fluidization). An initial bed contraction is observed, which then reaches a stable height. Bubbles lead to fluctuations in the bed height. They tend to rise very quickly with a velocity that is usually much higher than the interstitial gas velocity and churn up the solid bed causing back-mixing of the bed itself. Back-mixing can be detrimental since it favors the contribution of consecutive reactions, but it can also be useful for achieving better homogenization of the temperature in the bed and as well as promote mixing. The formation of bubbles in the case of oxidation reactions can be dangerous owing to the formation of flammable compositions within the bubble itself.

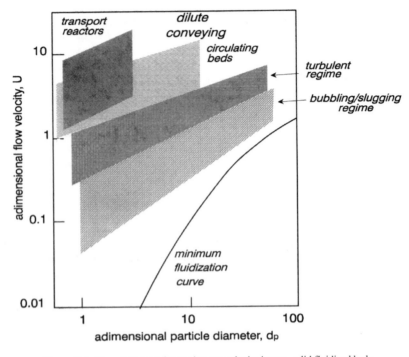

Figure 2.6. Flow regimes at increasing gas velocity in gas–solid fluidized beds.

4. With even greater gas flows the bubbles increase in size until the frontal diameter of the bubbles equals the diameter of the vessel at which point slugging sets in.
5. High-velocity-gas systems (turbulent regime and lean-phase with transport regimes) develop for very rapid gas flows, for which the bed density is much lower than with bubbling and slugging.

The oxidation of light alkane in a laboratory fluidized-bed reactor was studied recently by Bharadwaj and Schmidt[146,147] for ethane and methane transformation to syngas and propane, n-butane, and isobutane oxidehydrogenation to the corresponding alkenes. The use of these reactors operating autothermally at low residence times (in the turbulent regime), with a catalyst of Pt- or Rh-coated α-Al_2O_3 beads, led to remarkable selectivities for alkenes (i.e., > 60%) at high (> 80%) alkane conversion. High flow rates resulted in a thin extragranular boundary layer and thus enhanced mass transfer. The high selectivity achieved was explained by hypothesizing a total oxygen consumption close to the distributor and moderate gas back-mixing. Under these conditions the alkene formed at the very beginning of the catalytic bed did not undergo consecutive undesired combustion reactions.

A particular laboratory fluidized-bed reactor configuration (*in situ* redox fluidized-bed reactor) was proposed recently by Soler *et al.*[148] for alkane oxidative dehydrogenation. Owing to the specific type of hydrocarbon inlet inside the fluidized bed, the reactor operates like a two-region reactor, where the catalyst works in the virtual absence of oxygen in the upper part and the reduced catalyst is reoxidized in the bottom part. The catalyst is internally recirculated from one region to the other. In this way it is possible to have the advantage of separate hydrocarbon and oxygen interaction with the catalyst in a single reactor. It is claimed that better selectivities and productivities are possible with this reactor, although scale-up is probably difficult.

2.2.4. High-Gas-Velocity Systems: Circulating Fluidized-Bed Reactors

Since the 1940s circulating fluidized-bed reactors (CFlBRs) have been employed as alternatives to conventional fluidized-bed reactors for several gas–solid systems, such as in catalytic cracking, flue gas desulfurization, combustion of low-grade fuels (coal waste and biomass), pyrolysis, and gasification of solids. CFlBRs have been referred to over the years by a variety of names, including transfer-line reactors, transport-bed reactors, and high-velocity fluidized-bed reactors.

The main advantage of this reactor configuration lies in the possibility of achieving continuous catalyst reactivation by transport from the reaction vessel to the regeneration vessel. Other differences with respect to the fluidized bed concern the improved possibility of managing very fast, exothermic reactions.

In the petrochemical field, the most important industrial applications concern dehydrogenation and the recently developed process for the oxidation of *n*-butane to maleic anhydride. The latter was developed by DuPont, and went on-stream at the end of 1996 in Spain. The particular properties of this reactor type are in fact best exploited in selective oxidation reactions where a redox-type mechanism is operative. It is thus possible to decouple the two steps of the redox reaction in two separate reactors, and cause the hydrocarbon to interact with the catalyst in the absence of gas phase oxygen to yield the oxidized product (Figure 2.7).

The hydrodynamic regime in high-velocity two-phase (gas–solid) systems (Figure 2.8) may be one of three different types: (i) turbulent, (ii) fast fluidization, or (iii) pneumatic transport. Thus the main difference with respect to conventional fluidized beds concerns the gas velocity, which is usually in the range $3–10 \text{ m·s}^{-1}$, well beyond the bubbling and slugging regimes. At these velocities the solid (i.e., the catalyst or the powder to be treated) is carried over the top of the vessel. The turbulent regime may occur over a wide range of particle sizes and gas velocities.[149]

The regime occurring at a gas velocity higher than approximately $2–4 \text{ m·s}^{-1}$ is fast fluidization, which is typical in CFlBRs, as in FCC systems and in circulating

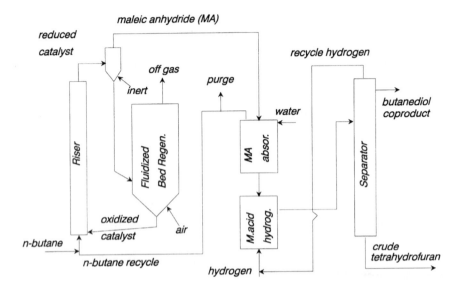

Figure 2.7. Simplified flow sheet of the CFIBR DuPont reactor for the oxidation of *n*-butane to maleic anhydride.

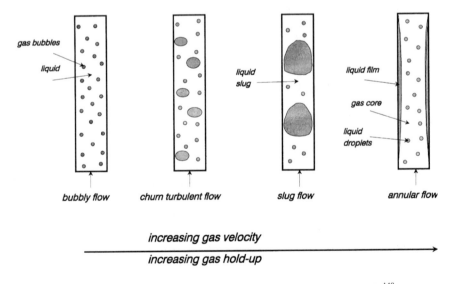

Figure 2.8. Hydrodynamic regimes in gas–solid high-velocity fluidized beds.[148]

fluidized-bed combustion systems.[150] The transition from turbulent to fast fluidization is not clear-cut, but is, rather, gradual, with a corresponding gradual increase in the bed void fraction. A large fraction of solid is entrained from the vessel top and has to be returned to it or fresh solid has to be added to maintain the solid inventory. The extensive overlapping between the fast-fluidization and pneumatic transport regimes (the latter generally occurring in riser reactors), is evident in Figure 2.6. The two regimes differ mainly in the fraction of solid; pneumatic transported-bed reactors are more dilute (1–5 vol% solid) than fast-fluidization reactors (2–15 vol% solid).

General advantages of the CFlBR in chemical processes, with respect to fixed-bed technology or conventional fluidized beds, are as follows:

1. Greater plant flexibility, since the two sections can be designed and optimized separately by choosing the best conditions for each, which also leads to a lower catalyst inventory.
2. Higher throughput, because of the higher gas velocity and the possibility of independently optimizing the kinetic conditions for the two stages (i.e., the two redox steps in the selective oxidation, or the reaction and the regeneration step in dehydrogenation or catalytic cracking).
3. Excellent intragrain and interphase heat and mass transfer, because of the fine solid particles used (typically <0.2 mm), and the high gas velocity relative to the solid; consequently, very high catalyst efficiencies can be achieved, and thermal gradients inside the particle are minimal.
4. Higher product concentration, which means fewer recovery problems.
5. The gas–solid contact is improved with respect to the bubbling/slugging regime.

On the other hand, there are disadvantages:

1. The need for a catalyst with special properties: high attrition resistance and high availability of active centers even in the absence of one reactant (i.e., in the decoupled redox reactions), and ready availability of bulk oxygen.
2. Significant uncertainty in scale-up, owing to radial and axial gradients in the solid, and the lack of a model that would enable prediction of geometric parameters and discontinuity at the inlet and outlet of the riser.
3. High energy costs associated with recirculation of the solid and high capital costs.

Dense riser transport reactors and fast-fluidization reactors exhibit limited axial gas dispersion, but considerable radial gradients, with the solid being concentrated close to the reactor wall (annular region). A core/annulus model has been proposed for CFlBRs, with one-dimensional upflow of gas and entrained solids in the dilute column core, a dense phase of downflow catalyst clusters in the thin annular region, with gas moving slowly upward or downward in the latter phase. This model has also been assumed in the DuPont process for n-butane oxidation,[151,152] even though some authors have suggested that it does not adequately describe the hydrodynamic regime in the DuPont reactor because that reactor is radially and axially homogeneous enough to be better represented by a homogeneous plug–flow model.[134]

In the DuPont process, the CFlBR technology is used for the synthesis of maleic anhydride by n-butane oxidation (Figure 2.7). The anhydride is then hydrogenated to tetrahydrofuran. The hydrocarbon is oxidized by the lattice oxygen of the catalyst, a vanadium/phosphorus mixed oxide,[153] in the riser reactor, which is then carried by the gas stream to the regeneration vessel, where it is exposed to air in order to restore the original oxidation state of vanadium.[133] Fresh n-butane, together with the recycled gas and the reoxidized catalyst, enters at the bottom of the riser. At the top of the riser the reduced catalyst is separated from the gas stream, and is then introduced into the regeneration vessel, a fluidized bed where it is oxidized with air before being recycled to the riser. Maleic anhydride is recovered from the off-gas by scrubbing with water. The rate of catalyst circulation is affected by the amount of active oxygen available on the catalyst surface. The catalyst circulation rate is approximately 0.5×10^3 kg·s^{-1} for a maleic anhydride plant with a capacity of 40×10^6 kg·year^{-1}.

2.2.5. Structured Catalysts and Reactors

The term "structured reactor" indicates a catalytic reactor in which the catalyst is not shaped in the usual form, as adopted for fixed-bed, fluidized-bed, or circulating fluidized-bed reactors (i.e., granules, pellets, extrudates, or microparticles). Structured catalysts are shaped by a metal or a ceramic structure (or both), which serves as a base for the deposition of the catalytically active component (wash-coating). They can be broadly classified into: (i) catalysts where no convective mass transfer occurs over a cross section of the reactor, and (ii) catalysts where convection occurs in the radial direction of the reactor.[154] To the first group belong monolithic catalysts (such as honeycomb structures for automotive exhaust gas treatment), while to the second belong corrugated-plate monoliths with cross-flow channels, parallel-plate monoliths, bead-string reactors, catalytic-gauzes, nets, and bundles. Monolithic reactors with permeable walls (membrane reactors) represent an intermediate situation, since radial flow is possible but occurs mainly by diffusion.

Important industrial applications of oxidation reactions which make use of structured catalysts are the following:

OXIDATION OF AMMONIA TO NO (AND THEN TO HNO_3)

$$NH_3 + \frac{5}{4} O_2 \rightarrow NO + \frac{3}{2} H_2O \qquad (2.11)$$

where the catalyst, discovered in the early 1900s by W. Ostwald, is made of Pt–10%Rh wires woven into gauze. The catalytic bed is arranged into several (5 to 20) layers of gauze, stacked one over the other to form a bed several millimeters deep.

ANDRUSSOW PROCESS OF METHANE AMMOXIDATION TO HCN

$$CH_4 + NH_3 + \frac{3}{2} O_2 \rightarrow HCN + 3H_2O \qquad (2.12)$$

where the catalyst is the same as for ammonia oxidation.

The main features of these processes are the very low residence time that is achieved (around 1 ms in both cases) and the adiabatic autothermal conditions. For this reason structured reactors are often referred to as *short-bed* reactors. There are similar features in the reaction of oxidative dehydrogenation of methanol to formaldehyde, but in this case the catalyst (made of silver needles) is not properly a "structured" one. The heat developed in these very exothermal reactions heats the catalyst surface to very high temperatures, e.g., 800°C for ammonia oxidation and 1100°C for methane ammoxidation. At such high temperatures the processes are usually under mass-transfer control. When the reaction is under kinetic control, a drawback may arise from the high sensitivity of the reaction to temperature and hence to the susceptibility to oscillation phenomena due to ignition–quench hysteresis.

In general, because of the adiabatic operation short-bed configurations are suitable for highly exothermic reactions and high operating temperatures. The main problem in such a reactor configuration is the necessity for rapid heat transfer via conduction and radiation from the end of the catalytic bed in the upward direction in order to heat the colder feedstock efficiently. Thus the feed rate must be properly controlled to avoid progressive cooling of the catalytic bed and extinction of the reaction.

These "extreme" conditions would seem suitable only for reactions where the control of selectivity is not a problem, but nevertheless have been proposed not only for the oxidative coupling of methane,[155] but also for reactions of partial oxidation of hydrocarbons.[156–161] The fundamental concept is to achieve very low residence times (on the order of milliseconds), under autothermal conditions at temperatures that can be as high as 1000°C. The feedstock usually does not contain diluents or ballast. Under these conditions the control of selectivity is not dictated by the catalyst properties, since the only catalysts that can operate under these conditions are noble metals, which are intrinsically nonselective, especially under these reaction conditions. The distribution of products is thus mainly dictated by the reaction parameters, and in particular by:

1. Hydrocarbon-to-oxygen ratio[159-161]: Referring to the stoichiometric ratio for total combustion of the hydrocarbon to CO_2 and H_2O ($\phi = 1$), excess O_2 leads to the prevailing formation of the latter compounds. For mixtures poorer in O_2, CO becomes the prevailing product, and conditions may exist at which CO and H_2 are preferentially produced (partial oxidation to syngas). For mixtures very poor in O_2, alkenes and oxygenates are the prevalent products. One example is shown in Figure 2.9,[159] which plots the expected distribution of products in hydrocarbon oxidation as a function of the hydrocarbon-to-oxygen ratio.

2. Residence time: The products of partial oxidation can easily undergo consecutive combustion reactions, especially when the reaction is aimed at the oxidative transformation of a less reactive reactant, such as an alkane. Under the conditions that are necessary to activate the alkane, the desired product can easily undergo degradation reactions. Therefore a short residence time (on the order of microseconds with gauzes and milliseconds with monoliths) allows the desired product to be carried out of the reactor quickly, and thus be saved from subsequent transformations. The catalysts are usually nonporous, as their effectiveness would be very low at such high temperatures in any case, and the active phase in the pores would be completely unexploited. Homogeneous decomposition reactions of the product can easily occur in the gas phase as well as in the extragranular layer at the interface between the catalyst and the fluid phase when laminar conditions develop, especially if the reaction is mass-transfer controlled. Quenching of the gaseous stream immediately after the catalytic zone may thus be fundamental to avoid a considerable contribution of undesired homogeneous reactions.

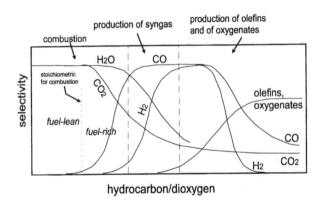

Figure 2.9. Qualitative distribution of products as a function of the hydrocarbon-to-oxygen ratio.[159]

3. Temperature at the catalyst surface and the contribution of homogeneous processes: The very high temperature of the catalyst surface (higher than in the gas phase, due to the existence of intergranular heat-transfer limitations) makes all the surface processes, including adsorption of reactants, surface reaction, and desorption of products or intermediates, very quick. This makes nonselective heterogeneous consecutive reactions over the intermediate at the adsorbed state less likely. The very short residence time and the cooler gas phase also make homogeneous degradation of unstable products less likely. It has been suggested that in many cases radical-chain homogeneous processes contribute as well. Possible radical intermediates that desorb from the catalyst surface may build up and start homogeneous reactions while diffusing through the film or in the bulk of the gas phase, possibly yielding products of selective oxidation. Indeed, there is evidence in the literature that in the oxyfunctionalization of alkanes mixed heterogeneous/homogeneous processes can lead to higher selectivity than completely heterogeneous processes.[162-168] For instance, in the case of ethane oxidehydrogenation it is generally accepted that the reaction mechanism involves the contribution of both heterogeneous and homogeneous reactions, the prevailing one being essentially a function of the reaction temperature. Figure 2.10 summarizes the possible reactions in ethane oxidehydrogenation.[169]

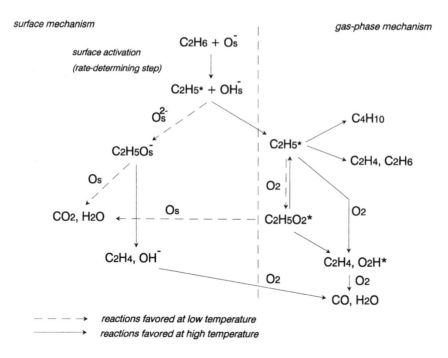

Figure 2.10. Possible reactions in ethane oxidehydrogenation.[169]

2.2.5.1. Monolith Reactors

In monolith reactors the catalysts are continuous and unitary structures.[170-176] Monolithic catalysts can be of different shapes and sizes, but are always made of many parallel passages, with the channels being circular, square, or hexagonal (honeycomb monoliths). The catalytically active component is coated in the form of a thin layer on the ceramic or metallic support, which constitutes the skeleton of the monolith structure. The walls can be either permeable or impervious to the passage of gases; in the former case radial mass transfer can occur by diffusion through the pores of the walls. In this case, the systems are referred to as "membrane reactors" (these systems will be analyzed in more detail later). In cross-flow monoliths one fluid (it can be either one of the reactants or a cooling fluid) flows in channels that are perpendicular to those in which the main fluid flows. Ceramic monoliths are usually adiabatic, showing low thermal conductivity, while metallic monoliths have a high conductivity and thus exhibit a radial heat transfer comparable or superior to that of packed beds.

Apart from automotive applications (catalytic mufflers), monolith reactors are used primarily for catalytic combustion, e.g., as afterburners of engine exhaust gases, for the incineration of off-gas in chemical plants, and for the combustion of fuels for boilers and turbines.

Monolith reactors solve the problem of pressure drop met in conventional fixed-bed reactors. The pressure drop can be up to two to three orders of magnitude less for gaseous feedstocks with respect to a fixed bed with catalyst particles of a size comparable to the channel width in monoliths. This of course leads to decreased costs for pumping of reactants. The pressure drop of the monolith channel can be calculated with good approximation by the Poiseuille equation:

$$\Delta P = 32 \, \mu_F L u_F / d_h^2 \tag{2.13}$$

where μ_F is the dynamic viscosity of the fluid, L is the reactor length, u_F is the linear interparticle fluid velocity, and d_h is the hydraulic diameter of the reactor ($d_h = 4$ flow-cross-sectional area/circumference).

The flow that develops in tubular monoliths is often laminar, being turbulent only in gas turbines, where it has a very high velocity (Re > 2000). However, notwithstanding the laminar flow, plug flow is approached because of the considerable radial diffusion occurring in the narrow channels. In foam monoliths the structure leads to remarkably better mixing than in extruded tubular monoliths, owing to the cellular structure.

The thin layer of catalytic component coating the monolith guarantees a shorter diffusion length for reactants and products with respect to catalytic particles in packed beds. This can be useful in reactions where the selectivity for the desired product may depend on the presence of diffusion phenomena.

Monoliths have been used increasingly in gas phase oxidation and combustion catalysis because of the high flow rates that can be achieved. The main problem that may be encountered concerns the exothermicity of the reaction, which can cause a considerable increase in the gas temperature owing to the adiabaticity of ceramic monoliths. The heat transfer can be improved by improving the heat conductivity of the gaseous feedstock (i.e., by adding steam). The high exothermicity of oxidation reactions can be better managed by the use of cross-flow monoliths. In this way, the heat of reaction can be used to raise the temperature of the reaction mixture or to carry out an endothermal reaction. Examples of oxidation reactions in monolithic reactors described in the literature are the oxidation of ammonia to nitrogen with a Co/α–Al$_2$O$_3$ catalyst,[177] oxidation of SO$_2$ with a Pt catalyst,[178] oxidation of polychlorinated biphenyls,[179] oxidative dehydrogenation of light alkanes,[180–185] and the partial oxidation of methane.[186–188]

A porous ceramic membrane, coated with the catalytically active component was used in the oxidehydrogenation of propane to propene.[184,185] All the reactants were fed in at the core side of the inorganic membrane, thus simulating a monolithic reactor. The reaction temperature was lower than 450°C, and residence times (calculated with respect to the catalyst volume) of a few milliseconds were used. As shown in Figure 2.11, which compares the selectivity for the alkene as a function

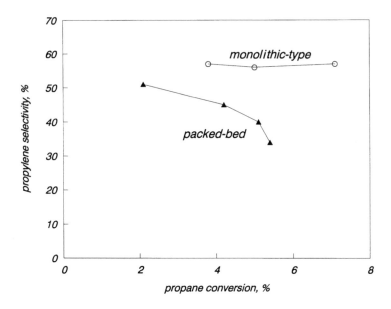

Figure 2.11. Selectivity to propene as a function of propane conversion in a monolith-type reactor and in a packed-bed reactor.[184]

of propane conversion in the monolith system developed and in a packed-bed reactor loaded with the same catalytic component but in particle form, a higher selectivity for propene was achieved in the former case. The improved selectivity was attributed to the smaller contribution of the consecutive reaction of combustion on propene, which could rapidly escape from the thin catalytic layer into the gas phase.

Results obtained by Schmidt and Goralski[189] in the oxidation of various hydrocarbons at millisecond residence times (either for selective oxidation of alkanes or for catalytic destruction of VOC) are summarized in Table 2.5. In all cases the reactor was made of a small-area (~0.1 m^2/g), porous monolith catalyst sealed in a quartz tube. Typically, a metal-coated ceramic foam 45 ppi α-Al_2O_3 monolith 10 mm thick and 18 mm in diameter was used. The advantage of the cellular structure of the foam with respect to the channel of the extruded monolith was better mixing. Indeed laminar flows usually develop in extruded monoliths. The metal was deposited from a metal salt solution; impregnation, calcination, and reduction led to the final catalytically active phase. In methane partial oxidation the best results were obtained with a rhodium-based catalyst,[190–192] while for oxidehydrogenation of higher alkanes platinum was found to give the best performance (rhodium produced mainly syngas). The main feature was the absence of carbon buildup into multilayers on the catalyst surface, even though solid graphite was the most favored product from a thermodynamic point of view. In practice all heterogeneous reactions occurred in the initial part of the monolith, and graphite formation was only observed in the region where no residual oxygen was present in the gas phase.

Ethane oxidation on Pt/Sn-coated alumina foam monoliths led to the formation of ethene with 65% selectivity (CO and CO_2 were the by-products), with more than 99% oxygen conversion. These were significantly affected by variations in residence time. The proposed mechanism consisted of the generation of an adsorbed $C_2H_5^{\bullet}$ intermediate species by reaction between ethane and adsorbed atomic oxygen, followed by β-elimination of $H^{\cdot}$ to generate ethene. A parallel route was the decomposition to CH_x fragments, thereafter transformed into carbon. The formation of the latter was postulated to be negligible owing to the presence of a

TABLE 2.5. Results Obtained in Millisecond Reactors for Oxidation of Hydrocarbons[189]

Reaction	Selectivity, %	Conversion, %	Yield, %
Methane to syngas	> 90	> 90	≈ 90
Ethane to ethene	≤ 70	≤ 90	≈ 60
Alkanes to alkenes	≤ 70	≤ 80	≤ 60
Toluene to CO_2	99.9	> 99	> 99

steam-reforming reaction:

$$C(s) + H_2O \rightarrow CO + H_2 \qquad (2.14)$$

All surface processes were very rapid, with very low surface coverage by all adsorbed species. The very quick desorption of ethene and its rapid escape from the reactor enabled the high selectivity for the alkene. Moreover, ethene could not readsorb on the following part of the monolith, which was covered with carbon and thus practically inactive.

Combustion of hydrocarbons could also be achieved with very high efficiency with the same reactor configuration, and with a fuel-lean feed composition. C1–C4 alkanes were burned more than 99.5% to CO_2 and H_2O with excess air. Methane was also added to increase the fuel content of the stream, sustain the reaction, and keep the catalyst in the ignited state.[193] Catalyst extinction for ethane combustion was found to occur at 1050°C, for a feedstock at room temperature and a residence time of 5 ms, and below 2.2% ethane in air (well below the stoichiometric content for combustion in air).

There might be a contribution of homogeneous reactions as well. Actually, the product distribution was very similar to that obtained in thermal pyrolysis. This indicates that either thermodynamic equilibrium is finally reached or that the heterogeneously assisted oxidation heated the fluid above the catalyst, and here thermal pyrolysis completed the conversion of the hydrocarbon. The authors argued that in most cases there might be a contribution due to homogeneous radical-chain propagation, but that the catalyst was nevertheless necessary to activate the hydrocarbon (heterogeneously initiated homogeneous reaction mechanism).

2.2.5.2. Membrane Reactors

Significant developments have been achieved in recent years in membrane reactor technology, where the capability of these materials in the selective permeation of particular components in a mixture is combined with a reactive application. A number of reviews have been published in recent years concerning uses for catalytic membrane reactors.[194–200] Different types of configurations can be envisaged in which the catalytically active component is deposited on the membrane by different techniques or the catalyst is in a chamber on the feed side contained by a membrane wall, in either a fixed or a fluidized bed.

The selective transport of oxygen in membranes can be classified according to the type of diffusion mechanism—in the form of atoms, ions, or molecules: (i) dense metal membranes, (ii) porous membranes, and (iii) solid-electrolyte membranes. The last class of membranes will be described in the following section, which is devoted to electrochemical reactors.

Silver membranes constitute the most important example of a dense metal membrane for the diffusion of atomic oxygen. Oxygen dissociates on one side of the membrane and the atoms dissolve in the metal and diffuse (permeate) toward the opposite side of the membrane, where they combine yielding oxygen, the limiting step being the desorption at the permeate side. Permeation can be increased by carrying out a chemical reaction at the permeate side, also due to the very high reactivity of the monatomic oxygen emerging at the membrane surface. Applications have been reported for the oxidative coupling of methane, for which it is claimed that operation is at temperatures lower than for conventional cofeeding reactors, and for the oxidation of alcohols to aldehydes.[201–203]

The transport mechanisms in porous membranes can be of different types, depending on the pore dimensions, the nature of the molecule, and operating conditions (temperature, pressure). The rate of permeation is much higher than for dense membranes, but the selectivity is rather poor. The transport mechanisms can be classified as follows:

1. Viscous flow (Poiseuille flow), which occurs when the mean pore diameter is larger than the mean free path of the molecules.
2. Knudsen-type diffusion, which occurs when the mean free path of the molecules increases and the collisions with pore walls become more frequent than collisions among molecules.
3. Surface diffusion, which occurs when one molecule adsorbs on the pore wall, and the preferential interaction of this molecule favors its diffusion through the porous structure.
4. Capillary condensation, which occurs at low temperature and small pore size.
5. Molecular sieving, which occurs when the pore diameter is small enough to sieve the molecules on the basis of their different dimensions.

Membrane reactors can be used with the aim of distributing oxygen along the reactor, thus simulating oxygen injection at different points.[204–206] Oxygen can be supplied through the reactor wall by using a porous ceramic membrane.[207,208] It was demonstrated that the gradual feeding of oxygen also limits the formation of hot spots, because the reaction is more diluted along the catalytic bed instead of being concentrated in the proximity of the reactor entrance, and hence runaway phenomena occur less often. Moreover, a controlled distribution of oxygen may kinetically favor the partial oxidation reactions instead of complete hydrocarbon combustion and keep the catalyst at a desired average oxidation level. The supply of the oxidant to the reactor medium is governed by mass-transport laws through porous membranes (or ionic diffusion laws through dense metal membranes), and these diffusion phenomena are much less dependent on temperature than chemical reactions. Hence, this system may limit hazards associated with thermal runs that sometimes occur in the reaction medium.

In general, however, results obtained with separate feeds of hydrocarbon and oxygen have not yet demonstrated significant advantages in terms of catalytic performance. In fact, in porous membranes effective control of reactant permeability can be achieved only in systems characterized by sieving effects. Moreover, in dense membranes for ionic diffusion of oxygen the rate of permeation is generally too low to be of practical interest.

2.2.6. Electrochemical Cells as Reactors

Fuel cells are devices used to convert the variation of free energy of a chemical reaction into electricity through electrochemical cell reactions. For instance, in the case of a H_2/O_2 fuel cell, the formation of H_2O from the two reactants generates electricity. The two reactions occurring at the anode and cathode, respectively, are:

$$H_2 \rightarrow 2\,H^+ + 2e^- \qquad (2.15)$$

$$\frac{1}{2}\,O_2 + 2H^+ + 2e^- \rightarrow H_2O \qquad (2.16)$$

with a net generation of H_2O. If, on the other hand, H_2O is generated by catalytic oxidation of H_2, the reaction can be divided into two electrochemical cell reactions.

The same can be done for catalytic oxidation of a hydrocarbon; thus the reaction can be decomposed into the formation of the oxygenated hydrocarbon at the anode and the reduction of oxygen to water at the cathode. The hydrocarbon–oxygen cell would consist of a H^+-conducting electrolyte held in a membrane as a separator[209] (see also Chapter 7).

The advantages of this system can be summarized as follows[210–217]:

1. The reaction rate and sometimes the selectivity can be controlled by varying the current in the external circuit.
2. Apart from the oxygenated product, electric power output also can be obtained.
3. Separation of the hydrocarbon and the oxygen leads to fewer hazards associated with the possible formation of potentially flammable mixtures.
4. Oxidation of the organic substrate can be carried out at room temperature.

A hydrocarbon–O_2 fuel cell has been used for the Wacker oxidation of ethene to acetaldehyde,[218] with a Pd-electrode at the anode (for the selective oxidation of ethene) and a Pt-electrode at the cathode to catalyze the electrochemical reduction of O_2. An aqueous solution of H_3PO_4 was chosen as the electrolyte. A gaseous mixture containing ethene, H_2O and He was fed in at the anode compartment, while O_2 was introduced at the cathode; the reaction temperature was 100°C. The reaction yielded acetaldehyde with more than 97% selectivity, current efficiency higher than 87%, and a 7% yield. It was possible to control the rate of oxidation by changing the potential between the electrodes (the reaction was stopped when the circuit was opened). The same authors also studied the oxidation of propene to acrolein.[219,220]

In some ways, this system simulates enzymes (such as Cytochrome P-450 or methane monoxygenase), where molecular oxygen is activated through reduction by an electron donor such as NADH, leading to the generation of a high-valence iron oxo species on a porphyrin cation radical. This center is also able to hydroxylate aromatics and alkanes. The H_2/O_2 fuel cell might activate molecular oxygen, generating active oxygen species such an $HO^\bullet$ on a properly chosen electrocatalyst, possibly able to hydroxylate aromatic and alkane substrates. Hydroxylation of alkanes and aromatics was carried out by passing these hydrocarbons through the cathode compartment during H_2/O_2 cell reactions.[221] Phenol was observed when benzene and Fe^{3+} ions were added together in an H_2O/HCl solution in the cathode compartment, in practice exploiting the classical Fenton chemistry. Propane could be oxidized to acetone with 65% selectivity in the gas phase at a carbon whisker cathode, at room temperature.[222]

A modification of this system involves the use of cell systems with solid oxide electrolytes, typically consisting of oxides with high ionic conductivity of O^{2-}. The use of solid-electrolyte membranes for the ionic diffusion of O^{2-} has been reviewed by various authors.[194,213,223,224] Typical solid electrolytes are ZrO_2, CeO_2, ThO_2, $Y_2O_3-ZrO_2$, and $SrCeO_2$. Superior permeabilities are exhibited by perovskite solids (La, Sr, Mn mixed oxides), which can transport oxygen at a rate of $10^{-11}-10^{-9}$ $mol\cdot cm^{-2}\cdot s^{-1}$ through a vacancy diffusion mechanism, orders of magnitude higher than for conventional solid electrolytes. The diffusion mechanism is dissociative chemisorption of oxygen, with ionization of atoms and transport of the latter through the crystalline lattice up to the opposite surface, where they lose their charge and again form oxygen. The driving force for the diffusion of the O^{2-} species can be either a difference in pressure at the two sides of the membrane or an electrical potential gradient obtained by electrodes deposited on the two sides of the membrane. The development of a specific ionic species at the membrane surface allows very specific oxidation reactions to be carried out, with a theoretically very high selectivity for specific products.

Another possible use of these ion conductors is to feed oxygen into a reaction medium while excluding other components. For instance, by using a dense PbO membrane in the form of a thin layer on a porous alumina tube, promoting ionic diffusion of oxygen, it is possible to feed air in at the shell side of the membrane and to allow oxygen to reach the reaction medium while excluding molecular nitrogen. In this way it is possible to overcome the problems associated with separation of nitrogen from the products and unconverted reactant in methane oxidative coupling.[225,226] The drawback of these systems is the low rate of permeation, and the resultant low productivity; permeation can be facilitated by operation at high temperature (higher than 600°C). In addition, one technological limitation involves the sealing of these membranes into devices that have to operate at high temperature. In general, the same drawback exists for fuel cell systems for hydrocarbon oxidation, where the rate of oxidation is too low for industrial application. Improvement of the system depends on the development of superior electrocatalysts and suitable electrolytes, but it might become of interest if very valuable fine chemicals could be synthesized.

2.3. AIR VERSUS OXYGEN PROCESSES

2.3.1. Advantages in the Use of Oxygen Instead of Air in Industrial Oxidation Processes

One of the most important aspects of the use of molecular oxygen concerns the choice between pure oxygen and air as the oxidizing agent. Air was always the preferred oxidant, but a number of oxidation processes, both in the liquid and the gas phases, have been modified over the years in order to allow for the use of pure oxygen. These changes have been driven by improvements in productivity and yield, while more recently revamping or modifications aimed at pure oxygen use have been undertaken in response to environmental constraints. It is likely that in the near future most industrial processes will be changed from air-based to oxygen-based.[227]

One of the major industrial applications of oxygen is in combustion processes. In fact the use of oxygen to enrich air has the following advantages:

1. An increase in oxygen concentration results in an increased rate of combustion, with a consequent increase in the adiabatic flame temperature. In some processes (production of ceramics and refractories), higher temperatures are desired since they lead to products of better quality.
2. The increase in oxygen concentration also leads to an improvement in heat transfer, and finally in heat utilization; however, it is possible to reduce fuel consumption while keeping the combustion rate with respect to the air–fuel flame constant. For instance, a 20% fuel savings can be reached using a 25% oxygen combustion mixture.

In refineries pure oxygen is used mainly in the form of oxygen-enriched air to debottleneck the catalytic crackers, by making the regeneration of deactivated catalysts through coke burn off more rapidly. Another important application is in the partial oxidation of heavy residues containing high levels of sulfur, asphaltenes, and metals. The product is syngas.

Analogous advantages are found with the use of oxygen instead of air in the chemical industry, which include energy savings and increased productivity, resulting in debottlenecking of plants. On the other hand, smaller reactors can be used for a fixed productivity. Smaller equipment is also needed in the section dedicated to product recovery and purification, owing to the lower volume of inert gas contained in the reactor outlet stream. The use of oxygen instead of air implies that the same partial pressure can be used with a much lower total pressure than with air, thus making it feasible to reduce the total pressure, with obvious energy advantages. Moreover, an increase in the reaction rate makes it possible to reach the same productivity while lowering the reaction temperature, with possible benefits from the selectivity point of view when several products are formed in the reaction. A lower content of nitrogen, a ballast with very poor thermal conduction properties, also allows better control of the temperature profile in the presence of strongly exothermal reactions.

An even more important benefit derives from the considerable decrease in polluting emissions into the atmosphere, as a consequence of the possibility of recycling spent gases when oxygen is used in place of air. Much less waste gas is produced, with energy savings during incineration. Other treatments are now possible, such as adsorption of some components or chemical treatments. In addition, the heating value of the stream is much higher than for the air-based process (the concentration of nitrogen is much lower, while that of hydrocarbons and carbon oxides is expected to be much higher). Therefore the purge stream instead of being treated can be used as a fuel to incinerate other wastes.

The following chemical processes make use of oxygen-enriched air or of oxygen, as an alternative to air:

- Partial oxidation of oil fractions and coke to synthesis gas.
- Oxidation of methanol to formaldehyde (either air or oxygen-enriched air).
- Oxidation of ethene to ethene oxide (either air or oxygen; new plants oxygen).
- Oxychlorination of ethene to 1,2-dichloroethane (either air or oxygen; new plants mostly oxygen).
- Oxidative acetoxylation of ethene to vinyl acetate (oxygen).
- Oxidation of n-butane to acetic acid (either air or oxygen).
- Oxidation of ethene to acetaldehyde (either air or oxygen).
- Oxidation of acetaldehyde to acetic anhydride (either air or oxygen).
- Ammoxidation of propene to acrylonitrile (oxygen-enriched air).
- Oxidation of cyclohexane to cyclohexanone (either air or oxygen).
- Oxidation of isobutane to *tert*-butylhydroperoxide (the latter is used for propene epoxidation) (oxygen).

Safety is one of the more serious problems that must be taken into account when considering the use of an oxygen-based oxidation process as an alternative to an air-based one. The absence of large amounts of ballast (as opposed to the air-based process) makes the formation of flammable mixtures more likely. Apart from the use of suitable instruments to continuously monitor the gas phase composition, fundamental aspects are knowing the flammability limits under reaction conditions (temperature and pressure), including the autoflammability temperature, a careful design of reactors and apparatus to avoid ignition sources, and the presence of safety valves and membranes to minimize the consequences of an explosion. The best way to mix components that might form flammable mixtures is inside a fluidized bed of a solid suspended in the gaseous stream. This is not possible with fixed beds, but fixed-bed reactors often operate under flammable conditions to reach higher productivities.

If new processes of alkane oxidation are developed in the future, they will likely operate under fuel-rich conditions, in order to obtain better selectivity for the desired product (an exception is the oxidation of n-butane to maleic anhydride, owing to the fact that at a high n-butane/oxygen ratio several by-products form that are not observed under fuel-lean conditions). Therefore there will be incomplete conversion of the hydrocarbon, and the unconverted alkane might possibly be recycled. This process configuration requires the use of oxygen to minimize the purge flow and atmospheric emissions.

Two very important examples of industrial gas phase oxidation where oxygen is used instead of air will be examined briefly, in order to make the advantages clear: (1) gas phase epoxidation of ethene, and (2) gas phase oxychlorination of ethene.

2.3.1.1. Synthesis of Ethene Epoxide

Epoxidation of ethene with air is achieved with two in-series reactors (three for the larger plants). This is necessary in order to avoid the significant loss of ethene that is left unconverted in the first reactor. In the first reactor the conversion is kept at about 40% to maintain high selectivity (higher than 70%), while in the next reactor (after separation of the ethene oxide produced) the overall conversion reaches 80% (95% in the third reactor), and a lower selectivity (close to 50%) is tolerated. In fact due to the consecutive reaction of ethene oxide combustion, the selectivity rapidly falls with increasing ethene conversion. A fraction of unconverted ethene exiting from the first main reactor is fed into the second reactor, while a fraction of it is recycled back to the main reactor. The loss of ethene that is contained in the purge stream corresponds to approximately 2–3% of the ethene fed into the plant.

The process that makes use of oxygen is single-stage, since the quantities of inerts entering the plant are very low when high-purity oxygen is used. Thus there is no need for additional reactors to recover ethene, because it is possible to recycle unconverted ethene. Typically, an ethene-to-oxygen ratio in the feed close to 3.0–3.5 is used; the oxygen concentration is kept at 9% to remain outside the flammable limits. In this process the ethene per-pass conversion is lower than 15% (typically around 11%), while the oxygen conversion is 35–45%, so there is a high selectivity (>80%) for ethene oxide. The overall ethene conversion is higher than 97%. The stream containing unconverted ethene after the ethene oxide absorber is recycled to the reactor. The purge stream leads to a loss of less than 1% of the ethene fed to the plant.

The oxygen-based process can operate from 10 to 40 mol% of ethene in the feed (usually between 25 and 30%). The oxygen concentration must be no higher than 9%. This high content of ethene leads to the formation of a mixture that is always above the upper limit of flammability under reaction conditions.

The ballast used in oxygen-based processes can be tailored for the reaction requirements. Properties that the ballast must possess are: (i) suitable thermal conductivity, (ii) capability to decrease the flammability zone to enable working at higher oxygen concentration (with a positive effect on the reaction rate), and (iii) absence of effect on the catalyst properties (or, better, a positive effect on catalytic performance). Nitrogen does not possess good heat-transfer properties, but does not negatively affect the catalyst surface properties. The amounts of argon and nitrogen in the feed (which enter the system as oxygen impurities) are kept lower than 10% through regulation of downstream purges. Carbon dioxide possesses good properties for heat removal, has the property of decreasing the flammability zone of ethene, and interacts with the catalyst showing a positive effect on selectivity at low concentration, but is a poison at high concentration (the optimal amount in the feed is less than 6–8%). Therefore usually only a part of the recycle gas is sent to the carbon dioxide absorber system, in order to have an optimal concentration of carbon dioxide in the reactor under stationary conditions. Methane (like ethane) has good heat removal properties, owing to the fact that it has greater heat capacity and thermal conductivity than nitrogen; concentrations of methane as high as 45–50% in the reactor are often reached by the addition of the hydrocarbon to the stream. Ethene in excess is a better ballast than nitrogen, because of its good properties of heat removal. In some cases mixtures of different ballasts are used.

Advantages in operation with oxygen with respect to operation with air are:

1. Increased yield and productivity of ethene oxide; this is because the reaction rate is first order with respect to oxygen partial pressure.
2. Increased selectivity for ethene oxide, due to the lower conversion of ethene that is now possible and to the higher concentration of ethene in the feed.
3. Smaller amount of vent gas and therefore less loss of ethene. The amount of ethene lost is a function of the: (i) purge flow rate, (ii) concentration of ethene in the feedstock (usually 20–30%), and (iii) conversion of ethene.

The vent stream depends on the quantity of inerts fed into the reactor: argon and nitrogen, present as impurities in the oxygen stream, in amounts that depend on the purity of the oxygen. Another inert is ethane, contained as an impurity in the makeup ethene stream. High-purity oxygen and ethene are thus required for oxygen-based processes.

4. Possibility of choosing the ideal ballast.
5. Lower costs associated with smaller equipment size in the reaction and recovery sections; no costs associated with air compression.

Apart from the cost of pure oxygen, disadvantages in the use of oxygen are the need for safety equipment and more elaborate mixing devices, and the costs for carbon dioxide removal by hot carbonate absorption (in the air-based process the carbon dioxide produced is purged together with nitrogen).

2.3.1.2. Synthesis of 1,2-Dichloroethane (DCE)

The use of oxygen instead of air as the oxidizing agent in oxychlorination goes back to the beginning of the 1960s. Shell, Monsanto, and PPG first used oxygen in their oxychlorination plants in the United States. These companies took advantage of on-site production of oxygen for other purposes, e.g., for acetylene production. Then, among the several different technologies developed during the 1960s, some were designed to make use of either oxidant, such as the Stauffer technology for oxychlorination in multitubular fixed-bed reactors.

In general, there are two different cases in which oxygen is used as the oxidizing agent instead of air: (i) when a large excess of ethene with respect to the stoichiometric amount is fed into the reactor, typically in fixed-bed technology, and recycle is aimed at converting the unconverted ethene, and (ii) when recycle is carried out to meet environmental constraints, and avoid emission of pollutants into the atmosphere. In the latter case, the residual unconverted ethene is minimal, since an almost stoichiometric composition is fed into the reactor. In both cases the use of air would lead to a too large concentration of inert gas (nitrogen), and the purge-to-recycle streams ratio would be excessively high.

The advantages of the use of oxygen have been examined in several papers.[228-232] The following are the most important points:

1. The stream size of the purge flow is from 20 to 100 times smaller than the vent stream of the air-based technology (depending essentially on the quantity of carbon oxides produced with the reaction, thus on the catalytic performance). When operation is carried out with a large excess of ethene with respect to the stoichiometric requirement (fixed-bed technology), the purge stream also has a very large concentration of ethene, and can be sent to the liquid phase direct chlorination section to further improve the yield for DCE. In this case, the gaseous vent from

the direct chlorination reactor becomes the only source of pollution. Alternatively, the catalytic or thermal incineration of the purge stream can be carried out more economically than with the larger and less concentrated vent stream of the air-based process. Typical vent compositions of the air-based and oxygen-based processes are given in Table 2.6 for fluid-bed processes (operation with almost stoichiometric feedstock). In the air-based process it is important to keep the oxygen concentration in the stream lower than 9%, the limit value of 5% being generally considered as a safe one. For oxygen-based processes a limit value of 2% for oxygen concentration in the recycled stream is usually considered safe.

2. When a large excess of ethene is used, the alkene is the major component in the gaseous stream entering the reactor. In this case, typically used in fixed-bed oxychlorination technology, the better heat-transfer properties (higher heat capacity) of ethene as compared to nitrogen (which is the major gaseous component in air-based processes) allows the hot-spot temperature and the average temperature in the reactor to be significantly lowered. This yields several advantages, among them: (i) improved heat transfer from the reaction medium to the heat-removal fluid; (ii) better selectivity owing to the smaller contribution of combustion reactions and less formation of chlorinated by-products such as trichloroethane (formed from vinyl chloride, itself formed by cracking of DCE); (iii) improved HCl conversion (for temperatures above 300°C in the hot spot the reaction becomes mass-transfer limited); and (iv) longer catalyst lifetime, owing to less active phase volatilization and less coke formation, which finally cause catalyst dusting.

3. As a consequence of the previous point, improved heat transfer allows operation to be carried out at increased capacity. In fact, the productivity is usually essentially limited by the heat transfer between the reaction medium and the cooling fluid. From another point of view, the same capacity as for the air-based process can be obtained with fewer tubes, thus with considerably reduced investment costs owing to the smaller reactor size. Finally, the DCE yield is increased with respect to ethene since the latter is completely converted.

TABLE 2.6. Vent Composition for Air-Based and Oxygen-Based Technologies

Component	Content, vol.%; flow rate, m^3/h	
	Air-based process	Oxygen-based process
Oxygen + argon	4–8; 400–2400	0.1–2.5; < 25
Ethene	0.1–0.8; 10–24	2–5; < 50
CO_x (CO_2/CO 3–4/1)	1–3; > 100–900	15–30; < 300
DCE and chlorinated comp.	0.02–0.2; 2–60	0.5–1; < 10
Molecular nitrogen	Rest	$\cong$ 600
Vent flow rate, m^3/h[a]	10,000–30,000	< 1000

[a]This corresponds to approximately 300 to 900 m^3/ton of VCM produced for a balanced plant.

4. Operating costs associated with the compression of oxygen are lower than costs for air compression. It is worth noting that the cost of the energy required for compression constitutes a significant portion of the cost for the oxidant supply. On the other hand, higher costs are obviously associated with the energy required for recycle compression.

5. It is claimed that capital investments reduced as a result of differences in the plant with respect to the air-based technology. In fact some units become superfluous; e.g., the absorption/stripping treatment on the purge stream for DCE recovery is not necessary in the oxygen case, since the volumetric flow rate is much lower than for the air-based process. Other parts are smaller in size (the compressor for oxygen, the incinerator). These savings will be possible for new plants, while for revamping of air-based processes the comparison of investments may become less advantageous. Investments associated with the use of oxygen are essentially related to the recycle compressor and to special piping to handle oxygen.

2.3.2. Use of Pure Oxygen in the Oxidation of Alkanes

The majority of reactions currently being studied for the oxidative transformation of alkanes to alkenes or to oxygenated compounds may require the use of pure oxygen instead of air as the oxidizing agent for the following reasons:

1. Often obtaining the best reaction conditions in terms of selectivity for the desired product (either alkene or oxygenated compound) implies the use of hydrocarbon-rich conditions, thus with oxygen as the limiting reactant. Under these conditions combustion reactions are minimized and the selectivity for the desired product is the highest possible. This is the case with propane ammoxidation to acrylonitrile, isobutane oxidation to methacrylic acid, oxidehydrogenation of ethane, and methane partial oxidation. Under these conditions a low alkane conversion is achieved, which implies the possible recycling of the unconverted reactant, and thus the use of pure oxygen as the oxidizing agent, in order to minimize the purge flow and maximize the concentration of fuel in the vent stream. With higher alkanes oxidehydrogenation, hydrocarbon-rich conditions leads to the formation of undesired by-products (unsaturated acids and other oxygenated compounds). Also in the case of n-butane and n-pentane oxidation, high alkane-to-oxygen ratios lead to the formation of several undesired by-products of partial oxidation and are therefore not suitable reaction conditions. In these cases, hydrocarbon-lean conditions are preferred.

2. The high exothermicity of the reactions of alkane oxidative transformation (in consideration of the large difference in energy between the reactants and the products, and of the extent of combustion usually present) makes the use of nitrogen as the ballast (and thus the use of air as the oxidizing agent) unsuitable, owing to

the poor characteristics of thermal conductivity. Excess alkane is a more suitable ballast for strongly exothermal reactions.

Recently, Mitsubishi announced that it is going to build a pilot plant for the ammoxidation of propane with oxygen or oxygen-enriched air as the oxidizing agent.[233] The BOC technology for selective hydrocarbon adsorption is used to recover more than 99% of unconverted propane from the reactor outlet stream. Propane is then compressed and recycled to the fluidized-bed reactor.

REFERENCES

1. A. Bielanski and J. Haber, *Oxygen in Catalysis*, Marcel Dekker: New York (1991).
2. J. Haber, in: *Heterogeneous Hydrocarbon Oxidation* (B.K. Warren and S.T. Oyama, eds.), American Chemical Society: Washington DC (1996), ACS Symp. Series 638, p. 20.
3. S.T. Oyama, B.K. Warren, and S.T. Oyama (eds.), ACS Symposium Series 638, American Chemical Society: Washington DC (1996), p. 2.
4. J.S. Lee and S.T. Oyama, *Catal. Rev.-Sci. Eng.* **30**, 249 (1988).
5. H.H. Kung, *Adv. Catal.* **40**, 1 (1994).
6. J.C. Vedrine, G. Coudurier, J.-M. Millet, *Catal. Today* **33**, 3 (1997).
7. J.C. Vedrine, J.-M. Millet, J.-C. Volta, *Catal. Today* **32**, 115 (1996).
8. G. Busca, E. Finocchio, G. Ramis, and G. Ricchiardi, *Catal. Today* **32**, 133 (1996).
9. F. Cavani and F. Trifirò, in: *Catalysis Vol. 11*, Royal Society of Chemistry: Cambridge 1994, p. 246.
10. S. Albonetti, F. Cavani, and F. Trifirò, *Catal. Rev.-Sci. Eng.* **38**, 413 (1996).
11. V.D. Sokolovskii, *Catal. Rev.-Sci. Eng.* **32**, 1 (1990).
12. B. Delmon, P. Ruiz, S.R.G. Carrazán, S. Korili, M.A. Vicente Rodriguez, and Z. Sobalik, in: *Catalysts in Petroleum Refining and Petrochemical Industries 1995* (M. Absi-Halabi, ed.), Elsevier Science: Amsterdam 1996, p. 1.
13. R. Burch and M.J. Hayes, *J. Molec. Catal. A: Chemical* **100**, 13 (1995).
14. L.D. Schmidt, M. Huff, and S.S. Bharadwaj, *Chem. Eng. Sci.* **49**, 3981 (1994).
15. H.H. Kung, P. Michalakos, L. Owens, M. Kung, P. Andersen, O. Owen, and I. Jahan, in: *Catalytic Selective Oxidation* (S.T. Oyama and J.W. Hightower, eds.), ACS Symposium Series 523, American Chemical Society: Washington DC (1993), p. 387.
16. F. Trifirò and F. Cavani, *Selective Partial Oxidation of Hydrocarbons and Related Oxidations*, Catalytica Studies Division: Mountain View, CA, No. 4193 SO (1994).

17. F. Trifirò and F. Cavani, *Oxidative Dehydrogenation and Alternative Dehydrogenation Processes*, Catalytica Studies Division: Mountain View, CA, No. 4192 OD (1993).
18. F. Cavani and F. Trifirò, *Catal. Today* **24**, 307 (1995).
19. F. Cavani and F. Trifirò, *Appl. Catal., A: General* **88**, 115 (1992).
20. E.A. Mamedov and V. Cortes Corberan, *Appl. Catal., A: General* **127**, 1 (1995).
21. E.A. Mamedov, *Appl. Catal., A: General* **116**, 49 (1994).
22. L.-T. Weng and B. Delmon, *Appl. Catal., A: General* **81**, 141 (1992).
23. G. Centi, in: *Elementary Reaction Steps in Heterogeneous Catalysis* (R.W. Joyner and R.A. Van Santen, eds.), NATO Symposium Series, Kluwer Academic Press: Dordrecht (1993), p. 93.
24. G. Centi, *Catal. Lett.* **22**, 53 (1993).
25. F. Cavani and F. Trifirò, *Catal. Today* **36**, 431 (1997).
26. J.L. Callahan and R.K. Grasselli, *AIChE J.* **9**, 755 (1963).
27. G. Centi (ed.), *Catal. Today* **16** (Special Issue on Vanadyl Pyrophosphate) (1993).
28. F. Cavani and F. Trifirò, in: *Preparation of Catalysts VI* (G. Poncelet, J. Martens, B. Delmon, P.A. Jacobs, and P. Grange, eds.), Studies in Surface Science and Catalysis Vol. 91, Elsevier Science: Amsterdam (1995), p. 1.
29. G.A. Foulds and J.A. Lapszewicz, in: *Catalysis Vol. 11*, Royal Chemical Society: Cambridge (1994), p. 412.
30. S.S. Bharadwaj and L.D. Schmidt, *Fuel Process. Techn.* **42**, 109 (1995).
31. M.A. Peña, J.P. Gómez, and J.L.G. Fierro, *Appl. Catal., A: General* **144**, 7 (1996).
32. M. Roy, M. Gubelmann-Bonneau, H. Ponceblanc, and J.-C. Volta, *Catal. Lett.* **42**, 93 (1996).
33. P. Barthe and G. Blanchard, *Fr. Patent* 90 12,519 (1990).
34. M. Merzouki, B. Taouk, L. Monceaux, E. Bordes, and P. Courtine, in: *New Developments in Selective Oxidation by Heterogeneous Catalysts* (P. Ruiz and B. Delmon, eds.), Studies in Surface Science and Catalysis Vol. 72, Elsevier Science: Amsterdam (1992), p. 165.
35. R.J. Blake and G.W. Roy, *US Patent* 3,657,367 (1972).
36. A.J. Magistro, *US Patent* 4,102,936 (1978).
37. *GB Patent Appl.* 9318507.2 (1993).
38. Y. Moro-oka and W. Ueda, in: *Catalysis Vol. 11*, Royal Chemical Society: Cambridge (1994), p. 223.
39. L.C. Glaeser, J.F. Brazdil, D.D. Suresh, D.A. Orndoff, and R.K. Grasselli, *US Patent* 4,788,173 (1988).
40. A.T. Guttmann, R.K. Grasselli, and J.F. Brazdil, *US Patent* 4,746,641 (1988).
41. S. Albonetti, G. Blanchard, P. Burattin, F. Cavani, and F. Trifirò, *EP Patent* 691,306 A1 (1995), *EP Patent* 420,025 A1 (1996), and *Fr. Patent Appl.* 95 11680 (1995).

42. R. Catani, G. Centi, F. Trifirò, and R.K. Grasselli, *Ind. Eng. Chem. Res.* **31**, 107 (1992).

43. A. Andersson, S.L.T. Andersson, G. Centi, R.K. Grasselli, M. Sanati, and F. Trifirò, in *New Frontiers in Catalysis* (L. Guczi, F. Solymosi, P. Tétény, eds.), Elsevier Science: Amsterdam (1993), p. 691.

44. V.D. Sokolovskii, A.A. Davydov, and O. Yu. Ovsitser, *Catal. Rev.-Sci. Eng.* **37**, 425 (1995).

45. F. Cavani, E. Etienne, M. Favaro, A. Galli, F. Trifirò, and G. Hecquet, *Catal. Lett.* **32**, 215 (1995).

46. G. Busca, F. Cavani, E. Etienne, E. Finocchio, A. Galli, G. Selleri, and F. Trifirò, *J. Molec. Catal.* **114**, 343 (1996).

47. F. Cavani, E. Etienne, G. Hecquet, G. Selleri, and F. Trifirò, in: *Catalysis of Organic Reactions* (R.E. Malz, ed.), Marcel Dekker: New York (1996), p. 107.

48. N. Mizuno, M. Tateishi, and M. Iwamoto, *Appl. Catal., A: General* **128**, L165 (1995).

49. N. Mizuno, W. Han, T. Kudo, and M. Iwamoto, in: *11th International Congress on Catalysis—40th Anniversary* (J.W. Hightower, W.N. Delgass, E. Iglesia, and A.T. Bell, eds.), Studies in Surface Science and Catalysis Vol. 101, Elsevier Science: Amsterdam (1996), p. 1001.

50. N. Mizuno, M, Tateishi, and M. Iwamoto, *J. Catal.* **163**, 87 (1996).

51. N. Mizuno, D.-J. Suh, W. Han, and T. Kudo, *J. Molec. Catal., A: Chemical* **114**, 309 (1996).

52. L. Jalowiecki-Duhamel, A. Monnier, Y. Barbaux, and G. Hecquet, *Catal. Today* **32**, 237 (1996).

53. W. Ueda, Y. Suzuki, W. Lee, and S. Imaoka, in: *11th International Congress on Catalysis—40th Anniversary* (J.W. Hightower, W.N. Delgass, E. Iglesia, and A.T. Bell, eds.), Studies in Surface Science and Catalysis Vol. 101, Elsevier Science: Amsterdam (1996), p. 1065.

54. H. Krieger and L.S. Kirch, *US Patent* 4,260,822 (1981).

55. D. Hönicke, K. Griesbaum, R. Augenstein, and Y. Yang, *Chem. Eng. Techn.* **59**, 222 (1987).

56. G. Centi, J. Lopez Nieto, D. Pinelli, and F. Trifirò, *Ind. Eng. Chem. Res.* **28**, 400 (1989).

57. G. Centi, J. Lopez Nieto, F. Ungarelli, and F. Trifirò, *Catal. Lett.* **4**, 309 (1990).

58. G. Calestani, F. Cavani, A. Duran, G. Mazzoni, G. Stefani, F. Trifirò, and P. Venturoli, in: *Science and Technology in Catalysis, 1994* (Y. Izumi, H. Arai, and M. Iwamoto, eds.), Kodansha: Tokyo, and Elsevier: Amsterdam (1995), p. 179.

59. S. Albonetti, F. Cavani, F. Trifirò, P. Venturoli, G. Calestani, M. Lopez Granados, and J.L.G. Fierro, *J. Catal.* **160**, 52 (1996).

60. F. Cavani, A. Colombo, F. Giuntoli, E. Gobbi, F. Trifirò, and P. Vazquez, *Catal. Today* **32**, 125 (1996).

61. F. Cavani, A. Colombo, F. Giuntoli, F. Trifirò, P. Vazquez, and P. Venturoli, in: *Advanced Catalysts and Nanostructured Materials* (W.R. Moser, ed.), Academic Press: New York (1996), p. 43.
62. F. Cavani, A. Colombo, F. Trifirò, M.T. Sananes Schulz, J.C. Volta, and G.J. Hutchings, *Catal. Lett.* **43**, 241 (1997).
63. D. Braks and D. Hönicke, in: *Proceedings DGMK Conference on Selective Oxidations in Petrochemistry*, DGMK: Hamburg (1992) p. 37.
64. V.A. Zazhigalov, J. Haber, J. Stoch, B.D. Mikhajluk, A.I. Pyatnitskaya, G.A. Komashko, and I.V. Bacherikova, *Catal. Lett.* **37**, 95 (1996).
65. S.A. Korili, P. Ruiz, and B. Delmon, in: *Heterogeneous Hydrocarbon Oxidation* (B.K. Warren and S.T. Oyama, eds.), ACS Symposium Series 638, American Chemical Society: Washington DC (1996), p. 192.
66. U.S. Ozkan, T.A. Harris, and B.T. Schilf, *Catal. Today* **33**, 57 (1997).
67. A.Kh. Mamedov, P.A. Shiryaev, D.P. Shashkin, and O.V. Krylov, in: *New Developments in Selective Oxidation* (G. Centi and F. Trifirò, eds.), Studies in Surface Science and Catalysis Vol. 55, Elsevier Science: Amsterdam (1990), p. 477.
68. L. Mendelovici and J.H. Lunsford, *J. Catal.* **94**, 37 (1985).
69. E. Iwamatsu, K. Aika, and T. Onishi, *Bull. Chem. Soc. Jpn* **59**, 1665 (1986).
70. A. Erdöhelyi and F. Solymosi, *Appl. Catal.* **39**, L11 (1988).
71. A. Erdöhelyi and F. Solymosi, *J. Catal.* **129**, 497 (1991).
72. S.N. Vereshchagin, N.N. Shishkina, and A.G. Anshits, in: *New Developments in Selective Oxidation* (G. Centi and F. Trifirò, eds.), Studies in Surface Science and Catalysis Vol. 55, Elsevier Science: Amsterdam (1990), p. 483.
73. A.G. Anshits, *Catal. Today* **13**, 495 (1992).
74. K. Aika, M. Isobe, K. Kido, T. Moriyama, and T. Onishi, *J. Chem. Soc., Faraday Trans. I* **83**, 3139 (1987).
75. R. Burch and C. Howitt, *Appl. Catal., A: General* **103**, 135 (1993).
76. E. Suzuki, K. Nakashiro, and Y. Ono, *Chem. Lett.* **6**, 953 (1988).
77. G. Panov, A. Kharitonov, and V. Sobolev, *Appl. Catal., A: General* **98**, 1 (1993).
78. M. Häfele, A. Reitzmann, E. Klemm, and G. Emig, in: *3rd World Congress on Oxidation Catalysis* (R.K. Grasselli, S.T. Oyama, A.M. Gaffney, and J.E. Lyons, eds.), Studies in Surface Science and Catalysis Vol. 110, Elsevier Science: Amsterdam (1997), p. 487.
79. M. Iwamoto, J. Hirata, K. Matzukami, and S. Kagawa, *J. Phys. Chem.* **87**, 903 (1983).
80. G.I. Panov, G.A. Sheveleva, A.S. Kharitonov, V.N. Romannikov, and L.A. Vostrikova, *Appl. Catal., A: General* **82**, 31 (1992).
81. R. Burch and C. Howitt, *Appl. Catal., A: General* **86**, 139 (1992).
82. M. Gubelmann and P.J. Tirel, *EP. Patent* 341165 (1989).
83. M. Gubelmann, J.M. Popa, and P.J. Tirel, *EP Patent* 406050 (1991).
84. M. Häfele, A. Reitzmann, D. Roppelt, and G. Emig, *Erdöl Erdgas Kohle* **12**, 512 (1996).

85. G.I. Panov, V.I. Sobolev, and A.S. Kharitonov, *J. Molec. Catal.* **61**, 85 (1990).
86. A.S. Kharitonov, G.I. Panov, K.G. Ione, V.N. Romannikov, G.A. Sheveleva, L.A. Vostrikova, and V.I. Sobolev, *US Patent* 5,110,995 (1992).
87. R. Burch and C. Howitt, *Appl. Catal, A: General* **106**, 167 (1993).
88. A.S. Kharitonov, G.A. Sheveleva, G.I. Panov, V.I. Sobolev, Y.A. Paukshtis, and V.N. Romannikov, *Appl. Catal., A: General* **98**, 33 (1993).
89. A.K. Uriarte, M.A. Rodkin, M.J. Gross, A.S. Kharitonov, and G.I. Panov, in: *3rd World Congress on Oxidation Catalysis* (R.K. Grasselli, S.T. Oyama, A.M. Gaffney, and J.E. Lyons, eds.), Studies in Surface Science and Catalysis Vol. 110, Elsevier Science: Amsterdam (1997), p. 857.
90. *Petrochemical News* **35**, 2 (1996).
91. A. Gervasini, G.C. Vezzoli, and V. Ragaini, *Catal. Today* **29**, 449 (1996).
92. W. Li and S.T. Oyama, in: *Heterogeneous Hydrocarbon Oxidation* (B.K. Warren and S.T. Oyama, eds.), ACS Symposium Series 638, American Chemical Society: Washington DC (1996), p. 364.
93. A. Naydenov and D. Mehandjiev, *Appl. Catal., A: General* **97**, 17 (1993).
94. W. Li and S.T. Oyama, in: *3rd World Congress on Oxidation Catalysis* (R.K. Grasselli, S.T. Oyama, A.M. Gaffney, and J.E. Lyons, eds.), Studies in Surface Science and Catalysis Vol. 110, Elsevier Science: Amsterdam (1997), p. 873.
95. M.G. Clerici and P. Ingallina, *Catal. Today* **41**, 351 (1998).
96. D.H.R. Barton and D. Doller, *Acc. Chem. Res.* **25**, 504 (1992).
97. D.H.R. Barton, F. Halley, N. Ozbalik, M. Schmitt, E. Young, and G. Balavoine, *J. Am. Chem. Soc.* **111**, 7144 (1989).
98. D.H.R. Barton, M.J. Gastiger, and W.B. Motherwell, *J. Chem. Soc., Chem. Comm.* **41** (1983).
99. U. Schuchardt, C.E.Z. Krahembaii, and W.A. Carvalho, *New J. Chem.* **15**, 955 (1991).
100. T. Yamada, T. Takai, O. Rhode, and T. Mukaiyama, *Bull. Chem. Soc. Jpn* **64**, 2109 (1991).
101. T. Yamada, K. Imagawa, T. Nagata, and T. Mukaiyama, *Chem. Lett.*, 2231 (1992).
102. O.A. Kholdeva, V.A. Grigoriev, G.M. Maksimov, M.A. Fedotov, A.V. Golovin, and K.I. Zamaraev, *J. Molec. Catal.* **114**, 123 (1996).
103. R. Neumann and A.M. Khenkin, *J. Chem. Soc., Chem. Comm.*, 2643 (1996).
104. S.I. Murahashi, Y. Oda, and T. Naota, *J. Am. Chem. Soc.* **114**, 7913 (1992).
105. C.L. Hill, A.M. Khenkin, M.S. Weeks, and Y. Hou, in: *Catalytic Selective Oxidation* (S.T. Oyama and J.W. Hightower, eds.), ACS Symposium Series 523, American Chemical Society: Washington DC (1993), p. 67.
106. C. Ferrini and H.W. Kouwenhoven, in: *Proceedings DGMK Conference on Selective Oxidation in Petrochemistry*, DGMK, Hamburg (1992), p. 205.
107. M.G. Clerici and G. Bellussi, *US Patent* 5,235,111 (1993).
108. M.G. Clerici, *Appl. Catal.* **68**, 249 (1991).

109. T. Tatsumi, K. Yuasa, and H. Tominaga, *J. Chem. Soc., Chem. Comm.*, 1446 (1992).

110. M. Lin and A. Sen, *J. Amer. Chem. Soc.* **114**, 7307 (1992).

111. T. Jintoku, K. Takaki, Y. Fujiwara, Y. Fuchita, and K. Hiraki, *Bull. Chem. Soc. Jpn* **63**, 438 (1990).

112. P.B. Weisz and J.S. Hicks, *Chem. Eng. Sci.* **17**, 265 (1962).

113. C. McGreavy and C.I. Adderley, *Chem. Eng. Sci.* **28**, 577 (1973).

114. D. Luss, in: *Chemical Reaction Theory—A Review*, Prentice-Hall: Englewood Cliffs, NJ (1977), p. 191.

115. V. Hlavacek, M. Kubicek, and M. Marek, *J. Catal.* **15**, 17 (1969).

116. V. Hlavacek and J. Votruba, in: *Chemical Reaction Theory—A Review*, Prentice-Hall: Englewood Cliffs, NJ (1977), p. 314.

117. H. Hoffmann and V. Hlavacek, *Chem. Eng. Sci.* **25**, 173 (1970).

118. K.R. Westerterp, W.P.M. van Swaaij, and A.A.C.M. Beenackers, *Chemical Reactor Design and Operation*, John Wiley & Sons, New York, (1983).

119. G.F. Froment, *Analysis and Design of Fixed Bed Catalytic Reactors*, Advances in Chemistry Series 109, American Chemical Society: Washington DC (1972).

120. O. Bilous and N.R. Amundson, *AIChE J.* **2**, 117 (1956).

121. C.H. Barkelew, *Chem. Eng. Prog. Symp. Ser.* **55**, 37 (1959).

122. R.J. van Welsenaere and G.F. Froment, *Chem. Eng. Sci.* **25**, 1503 (1970).

123. M. Morbidelli and A. Varma, *AIChE J.* **28**, 705 (1982).

124. J.B. Agnew and O.E. Potter, *Trans. Inst. Chem. Eng.* **44**, T216 (1966).

125. G. Emig, H. Hofmann, V. Hoffmann, and V. Fiand, *Chem. Eng. Sci.* **35**, 249 (1980).

126. G.F. Froment, *Ind. Eng. Chem.* **59**, 18 (1967).

127. J.C. Pirkle and I.E. Wachs, *Chem. Eng. Progr.* **29**, (1987).

128. G. Eigenberger, *Chem. Eng. Process.* **18**, 55 (1984).

129. J.N. Papageorgiu and G.F. Froment, *Chem. Eng. Sci.* **51**, 2091 (1996).

130. Y.S. Matros, *Chem. Eng. Sci.* **45**, 2097 (1990).

131. Y.S. Matros and G.A. Bunimovich, *Catal. Rev.-Sci. Eng.* **38**, 1 (1996).

132. R.M. Contractor, in: *Circulating Fluidized Bed Technology II* (P. Basu and J.F. Large, eds.), Pergamon Press: Toronto (1988), p. 467.

133. R.M. Contractor and A.W. Sleight, *Catal. Today* **1**, 587 (1987).

134. R.M. Contractor, H.E. Bergna, U. Chowdhry, and A.W. Sleight, in: *Fluidization VI* (J.R. Grace, L.W. Shemilt, and M.A. Bergognou, eds.), Engineering Foundation: New York (1989), p. 589.

135. R.M. Contractor, D.I. Garnett, H.S. Horowitz, H.E. Bergna, G.S. Patience, J.T. Schwartz, and G.M. Sisler, in: *New Developments in Selective Oxidation II* (V. Corberan and S. Vic Bellon, eds.), Studies in Surface Science and Catalysis Vol. 82, Elsevier Science: Amsterdam (1994), p. 233.

136. E. Kesteman, M. Merzouki, B. Taouk, E. Bordes, and R. Contractor, in: *Preparation of Catalysts VI* (G. Poncelet, J. Martens, B. Delmon, P.A. Jacobs, and P. Grange, eds.), Studies in Surface Science and Catalysis Vol. 91, Elsevier Science: Amsterdam (1995), p. 707.

137. J.A. Sofranko, J.J. Leonard, and C.A. Jones, *J. Catal.* **103**, 302 (1987).

138. R.H. Kahney and T.D. McMinn, *US Patent* 4,000,178 (1976).

139. J.H. Edwards, K.T. Do, and R.J. Tyler, in: *Natural Gas Conversion* (A. Holman, K.J. Jens, and S. Kolboe, eds.), Studies in Surface Science and Catalysis Vol. 61, Elsevier Science: Amsterdam (1991), p. 489.

140. C. Raimbault and C.J. Cameron, in: *Natural Gas Conversion* (A. Holman, K.J. Jens, and S. Kolboe, eds.), Studies in Surface Science and Catalysis Vol. 61, Elsevier Science: Amsterdam (1991), p. 479.

141. T. Imai and R.J. Schmidt, *US Patent* 4,886,928 (1989).

142. R.E. Reitmeier, F.D. Mayfield, and J.H. Mayes, *US Patent* 3,437,703 (1969).

143. J. Romatier, M. Bentham, T. Foley, and J.A. Valentine, in: *DeWitt Petrochemical Review*, DeWitt: Houston (1992), K1.

144. F. Cavani and F. Trifirò, *Catal. Today* **36**, 431 (1997).

145. F.M. Dautzenberg, J.C. Schlatter, J.M. Fox, J.R. Rostrup-Nielsen, and L.J. Christiansen, *Catal. Today* **13**, 503 (1992).

146. S.S. Bharadwaj and L.D. Schmidt, *J. Catal.* **146**, 11 (1994).

147. S.S. Bharadwaj and L.D. Schmidt, *J. Catal.* **155**, 403 (1995).

148. J. Soler, J.M. Lòpez Nieto, J. Herguido, M. Menéndez, and J. Santamaria, *Catal. Lett.* **50**, 25 (1998).

149. J.R. Grace, *Chem. Eng. Sci.* **45**, 1953 (1990).

150. R.J. Dry and R.D. La Nauze, *Chem. Eng. Progr.* **7**, 31 (1990).

151. T.S. Pugsley, G.S. Patience, F. Berruti, and J. Chaouki, *Ind. Eng. Chem. Res.* **31**, 2652 (1992).

152. G.S. Patience and J. Chaouki, *Chem. Eng. Sci.* **48**, 3195 (1993).

153. F. Cavani and F. Trifirò, *CHEMTECH* **24**, 18 (1994).

154. A. Cybulski and J.A. Mouljin, (eds.), *Structured Catalysts and Reactors*, Marcel Dekker: New York (1997).

155. D.W. Leyshon, in: *Natural Gas Conversion* (A. Holman, K.J. Jens, and S. Kolboe, eds.), Studies in Surface Science and Catalysis Vol. 61, Elsevier Science: Amsterdam (1991), p. 497.

156. A.T. Ashcroft, A.K. Cheetham, J.S. Foord, M.L.H. Green, C.P. Grey, A.J. Murrell, and P.D.F. Vernon, *Nature* **344**, 319 (1990).

157. E. Morales and J.H. Lunsford, *J. Catal.* **118**, 255 (1989).

158. V.R. Choudhary, B.S. Uphade, and A.A. Belhekar, *J. Catal.* **163**, 312 (1996).

159. V.R. Choudhary, B. Prabhakar, and A.M. Rajput, *J. Catal.* **157**, 752 (1995).

160. L.D. Schmidt, M. Huff, and S.S. Bharadwaj, *Chem. Eng. Science* **49**, 3981 (1994).

161. D.AQ. Goetsch, P.M. Witt, and L.D. Schmidt, in: *Heterogeneous Hydrocarbon Oxidation* (B.K. Warren and S.T. Oyama, eds.), ACS Symposium Series 638, American Chemical Society: Washington DC (1996), p. 124.

162. R. Burch and E.M. Crabb, *Appl. Catal.* **100**, 111 (1993).

163. M.Yu. Sinev, L.Ya. Margolis, and V.N. Korchak, *Russ. Chem. Rev.* **64**, 349 (1995).

164. M.Yu. Sinev, *Catal. Today* **24**, 389 (1995).

165. D.J. Driscoll, K.D. Campbell, and J.H. Lunsford, *Adv. Catal.* **35**, 139 (1987).

166. J.H. Lunsford, *Langmuir* **5**, 12 (1989).

167. O.V. Krylov, *Catal. Today* **18**, 209 (1993).

168. D.J. Driscoll, M. Wilson, J.-X. Wang, and J.H. Lunsford, *J. Am. Chem. Soc.* **107**, 58 (1985).

169. F. Cavani and F. Trifirò, *Catal. Today* **24**, 307 (1995).

170. J.H.B.J. Hoebonk, G.B. Marin, in: *Structured Catalysts and Reactors* (A. Cybulski and J.A. Moulijn, eds.), Marcel Dekker: New York (1997), p. 209.

171. A. Cybulski and J.A. Moulijn, *Catal. Rev.-Sci. Eng.* **36**, 179 (1994).

172. S. Irandoust and B. Andersson, *Catal. Rev.-Sci. Eng.* **30**, 341 (1988).

173. J.P. Deluca, L.E. Campbell, in: *Advanced Materials in Catalysis* (J.J. Burton and R.L. Garten, eds.), Academic Press: New York (1977), p. 293.

174. L.D. Pfefferle and W.C. Pfefferle, *Catal. Rev.-Sci. Eng.* **29**, 219 (1987).

175. R. Prasad, L.A. Kennedy, and E. Ruckenstein, *Catal. Rev.-Sci. Eng.* **26**, 1 (1984).

176. D.L. Trimm, *Appl. Catal.* **7**, 249 (1983).

177. B. Bernauer, A. Simecek, and J. Vosolsobe, *Collect. Czech. Chem. Comm.* **47**, 2087 (1982).

178. E. Zabar and M. Sheintuch, *Chem. Eng. Comm.* **16**, 313 (1982).

179. P. Subbana, H. Greene, and F. Desai, *Environ. Sci. Technol.* **22**, 557 (1988).

180. J.J.H.M. Font Freide, M.J. Howard, and T.A. Lomas, *EP Patent* 3,322,289 A2 (1989).

181. M. Huff and L.D. Schmidt, *J. Phys. Chem.* **97**, 11815 (1993).

182. M. Huff and L.D. Schmidt, *J. Catal.* **149**, 127 (1994).

183. M. Huff and L.D. Schmidt, *J. Catal.* **155**, 82 (1995).

184. G. Capannelli, E. Carosini, F. Cavani, O. Monticelli, and F. Trifirò, *Chem. Eng. Sci.* **51**, 1817 (1996).

185. G. Capannelli, E. Carosini, F. Cavani, O. Monticelli, and F. Trifirò, *Catal. Lett.* **39**, 241 (1996).

186. P.M. Torniainen, X. Chu, and L.D. Schmidt, *J. Catal.* **146**, 1 (1994).

187. D.A. Hickman, E.A. Haupfear, and L.D. Schmidt, *Catal. Lett.* **17**, 223 (1993).

188. D.A. Hickman and L.D. Schmidt, *J. Catal.* **138**, 267 (1992).

189. L.D. Schmidt and C.T. Goralski, in: *3rd World Congress on Oxidation Catalysis* (R.K. Grasselli, S.T. Oyama, A.M. Gaffney, and J.E. Lyons, eds.), Studies in Surface Science and Catalysis Vol. 110, Elsevier Science: Amsterdam (1997), p. 491.

190. D.A. Hickman and L.D. Schmidt, *AIChE J.* **39**, 1164 (1993).

191. P. Witt and L.D. Schmidt, *J. Catal.* **163**, 465 (1996).

192. A. Dietz III and L.D. Schmidt, *Catal. Lett.* **33**, 15 (1995).

193. C.T. Goralski and L.D. Schmidt, *Catal. Lett.* **42**, 47 (1996).

194. G. Saracco and V. Specchia, *Catal. Rev.-Sci. Eng.* **36**, 305 (1994).

195. J.N. Armor, *Appl. Catal.* **49**, 1 (1989).

196. J.N. Armor, *CHEMTECH* **22**, 557 (1992).

197. J.L. Falconer, R.D. Noble, and D.P. Sperry, in: *The Handbook of Membrane Separations* (S.A. Stern and R.D. Noble, eds.), Marcel Dekker: New York (1993).

198. J. Shu, B.P.A. Grandjean, A. van Neste, and S. Kaliaguine, *Can. J. Chem. Eng.* **69**, 1036 (1991).

199. H.P. Hsieh, *Catal. Rev.-Sci. Eng.* **33**, 1 (1991).

200. M. Stoukides, *Ind. Eng. Chem. Res.* **27**, 1745 (1988).

201. V.M. Gryaznov, V.I. Vedernikov, and S.G. Gul'yanova, *Kinet. Catal.* **26**, 129 (1986).

202. A.G. Anshits, A.N. Shigapov, S.N. Vereshchagin, and V.N. Shevnin, *Kinet. Catal.* **30**, 1103 (1989).

203. V.M. Gryaznov, S.G. Gul'yanova, and Yu.M. Serov, *Russ. Chem. Rev.* **58**, 35 (1989).

204. J.W. Veldsink, R.M.J. van Damme, G.F. Versteeg, and W.P.M. van Swaaij, *Chem. Eng. Sci.* **47**, 2939 (1992).

205. D. Lafarga, J. Santamaria, and M. Menendez, *Chem. Eng. Sci.* **49**, 2005 (1994).

206. M.P. Harold, V.T. Zaspalis, K. Keizer, and A.J. Burggraaf, *5th Annual Meeting of the NAMS*, Lexington, KY, (1992), paper 11B.

207. J. Coronas, M. Menendez, and J. Santamaria, *J. Loss Prev. Proc. Ind.* **8**, 97 (1995).

208. J. Coronas, M. Menendez, and J. Santamaria, *Chem. Eng. Sci.* **49**, 4749 (1994).

209. K. Otsuka and I. Yamanaka, *Catal. Today* **41**, 311 (1998).

210. C.G. Vayenas and R.D. Farr, *Science* **208**, 593 (1980).

211. C.G. Vayenas, S.I. Bebelis, and C.C. Kyriazis, *CHEMTECH* **7**, 422 (1991).

212. M. Stoukides, *Ind. Eng. Chem. Research* **27**, 1745 (1988).

213. D. Eng and M. Stoukides, *Catal. Rev., Sci. Eng.* **33**, 375 (1991).

214. K. Otsuka, S. Yokoyama, and A. Morikawa, *Chem. Lett.* **319** (1985).

215. K. Otsuka, K. Suga, and I. Yamanaka, *Catal. Today* **6**, 587 (1990).

216. T. Hayakawa, T. Tsunoda, H. Orita, T. Kameyama, H. Takahasi, K. Takehira, and K. Fukuda, *Chem. Comm.*, 961 (1986).

217. T. Hayakawa, K. Kito, A. York, T. Tsunoda, K. Suzuki, M. Shimizu, and K. Takehira, *J. Electrochem. Soc.* **144**, 1 (1997).

218. K. Otsuka, Y. Shimizu, and I. Yamanaka, *J. Electrochem. Soc.* **137**, 2076 (1990).

219. K. Otsuka, Y. Shimizu, I. Yamanaka, and T. Komatsu, *Catal. Lett.* **3**, 365 (1989).

220. K. Otsuka, K. Ishizuka, I. Yamanaka, and M. Hatano, *J. Electrochem. Soc.* **138**, 3176 (1991).

221. K. Otsuka and I. Yamanaka, *Electrochim. Acta* **35**, 319 (1990).

222. Q. Zhang and K. Otsuka, *Chem. Lett.* **363** (1997).

223. A. Clearfield, *Chem. Rev.* **88**, 125 (1988).
224. T.M. Gur, A. Belzner, and R.A. Huggins, *J. Membr. Sci.* **75**, 151 (1992).
225. K. Omata, S. Hashimoto, H. Tominaga, and K. Fujimoto, *Appl. Catal.* **52**, L1 (1989).
226. K. Fujimoto, K. Asami, K. Omata, and S. Hashimoto, in: *Natural Gas Conversion* (A. Holman, K.J. Jens, and S. Kolboe, eds.), Studies in Surface Science and Catalysis Vol. 61, Elsevier Science: Amsterdam (1991), p. 525.
227. R. Ramachandran and D. MacLean, *Techn. Magazine* **6**, 43 (1994).
228. R.P. Burke and R. Miller, *Chem. Week* **8**, 93 (1964).
229. R.G. Markeloff, *Hydroc. Proc.* **11**, 91 (1984).
230. W.E. Wimer and R.E. Feathers, *Hydroc. Proc.* **3**, 81 (1976).
231. P. Reich, *Hydroc. Proc.* **3**, 85 (1976).
232. R.W. McPherson, C.M. Starks, and G.J. Fryar, *Hydroc. Proc.* **3**, 75 (1979).
233. T. Ihara, A. Kayou, and H. Kameo, Absts. *3rd Tokyo Conference on Advanced Catalytic Science and Technology*, Catalysis Society of Japan: Tokyo (1998), p. 114.

NEW TECHNOLOGICAL AND INDUSTRIAL OPPORTUNITIES
Examples

3.1. INTRODUCTION

Several scientific and technological developments as well as new demands for selective oxidation processes have significantly modified the outlines of this field compared to a decade or two ago.

Integration between catalytic science and reaction engineering has had a marked effect on recent developments in selective oxidation and will continue to have a positive influence for the next few years. New types of reactors have been developed and others have been adapted for selective oxidation reactions (see Chapter 2) leading to:

1. Better use or control of the heat of reaction and of homogeneous–heterogeneous processes (adiabatic monolith reactors, combined reaction and separation, structured beds).
2. Full efficient use of the unusual potential of selective oxidation catalysts (separate reduction and reoxidation stages, activity in ultrashort contact time, operation with metastable catalysts and under dynamic catalyst conditions, e.g., using reverse flow).
3. Integration of the catalytic step in a more complicated network of functions to be carried out by a system (several reactions in the same reactor, reaction and separation, combustion and production of energy).

New catalytic materials have been developed, and recent advances in their characterization have made it possible to clarify the key features of their action mechanisms and thus to tune their activities and develop innovative materials. New types of both heterogeneous and homogeneous catalysts have been developed. Particularly remarkable are the development and/or improvement of: (i) a series of new mixed oxides for the selective oxidation or oxidative dehydrogenation of light alkanes, (ii) titanium silicalite and related selective oxidation processes in the liquid phase and from this a long series of transition metal–containing microporous materials for the selective oxidation of a wide range of substrates, (iii) a series of new catalysts for the synthesis of fine chemicals by selective oxidation processes, (iv) metalloporphyrin complexes that simulate the action of enzymatic monoxygenases, and (v) multifunctional heteropoly compounds that can be readily tailored for specific applications.

New processes have also appeared in the last 10 years, such as[1]:

- Synthesis of methylmethacrylate from isobutene (Asahi Chemical Industries).
- Oxidation of methylal ($CH_3O–CH_2–OCH_3$) to formaldehyde on iron–molybdenum oxides.
- Selective gas phase oxidation of ethene to acetic acid (Showa Denko).
- Oxidation of 4-methoxy toluene to *p*-methoxybenzaldehyde over alkali-doped supported vanadium catalysts (Nippon Shokubai and Kagaku Kogyo).
- Oxidative dehydrogenation of methylformamide to methyl isocyanate (for *in situ* use) (Dupont).
- Direct oxidation of benzene to phenol over platinum-doped vanadium oxide on silica (Tosoh Co.).
- Tetrahydrofuran synthesis from maleic anhydride via *n*-butane oxidation to maleic anhydride in riser reactors (Dupont).
- Acetoxylation of butadiene as the intermediate to a new route to 1,4 butane-diol and tetrahydrofuran on palladium/telleurium catalysts (Mitsubishi Kasei).
- Caprolactam synthesis from cyclohexanone via intermediate synthesis of cyclohexanone oxime avoiding formation of ammonium sulfate (EniChem).
- Propene oxide synthesis from propene using the TS-1/H_2O_2 process, also with *in situ* production of H_2O_2 (EniChem).

All these developments are evidence of the dynamism and inventiveness in the field of selective oxidation and indicate clearly that this area of research is at the threshold of a innovative period with relevance for other forms of catalysis. There are several new technological and industrial opportunities for selective oxidation. In this chapter some examples of new processes and catalysts for selective oxidation will be discussed in order to outline general aspects and possibilities, which will then be discussed in more detail in subsequent chapters. The discussion will focus primarily on the use of H_2O_2 as the selective oxidizing agent. Some of the innovative approaches being developed for using new types of oxidizing agents and new ways to activate the least costly of them (O_2) will be surveyed in Chapter 7.

A second group of new opportunities for selective oxidation processes derives from the use of new types of substrates, in particular alkanes. This area has seen more innovations in the field than any other and the new technologies that have been developed will be discussed in detail in Chapters 4 and 5.

A brief outline of catalytic systems that have recently been subjected to particular attention concludes this chapter. Other examples will be described in Chapter 6.

3.2 EXAMPLES OF OPPORTUNITIES FOR NEW OXIDATION PROCESSES

3.2.1. Selective Oxidation for Fine Chemicals and Pharmaceuticals

Most of the reactions for the synthesis of fine chemicals and pharmaceuticals involve one or more steps of oxidation of organic functional groups. However most oxidation reactions use inorganic oxidants such as permanganates and chromium oxide, and their disposal is becoming environmentally unacceptable. In the production of fine chemicals and pharmaceuticals the ratio between waste and product production may reach values of 50–100 by weight, clear indication of the necessity to develop cleaner processes for organic synthesis involving catalytic steps and solid catalysts.

For the synthesis of complex molecules of interest for fine chemicals by selective oxidation using solid catalysts, many studies are currently focusing on:

1. Development of transition metal–containing microporous materials using hydrogen peroxide and organic peroxides as primary oxidants.[2–7]
2. Development of noble metal catalysts using oxygen as the oxidizing agent.[2,8,9]
3. The bioinorganic approach, i.e., oxidation reactions mimicking biological oxidative processes.[10,11]

3.2.1.1. Oxidation with Hydrogen Peroxide and Organic Peroxide

The activity of oxides such as MoO_3 and WO_3 (bulk oxides or supported on silica) in the monoxygenation of organic substrates (e.g., propene to propene oxide) with alkyl hydroperoxide is known,[12,13] although the activity is probably related to the formation of soluble peroxo complexes. However, Ti(IV) ions on silica have been shown to be truly heterogeneous catalysts in the epoxidation of double bonds with ethylbenzene hydroperoxides.[12]

The limitations in the use of organic peroxides are their cost and the formation of coproducts. Thus the discovery that titanium ions incorporated in the framework of a silica zeolite (TS-1) show a high stable activity in epoxidation reactions, using

H_2O_2 instead of an organic peroxide[14] has stimulated research on the extension of the use of transition metal–containing microporous materials for selective oxidation reactions using hydrogen peroxide, although in most cases the critical problem of metal leaching must be solved before application.[6]

Three major applications of titanium silicalite, namely alkene epoxidation, cyclohexanone ammoximation, and phenol hydroxylation will be discussed later. Here we relate only to other types of reactions of possible interest for the synthesis of fine chemicals and pharmaceuticals.

Titanium silicalite can be synthesized in two crystalline forms, TS-1 and TS-2, although the former is the one more commonly used. TS-1 has a crystalline structure similar to ZSM-5 with a tridimensional channel structure: parallel channels in one direction, and zigzag type channels perpendicular to them. The opening of the channels is 0.55–0.60 nm.[15] Titanium ions are incorporated in the zeolite framework in partial substitution for silica ions. They are all accessible to reactants but isolated from each other, so side reactions owing to the presence of Ti–O–Ti units are avoided. The presence of extraframework titanium oxide, which causes such types of side reactivity also must be avoided because, e.g., it catalyzes the decomposition of hydrogen peroxide.

Several catalytic reactions of interest for organic synthesis are catalyzed by the TS-1/H_2O_2 system. Secondary alcohols are very selectively converted to the corresponding ketones (with selectivity close to 95%),[16] while primary alcohols are converted to the corresponding aldehydes and ketones.[16,17] The reactions occur in the 50–70°C temperature range and the order of reactivity with respect to the position of the carbon atom in the hydrocarbon chain to which the OH group is linked is $\beta > \alpha > \gamma$.[16,18] The difference in reactivity between the β and γ positions is related to constraints in the transition states inside the zeolite cages and further evidences the differences in the reactivity when the active sites are located inside a microporous environment, which creates sterical constraints that are not present in the case of homogeneous catalysts.

The TS-1/H_2O_2 system catalyzes the epoxidation of alkenes at near-to-room temperatures and the reaction is especially fast and selective when the double bond does not have near-neighbor groups withdrawing electrons.[19] Epoxidation of terminal double bonds is slower than that of internal double bonds, but when a primary or secondary alcohol is present along with a terminal double bond, the oxidation proceeds selectively at the double bond. For example, the reaction of allyl alcohol at 40°C with hydrogen peroxide in the presence of TS-1 gives an 85% yield of the epoxide and minimal formation of acrolein or acrylic acid,[20] while in the gas phase over various vanadium and molybdenum oxides or mixed oxides the allyl alcohol rapidly converts to acrolein at low reaction temperatures (see Chapter 8). This points out the differences in the types of products and reaction mechanisms between TS-1 and mixed oxides.

With more bulky and substituted unsaturated alcohols, however, there is oxidation of both the double bond and the alcohols, again for steric reasons. The selectivity is determined by the degree of substitution on the double bond and the carbinol carbon.[21] Increasing the steric hindrance around the double bond increases the degree of alcohol oxidation. For example, 3-penten-2-ol oxidation with TS-1/H_2O_2 gives the epoxy alcohol and unsaturated ketone is nearly equal amounts.

Other examples of reactions made by the TS-1/H_2O_2 system include phenol hydroxylation,[22] sulfoxidation,[23] ammoximation of aldehydes and ketones with NH_3/H_2O_2,[24] oxidation of primary[25] and secondary amines[26] to oximes and hydroxylamines, respectively, and hydroxylation of alkanes to secondary alcohols and alkanes.[27]

The reactivity of TS-1 is due to site-isolation of tetrahedral titanium(IV) in a hydrophobic environment. The hydrophobic character of the silicalite provides for selective adsorption of apolar substrates and relatively apolar H_2O_2 from the aqueous reaction mixtures. The titanium(IV) acts as a Lewis acid, thereby decreasing the electron density in the coordinated hydroperoxide and promoting nucleophilic attack of the hydrocarbon on the O–O bond. The electrophilicity of this species is further increased by coordination of the alcohol solvent (usually methanol, see Chapter 7) to the titanium hydroperoxo species forming a five-member peracid-type species, in which oxygen transfer is facilitated.

A disadvantage of TS-1 is its limited pore size. Molecules with a cross section greater than about 0.60 nm cannot enter the channel and thus TS-1 is essentially inert in the H_2O_2 oxidation of large molecules. For this reason, in recent years much research activity has been focused on the development of analogous Ti-containing microporous materials such as Ti-β, Ti-UTD, TAPSO-5, Ti-containing mesoporous materials (Ti-MCM-41).[28–31] However, differently from TS-1, titanium ions in these kinds of new titanium-containing zeolite-like materials are not stable against metal leaching during the reaction conditions of catalytic oxidation with hydrogen peroxide.[6,32] Primarily because formation of the active titanium hydroperoxide species requires a change in coordination of the titanium ion from tetrahedral to octahedral, thus breaking of some Ti–O(zeolite) bonds. After the reaction of monoxygen transfer, the rate at which this titanium species closes again to reform the stable tetrahedral species must be faster than the rate of further hydrolysis of the remaining Ti–O(zeolite) bonds, which cause the solvation of the titanium ion and thus its leaching from the zeolite structure. TS-1 has unique characteristics of stability in this sense, related to the peculiar local environment of the titanium ions. All other kinds of Ti-containing micro- and mesoporous materials developed up to now show a faster rate of hydrolysis compared to the rate of octahedral to tetrahedral closure and thus the rate of leaching of titanium ions is a rapid one.

These samples and analogous systems containing different transition metals such as Co, Sn, V, Cr, and Zr, are more stable in organic-type solvents, which, while not suitable for H_2O_2, can be used with organic peroxides such as *tert*-butyl hydroperoxide (TBHP). It thus may be concluded that while H_2O_2 seems to be the best choice for an oxidizing agent for cleaner oxidation processes of interest for the synthesis of fine chemicals and pharmaceuticals, the limitation on the use of TS-1 (due to stability) and thus to small molecules makes it unsuitable for this kind of application. Reactions of monoxygenation of interest for the synthesis of fine chemicals and pharmaceuticals thus require the use of organic peroxides such as TBHP, organic solvents, and transition metals anchored on zeolite-like materials with large pores. It must be noted, however, that the possible shape selectivity effects are usually lost when materials with very large pores such as mesoporous MCM-41 are used.

Generally pore restrictions preclude the use of TBHP in the oxidation of organic compounds using TS-1 owing to severe steric constraints in the transition state for oxygen transfer, but the difficulty can be resolved by using large-pore molecular sieves, and Ti-β, Ti mesoporous silicates (HMS), and Ti-MCM-41 are active catalysts for oxidation with TBHP.[33–35] The Brönsted acidity in a Ti-substituted zeolite can also be an asset to the catalytic activity, for instance, in the conversion of linalool (di-3, 7-dimethyl-3-hydroxy-1,6-octadiene) to cyclic ethers. The bifunctional (Ti-β or Ti-MCM-41) catalyst first epoxidizes the electron-rich double bond at a titanium site, which is then followed by acid-catalyzed rearrangement at an aluminum site. The observed conversion of linalool to cyclic hydroxy ethers closely resembles that catalyzed by epoxidase enzymes *in vivo*.

Success with TS-1 has also stimulated investigations of zeolite-like materials containing other metals such as Zr, Sn, V, Cr with properties similar to Ti, but which can catalyze different types of reactions.[4,6,36–46] For example, vanadium can, depending on the substrate used, react via an oxometal or a peroxometal pathway. Chromium also reacts via oxometal pathways and is a useful catalyst for allylic and benzylic oxidations. Because the active oxidant in oxometal pathways does not contain the alkylperoxo group, oxidation with TBHP is not subject to the same steric constraints as in peroxometal pathways. Thus, CrS-1 (an analogue of TS-1, but containing chromium) was shown to be an active and recyclable catalyst for the oxidation of benzyl alcohol and ethylbenzene with TBHP to benzaldehyde/benzoic acid and acetophenone, respectively.[6,32]

Another approach to designing shape-selective solid catalysts is to use the transition metal as the pillaring component of pillared clays.[47] Chromium-pillared montmorillonite (Cr-PILC) is an effective catalyst for the selective oxidation with TBHP of primary aliphatic and aromatic alcohols to the corresponding aldehydes, while secondary alcohols are selectively oxidized in the presence of a primary hydroxy group of a diol to give keto alcohols with selectivities higher than 90%.[48] Cr-PILC also catalyzes the benzylic oxidation of acylmethylene to the corresponding carbonyl compound.[49] For example, tetralin is oxidized to α-tetralone, bibenzyl to desoxybenzoin, *p*-isopropylethylbenzene to *p*-isopropylacetophenone.

Several other reactions useful in synthetic organic chemistry are catalyzed by Cr-PILC; it can be useful, e.g., in the oxidative deprotection of allyl and benzyl ethers and amines.[49,50] Furthermore, different elements such as V, Ti, and Zr, can also be pillared in the clay, thus obtaining different kinds of reactivities in a manner similar to those with zeolite-like materials containing transition metals.

These examples, although not comprehensive regarding the large body of research in these areas in the last decade, illustrate out the wide range of synthetic possibilities offered by the use of solid catalysts based on transition metals in micro- and mesoporous materials or clays. The activity and selectivity can be very well tuned based on the choice of transition metal and reaction conditions. However, it must be noted that the more interesting and cleaner reactant, i.e., hydrogen peroxide, can only be utilized effectively with the TS-1 catalyst and for small substrates, since the titanium and other transition metals are not stably anchored on other kinds of materials in the presence of 30% hydrogen peroxide. These other materials require the use of organic solvents and organic peroxides as oxidants. Nevertheless, a wide range of applications is possible with these systems.

3.2.1.2. Oxidation with Molecular Oxygen and Noble Metal-Based Catalysts

Oxidation with molecular oxygen using noble metal-based catalysts is an important new route for the conversion of raw materials of interest and this possibility is discussed here for the oxidation of carbohydrates and carbohydrate derivatives.[8]

Although the use of carbohydrates as chemical feedstock is limited compared to the agricultural production, their use is expanding: (i) as raw materials for the synthesis of, e.g., biodegradable polymers, surface active agents, and colloidal agents; and (ii) as chemicals in, e.g., biotechnology, production of paper and boards, and thickeners. In many cases, however, chemical modification of the raw carbohydrates is required in order to obtain the desired chemical and physical properties. Oxidation is one of the useful techniques for such required modifications.[9] Up to now, mainly stoichiometric oxidation procedures such as oxidation of starch with sodium hypochlorite or periodate to give dicarboxy- or dialdehyde starch,[51] or production of oxalic acid from starch or cellulose by oxidation with nitric acid[52] have been used in industry, but new cleaner and environmentally compatible catalytic processes are available for many of these transformations:

1. Oxidative cleavage with hydroperoxides in the presence of homogeneous or heterogeneous transition metal–containing catalysts (Fe^{3+}, W^{6+}, V^{5+}, Ti^{4+}).
2. Metal-catalyzed oxidation with O_2 using heterogeneous noble metal–based catalysts.

Notwithstanding the higher catalyst cost, the second method using O_2 instead of the much more costly hydroperoxides is clearly preferable for larger-scale applications. On the other hand, in several cases both methods are preferred over stoichiometric oxidation. Table 3.1 summarizes the advantages and disadvantages of stoichiometric and catalytic processes of oxidation of carbohydrate raw materials.[8]

Noble metal oxidation is well suited for the selective conversion of primary hydroxyl or aldehyde groups to carboxylic acid. The oxidation can be viewed as an oxidative dehydrogenation of the substrate by the noble metal, followed by oxidation of the adsorbed hydrogen atoms.[53] The dehydrogenation step probably occurs by a concerted mechanism of deprotonation of the hydroxyl group and hydride transfer from the carbon atom to the noble metal surface.

Oxygen, although required to oxidize the adsorbed hydrogen, also acts as a poison for the noble metal. The oxygen tolerance depends on the type of noble metal, the reaction conditions (temperature, pH, etc.), and the type of substrate. Deactivation of the catalyst by oxygen proceeds in two or three stages. First the noble metal surface will be occupied by chemisorbed oxygen (up to a surface coverage of about 0.25 of the monolayer). This dissociatively chemisorbed oxygen can be easily removed by simple feeding of an inert gas. After prolonged exposure to oxygen, the surface oxygen migrates into the noble metal lattice forming an oxide layer or even noble metal oxide crystallites. In this case, reactivation requires the use of reducing agents such as hydrogen.

We will discuss here three examples of oxidation of carbohydrates and their derivatives[8]: (i) oxidation of monosaccharides, (ii) oxidation of di- and oligosaccharides, and (iii) oxidation of 5-hydroxymethylfurfural.

Of the first class of reactions, the oxidation of glucose is the most representative example of the conversion of unprotected sugars, and several different research

TABLE 3.1. Advantages and Disadvantages of Stoichiometric and Catalytic Processes of Oxidation of Carbohydrate Raw Materials[8]

| | Stoichiometric | Catalytic | |
		Heterogeneous	Homogeneous
Advantages	Often highly selective	Easy separation of dissolved products	Mild conditions
	Suitable for specific reaction	Catalyst recycling not difficult	Applicable to solid substrates
			Good heat transfer
Disadvantages	Often by-products (salts)	Not applicable to solid substrates	Difficult separation of dissolved products
	Expensive	Heat-transfer problems may be encountered	No continuous processing possible

groups have studied it.[54] The primary oxidation products of glucose are gluconic and glucaric acid, and from gluconic acid, 2-keto-gluconic acid and glucuronic acid (Scheme 3.1). The oxidation toward gluconic acid can be made very selectively.[55] Hattori[56] showed Pd/C to be a very active and selective catalyst, and more recently attention has been focused on the promotion of this type of catalyst by Bi or other components to further improve the selectivity.[57] Using the promoted catalysts, the reaction is fully competitive with the biocatalytic process.

Palladium is preferred over platinum, notwithstanding the fact that the latter has a higher tolerance toward oxygen (thus is less sensitive to deactivation), because the passivation of the catalyst is one of the keys to high selectivity for gluconic acid. In fact, owing to the higher rate of oxidation of the glucose as opposed to further conversion of the gluconic acid, at the end of the reaction (where gluconic acid concentration is higher), the Pd-based catalyst is deactivated by oxygen and unable to further convert the gluconic acid, whereas Pt-based catalysts maintain a residual activity and thus give a lower selectivity.

Another interesting aspect of the reactivity of these catalysts is the role of the interaction of the substrate and the catalyst. This is a controlling aspect of the

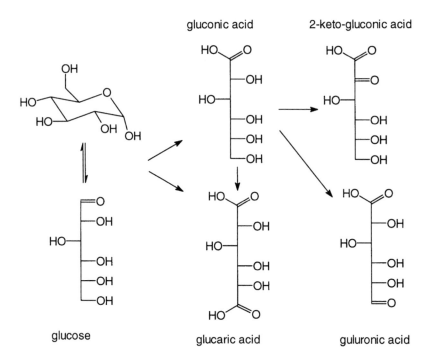

Scheme 3.1. Oxidation products of glucose obtained by noble metal–catalyzed oxidation.[8]

reactivity and can explain why the oxidation of glucose or gluconic acid toward glucaric acid is only possible with Pt catalysts and not with Pd catalysts.[8] Selectivity, however, is only about 70% owing to the formation of by-products such as oxalic acid.

Another carbohydrate can be used as a hydrogen acceptor instead of oxygen. At pH > 12 Pt and Rh effectively catalyze the conversion of glucose to gluconic acid in the presence of fructose, which transforms to sorbitol and mannitol.[58]

In protected sugars the scope of the oxidation is to convert the primary alcohol groups, while leaving the anomeric center intact. For example, 1-O-methyl glucopyranoside can be converted to 1-O-methyl glucuronic acid over Pt/C catalysts,[59] although the selectivity is not very high (about 70%). A similar reaction is the oxidation of 1-O-alkyl glucosides (alkyl = C8, C10, and C12) to 1-O-alkyl glucuronides, products interesting as special anionic–nonionic surfactants (Scheme 3.2a).[8,60] Using Pt or Pd catalysts with low noble metal dispersion, the selectivity is about 85%.

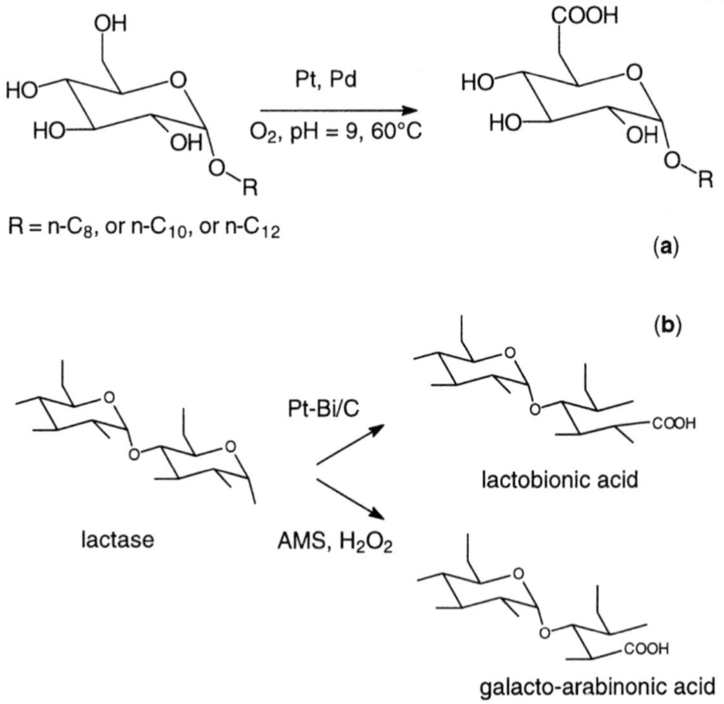

Scheme 3.2. Oxidation of 1–O–alkyl–α-D-glucopyranoside to the corresponding glucuronic acid (a), and oxidation of lactose to lactobionic acid and galacto-arabinonic acid (b).[8]

The oxidation of di- and oligosaccharides can also be devised in reducing (unprotected aldoses) and nonreducing (protected or keto-) sugars analogously to monosaccharides. For the first class of products the reactivity is also similar to that of monosaccharides. For example, lactose can be selectively converted (up to 95% selectivity) using Pd–Bi/C catalysts (Scheme 3.2b).[61] Using H_2O_2 in the presence of anthraquinone monosulfonate (AMS) instead, results in a selective decarboxylation with formation of galactoarabinonate.

The oxidation of oligosaccharides is also commercially interesting, because, e.g., the oxidation of β-cyclodextrin at the C6 position provides chemicals that can be used as cation complexing agents. Pt catalysts can be used for this reaction, although the selectivity is not good.[62]

5-Hydroxymethylfurfural (HMF), which can be obtained by acid-catalyzed dehydration of carboxydrates such as fructose,[63] is also an interesting reactant of the oxidation using noble metal catalysts, because several commercially interesting chemicals can be obtained. For example, hydroxymethyl-2-furancarboxylic acid (HFCA) and 2,5-furandicarboxylic acid (FDCA) are useful monomers for the production of polyesters, and 2,5-furandicarboxaldehyde (FDC) can be used in photochromic and conducting materials.

FDCA can be obtained in high yields using noble metal catalysts and air, whereas HFCA can be selectively formed using Ag_2O/CuO as catalysts with oxygen as the oxidant.[64] Various noble metals are active in the oxidation of HMF to FDCA via intermediate formation of 5-formyl-2-furancarboxylic acid.[65] None of the catalysts is poisoned by oxygen, probably owing to the strong adsorption of the aromatic furan ring, which inhibits oxygen chemisorption and deactivation. In this case, the reactant itself has a self-protective function of the catalytic activity.

These general indications on the reactivity of carboxylates in the oxidation by noble metals in the presence of air, although not exhausting the interesting chemistry of the system, illustrate some of the potentialities and difficulties of this approach, which, apart from its intrinsic commercial interest, can be viewed as a model for a broader use of this catalytic methodology in the selective conversion of highly functionalized organic substrates.

3.2.1.3. Bioinorganic-Type Oxidation

Molecular oxygen is the cheapest and most attractive oxidant for the chemical industry, but it is difficult to find effective catalysts to use with it directly under mild conditions, as noted repeatedly throughout this book. However, there is a group of enzymes (monooxygenases) that catalyze reactions such as epoxidation of alkenes, hydroxylation of aromatics, and selective conversion of lower alkanes quite selectively under mild conditions and molecular oxygen as the oxidant. A well-known example is Cytochrome P-450. An active and very promising line of research is to mimic the active sites and reactions of monoxygenase in order to develop novel catalysts and oxidation reactions with molecular oxygen.[10,11]

Enzymes that activate dioxygen to oxidize organic substrates can be classified into two classes according to the mode of oxygen incorporation (Scheme 3.3):

1. Dioxygenase incorporates two oxygen atoms of a dioxygen molecule into the substrate.
2. Monoxygenase transfers only one oxygen atom; the latter usually requires NAD(P)H as a cofactor and the other oxygen atom in dioxygen is converted to water.

Reactions catalyzed by monoxygenases include epoxidation of the C–C unsaturated bond and hydroxylation of the aromatic ring or C–H bond in saturated hydrocarbons, both of which are difficult to achieve with conventional catalytic systems. Dioxygenase also catalyses very interesting reactions including regio- and stereoselective oxidation reactions as well as reactions producing very unstable products such as prostaglandin. Iron and copper are the main active elements of these enzymes. The reaction mechanism, e.g., for the most studied Cytochrome P-450 enzyme, involves the formation of a high-valence iron oxo species resulting from the heterolytic O–O bond cleavage of the coordinated dioxygen molecule.[66,67] The principal aspect of the reaction scheme is the reductive activation of dioxygen with two electrons and protons from NADPH associated with effective electron and proton transfer systems. The high-valence iron oxo formed species is active enough to induce electrophilic addition to unsaturated C–C bonds and direct oxidation of saturated hydrocarbons. In anaerobic conditions, but in the presence of oxidants such as iodosylbenzene, alkylhydroperoxide, hydrogen peroxide, or peracid, Cytochrome P-450 also catalyzes these reactions ("shunt path" mechanism, see Scheme 3.4).[66]

A huge number of catalyst systems (metal porphyrin, metal complexes, heteropoly compounds, transition metal–containing microporous materials) for alkene epoxidation have been reported on the basis of the shunt path concept of Cytochrome P-450. For example, many transition metal porphyrins such as manganese,[68,69] chromium,[70] ruthenium,[71] and osmium complexes[72] are effective for alkene epoxidation.

Because of the oxidative degradation of the metalloporphyrin skeleton under the reaction conditions, there have been many attempts to prepare robust porphyrins, for instance, by adding electronegative groups on the pyrrole ring of the porphyrin. A maximum turnover number higher than 200,000 has been reported for the epoxidation of α-methylstyrene with 5,10,15,20-tetrakis (2′,6′-difluorophenyl)porphyrin.[73] Organic and inorganic matrices can also be utilized to support active transition metal cations, the latter with the advantage of higher

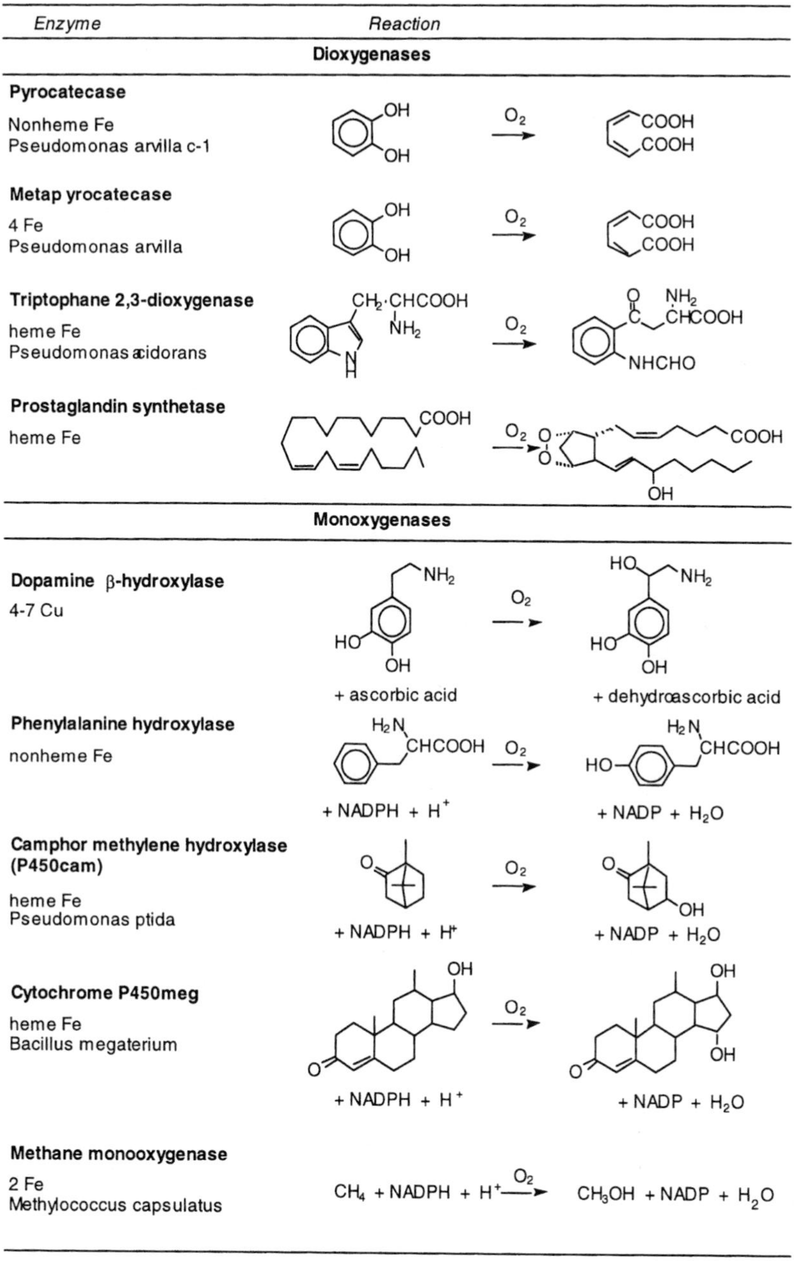
Enzyme
Reaction
Dioxygenases
Pyrocatecase
Nonheme Fe
Pseudomonas arvilla c-1
OH
OH
O2
COOH
COOH
Metap yrocatecase
4 Fe
Pseudomonas arvilla
OH
OH
O2
COOH
COOH
Triptophane 2,3-dioxygenase
heme Fe
Pseudomonas acidorans
CH2·CHCOOH
NH2
N
H
O2
O
C
NH2
CHCOOH
NHCHO
Prostaglandin synthetase
heme Fe
COOH
O2
O
O
COOH
OH
Monoxygenases
Dopamine β-hydroxylase
4-7 Cu
NH2
HO
OH
O2
HO
NH2
HO
OH
+ ascorbic acid
+ dehydroascorbic acid
Phenylalanine hydroxylase
nonheme Fe
H2N
CHCOOH
O2
HO
H2N
CHCOOH
+ NADPH + H+
+ NADP + H2O
Camphor methylene hydroxylase
(P450cam)
heme Fe
Pseudomonas ptida
O
O2
O
OH
+ NADPH + H+
+ NADP + H2O
Cytochrome P450meg
heme Fe
Bacillus megaterium
OH
O
O2
OH
O
OH
+ NADPH + H+
+ NADP + H2O
Methane monooxygenase
2 Fe
Methylococcus capsulatus
CH4 + NADPH + H+
O2
CH3OH + NADP + H2O

Scheme 3.3. Dioxygenase and monoxygenase enzymes that are able to catalyze reactions at room temperature. Elaborated from Moro-oka and Akita.[10]

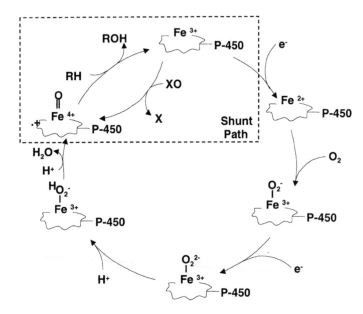

Scheme 3.4. Reaction mechanism for hydrocarbon oxidation with Cytochrome P-450. Elaborated from Moro-oka and Akita.[10]

chemical and thermal stability. A well-known example is titanium silicate (see also the following section).[74] Heteropoly compounds combined with phase transfer catalysts or not are also active in the oxidation with hydrogen peroxide.[75,76]

The main disadvantage of reactions mimicking the "shunt path" oxidation mechanism is the consumption of expensive oxidants, which limits their use for the production of bulk chemicals, whereas their use for the synthesis of more expensive fine chemicals (e.g., regio- or stereoselective reactions) is very interesting.

The biomimetic approach has been used extensively to realize this goal of regio- or stereoselective reactions, especially in trying to achieve steric control analogous to that for enzymes. For example, regioselective hydroxylation of n-alkane can be obtained by using hindered porphyrin manganese complexes.[77] Similarly, iron phthalocyanine inside the cavity of Y-zeolite has been used for the terminal oxidation of n-alkane.[78,79] Asymmetric epoxidation of prochiral olefin using chiral porphyrin complexes has also been studied.[80–83]

More difficult to mimic is the action of Cytochrome P-450 in oxidation using dioxygen. The ruthenium(II) porphyrin complex is oxidized by molecular oxygen to the di-oxo ruthenium(VI) complex, which oxidizes tetramethylethylene to the epoxide at room temperature, reforming the starting ruthenium(II) complex.[71] However, the turnover frequency is very low.

Effective catalysts for oxidation using dioxygen have not yet been developed. However, an alternative possibility is to use a reducing agent as coreactant. The aerobic oxidation of saturated C–H bonds and unsaturated C–C bonds with an iron complex catalyst in the presence of a reducing agent (ascorbic acid) has been known for a long time.[84] Barton and Doller[85] developed an aerobic oxidation using zinc powder as a reducing agent in acidic conditions (the iron complex / pyridine / acetic acid / zinc powder is well known as the Gif system). Saturated hydrocarbon is oxidized to the corresponding alcohol or ketone by aerobic oxidation at room temperature.

The consumption of NAD(P)H, ascorbic acid, or zinc powder may not be practical in industrial syntheses. The challenge is to find a way to use cheaper reducing agents such as hydrogen, carbon monoxide, and aldehyde. Aerobic epoxidation of olefins with molecular hydrogen was first demonstrated by Tabushi et al.[86,87] using a manganese tetraphenylporphyrin complex. Geraniol and related olefins were epoxidized regioselectively by molecular oxygen. More effective epoxidation was developed by Muhaiyama and Yamada[88] and Murahashi[11] using acetaldehyde as the reducing agent. The system is applicable to most kinds of alkenes. Hydroxylation of aromatic hydrocarbons to phenols was studied by Sasaki et al. using the system $Cu(I)/O_2/H_2SO_4$.[89] Cu(I) is converted to Cu(II) stoichiometrically, but the reaction proceeds catalytically when Cu(II) is reduced to Cu(I) by molecular hydrogen in the presence of a palladium cocatalyst.[90] The most effective hydroxylation of aromatic hydrocarbons with molecular oxygen and hydrogen was recently developed by Miyake et al.[91] Benzene is selectively oxidized to phenol on a Pt–V_2O_5 / SiO_2 catalyst. The reaction is carried out in both the liquid phase in the presence of acetic acid at 20–60°C and the gas phase at elevated temperatures,[92] but the critical point is the consumption of hydrogen in the side reaction to water. Some 7–8 mol of hydrogen are consumed to form 1 mol of phenol. The challenge for the future is to reduce hydrogen consumption, but nevertheless the reductive activation of oxygen by hydrogen may be the basis for new industrial oxidation processes.

A number of enzymes having a binuclear active center (e.g., methane monoxygenase having a binuclear iron site and tyrosinase having a dicopper center) are also known, and thus the interesting possibility of mimicking their action has been extensively investigated. Binuclear sites are also found in oxygen carriers such as hemerythrin and hemocyanin. Some copper, iron, and cobalt binuclear dioxygen complexes mimicking the characteristics of the active centers of hemocyanin and tyrosinase have been developed,[93–97] and it is expected that new types of dioxygen activation will be possible using these binuclear complexes. Only preliminary data are available, but aerobic oxidation based on the catalysts having a multimetal center will probably be a key direction of research to develop new fields of oxidation in the near future.

The bioinorganic approach to selective oxidation is thus one of the innovative research directions with several possibilities for industrial application in highly sophisticated chemical syntheses for the production of high-value chemicals.

3.2.2. New Catalytic Processes for Bulk Chemicals Using Hydrogen Peroxide

Hydrogen peroxide is the most environmentally acceptable, low-price monoxygenation agent. However, its commercial application has been limited by several factors, including:

1. The need for a suitable solvent for both hydrogen peroxide and the organic substrate to be oxidized.
2. The formation of side products through hydrolysis of the primary product with hydrogen peroxide.
3. The hazards associated with handling concentrated or anhydrous hydrogen peroxide.
4. The ready decomposition of hydrogen peroxide.
5. The inhibiting effect of water on reaction kinetics; however, when water is removed from the reaction medium, e.g., by azeotropic or catalytic distillation and using a solvent such as a polyether, it is technically very difficult to avoid the possibility of explosions of the anhydrous hydrogen peroxide.

Despite these problems, two catalytic processes that use hydrogen peroxide as an oxidant have become well established: (i) liquid phase catalytic oxidation using titanium silicalite or other metal-containing microporous materials,[3,32] and (ii) two-phase catalysis using heteropoly compounds.[98] As discussed in the previous section, as well as later in Chapter 7, hydrogen peroxide is a monoxygenation agent of increasing importance in the production of high-value chemicals, but it has limitations associated with the problems of finding effective and stable catalysts to use with large and complex molecules. However, some processes have been developed for the production of bulk chemicals that are very interesting for the possibility of reducing process complexity and its environmental impact[99]:

- Epoxidation of alkenes.
- Ammoxidation of cyclohexanone with ammonia.
- Hydroxylation of phenol to produce hydroquinone and catechol.

The last is already a commercial process; the cyclohexanone ammoxidation has been developed up to the stage of a demonstration unit, whereas the first process (propene epoxidation, in particular) will probably be commercialized in the near future, depending on the cost of hydrogen peroxide and/or the possibility of its production being integrated in the plant. All three processes have the common characteristic of using titanium silicalite as the catalyst (see Section 3.3).

The possibility of using dioxygen instead of hydrogen peroxide in these processes is certainly attractive, but results obtained to date are still far from commercialization. It may be useful, however, to cite some of these alternative directions studied with the aim of carrying out the above reactions using molecular oxygen as the oxidant:

1. The synthesis of epoxides with molecular oxygen in the presence of hydrogen (in situ generation of hydrogen peroxide) over titanium silicalite using an organic redox complex[100,101] (see also Chapter 7), over noble metal–doped titanium silicalite,[102,103] or gold nanoparticles over titanium silicalite or titania.[104]
2. The synthesis of epoxides with molecular oxygen in the presence of molten salts.[105]
3. The ammoximation of cyclohexanone with ammonia and molecular oxygen in the presence of amorphous silica as the catalyst.[106–108]

3.2.2.1. Alkene Epoxidation Reactions

Epoxides, with the exception of ethene oxide, are generally produced by one of three methods: (i) the chlorohydrin route, (ii) catalytic epoxidation with hydroperoxide, or (iii) the organic peracid route.

The chlorohydrin route involves the formation of a chlorohydrin by reaction of an alkene and hypochlorous acid, which is generated by the reaction of chlorine and water. The resulting chlorohydrin is dehydrochlorinated in the presence of calcium oxide to give the epoxide and calcium chloride. The route suffers from: (i) the use of chlorine (a very hazardous reagent, which is also expensive and corrosive), (ii) the production of undesirable chlorinated by-products and large amounts of aqueous wastes that require costly treatment, and (iii) the formation of large quantities of calcium chloride.

In catalytic epoxidation with hydroperoxide, epoxides are obtained from alkenes using tert-butyl hydroperoxide or ethylbenzene hydroperoxide. Selectivity is very high. In the homogeneous phase, the catalyst is a molybdenum complex. In the heterogeneous phase, the catalyst is titania/silica or other transition metal–containing microporous materials. The corresponding alcohol is formed in stoichiometric amounts. The economic viability of this multistep, capital-intensive process depends on the transformation of the coproduct (alcohol) to a relatively high-value product. In particular, MTBE is produced from tert-butanol by reaction with methanol, and styrene is produced from methylphenylcarbinol by dehydration. However, the market requirements for the two coproducts (the epoxide and the products of hydroperoxide transformation) are often different.

The organic peracid route produces epoxides by reaction of a peracid with alkenes. Stoichiometric amounts of the corresponding acid are coproduced. Commercial applications of this route are limited by the hazards associated with the peracid, high costs, and the low value of the by-product acid.

The disadvantages associated with these three processes have engendered continued research into alternative epoxidation routes of which the two most promising are: (i) epoxidation with aqueous hydrogen peroxide, which is a safer oxidant than the oxidants described above and which produces only water as a by-product; and (ii) epoxidation with molecular oxygen (air), which is the cheapest oxidant, but requires particular conditions to be activated (molten salts) or the use of a reductant such as hydrogen for *in situ* generation of hydrogen peroxide.

The key features of the alkene epoxidation process using titanium silicalite are[14]:

1. A titanium silicalite catalyst (TS-1) that is isostructural with the silicalite-1 catalyst. In TS-1, titanium substitutes for silicon in the framework of the zeolite.[109] The general formula of the catalyst is $(TiO_2)_x(SiO_2)_{1-x}$, where $0.01 \leq x \leq 0.025$.
2. A polar solvent that can dissolve both the hydrogen peroxide and the alkene. Potential solvents include alcohols (preferably methanol or *tert*-butanol), ketones (preferably acetone), acids (preferably acetic acid and propionic acid), ethers, and glycols.
3. The use of hydrogen peroxide in concentrations from 10 to 70 wt % at 10 to 70°C and 1 to 20 atm. The low-temperature limit is determined by the low activity, whereas the higher-temperature limit is determined by the formation of by-products and decomposition of hydrogen peroxide. High pressures are necessary to dissolve the low-boiling-temperature alkenes.
4. The use of slurry or trickle-bed reactors.

Table 3.2 summarizes the catalytic performance and experimental conditions for the oxidation of several alkenes of possible commercial interest.[110] The major by-products observed for the different alkenes are those derived from the consecutive transformation of the oxirane ring. No products of allylic oxidation were observed. The most interesting reaction from the commercial point of view is propene oxide synthesis. Table 3.3 summarizes the catalytic performance and typical experimental conditions for propene oxidation.[14] The typical by-products of propene epoxidation are shown in Scheme 3.5.[19]

The stability, activity, and selectivity of titanium silicalite in alkene epoxidation depends on many factors, including the crystal size of the zeolite, deactivation with time-on-stream, solvent choice, and the use of promoters. For example, in 1-butene epoxidation to 1,2-epoxybutane, two TS-1 samples with different crystal

TABLE 3.2. Catalytic Performances of Titanium Silicalite in the Epoxidation of Alkenes of Potential Industrial Interest

[The conversion of hydrogen peroxide is almost 100%, whereas the conversion of alkene is usually less than 50%; the selectivity to propene oxide is almost total (98%) with respect to hydrogen peroxide when small amounts of bases are added].[110]

Alkene	Temperature, °C	Yield epoxide, %[a]	Selectivity for epoxide, %[b]	Product
Propene	50	90	98	Propene oxide
Butene	50	90	95	Butene oxide
Dodecene	70	60	90	Dodecene oxide
Allyl alcohol	70	75	75	Glycinol
Allyl chloride	65	85	95	Epichlorohydrin
Styrene	65	80	95	Phenyl acetaldehyde
Butadiene	20	85	85	Butadiene monoepoxide

[a]Yield on hydrogen peroxide basis.
[b]Selectivity on an alkene basis.

sizes (0.1–0.3 and 2–3 μms, respectively) show similar initial activity, but a more rapid rate of deactivation of the sample with the larger size crystals. Thus, after about 1 h of reaction the yield of epoxide for the sample with smaller dimensions is about two-fold higher.[22] Deactivation is due to the blockage of zeolite channels by slowly diffusing epoxide solvolysis products.

Catalyst deactivation occurs when the zeolite is used in epoxidation reactions in consecutive cycles. However, deactivation can be eliminated by selective extraction after the catalytic reaction and subsequent oxidation, because it is mainly due to the formation of polyethers. The solvent also plays a critical role. Methanol is preferable not only for the higher productivity (Table 3.3), but also because it is relatively chemically inert (it is only slightly oxidized to formaldehyde) and reduces the rate of deactivation. In fact, using *tert*-butyl alcohol or ketones, e.g., the formation of hydroperoxides enhances the rate of deactivation.

The addition of other components to the catalyst, e.g., the addition of silica to increase its mechanical strength, enhances the rate of deactivation and causes a

TABLE 3.3. Epoxidation of Propene with Various Solvents

[a continuous flow reactor loaded with 3.5 g of catalyst is used][14]

Solvent	Temperature, °C	H_2O_2/alkene molar ratio	Product, kg/h	H_2O_2 conversion, %	Epoxide selectivity, %	Glycol selectivity, %	Others[a] selectivity, %
Methanol	15	0.68	4.15	98	88.5	1.5	10
Acetone	15	0.55	1.55	90	92	6	2
tert-Butanol	20	0.55	2.20	85	96	4	0

[a] Glycol monomethyl ether and glycol ketal.

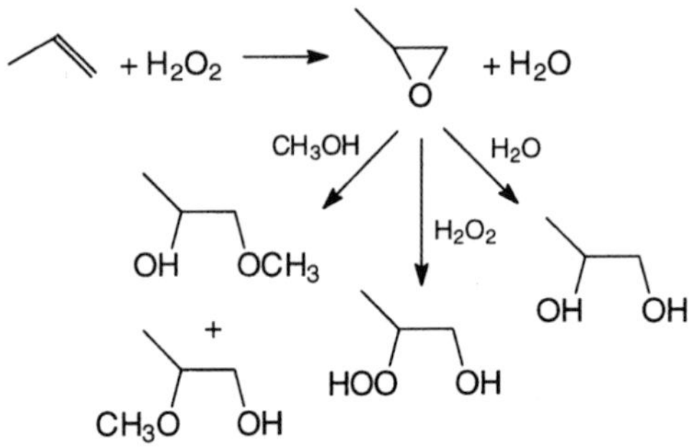

Scheme 3.5. Products of the oxidation of propene with hydrogen peroxide.[19]

lowering of the selectivity. In the case of silica, this is due to the presence of residual acidity in the silica (free silanol groups owing to defects), which catalyze opening of the epoxide ring. In fact, the addition of small quantities of bases enhances the selectivity (Table 3.4),[110] although higher concentrations of bases decrease the activity.

The general scheme of the process, e.g., for the synthesis of epichlorohydrin by epoxidation of allyl chloride with hydrogen peroxide, is shown in Fig. 3.1. The main features of the process are the following:

- Mild reaction conditions (50°C, 1 atm), which thus do not require special costly apparatus.
- Use of methanol as the solvent and cocatalysts (see Chapter 7).
- High hydrogen peroxide conversion.
- Absence of side products other than glycols or glycol ethers.
- Easy recycling of the solvent and unreacted alkene.

Recent work in alkene epoxidation involves the search for processes in which hydrogen peroxide is generated either *in situ* or *ex situ* but integrated in the process (see also Chapter 7).[22,100,101] The economic advantages of these two alternative

TABLE 3.4. Effect of Catalyst Nature on the Behavior in Propene Epoxidation [reaction conditions: 50°C, propene pressure : 8 atm, peroxide concentration = 0.99 mol/kg, catalyst = 8.2 g/kg, reaction time = 70 min][110]

Catalyst	H_2O_2 conversion, %	Epoxide selectivity, %	Glycol selectivity, %
TS-1	96	87	11
TS-1/SiO$_2$ 90%	96	85	13
TS-1/SiO$_2$ 90% + Na[a]	97	97	3

[a]Sample pretreated with sodium acetate.

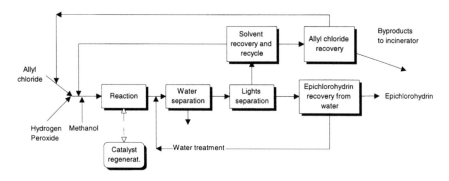

Figure 3.1. Block diagram for the process of epichlorohydrin synthesis by oxidation of allyl chloride with hydrogen peroxide.

processes result from the elimination of an expensive hydrogen peroxide purification unit, reduction of the costs and risks of hydrogen peroxide storage, higher flexibility in production, reduced capital costs, and easier scale-up and scale-down.

The classical alkylanthraquinone system can be used as a redox couple to generate the hydrogen peroxide *in situ* (Scheme 3.6), because titanium silicalite is not affected by the polynuclear compounds forming the redox couple, which, on the other hand are too large to enter the zeolite cavities and thus they are not oxidized. Key features of the process are[100,101]:

1. Use of titanium silicalite, $(TiO_2)_x(SiO_2)_{1-x}$, where $0.0001 \leq x \leq 0.04$, prepared according to Taramasso *et al.*[109]: Part of the titanium can be replaced by elements such as aluminum, iron, or gallium.

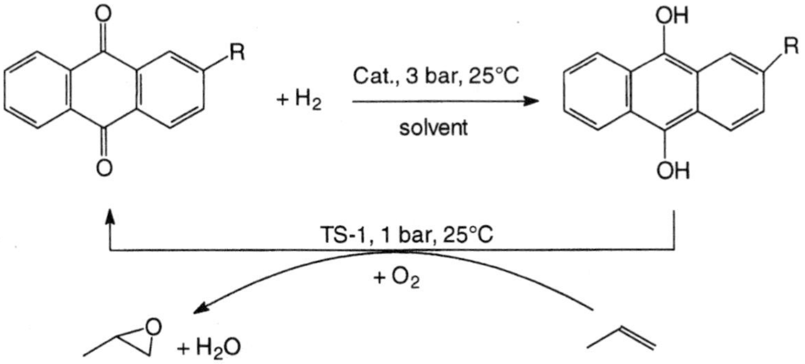

Scheme 3.6. Synthesis of propene oxide by oxidation of propene with an *in situ* coupled reaction for hydrogen peroxide generation.[22]

2. Use of a redox couple consisting of an alkylanthraquinone and an alkylan-thrahydroquinone, preferably 2-*sec*-butylanthraquinone: Steric hindrance prevents the alkylanthraquinone component of the redox couple from entering the zeolite cages. Also, the titanium silicalite does not influence the oxidation reaction of the alkylanthrahydroquinone.

3. Use of a mixture of solvents: One or more aromatic or alkylaromatic hydrocarbons, preferably 2-methylnaphthalene, are used to dissolve the anthraquinone. One or more polar organic compounds with a boiling temperature of 150–350°C, preferably diisobutylketone, or diisobutylcar-binol, methylcyclohexylacetate, dimethylacetate, or dibutylphthalate are used to dissolve the anthrahydroquinone. Finally, a low-molecular-weight alcohol, preferably methanol, is used to produce the active intermediate species.[22]

4. Mild reaction conditions: The reaction temperature is 0–60°C and the reaction pressure is 1–20 atm. Higher pressures enhance the dissolution of the alkene.

However, the optimal experimental conditions for the generation of hydrogen peroxide are not the same as those for the epoxidation reaction. For example, it is necessary to dissolve the redox couple in large quantities of solvent to achieve acceptable hydrogen peroxide productivities.

Figure 3.2 (top) shows the process flowsheet for epoxidation with *in situ* generation of hydrogen peroxide.[22,101] Methanol and propene are fed to reactor R-1, together with air and a solution of ethylanthrahydroquinone. Two streams leave the reactor, one gaseous and one liquid. The gaseous stream contains essentially residual nitrogen and traces of other gases. The liquid stream contains the product, together with dissolved unreacted propene, ethylanthraquinone, and the solvent. This liquid stream is directed to distillation column C-1, where propene oxide is separated from propene and from the other heavier compounds. A grid helps to retain the catalyst in the reactor R-1. The bottom fraction from the distillation column is sent to a separator, where the organic layer that contains the ethylan-thraquinone is introduced into the hydrogenation reactor. The aqueous layer, which contains methanol, is sent to distillation column C-3, where methanol is recovered from the top and recycled. The bottom fraction is sent to distillation column C-4, where water produced in the epoxidation step is eliminated and the by-products (glycols) are separated. The process can also be used for other types of reactions catalyzed by TS-1, such as alkane hydroxylation.[111]

The alternative and preferable solution, even though it requires a large invest-ment, is integration of the hydrogen peroxide production with the process of epoxidation (Figure 3.2, bottom).[101] The key feature of this process is the use of a solvent consisting of 20–60% by weight of methanol in water. This solvent, which is used to extract hydrogen peroxide from the organic solvent of the redox process,

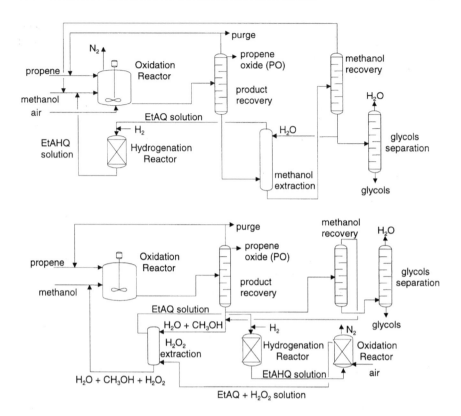

Figure 3.2. Synthesis of propene oxide via oxidation of propene with hydrogen peroxide generated *in situ* (top).[101] Integrated process for propene oxide synthesis using hydrogen peroxide (bottom).[22]

is also the solvent for the epoxidation reaction. The advantage of this integrated process over a conventional one in which hydrogen peroxide is produced in a separate unit is the elimination of the expensive hydrogen peroxide purification and concentration steps. The advantage with respect to *in situ* generation is the flexibility of the reaction configuration, with the optimization of the reaction conditions and especially of the nature of the solvent, which has a considerable influence on the reaction rates. The overall process flowsheet is analogous to that of the *in situ* process, but three reactors are required as well as a more complex reaction process.

Recent developments in this area are the substitution of the anthraquinone/anthrahydroquinone redox couple to generate hydrogen peroxide with a direct catalytic process of generation from hydrogen and oxygen using supported palladium catalysts (see Chapter 7). Another development area for this case is the direct *in situ* generation and use of hydrogen peroxide, but to date the integrated process is still preferred. The generation of H_2O_2 from H_2/O_2 has the advantages of cleaner production, elimination of carcinogic-type reactants and solvents, and better flexibility in operation.

Although the process of epoxidation with the TS-1 catalyst is the more promising and commercialization is expected in the near future, when the cost of generating hydrogen peroxide becomes more competitive, two alternative processes will have to be considered: (i) two-phase catalytic epoxidation in the presence of W^{6+} heteroanions, and (ii) epoxidation with molecular oxygen in the presence of molten salts.

Water-insoluble, nonfunctionalized, open-chain or cyclic alkenes can be epoxidized with diluted hydrogen peroxide (15%) at high selectivity using an organic solvent and W^{6+} heteroanions as catalysts.[88,112–115] Key features of this epoxidation route are:

1. Use of a peroxotungsten complex as the catalyst: The peroxotungsten complex is formed in an aqueous medium in the presence of hydrogen peroxide and mixtures of tungstic acid with phosphorus or arsenic acid. The formula of the active form of the catalyst is $Q^+_3\{XO_4[W(O)(O_2)_2]_4\}^{3-}$, where X = P or As. Q^+ is a lipophilic, long-chain "onium" group that helps in the transfer of the active complex from the aqueous phase to the organic phase. Q^+ can be $[(n\text{-}C_4H_9)_4N]^+$, $[(n\text{-}C_6H_{13})_4N]^+$, or $[(n\text{-}C_8H_{17})_3NCH_3]^+$.

2. Use of a biphasic system consisting of a hydrogen peroxide solution, the alkene, and a solvent (usually 1,2-dichloroethane or benzene) that is not soluble in water: The epoxidation is carried out at 60–70°C in a stirred-tank reactor.

High yields in epoxide are possible especially using unactivated, monosubstituted alkenes. As a reference, 85% yield in epichlorohydrin from allyl chloride can be obtained.[88] High yields are obtained, even though the aqueous layer is acidic, owing to the biphasic nature of the reaction medium. Pilot plant tests have confirmed the interest in this route,[88] especially for high-value products. A 100 ton/year plant has been built to produce isobutyl 3,4-epoxybutanoate, a key intermediate in the synthesis of an active nontropic agent (Oxiracetam).

The epoxidation with molecular oxygen in the presence of molten salts has been developed by researchers at Olin Corporation.[105,116–121] Although it has been claimed that the salt acts as a catalyst, the oxidation is probably not catalytic. Rather, the molten salt probably serves to remove the heat of reaction, allowing better control of the temperature, and thus allowing the very exothermic reaction to run more smoothly. Key features of the Olin process are:

1. Propene oxidation at a concentration of about 50%, with an oxygen deficiency (<10%) and inert gas (nitrogen and carbon oxides): The reaction proceeds at 8–300 psia, and 200–300°C, with relatively long contact times (10–100 s)

2. Use of a molten salt consisting of a mixture of $LiNO_3$, $NaNO_3$, and KNO_3: The composition of the mixture determines the reaction temperature. Care must be taken to avoid the formation of crusts of salt inside the reactor, because such crusts can create local overheating of the reactants and cause reaction runaway.

3. Flexibility in introducing gas reactants: Gas reactants may be (i) bubbled through a stirred-tank reactor filled with molten salt, (ii) moved counter-currently in a tower through a downflow of molten salt, or (iii) injected into a loop through which the molten salt circulates.

4. Introduction, optionally, of a true catalyst (cocatalyst in the Olin patents) into the molten salt: The catalyst, which may be based on platinum or sodium hydroxide, enables the reaction to proceed at a lower pressure.

5. Recycling of unreacted propene, together with part of the by-product gas (carbon oxides). The by-product acetaldehyde can also be recycled with the unreacted propene.

Epoxide selectivity in the 41–68% range (depending on the reaction conditions) and propene conversion in the 5–15% range were found.[105,116–121] Although selectivities and yield are much lower than in the processes using hydrogen peroxide, the use of oxygen instead of hydrogen peroxide can compensate for the poorer performances. A critical problem, however, is the wide range of by-products (Table 3.5) which makes purification of the product expensive.

3.2.2.2. Cyclohexanone Ammoximation

Cyclohexanone oxime, an intermediate in the production of caprolactam (monomer for Nylon 6) is manufactured using a number of different processes, all

TABLE 3.5. Product Distribution in the Oxidation of Propene in the Presence of Molten Salts
[reaction conditions: 53% propene, 8% oxygen, 39% nitrogen, flowrate = 2 K/min, pressure = 358 psig, temperature = 225°C, per pass conversion of propene in cocur-rent reactor = 9.7%, oxygen conversion = 84%][116]

Product	Selectivity, %	Product	Selectivity, %
CO	8.9	Ethanol	0.01
CO_2	13.5	Methyl acetate	0.01
CH_4	0.02	Propene oxide	56.7
HCHO	0.3	Acrolein	2.5
CH_3OH	0.5	Acetone	1.8
$CH_2=CH_2$	0.04	Allyl alcohol	1.1
CH_3-CH_3	0.16	Propene glycol	1.5
CH_3CHO	11.7	1,5-hexadiene	0.2
CH_3OCOCH_3	0.3	Other	0.4

of which suffer from similar disadvantages. For example, the Rashig process uses the following reactions to produce hydroxylamine:

$$(NH_4)_2CO_3 + N_2O_3 \rightarrow 2NH_4NO_2 + CO_2 \tag{3.1}$$

$$(NH_4)_2CO_3 + SO_2 \rightarrow (NH_4)_2SO_3 + CO_2 \tag{3.2}$$

$$2NH_4NO_2 + (NH_4)_2SO_3 + 3SO_2 + H_2O \rightarrow 2HON(SO_2NH_4)_2 \tag{3.3}$$

$$2HON(SO_2NH_4)_2 + 2H_2O \rightarrow NH_2OH \cdot \tfrac{1}{2}H_2SO_4 + (NH_4)_2SO_4 + \tfrac{1}{2}H_2SO_4 \tag{3.4}$$

The $NH_2OH \cdot 1/2H_2SO_4$ is then reacted with cyclohexanone to yield the oxime and coproduct $(NH_4)_2SO_4$ and H_2O. The disadvantages of this process are its complexity, the formation of huge quantities of low-value ammonium sulfate, and the sulfur dioxide and nitrogen oxide emissions. Other commercial processes for the production of cyclohexanone oxime suffer from the same disadvantages.

Three new catalytic routes to cyclohexanone oxime have been proposed: (1) the Montedipe (now EniChem) route, (2) the Tagosei route, and (3) the Allied Chemical route. Each of these possibilities uses a one-step ammoximation reaction of the type[108]:

$$\text{Cyclohexanone} + NH_3 + (O) - (\text{catalyst}) \rightarrow \text{cyclohexanone oxime} + H_2O \tag{3.5}$$

where (O) is either hydrogen peroxide or molecular oxygen.

In the first route, ammoximation is carried out in the liquid phase at 60–95°C, with hydrogen peroxide, using titanium silicalite as the catalyst.[122] Yields of 95% (on a cyclohexanone base) and 90% (on a hydrogen peroxide base) have been obtained. The process has been developed to the stage of a demonstration unit.

In the Tagosei route, ammoximation is carried out at 30°C, with hydrogen peroxide, using a catalyst based on heteropoly compounds of tungsten and phosphorus. Cyclohexanone oxime yields of 94% (on a cyclohexanone base) and 64% (on a hydrogen peroxide base) have been obtained.[123] Commercialization of this route has been hindered by the difficulty of separating the catalyst from the product, the instability of the catalyst, and the high level of hydrogen peroxide decomposition.

In the Allied Chemical process, ammoximation is carried out in the vapor phase at 200°C with air, using silica as the catalyst.[108] Cyclohexanone oxime yields of 45% (on a cyclohexanone base) have been obtained. Low oxime selectivity and rapid catalyst deactivation have prevented commercialization of this process to date.

The ammoximation process with hydrogen peroxide over a titanium silicalite catalyst (EniChem process) is the most promising of these three routes and will probably replace the conventional processes in the near future because of its much lower environmental impact and production of waste by-products, especially when the second stage (Beckmann transposition) is carried out by a gas phase catalytic process using weakly acid oxide catalysts. There is also much research in this direction both in EniChem and by Japanese researchers, and thus in the near future it will be possible to have a two-stage process from cyclohexanone to caprolactam without waste production (in the conventional process the ratio between ammonium sulfate and caprolactam, e.g., is about 8–10).

The key features of the EniChem process for cyclohexanone ammoximation are[122]:

1. A titanium silicalite catalyst with Si/Ti > 30, prepared according to the procedure outlined by Roffia et al.[124]
2. Use of a dilute aqueous solution of hydrogen peroxide.
3. Use of tert-butanol as the solvent for hydrogen peroxide, reagents, and products, or the use of a biphasic system consisting of water and a hydrophobic solvent, such as toluene.
4. Operation in the liquid phase, either in a slurry or in a trickle-bed reactor, at 60–95°C, with gaseous ammonia or an aqueous solution of ammonia.

This process uses a continuous reactor configuration.[124] An aqueous solution of hydrogen peroxide is fed into the reactor, together with the aqueous recycle and fresh cyclohexanone. Cyclohexanone, in a solution of tert-butanol, is fed into the reactor separately, and gaseous ammonia or an aqueous solution of ammonia is introduced at a third inlet. The tank reactor contains a blade stirrer and heating jacket fed with steam. An automatic valve controls both the flow rate at the exit and the level of solution inside the reactor. A grid at the bottom of the reactor helps to retain the catalyst. An extraction device separates the oxime, through the use of a hydrophobic solvent such as toluene, from an aqueous layer that contains unreacted cyclohexanone, ammonia, and traces of oxime. The unreacted components are recycled. After 28 h, the ketone conversion is nearly complete and the oxime yield is 98.3% (on a ketone basis).

The characteristics of the process are:

- High selectivity of cyclohexanone to cyclohexanone oxime (> 95%).
- High yield of oxime on a hydrogen peroxide basis (90–95%).
- Very low catalyst consumption.
- Easy recovery of oxime from the reactor effluent.
- High process reliability from an environmental standpoint (no nitrogen oxide and sulfur oxide emissions).

Several factors affect oxime yields, including the nature of the catalyst, catalyst pretreatment, catalyst deactivation, noncatalytic oxidation, and the nature of the solvent.[125,126] Table 3.6 shows the role of the catalyst in the ammoximation reaction. High conversion is obtained without a catalyst, but no oxime is produced. A catalyst based on titanium supported on amorphous silica gives interesting oxime selectivities, but the catalyst activity is much lower than that of titanium silicalite. Pretreatment of the catalyst with sulfuric acid and hydrogen peroxide improves selectivity.[124]

Three types of catalyst deactivation have been observed: (i) dissolution of silica, (ii) loss of titanium, and (iii) occlusion of zeolite channels by organic compounds formed in side reactions.[127] Dissolution of silica and loss of titanium are both caused by ammonia. Deactivation by occlusion of zeolite channels can be partially eliminated by washing the catalyst with *tert*-butanol.

Noncatalytic oxidation can be minimized by increasing the reaction rate of the catalytic reaction. The reaction rate has been increased by operating at high catalyst concentrations and high temperatures.[126] For example, when the reaction temperature is increased from 60° to 80°C, the oxime selectivity increases from 87.0 to 98.8%, but at 80°C the selectivity decreases to 86.0%. The role of the solvent is also important.[126] For instance, if benzene is used instead of a water/n-butanol mixture, the selectivity to oxime drops from 99.5 to 95.0%.

The process flowsheet is shown in Figure 3.3 for the part regarding the oxime synthesis section. Cyclohexanone, fresh and recycled ammonia, and an azeotropic water/solvent mixture are fed into the bottom of the reactor. Hydrogen peroxide is fed into the top of the reactor to keep organic compounds from reaching the

TABLE 3.6. Ammoximation of Cyclohexanone over Various Catalysts
[reaction conditions: solvent = water/*tert*-butanol, catalyst concentration = 2 wt%, temperature = 80°C, ammonia/H_2O_2 molar ratio = 2.0, reaction time = 1.5 h][126]

Catalyst	Ti, %	H_2O_2/CH^a molar ratio	CH^a conversion, %	Oxime Selectivity,[b] %	Oxime yield,[c] %
None	—	1.07	53.7	0.6	0.3
Amorphous SiO$_2$	0	1.03	55.7	1.3	0.7
Silicalite	0	1.09	59.4	0.5	0.3
H-ZSM5	0	1.08	53.9	0.9	0.4
TiO$_2$/SiO$_2$	1.5	1.04	49.3	9.3	4.4
TiO$_2$/SiO$_2$	9.8	1.06	66.8	85.9	54.0
TS-1^e	1.5	1.05	99.9	98.2	93.2

^{a}CH = cyclohexanone.
^{b}Selectivity on a cyclohexanone basis.
^{c}Yield on a hydrogen peroxide basis.
^{d}Reaction time = 5 h.
^{e}Titanium silicalite.

hydrogen peroxide tank and creating explosive conditions. A device in the reactor separates the catalyst from the organic phase by filtration. The outlet stream is sent first to a tank and then to a column for solvent recovery. The low-boiling azeotropic water/solvent mixture is recycled to the reactor. The gas phase, which is rich in ammonia, is compressed and also recycled. A 1:1 by weight water/oxime mixture is recovered at the bottom of the column. A 40% aqueous solution of ammonium sulfate is added to the water/oxime mixture, and the resulting solution is sent to a settler, where two phases (one aqueous and one organic) are obtained. The organic phase, which contains cyclohexanone oxime and 5–6% water is sent to the Beckmann rearrangement reactor.

Although the process allows very high yields and selectivities for oxime, the critical factor remains the cost of hydrogen peroxide. Thus there have been several attempts to avoid its direct use, either by generating it *in situ* or by eliminating the need for it altogether.

An interesting development in this area is the process developed by EniChem for using cyclohexanol as the raw material and generating hydrogen peroxide directly *in situ*.[128] Key features of the proposed process are:

1. Autoxidation of a secondary alcohol corresponding to the required ketoxime (e.g., cyclohexanol for the cyclohexanone oxime). The autoxidation is carried out in the liquid phase with molecular oxygen. The reaction, which is run in the presence of a radical initiator (azo-bis-isobutyronitrile), produces the ketone and hydrogen peroxide simultaneously.
2. Direct feeding of the autoxidation reactor outlet stream to the ammoximation reactor, which contains the titanium silicalite and an aqueous solution of ammonia.

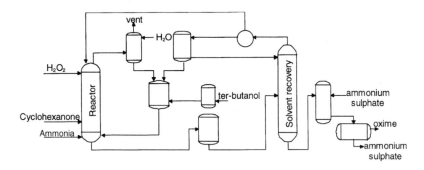

Figure 3.3. Process flow for cyclohexanone ammoximation.

The need for costly separation of impurities, such as carboxylic acids, is avoided. The direct integration of the two reactions is possible because the impurities present in the reactor outlet stream have no influence on the ammoximation stage. The critical problem, however, is the level of purity of the oxime obtained.

An alternative possibility is to generate hydrogen peroxide inside the ammoximation reactor via the reaction mechanism shown in Scheme 3.7.[129] The hydrogen peroxide formed *in situ* reacts with cyclohexanone and ammonia over a titanium silicalite catalyst. The cyclohexanone conversion is quite high (80–90%), but the maximum in oxime selectivity is of the order of 45% (using diglyme as the solvent). Furthermore, hydrogen peroxide is formed with the coproduct decahydrophenazine. Thus commercialization of this process is unlikely.

The possibility of avoiding the need for hydrogen peroxide and using molecular oxygen directly in a gas phase process has been explored mainly by Allied Chemical researchers[106,130–135] and also studied extensively in an academic framework.[136–145]

Cyclohexanone is converted in the presence of oxygen and ammonia to the oxime over amorphous silica or silica/alumina catalysts. The characteristic feature of the reaction is that the catalytic behavior changes rapidly with time-on-stream.[139,140] During the first 10–15 h, both the cyclohexanone conversion and the oxime yield increase, reaching values of about 70% conversion and 30% yield. For longer times-on-stream, however, both conversion and yield decrease, the latter becoming nearly negligible in 40–50 h of time-on-stream. At the end, the formation

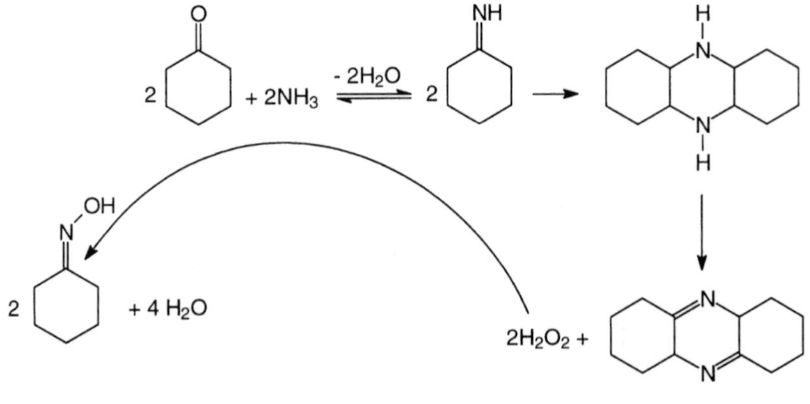

Scheme 3.7. Reaction mechanism for cyclohexanone ammoximation using hydrogen peroxide generated *in situ*.[129]

of imine increases, reaching yields of about 5%. Formation of tars over the catalyst surface is a continuous process reaching values of about 3 g of tars/g catalyst after about 50 h. These results clearly show the disadvantages of this process, namely poor oxime selectivity and rapid catalyst deactivation, due especially to the formation of heavy by-products remaining on the catalyst surface.

Scheme 3.8 illustrates the simplified reaction network.[136,139,140] It has been suggested that the initial reaction is the formation of cyclohexanone imine on the weak acid sites of silica, followed by three parallel reactions that lead to the oxime and by-products. The rates of these three parallel reactions depend on the gas composition. The formation of oxime is the only reaction that depends strongly on ammonia partial pressure, which is why the optimal conditions for high selectivity require a very large excess of ammonia, another drawback of the process. The formation of aldol condensation products does not depend on oxygen partial pressure. However, operating at high oxygen partial pressure does not increase selectivity because the formation rate of tars also increases.

3.2.2.3. Hydroxylation of Phenols

All three dihydroxybenzene isomers are used commercially as intermediates, and several different technologies have been developed for their synthesis. In the Eastman process for the production of hydroquinone (the para isomer), aniline is stoichiometrically oxidized to p-quinone using MnO_2 in an acidic medium. The quinone is then reduced to hydroquinone with Fe^{2+}. This process was in industrial operation for many years despite the large quantities of by-products and the expensive waste treatment required, but is no longer used, and Eastman is now employing the newer technology based on p-diisopropylbenzene.

Mitsui and Sumitomo use processes that produce either hydroquinone or resorcinol (the meta isomer). The raw material, p-diisopropylbenzene or m-diiso-

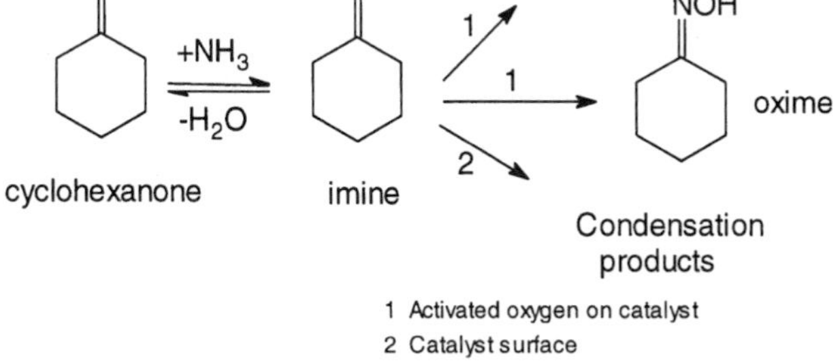

1 Activated oxygen on catalyst
2 Catalyst surface

Scheme 3.8. Simplified reaction scheme for gas phase cyclohexanone ammoximation.[139]

propylbenzene, is oxidized to the dihydroperoxide with molecular oxygen. Decomposition of the hydroperoxide leads to the formation of diphenols and coproduced acetone. The similar Signal/Goodyear process produces hydroquinone from the para isomer.

Three processes developed in Europe (the Rhône-Poulenc, the Brichima, and the EniChem processes) are also currently used for the synthesis of mixtures of hydroquinone and catechol (the ortho isomer). All three processes use hydrogen peroxide as the oxidizing agent. Of the more than 50,000 tons/year of diphenols produced in the world, approximately half is produced using these three processes.

The Rhône-Poulenc process[146,147] operates at 90°C in an acidic medium, using concentrated hydrogen peroxide as the oxidant and a phenol/H_2O_2 molar ratio of 20. Perchloric acid is used as the catalyst, and phosphoric acid is used as a masking agent to prevent side reactions such as the degradation of the product via radical side reactions. The mechanism involves a peroxonium ion intermediate [HOO-$(H_2)^+$], which is considered to be the true oxidizing agent.

The Brichima process[148,149] is based on a Fenton-like radical reaction that uses ferrocene or cobalt salts as catalysts. The mechanism proceeds via the decomposition of hydrogen peroxide into radicals.

The EniChem synthesis process uses TS-1[109] as the catalyst. A 10,000 tons/year plant for the production of the p- and o-isomers began operation at Ravenna, Italy, in 1986.[150,151] The performances of the three hydrogen peroxide–based technologies are compared in Table 3.7.[99,152] The process has recently been acquired by Borregaard Italia.

A fourth, Japanese process is used by Ube Industries for the production of catechol and hydroquinone, in which phenol is hydroxylated by ketone peroxides in the presence of an acid catalyst.[153] Ketone peroxides, formed in situ from a ketone and hydrogen peroxide, react electrophilically with phenol to yield the ortho and para isomers in a 1:1 ratio. A small amount of sulfuric acid is added as the catalyst. The phenol conversion is less than 5%, and the selectivity to dihydroxybenzenes is approximately 90%.

TABLE 3.7. Catalytic Performance of Three Processes for Catechol and Hydroquinone Synthesis[99,152]

Process	Phenol conversion, %	Diphenol yield,[a] %	Diphenol selectivity,[b] %	Catechol/ hydroquinone	Tars selectivity,[b] %
Rhône-Poulenc	5	70	90	1.4–1.5	10
Brichima	9–10	50	80	2–2.3	20
EniChem	25–30	84	90–94	1	6–10

[a]Yield on a hydrogen peroxide basis.
[b]Selectivity with respect to phenol.

It can be clearly seen that the EniChem process is significantly better than other technologies using hydrogen peroxide as regards conversion, selectivity and less formation of tars, and in comparison with the older technologies, as well as in terms of environmental impact. This process is the first example of commercial application of an oxidation reaction using a solid zeolite material and represents one of the most innovative applications developed in recent years in the field of heterogeneous catalysis. The key features of the EniChem synthesis for phenol hydroxylation with hydrogen peroxide are:

1. Use of titanium silicalite as the catalyst: This heterogeneous, noncorrosive catalyst minimizes the problems of corrosion, handling, and separation of the catalyst from the products.
2. High selectivity with respect to phenol, which is attributable to low tar formation.
3. Ability to operate at higher phenol conversions than can be achieved with other processes, while keeping tar formation low and selectivity on both the hydrogen peroxide and phenol basis high: The higher phenol conversion is possible because the consecutive hydroxylation reactions of the diphenols, which are more readily oxidized than phenol itself, are sterically hindered inside the zeolite cavities. In other hydroxylation processes, an increase in phenol conversion, and thus an increase in the hydrogen peroxide to phenol ratio, leads to an increase in tar formation and hydrogen peroxide decomposition.[154] Higher phenol conversion also reduces the recycling of unreacted phenol.
4. High yields on a hydrogen peroxide base: The selective conversion of hydrogen peroxide is very rapid, which minimizes its decomposition, and because the mechanism does not proceed via its decomposition, hydrogen peroxide is used efficiently and coupling of intermediate organic radicals is avoided.
5. Lower catechol to hydroquinone ratios, with increased formation of the preferred hydroquinone.
6. Possibility of adjusting the operating conditions in order to change the relative proportions of the two products according to market demand.
7. Absence of a requirement for an organic solvent in the separation of water and unreacted phenol.
8. Low energy consumption, combined with high product quality.

Typical operating conditions and performances of the EniChem synthesis are shown in Scheme 3.9.[99] Figure 3.4 illustrates a simplified process flow diagram for the EniChem synthesis. The reaction mixture, which consists of phenol, water, and the solvent, is loaded into a batch reactor. The catalyst (3% wt) is added, and the

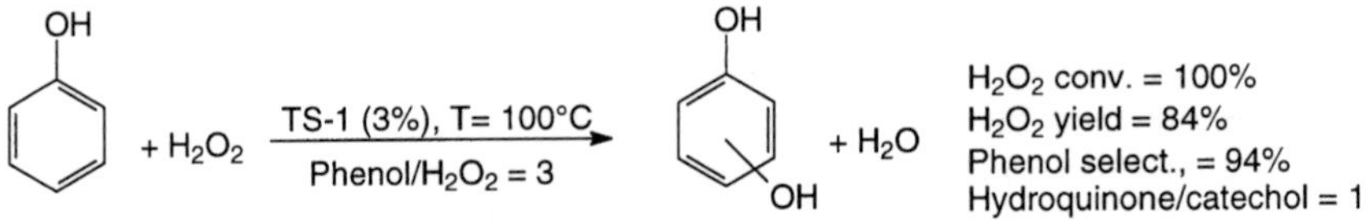

OH
$+ H_2O_2$ $\xrightarrow{\text{TS-1 (3\%), T= 100°C}}{\text{Phenol/H}_2\text{O}_2 = 3}$ OH ... OH $+ H_2O$

H_2O_2 conv. = 100%
H_2O_2 yield = 84%
Phenol select., = 94%
Hydroquinone/catechol = 1

Scheme 3.9. Typical performances in phenol hydroxylation catalyzed by the TS-1 catalyst.[99]

mixture is heated to close to 100°C, at which time hydrogen peroxide is fed into the reactor at a specified feed rate. The reaction temperature is maintained by refluxing the solvent. The overall phenol/hydrogen peroxide ratio is approximately 3. The reactor is then unloaded, and the catalyst is separated from the liquid mixture by filtration and sent to the regeneration section. The purification unit consists of a distillation column for solvent separation (the solvent is then recycled) and a second column for the separation of phenol and water. Phenol is recovered in a demixer, where a phenol-rich phase and a water-rich phase are separated. The phenol-rich phase is recycled, while the water-rich phase is treated to recover the phenol. A thin-film evaporator is used to separate catechol and hydroquinone from the tars. A third column separates catechol, and a fourth separates hydroquinone.

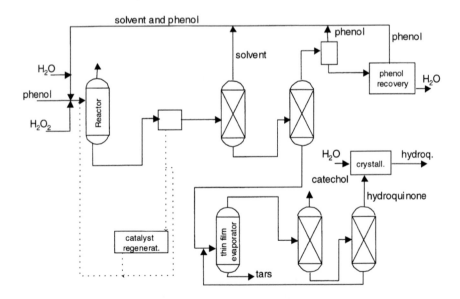

Figure 3.4. Simplified flow sheet of the phenol hydroxylation process.

Both the third and fourth columns operate under vacuum. Finally, hydroquinone is crystallized to give a photograde product.

Several factors influence the catalytic performance, including crystallite size, nature of the binder, titanium content (and especially the presence of extraframework titanium ions), nature of the solvent, and hydrogen peroxide to phenol ratio.

The crystal size of the TS-1 catalyst particles has a considerable effect on the catalytic behavior.[155,156] The best catalytic results are obtained using crystallites with a size of 0.2–0.3 μm, as larger sizes reduce the selectivity for diphenols. For example, increasing the crystal size from 0.5 to 5 μm almost halves the diphenol yield (49% vs. 81%) on a hydrogen peroxide base and increases the selectivity for tars (19% vs. 5%). This effect is due to consecutive oxidation reactions of the diphenols and the presence of diffusional limitations for reactant access to internal sites, but is not associated with the role of the external surfaces of the zeolite crystallites.[156] Crystal size also affects the distribution of the diphenol isomers and, hence, the ortho-to-para ratio. These effects are due to a combination of shape selectivity (more relevant in the larger crystals) and the increased rates of consecutive reactions that favor tar formation from cathecol.[19]

The role of the binder is also interesting and illustrates a typical problem encountered in industrial catalyst development. The first patents on phenol hydroxylation over TS-1 claimed the use of unmodified crystallites rather than crystallites agglomerated into larger particles through the use of binders.[150] Unmodified crystallites were claimed because the use of conventional nonacidic binders such as Ludox silica or silicates, led to a deterioration in catalytic performance with only a relatively small increase in mechanical resistance.[151] However, the use of such small particles led to problems in catalyst recovery by filtration, and the severe regeneration conditions and the catalyst handling procedures required that particles with superior mechanical resistance be used. The high cost of the zeolite material, which has a measurable impact on the production cost of diphenols, necessitates maximization of zeolite recovery and recycling.

The binder problem was solved by forming a thin layer of oligomeric silica (10–30% binder used) to connect all the crystallites through Si–O–Si bridges.[151] The silica layer is prepared from an aqueous solution of silica and tetraalkylammonium hydroxide obtained by hydrolyzing tetraethylorthosilicate under alkaline conditions at 60°C. Under these conditions, the silica is present in an oligomeric form. Small crystals of titanium silicalite are then dispersed in this solution, and under alkaline conditions stable chemical bonds form between the surface of the crystals and the oligomeric silicates. The resulting slurry is spray-dried, which favors cross-linking of the crystallites, and the particles are calcined. The final aggregates are 5–100 μm in size with an average size of 20 μm. This procedure allows highly efficient recovery of the catalyst from the liquid reaction medium, with only a slight penalty in catalyst performance.

Owing to the presence of several competitive reactions[157] and the unique catalytic role of framework titanium in catalyzing the selective reaction, the amount of titanium in the TS-1 catalyst is clearly determinative. Decreasing it linearly decreases the yield of diphenols and increases that of tars. However, the limiting quantity of titanium corresponds to the maximum that can substitutionally enter the silicalite framework [this value corresponds to $x = 0.025$ in the $(TiO_2)_x(SiO_2)_{1-x}$ general formula],[158] because extraframework titanium ions have a marked negative effect catalyzing side hydrogen peroxide decomposition. Commercial TS-1 thus has the general composition $Si_{40}TiO_{82}$.

The solvent, similar to the cases discussed earlier, is also of fundamental importance for two main reasons: first, the solvent influences the temperature at which the reaction is carried out, and, second, because of the organophilic nature of the titanium silicate, the diffusion of the solvent within the zeolite pores depends on its physicochemical features. Thus the concentration of reactants within the pores of the zeolite depends on the nature of the solvent.[157,159–161] Acetone or methanol or their mixtures with water (water, however, also forms during the reaction) are the preferred solvents.[162] The nature and composition of the solvent also influences the catechol to hydroquinone ratio[160] and selectivity to the para product.[161]

The H_2O_2/phenol ratio also has an effect on the diphenol formation, which increases as that ratio decreases, because the larger quantity of hydrogen peroxide favors further oxidation of the diphenols and formation of tars, besides increasing the rate of hydrogen peroxide side decomposition. Thus the H_2O_2/phenol ratio, on a weight basis, should be maintained at less than 0.1

3.3. EXAMPLES OF NEW CATALYTIC SYSTEMS

Both heterogeneous, gas phase oxidation and homogeneous liquid phase oxidation are used in petrochemistry (see Chapter 1). Liquid phase oxidation reactions are also usually employed for the synthesis of fine chemicals, which are often complex molecules with limited thermal stability. There is also increasing interest in the development of heterogeneous catalysts to be utilized in liquid phase oxidation, as an alternative to the dissolved metal salts generally used. One important example is titanium silicalite with other microporous materials containing transition metals (see Chapter 6). Heterogeneous systems have the advantages of an easier separation from the product mixture and the possibility of achieving continuous processing. On the other hand, homogeneous oxidation reactions usually allow for more efficient heat transfer and milder operating conditions. A third new area of development is that represented by metal complexes or metallo–organo complexes entrapped in constrained environments[32] such as the cavities of zeolites or organic membranes (see Chapter 6).

Recently, many patents and scientific publications have dealt with the applications of some particular classes of catalysts for selective oxidation reactions. Among these, attention has been focused on systems that simulate enzymatic action in the monoxygenation of organic substrates, and upon molecular-type solids, such as polyoxometallates, where the high structural flexibility and the possibility of modulating (within certain limits) the composition while maintaining the structure intact enables the practical application of concepts such as catalyst tailoring, isolation of active sites, and multifunctionality. These materials can be considered molecular solids and in fact have been used in both the liquid and solid phase, although their limited thermal stability is their primary drawback as heterogeneous catalysts.

In gas phase oxidations the only feasible oxidizing agent is molecular oxygen, but often the true oxygen species to be inserted into the organic substrate is oxygen deriving from oxide catalysts. At a molecular level all oxygen atoms are potentially useful for the formation of the active oxidizing species, and the nature of the latter (in terms of reactivity toward the substrate, and therefore of stability and lifetime) is only affected by the physicochemical properties of the oxide. This mechanism implies, first, that the reaction rate as well as the distribution of products is very strongly affected by the nature and chemical–physical properties of the catalyst, and, second, that it seldom happens that a catalyst that is active and selective in a reaction of hydrocarbon oxidation works for another substrate. Some reactions are only catalyzed by a very specific composition or structure, while others are catalyzed by many different types of catalysts (facile reactions). Typical is the case of the catalyst for n-butane oxidation to maleic anhydride, $(VO)_2P_2O_7$ (see Chapter 4), which is the only system that works in this reaction. On the other hand, paraffin oxidehydrogenation to olefins and o-xylene oxidation to phthalic anhydride are classified as facile reactions because they are catalyzed (of course, with different levels of performance in terms of activity and selectivity) by several types of catalysts. Thus different factors determine the selection of catalytic materials and parameters to be studied for the development of catalysts. For the second class of facile reactions research must focus primarily on optimization, based on the analysis of the reaction mechanism (see Chapter 8) and the reaction kinetics. While for the first class the selective reaction by tuning surface properties still requires a very extensive search for innovative materials to perform the reaction selectively, although catalyst design driven by the analysis of the reaction mechanism remains a useful guideline. Moreover, for the first class of materials, innovation derives from an integration of reaction engineering and catalytic material development.

A useful direction for research in new materials also derives from investigations on the fundamental concepts necessary to understand and generalize catalytic behavior in selective oxidation reactions catalyzed by metal oxides. The different concepts that have been developed—structure sensitivity, synergy between mixed oxides, the monolayer of metal oxides, and catalyst polyfunctionality—offer several clues for designing innovative materials. The development of these concepts has also stimulated the study of solid state chemistry and has further underscored the importance of surface and interface phenomena in catalytic phenomenology.

In homogeneous oxidation, the search for a suitable catalytic system refers to the use of metals able to catalyze the transfer of oxygen from the oxygen donor to the organic substrate. The mechanism by which this transfer may occur, in the absence of autoxidation, involves the heterolytic activation of the oxidant (e.g., an alkylhydroperoxide) by the metal center and the transfer of the activated oxygen to the substrate. The role of the metal center is to bind the substrate and to act as an electrophilic center. This kind of mechanism operates, e.g., in the homogeneous asymmetric Sharpless epoxidation technology,[163] and in the ARCO homogeneous epoxidation of propylene with alkylhydroperoxide. The latter is catalyzed by Mo^{6+} compounds, and the reaction proceeds via a peroxometal mechanism, with a rate-limiting oxygen transfer to the olefin. The same mechanism also operates in alkene epoxidation by hydrogen peroxide catalyzed by titanium silicalite.

The second mechanism involves the oxidation of a substrate by a metal ion, which is reduced in the process and subsequently regenerated by reaction with molecular oxygen. That is the case of the Pd^{2+}-catalyzed Wacker oxidation of alkenes, where the metal is in the form of a soluble salt, and the alkene epoxidation and alkane oxygenation with metal porphyrin-based catalysts, where the oxidizing agent is a high-valence oxometal intermediate species. It has recently been reported that some systems (halogenated metalloporphyrin complexes and transition metal–substituted polyoxometallates) can catalyze the liquid phase oxidation of organic substrates with molecular oxygen without the use of coreductants. This could represent an interesting innovation, when the mechanism through which they operate can be elucidated. Indeed, it is not often clear whether a catalytic action is really at work or the catalyst is only needed to initiate the reaction, which then proceeds via the classical free radical autoxidation route. Thus it is often difficult in these cases to decide whether or not the proposed materials are really innovative classes of catalysts for selective oxidation.

In the following section we briefly describe the catalytic systems for liquid and gas phase oxidation that have recently received particular attention or that look particularly interesting in terms of potential industrial exploitation.

3.3.1. Metalloporphyrin Complexes

Metalloporphyrins have been used in attempts to simulate the action of enzymatic monoxygenases. These porphyrin systems catalyze C–H bond hydroxylation and double-bond epoxidation using a single oxygen-atom transfer agent such as iodosylbenzene, hypochlorite, or organic or inorganic peroxide. They can also operate with hydrogen peroxide or alkylhydroperoxide, although the peroxides usually dismutate, resulting in poor oxygen transfer to the organic substrate. Although molecular oxygen can also be used, stoichiometric coreductants are needed (see Section 3.2.1.1). It has recently been reported that polyhalogenated iron porphyrins catalyze the oxidation of light n-alkanes into alcohols and ketones using molecular oxygen without coreductants.[164,165]

The halogenation of metallotetraarylporphyrins makes them more resistant toward oxidative degradation. Halogenation also affects the redox potential of the metal, which is usually iron or manganese, making the metal more active toward the substrate.[165,166] Hill *et al.*[167] suggested, however, that the higher the redox potential of the transition metal ion, the less well the ion can coordinate molecular oxygen or other oxygen donors.

The substrate oxidation proceeds as

$$M^{(n+2)+}O + R \rightarrow M^{n+} + RO \tag{3.6}$$

and the metal is reoxidized as follows:

$$M^{n+} + O_2 \rightarrow M^{(n+1)+}O_2 \tag{3.7}$$

$$M^{(n+1)+}O_2 + M^{n+} \rightarrow M^{(n+1)+}OOM^{(n+1)+} \tag{3.8}$$

$$M^{(n+1)+}OOM^{(n+1)+} \rightarrow 2M^{(n+2)+}O \tag{3.9}$$

The actual mechanism probably consists of the initial formation of alkyl hydroperoxides by oxidation of the organic substrate, followed by an autoxidation mechanism.

The limitations of metalloporphyrin systems are as follows[167]:

1. The catalyst is unstable under reaction conditions. Organic ligands are often unstable under oxidation and can suffer extensive degradation. More robust metalloporphyrins have been obtained by halogenating the ligands.
2. The active metal peroxo species is deactivated by conversion to oxo-bridged dimers or oligomers.
3. Side reactions lead to species with the intermediate oxidation state $M^{(n+1)+}$. These species may not be able to reenter the redox cycle, and thus the cycle terminates. Analogously, inactive species can be formed if a

reactant or intermediate binds to metalloporphyrin to give a coordinated saturated adduct that is inactive under the reaction conditions.
4. High substrate-to-oxidant ratios are required to minimize side reactions.
5. Turnover numbers are low.
6. Metalloporphyrin systems are expensive.

All these factors limit the potential use of metalloporphyrin systems to a few, rather unusual applications, although they remain of considerable interest for more sophisticated organic synthesis to high-value products. A further area of particular interest involves overcoming of their limitations by including them in suitable organic or inorganic matrices, which makes them easier to use and especially more stable under reaction conditions (see Chapter 6).

3.3.2. Polyoxometallates

Isopoly acids, heteropoly acids, and their salts are molecular-type compounds that have polyoxoanions as the structural unit. The most studied of these crystalline structures are the Keggin-type heteropoly compounds. These compounds, which exhibit the highest thermal structural stability and find the widest applications as both homogeneous and heterogeneous catalysts, have the general formula $(XM_{12}O_{40})^{3-}$, where X=Si, Ge, P, or As and M=Mo^{6+} or W^{6+}. Metals in peripheral positions can be partially substituted by V^{5+} ions without loss of structural integrity. Similarly, the anionic charge can be compensated for by protons, alkali and alkaline earth metals, or transition metal ions such as Fe^{2+}, Co^{2+}, and Ni^{2+}. These changes in composition produce changes in thermal structural stability and oxidative properties. Keggin compounds are both strong Brönsted acids and multielectron oxidants. Because they are potentially multifunctional, they are particularly interesting for the activation of organic substrates such as alkanes.

The most interesting of the various catalytic applications of heteropoly compounds from an industrial standpoint is the synthesis of methacrolein and methacrylic acid by direct gas phase oxidation of isobutane. Heteropoly compounds also catalyze the oxidative dehydrogenation of isobutyric acid to methacrylic acid and the oxidation of methacrolein to methacrylic acid. However, the multielectron functionalization of isobutane requires active centers that can perform different kinds of oxidative attacks. The catalysts claimed for the oxidation of isobutane are Keggin-type 12-molybdophosphates—alkali metals that contain vanadium in anionic positions and small amounts of copper in cationic positions. Processes for the oxidation of isobutane and the composition and performance of the required catalysts are discussed in Chapter 5.

Transition metal–substituted polyoxometallate (TMSP) systems have been prepared from 12-molybdophosphoric acid, tungstophosphoric acid, or their salts by substitution of one or more peripheral molybdenum or tungsten atoms in the anion with transition metal ions.[164,168-172] Transition metal ions in cationic positions are substituted by simple ion-exchange of the acid form or by direct precipitation of insoluble salts. The replacement of molybdenum or tungsten with metal ions other than vanadium is more difficult and requires special preparation procedures.[173,174]

TMSP systems are interesting in that they represent an inorganic analog of the metalloporphyrins in which the oxidizing centers are isolated inside a matrix that is stable to oxidative degradation under reaction conditions. They have significant potential as selective catalysts for liquid phase oxidation using hydrogen peroxide or molecular oxygen, although the use of these systems in the liquid phase with molecular oxygen seems to be limited primarily to applications that require either coreductants[175] or photochemical catalyst activation.[176] TMSP materials, used with palladium, are also active with molecular oxygen in Wacker-type catalysis.[177] In recent patents, Lyons et al.[164,165] claimed that transition metal–trisubstituted polyoxometallates catalyze the liquid phase oxidation of light alkanes with molecular oxygen, although the real catalytic action of these systems has yet to be demonstrated. Improving the thermal stability of TMSP compounds through the use of supports or by better neutralization could make these compounds useful for heterogeneous gas phase catalytic applications.

Interest has also been focused recently on ruthenium, rhodium, and iridium supported heteropoly compounds.[178] These systems, developed by Finke et al.,[179] show high activity in cyclohexene oxidation with oxygen. The metal ions in the complexes are not incorporated in the anion, but may be located on the anion surface. However, few characterization methods can distinguish between metal ions supported on heteropoly compounds, heteropoly compounds that have metals in cationic positions, and transition metal–substituted heteropoly compounds. Although, few catalytic results have been reported for these systems, the possibility of preparing compounds with a broad range of compositions suggests that much attention will be devoted to the study and characterization of these metal-modified heteropoly compounds in the near future.

Further aspects of polyoxometallates as catalysts for liquid and gas phase reactions will be discussed in Chapter 6.

3.3.3. Supported Metals

Palladium and platinum have long been known to catalyze the oxidation of primary alcohols and aldehydes, which is then followed by the formation of carboxylic acids in aqueous solution. The reaction proceeds via dehydrogenation (Section 3.2.1.2), followed by oxidation of the metal hydride by oxygen. More recently, liquid phase oxidative dehydrogenation reactions using supported noble metals have been used for the oxidation of alcohols and vicinal diols.[180] Several groups have also studied the oxidation of carbohydrates (Section 3.2.1.2).[181,182] In all these reactions, the catalysts suffer from rapid deactivation in the presence of molecular oxygen. The deactivation is attributed to the migration of adsorbed hydroxyl species into the platinum lattice. The adsorbed hydroxyl species are formed by reaction between adsorbed oxygen atoms and water. Catalyst deactivation can be reduced by using low oxygen partial pressures, low stirring speeds, and large catalyst particles in which oxygen diffusion is limited. While there may be applications for supported metal catalysts in the synthesis of fine chemicals, their use will probably be limited by deactivation problems, although several attempts have been made to overcome or minimize this problem (Section 3.2.1.2). Although no large-scale processes based on noble metal–catalyzed liquid phase oxidation are in operation at the moment, the increasing number of patents in this field is evidence of the industrial interest and thus an expanding number of applications can be expected in the near future.

3.3.4. Isomorphically Substituted Molecular Sieves

In a number of zeolites and silicalites, silicon or aluminum atoms in the framework can be isomorphically replaced with transition metal ions (see Table 3.8 for selected examples, of the continuously increasing number of new materials developed; see also Chapter 6).[183] The resulting isomorphically replaced molecular sieves contain isolated oxidizing centers located inside an inert and stable matrix that is characterized by uniform, well-defined cavities and channels. Thus, the advantage of the shape-selectivity of the zeolite and the presence of isolated oxidizing centers are combined to attain high selectivity in oxidation reactions without overoxidation.[184]

The metal ions in these isomorphically substituted molecular sieves are characterized by unique reactivity because their oxidizability and reducibility are strongly affected by the environment and because the physicochemical properties

TABLE 3.8. Structural Characteristics of Micro- and Mesoporous Materials and Examples of Incorporated Transition Metals[183]

Structure type	Isotopic framework structure	Ring number	Pore size, Å	Dimension	Metal incorporated
MFI	ZSM-5	10	5.6,5.4	3	Ti, Zr, V, Cr
MEL	ZSM-11	10	5.1,5.5	3	Ti, V
ZSM-48	ZSM-48	10	5.4,4.1	1	Ti
FAU	Y,X	12	7.4,7.4	3	Ti, Fe
MOR	Mordenite	12	6.7,7.0	2	Ti, Fe
BEA	Beta	12	7.6,6.4	3	Ti, Fe
MCM-41	MCM-41	—	40–100	1	Ti, Fe
AEL	AlPO$_4$-11	10	6.3,3.9	1	Co, Mn, Cr, V, Ti
AFI	AlPO$_4$-5	12	7.3	1	Co, Mn, Cr, V, Ti

of the isolated ions are quite different from those of the corresponding bulk metal oxide. Indeed, the unique reactivity of the metals in these systems is due to a mechanism of oxygen transfer via peroxo intermediates that is quite different from the redox mechanism in the corresponding transition metal bulk oxides.

These isomorphically substituted molecular sieve catalysts represent a valid alternative to metal ion–catalyzed oxidation reactions in the liquid phase. The molecular sieve provides a robust system that is easily recovered from the reaction medium and in which the active center is stabilized. However, as discussed in Section 3.2.1.1, titanium silicalite is the only real example of a system in which the transition metal is resistant to leaching in the presence of aqueous solutions of hydrogen peroxide, the most interesting oxidant for these catalysts. Nevertheless, they show much higher stability in organic media and thus they are a very interesting class of materials using organic peroxide reagents.

Titanium silicalite (TS-1) was the first example of this class of materials,[109] developed and then used commercially for liquid phase oxidation reactions. Since then, titanium has been incorporated into other molecular sieves (Table 3.8), although it is much less resistant to leaching. As discussed in this chapter and in Chapter 6, titanium and other transition metal ions incorporated in micro- and mesoporous materials represent one of the more significant developments in the field of selective oxidation, although the use of systems other than the TS-1 catalyst has not yet been developed on an industrial scale, primarily because of the cited problems of catalyst deactivation. Improvement in this area will open a wide range of reactions that cannot currently be done by TS-1, due either to steric limitations to diffusion of the reactants or to the inability of titanium to perform such a reaction (e.g., oxidation of primary carbon in the side-chain position to the aromatic ring).

A further area under initial development is that of transition metal ions or complexes anchored, or tethered, to the zeo-type material (micro- or mesoporous material) or generically in extraframework positions. In some cases, the complex may have only weak types of bonds with the zeo-type material, but owing to the different characteristics of hydrophobicity/hydrophilicity of the micro- or mesoporous material from the solvent, the complex is preferably adsorbed on the latter and thus can be easily recovered (adsorbed on the zeo-type material) by filtration at low temperature after the catalytic reaction.

3.3.5. Redox Pillared Clays

Smectite cationic clays and layered double hydroxide clays can be pillared with redox metal ions to give materials with interesting oxidation properties.[180,185] For example, intercalation of montmorillonite with vanadium ions[186] gives heterogeneous catalysts that are active in the epoxidation of allylic alcohols with alkylhydroperoxides. The activity of these catalysts is comparable to that of vanadium salts. In addition, the pillared clays exhibit interesting regioselectivity effects.

Also of interest is the pillaring of anionic clays by heteropolyoxoanions and isopolyoxoanions.[187-190] The many oxygen atoms and the large negative charge carried in the polyoxometallate anions result in pore structures with optimum vertical expansion of the layered double-hydroxide galleries and lateral separation of the pillars.[187] However, intercalation does not lead to a substantial improvement in the structural stability of the heteropolyanions, and the materials have been used successfully only for liquid phase alkene epoxidation and photolytic isopropanol oxidation. The oxidation that occurs in the gallery region of clays is constrained by steric restrictions. Thus these materials may function as shape-selective catalysts in the same way as zeolites, but with considerable possibility of fine tuning the dimensions of the galleries by modification of the pillars.

Furthermore, in these materials, it is relatively easier to create multifunctional systems by intercalating or pillaring different metal oxides, ions, polyoxoanions, and complexes. It is also possible to have mixed organic–inorganic systems. These materials thus show a very wide range of possibilities for the creation of supramolecular and polyfunctional catalysts, but their characteristics must be better understood if the opportunities for catalysis that they can offer are to be maximized.

3.3.6. Phase-Transfer Catalysts

Phase-transfer catalysis can be used for oxidation when either the oxygen donor or the catalyst is insoluble in the organic solvent in which the organic substrate to be oxidized is dissolved. For example, an ionic water-soluble catalyst can be transferred to an organic phase as an organic-soluble quaternary ammonium salt. Recently reported examples of applications of phase-transfer catalysis in oxidation reactions include oxidative cleavage of alkenes, oxidation of methylbenzenes to the corresponding acids, and oxidation of alcohols.[191]

Venturello et al.[192] reported the first example of catalytic epoxidation of alkenes in a two-phase system using hydrogen peroxide as the oxidizing agent. Heteropolyanions, dissolved in organic solvents by neutralization with cetylpyridinium cations, have been used as efficient epoxidation catalysts for oxidation of secondary alcohols to ketones and for oxidative cleavage of alkenes and vicinal diols (see also Section 3.2.2.1)

For the production of high-value chemicals, phase-transfer catalysts offer the possibility of combining the characteristics of the selectivity of homogeneous catalysts to reduce problems in catalyst recycling and separation of products, and thus offer a further opportunity to develop better catalytic organic syntheses of industrial interest.

3.3.7. Guest Oxide Nanoparticles within Host Zeo-Type Materials

A new area of solid catalysts which potentially offer many opportunities for selective oxidation is that of guest oxides or mixed oxides with host micro- or mesoporous materials.[193,194]

Oxide nanocrystals often possess quite different and particular reactivity properties in comparison to the corresponding bulk oxide,[195] as indicated by differences in, e.g., the crystalline habits, morphology, bulk oxygen transport properties, and electronic properties. Various methodologies have been developed to synthesize these oxide nanocrystals,[195] but it is usually very difficult to control the characteristics of the nanocrystals obtained and especially to avoid recrystallization during catalytic runs.

An interesting possibility is to stabilize these oxide crystals within a solid ordered matrix such as a zeolite, in which case it can be expected that the growth and crystalline habit of the guest nanoxide are influenced by the host zeolite (Figure 3.5). Additional factors that contribute to changing the reactivity of the oxide, include the presence of a strong electrostatic field inside the zeolite channels, the possibility of synergetic cooperation between isolated transition metal ions and oxide nanoparticles, and the possible presence of shape-selectivity effects. However, while the study of the characteristics and reactivity of guest oxides inside ordered crystalline host matrices (zeolite, in particular) offers an interesting new opportunity, the groundwork on the method of preparation of these solids has yet to be developed.

Especially critical problems in the synthesis of catalytic active oxides within ordered matrices are those of preparation,[193,194] such as: (i) avoiding external deposition on zeolite crystals, (ii) avoiding deposition of the oxide in a way so as

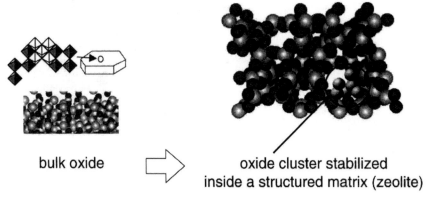

bulk oxide ⟹ oxide cluster stabilized inside a structured matrix (zeolite)

Figure 3.5. Schematic drawing of the change in characteristics of bulk oxide particles when the oxide is inserted in a host zeolite-type material.

to inhibit diffusion of reactants or reaction products, (iii) avoiding the nonselective catalytic role of the active sites already present in the zeolite matrix (especially acid sites), (iv) controlling the effective local structure and composition of nanoxide particles within the zeolite, and (v) deposition of a sufficient amount of active components within the zeolite pores to have reasonably high reaction rates.

The first attempts were made by inclusion of vanadium oxide[196,197] and copper oxide[194] nanoparticles in pentasyl-type zeolites. The oxide particles within the host zeolites showed different reactivity characteristics than the corresponding bulk oxides [e.g., higher turnover number for vanadium oxide nanoparticles within a host pentasyl-type TS-1 structure (see Figure 3.6); with a decreasing SiO_2/V_2O_3 ratio, larger vanadium oxide particles form] and there is better thermal stability than when oxide nanoparticles are obtained by other methods, but the problem of controlling the characteristics of the oxide particles within the host zeolite was not solved.

Recently it has been reported that Fe–Mo/DBH zeolites (where DBH indicates a deboronated borosilicate in the H form) prepared by chemical vapor deposition show interesting reactivity and selectivity characteristics in the selective oxidation of p-xylene to the corresponding aromatic aldehydes.[198-202] The preparation of

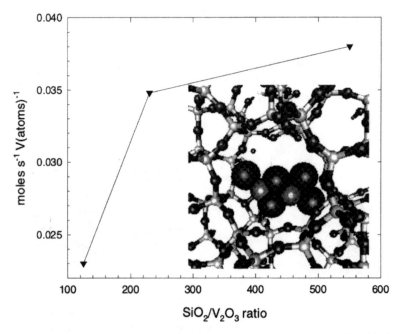

Figure 3.6. Dependence of the rate of propene formation at 500°C by propane oxidative dehydrogenation from the SiO_2/V_2O_3 ratio in vanadium oxide nanoparticles within silicalite-1 microporous materials (inset: model of the structure of vanadium oxide nanoparticles).

these catalysts is very complex, and control of all the parameters is very difficult. Recent studies[193] have shown that the creation of active and selective iron–molybdenum mixed oxide nanoparticles within the host zeolite involves: (i) the creation of defects in the host zeolite where iron ions are introduced (alternatively, iron ions can be in framework positions, but their distribution is less favorably controlled), and (ii) the reaction of molybdenum with these iron ions as well other iron oxide particles in extraframework positions to create a mixed oxide analogous to $Fe_2(MoO_4)_3$, but with controlled deformations/distortions in the bonds owing to the anchoring of some of the iron atoms constituting the iron–molybdate nanoparticle to the guest zeolite framework. The model proposed for the structure is shown in Figure 3.7.

These results indicate that the zeolite host matrix offers not only the possibility of orienting the growth of the guest (mixed) oxide by steric limitations (e.g., a chain-type copper oxide species forms inside ZSM-5, instead of tridimensional-like particles[194]), but also the additional possibility of creating further modifications by partial anchoring of some of the ions of the guest oxide to the zeolite framework. Thus these materials provide new possibilities for creating (mixed) oxides that will be unstable without an assisting rigid and porous matrix and they have the classical known characteristic of control of the reactions owing to steric limitations in restricted reaction environments (shape-selectivity).

The study of this class of materials, which is at an early stage, thus offers many possibilities for the development of very interesting new catalytic materials for selective oxidation reactions.

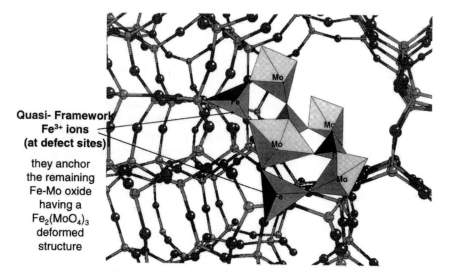

Quasi- Framework
Fe³⁺ ions
(at defect sites)

they anchor
the remaining
Fe-Mo oxide
having a
$Fe_2(MoO_4)_3$
deformed
structure

Figure 3.7. Model of the structure of iron–molybdate nanoparticles within a host boralite zeolite.[193]

3.4. CONCLUSIONS

The academic and industrial research in the field of selective oxidation has been quite active in the last two decades and several new technological and industrial opportunities have been developed with the common characteristics of process simplification and reduced environmental impact, as well as reduction of production costs.

In this chapter the discussion centered on the use of hydrogen peroxide as a new, environmentally acceptable reactant and the various opportunities offered by its use, some of them already commercially employed, and several others still under development. Its use extends from syntheses of interest for fine chemicals to the production of bulk chemicals such as propene oxide and cyclohexanone oxime. However, discussion of the possibility of using this reactant also offered the opportunity to discuss alternative solutions, the trends in searching for ways to reduce the critical problem of the cost of hydrogen peroxide, and the limitations in the use of this reactant and which alternative monoxygenation reactants can be used, especially for the synthesis of high-value products.

New opportunities are also offered by the use of new hydrocarbon raw materials such as alkanes, but these will be discussed in subsequent chapters. However, this chapter did contain a brief discussion to illustrate some classes of catalytic materials that have received increasing attention in the last decade and thus offer new possibilities for the development of new kinds of catalytic reactions of selective oxidation. Of course, as the development of new catalytic materials is the basis for innovation in catalysis, the discussion of specific examples of new catalytic materials of potential interest can be endless and is outside the scope of this book. Discussion is thus limited to showing the general characteristics of selected classes of materials for catalytic oxidation. Several of them will then be discussed in more detail in other later chapters.

In order to highlight the critical problems in process development and use them as examples and guidelines for future research, discussion was centered on an analysis of alternative process options and more specific data on the catalytic behavior of particular importance for process development. A general discussion of the flowsheet of processes, when relevant, was also added, because catalyst development cannot ignore the technical aspects of process development.

REFERENCES

1. D. Delmon, in: *3rd World Congress on Oxidation Catalysis* (R.K. Grasselli, S.T. Oyama, A.M. Gaffney, and J.E. Lyons, eds.), Studies in Surface Science and Catalysis Vol. 110, Elsevier Science: Amsterdam (1997), p. 43.
2. R.L. Augustine, *Heterogeneous Catalysis for the Synthetic Chemist*, Marcel Dekker: New York (1996), Ch. 21.

3. R.A. Sheldon and J. Dakka, *Catal. Today* **19**, 215 (1994).
4. G. Bellussi and M.S. Rigutto, in: *Advanced Zeolite Science and Applications* (J.C. Jansen, M. Stöcker, H.G. Karge, and J. Weitkamp, eds.), Studies in Surface Science and Catalysis Vol. 85, Elsevier Science: Amsterdam (1994), Ch. 6, p. 177.
5. R.A. Sheldon and J.K. Kochi, *Metal-Catalyzed Oxidations of Organic Compounds*, Academic Press: New York (1981).
6. R.A. Sheldon, in: *3rd World Congress on Oxidation Catalysis* (R.K. Grasselli, S.T. Oyama, A.M. Gaffney, and J.E. Lyon, eds.), Studies in Surface Science and Catalysis Vol. 110, Elsevier Science: Amsterdam (1997), p. 151.
7. R.A. Sheldon, *J. Mol. Catal. A: Chemical* **107**, 75 (1996).
8. P. Vinke, D. de Wit, A.T.J. de Goede, and H. van Bekkum, in: *New Developments in Selective Oxidation by Heterogeneous Catalysts* (P. Ruiz and B. Delmon, eds.), Studies in Surface Science and Catalysis Vol. 72, Elsevier Science: Amsterdam (1992), p. 1.
9. H. van Bekkum, in: *Carbohydrates as Organic Raw Materials* (F.W. Lichtenthaler, ed.), VCH: Weinheim (1991), p. 289.
10. Y. Moro-oka and M. Akita, *Catal. Today* **41**, 327 (1998).
11. S.-I. Murahashi and N. Komiya, *Catal. Today* **41**, 339 (1998).
12. R.A. Sheldon, in: *Aspects of Homogeneous Catalysis* Vol. 4 (R. Ugo, ed.), Reidel: Dordrecht, (1981), p. 1.
13. J. Sobczak and J.J. Ziolkowski, *React. Kinet. Catal. Lett.* **11**, 3583 (1985).
14. C. Neri, B. Anfossi, A. Esposito, and F. Buonomo, *EP Patent* 100,119 B1 (1983).
15. R. Millini, E.P. Massara, G. Perego, and G. Bellussi, *J. Catal.* **137**, 497 (1992).
16. A.J.H.P. van der Pol and J.H.C. van Hooff, *Appl. Catal. A* **106**, 97 (1993).
17. F. Maspero and U. Romano, *J. Catal.* **146**, 476 (1994).
18. U. Romano, A. Esposito, F. Maspero, C. Neri, and M.G. Clerici, *Chim. Ind. (Milan)* **72**, 610 (1990).
19. U. Romano, A. Esposito, F. Maspero, C. Neri, and M.G. Clerici, in: *New Developments in Selective Oxidation* (G. Centi and F. Trifirò eds.), Studies in Surface Science and Catalysis Vol. 55, Elsevier Science: Amsterdam (1990), p. 33.
20. C. Neri and F. Buonomo, *EP. Patent* 102,097 (1984).
21. T. Tasumi, M. Yako, N. Nakajura, Y. Yuhara, and H. Tominaga, *J. Mol. Catal.* **78**, L41 (1993).
22. M.G. Clerici and P. Ingallina, *J. Catal.* **140**, 71 (1993).
23. V. Hulea, P. Moreau, and F. di Renzo, *J. Mol. Catal. A: Chemical* **111**, 325 (1996).
24. A. Zecchina, G. Spoto, S. Bordiga, F. Geobaldo, G. Petrini, G. Leofanti, M. Padovan, M. Mantegazza, and P. Roffia, in: *New Frontiers in Catalysis* (L. Guczi, F. Solymosi, P. Tétényi, eds.), Studies in Surface Science and Catalysis Vol. 75, Elsevier Science: Amsterdam (1993), p. 719.
25. J.S. Reddy and P.A. Jacobs, *J. Chem. Soc. Perkin Trans. I* **22**, 2665 (1993).
26. J.S. Reddy and P.A. Jacobs, *Catal. Lett.* **37**, 213 (1996).
27. M.G. Clerici, *Appl. Catal.* **68**, 249 (1991).
28. T. Blasco, M.A. Camblor, A. Corma, and J. Pérez-Pariente, *J. Am. Chem. Soc.* **115**, 11806 (1993).

29. A. Tuel, *Zeolites* **15**, 228 (1995).

30. P.T. Tanev, M. Chibwe, and T.J. Pinnavaia, *Nature* **368**, 321 (1994).

31. A. Corma, M.A. Camblor, P. Esteve, A. Martínez, and J. Pérez-Pariente, *J. Catal.* **145**, 151 (1994).

32. R.A. Sheldon, I.W.C.E. Arends, and H.E.B. Lempers, *Catal. Today* **41**, 387 (1998).

33. T. Sato, J. Dakka, and R.A. Sheldon, *J. Chem. Soc., Chem Commun.* 1887 (1994).

34. A. Tuel, S. Gontier, and R. Teissier, *J. Chem. Soc., Chem. Commun.* 651 (1996).

35. A. Corma, M. Iglesias, and F. Sanchéz, *J. Chem. Soc., Chem. Commun.* 1653 (1995).

36. P.R. Hari Prasad Rao and A.V. Ramaswami, *J. Chem. Soc. Chem. Comm.* 1245 (1992).

37. M.K. Dongare, P. Singh, P.P. Moghe, and T. Ratnasamy, *Zeolites* **11**, 690 (1992).

38. N. Ulagappan and C.N.R. Rao, *J. Chem. Soc., Chem. Commun.* 1047 (1996).

39. W. Zhang and T.J. Pinnavaia, *Catal. Lett.* **38**, 261 (1996).

40. W. Zhang, J. Wang, P.T. Tanev, and T.J. Pinnavaia, *J. Chem. Soc., Chem. Commun.*, 979 (1996).

41. J.S. Reddy and A. Sayari, *J. Chem. Soc., Chem Commun.*, 2231 (1995).

42. B.I. Whittington and J.R. Anderson, *J. Phys. Chem.* **97**, 1032 (1993).

43. M.S. Rigutto and H. van Bekkum, *J. Mol. Catal.* **81**, 77 (1993).

44. M.J. Haanepen and J.H.C. van Hooff, *Appl. Catal. A: General* **152**, 183 (1997).

45. N.K. Mal, V. Ramaswamy, S. Ganapathy, and A.V. Ramaswamy, *J. Chem. Soc., Chem. Commun.* 1933 (1994).

46. T.K. Das, K. Chaudhari, A.J. Chandwadhar, and S. Sivasankar, *J. Chem. Soc., Chem. Commun.* 2495 (1995).

47. F. Figueras, *Catal. Rev.-Sci. Eng.* **30**, 457 (1988).

48. B.M. Chaudary, A. Durga Prasad, and V.L.K. Valli, *Tetrahedron Lett.* **31**, 57885 (1990).

49. B.M. Chaudary, A. Durga Prasad, V. Bhuma, and V. Swapna, *J. Org. Chem.* **57**, 5841 (1992).

50. B.M. Chaudary, A. Durga Prasad, V. Swapna, V.L.K. Valli, and V. Bhuma, *Tetrahedron* **48**, 953 (1992).

51. M. Floor, A.P.G. Kieboom, and H. van Bekkum, *Starch* **41**, 348 (1989).

52. A. Sattar, D. Muhammad, M. Ashraf, S.A. Khan, and M.K. Bhatty, *Pak. J. Sci. Ind. Res.* **31**, 745 (1988).

53. K. Heyns and H. Paulsen, *Angew. Chem.* **69**, 600 (1957).

54. H. Röper, in: *Carbohydrates as Organic Raw Materials* (F.W. Lichtenthaler, ed.), VCH: Weinheim (1991), p. 267.

55. J.M.H. Dirkx and H.S. van der Baan, *J. Catal.* **67**, 1 (1981).

56. K. Hattori, *JP Patent* 784,0713 (1978).

57. K. Deller, H. Krause, E. Peldszus, and B. Despeyroux, *DE Patent* 382,3301 (1989).

58. A.J. van Hengstum, A.P.G. Kieboom, and H. van Bekkum, *Starch* **36**, 317 (1984).

59. Y. Shuurman, B.F.M. Kuster, K. van der Wiele, and G.B. Marin, in: *New Developments in Selective Oxidation by Heterogeneous Catalysts* (P. Ruiz and B. Delmon,

eds.), Studies in Surface Science and Catalysis Vol. 72, Elsevier Science: Amsterdam (1992), p. 43.

60. N. Ripke, J. Thiem, and Th. Böcker, EP *Patent*, EP 032,6673 (1988).

61. H.E. Hendriks, B.F.M. Kuster, and G.B. Marin, *Carbohydr. Res.* **204**, 121 (1990).

62. B. Casu, G. Scovenna, A.J. Cifonelli, and A.S. Perlin, *Carbohydr. Res.* **63**, 13 (1978).

63. B.F.M. Kuster, *Starch* **42**, 314 (1990).

64. B.W. Lew, *US Patent* 332,6944 (1967).

65. P. Vinke, H.E. van Dam, and H. van Bekkum, in: *New Developments in Selective Oxidation* (G. Centi and F. Trifirò, eds.), Studies in Surface Science and Catalysis Vol. 55, Elsevier Science: Amsterdam (1990), p. 147.

66. J.T. Groves, T.E. Nemo, and R. S.Meyers, *J. Am. Chem. Soc.* **101**, 1032 (1979).

67. J.T. Groves and Y. Watanabe, *J. Am. Chem. Soc.* **110**, 8443 (1988).

68. C.L. Hill and B.C. Schardt, *J. Am. Chem. Soc.* **102**, 6374 (1980).

69. J.T. Groves, W.J. Kruper, Jr., and R.C. Hauschalter, *J. Am. Chem. Soc.* **102**, 6375 (1980).

70. J.T. Groves and W.J. Kruper, *J. Am. Chem. Soc.* **101**, 7613 (1979).

71. J.T. Groves and R. Quinn, *J. Am. Chem. Soc.* **107**, 5790 (1985).

72. C.M. Che and W.C. Chung, *J. Chem. Soc., Chem. Commun.*, 386 (1986).

73. S. Takagi, E. Takahashi, T.K. Miyamoto, and Y. Sasaki, *Chem. Lett.* 1275 (1986).

74. B. Notari, *Catal. Today* **18**, 163 (1993).

75. Y. Ishii, K. Yamayaki, T. Yoshida, and T. Ura, *J. Org. Chem.* **52**, 1868 (1987).

76. M. Misono, N. Mizuno, K. Inumaru, G. Koyano, and X.H. Lu, in: *3rd World Congress on Oxidation Catalysis* (R.K. Grasselli, S.T. Oyama, A.M. Gaffney, and J.E. Lyon, eds.), Studies in Surface Science and Catalysis Vol. 110, Elsevier Science: Amsterdam (1997), p. 35.

77. B.R. Cook, T.J. Reinert, and K.S. Suslick, *J. Am. Chem. Soc.* **108**, 7281 (1986).

78. N. Herron, G.D. Stucky, and C.A. Tolman, *J. Chem. Soc., Chem. Commun.*, 1521 (1986).

79. N. Herron and C.A. Tolman, *J. Am. Chem. Soc.* **109**, 2837 (1987).

80. D. Mansuy, P. Rattioni, J.P. Renard, and P. Guerin, *J. Chem. Soc., Chem. Commun.*, 155 (1985).

81. Y. Naruta, F. Tani, N. Ishihara, and K. Maruyama, *J. Am. Chem. Soc.* **113**, 6865 (1992).

82. W. Zhang, J.L. Loebach, S.R. Wilson, and E.N. Jacobsen, *J. Am. Chem. Soc.* **112**, 2801 (1990).

83. T. Katsuki, *J. Mol. Catal. A: Chemical* **113**, 87 (1996).

84. S. Udenfriend, C.T. Clark, J. Axelrod, and B.B. Brodie, *J. Biol. Chem.* **208**, 731 (1954).

85. D.H.R. Barton and D. Doller, *Acc. Chem. Res.* **25**, 504 (1992).

86. I. Tabushi and N. Koga, *J. Am. Chem. Soc.* **101**, 6456 (1979).

87. I. Tabushi and A. Yazaki, *J. Am. Chem. Soc.* **103**, 7371 (1981).

88. T. Mukaiyama and T. Yamada, *Bull. Chem. Soc. Jpn* **68**, 17 (1995).

89. A. Kunai, S. Hara, S. Ito, and K. Sasaki, *J. Org. Chem.* **51**, 3471 (1986).

90. A. Kunai, T. Wani, Y. Uehara, F. Iwasaki, Y. Kuroda, S. Ito, and K. Sasaki, *Bull. Chem. Soc. Jpn* **67**, 2613 (1989).

91. T. Miyake, M. Hamada, Y. Sasaki, and M. Oguri, *Appl. Catal., A* **131**, 33 (1995).

92. M. Hamada, H. Niwa, M. Oguri, and T. Miyake, *JP Patent* 179383 (1995).

93. N. Kitajima, K. Fujisawa, C. Fujimoto, Y. Moro-oka, S. Hashimoto, T. Kitagawa, T. Toriumi, K. Tatsumi, and A. Nakamura, *J. Am. Chem. Soc.* **114**, 1277 (1992).

94. N. Kitajima and Y. Moro-oka, *Chem. Rev.* **94**, 737 (1994).

95. N. Kitajima, N. Tamura, H. Amagai, H. Fukui, Y. Moro-oka, Y. Mizutani, T. Kitagawa, R. Mathur, K. Heerwegh, C.A. Reed, C.R. Ranfall, L. Que, Jr., and K. Tatsumi, *J. Am. Chem. Soc.* **116**, 9071 (1994).

96. T. Ohkubo, H. Sugimoto, T. Nagayama, H. Masuda, T. Sayo, K. Tanaka, Y. Maeda, H. Okawa, Y. Hayashi, A. Uehara, and M. Suzuki, *J. Am. Chem. Soc.* **118**, 701 (1996).

97. S. Hikichi, H. Komatsuzaki, N. Kitajima, M. Akita, M. Mukai, T. Kitagawa, and Y. Moro-oka, *Inorg. Chem.* **36**, 266 (1997).

98. C. Venturello, *Chim. Ind. (Milan)* **75**, 283 (1993).

99. F. Maspero, *Chim. Ind. (Milan)* **75**, 291 (1993).

100. M.G. Clerici and P. Ingallina, *EP Patent* 202,362 (1992).

101. M.G. Clerici and P. Ingallina, *EP Patent* 203,625 (1992).

102. W. Laufer, R. Meiers, and W.F. Hölderich, *Proceedings 12th International Zeolite Conference*, Baltimore (US) July 1998, B-8.

103. T. Tasumi, K. Yuasa, and H. Tominaga, *J. Chem. Soc. Chem. Comm.* 1446 (1992).

104. M. Haruta, in: *3rd World Congress on Oxidation Catalysis* (R.K. Grasselli, S.T. Oyama, A.M. Gaffney, and J.E. Lyon, eds.), Studies in Surface Science and Catalysis Vol. 110, Elsevier Science: Amsterdam (1997), p. 123.

105. B.T. Pennington, *US Patent* 4,885,374 (1989).

106. J.N. Armor, *J. Catal.* **70**, 72 (1981).

107. D. Dreoni, D. Pinelli, and F. Trifirò, in: *New Developments in Selective Oxidation by Heterogeneous Catalysts* (P. Ruiz and B. Delmon, eds.), Studies in Surface Science and Catalysis. Vol. 72, Elsevier Science: Amsterdam (1992), p. 109.

108. J.N. Armor, in: *Catalysis in Organic Chemistry*, Vol. 18, (J. Kosak, ed.), Marcel Dekker: New York (1984) p. 509.

109. M. Taramasso, G. Perego, and N. Notari, *US Patent* 4,410,501 (1983).

110. M.G. Clerici and P. Ingallina, *IT Patent* 003153 A/91 (1991).

111. M.G. Clerici and G. Bellussi, *IT Patent* 21158 A/90 (1990).

112. C. Venturello and R. D'Aloisio, *J. Org. Chem.* **53**, 1553 (1988).

113. C. Venturello, R. D'Aloisio, and M. Ricci, *EP Patent* 109,273 (1984).

114. C. Venturello, R. D'Aloisio, J.C. Bart, and M. Ricci, *J. Mol. Catal.* **32**, 107 (1985).

115. C. Venturello, E. Alneri, and G. Lana, *DE Patent* 3,027,349 (1981).

116. T.B. Pennington and M.C. Fullington, *US Patent* 4,943,643 (1990).

117. T.B. Pennington and M.C. Fullington, *US Patent* 5,117,011 (1992).

118. M.C. Fullington and B. Pennington, *WO Patent*, 92/09588 (1992).

119. J.L. Meyer and T.B. Pennington, *US Patent* 4,992,567 (1990).

120. J.L. Meyer and T.B. Pennington, *WO Patent* 90/15054 (1990).

121. J.L. Meyer, P.J. Craney, and T.B. Pennington, *WO Patent* 90/15053 (1990).

122. P. Roffia, M. Padovan, E. Moretti, and G. De Alberti, *EP Patent* 208,311 A2 (1986).

123. S. Tsuda, *Chem. Econ. Eng. Rev.* 39 (1970).
124. P. Roffia, M. Padovan, G. Leofanti, M. Mantegazza, G. De Alberti, and G.R. Tauszik, *US Patent* 4,794,198 (1988).
125. P. Roffia, G. Leofanti, A. Cesana, M. Mantegazza, M. Padovan, G. Petrini, S. Tonti, P. Gervasutti, and R. Varagnolo, *Chim. Ind. (Milan)* **72**, 598 (1990).
126. P. Roffia, G. Leofanti, A. Cesana, M. Mantegazza, M. Padovan, G. Petrini, S. Tonti, and P. Gervasutti, in *New Developments in Selective Oxidation* (G. Centi, and F. Trifirò, eds.), Studies in Surface Science and Catalysis Vol. 55, Elsevier Science: Amsterdam (1990), p. 43.
127. G. Petrini, A. Cesana, G. De Alberti, F. Genoni, G. Leofanti, M. Padovan, G. Paparatto, and P. Roffia, in: *Catalyst Deactivation 1991* (C.H. Bartholomew and J.B. Butt, eds.), Studies in Surface Science and Catalysis Vol. 68, Elsevier Science: Amsterdam (1991), p. 761.
128. P. Roffia, G. Paparatto, A. Cesana, and G. Tauszik, *US Patent* 4,894,478 (1990).
129. G. Paparatto, G. Petrini, and P. Roffia, in: *Proceedings 9th International Zeolite Conference* (Montreal, 1992), R. van Ballmoos, J.B. Higgins, and M.M. Treacy, eds., Butterworth-Heinemann: Boston (1993).
130. J.N. Armor, *US Patent* 4,163,756 (1979).
131. J.N. Armor, *US Patent* 4,281,194 (1971).
132. J.N. Armor and P. Zambri, *J. Catal.* **73**, 57 (1982).
133. J.N. Armor, *J. Am. Chem. Soc.* **102**, 1453 (1980).
134. J.N. Armor, E.J. Carlson, S. Soled, W.D. Conner, A. Laverick, B. De Rites, and W. Gates, *J. Catal.* **70**, 84 (1981).
135. J.N. Armor, P.M. Zambri, and R. Leming, *J. Catal.* **72**, 66 (1982).
136. G. Fornasari and F. Trifirò, *Catal. Today* **41**, 443 (1998).
137. D.P. Dreoni, D. Pinelli, F. Trifirò, Z. Tvaruzkova, H. Habersberger, and P. Jiru, in: *New Frontiers in Catalysis* (L. Guczi, F. Solymosi, and P. Tétényi, eds.), Studies in Surface Science and Catalysis Vol. 75, Part C, Elsevier Science: Amsterdam (1993), p. 2011.
138. D.P. Dreoni, D. Pinelli, and F. Trifirò, in: *New Developments in Selective Oxidation by Heterogeneous Catalysts* (P. Ruiz and B. Delmon, eds.), Studies in Surface Science and Catalysis Vol. 72, Elsevier Science: Amsterdam (1992), p. 109.
139. D.P. Dreoni, D. Pinelli, and F. Trifirò, *J. Mol. Catal.* **69**, 171 (1991).
140. E. Pieri, D. Pinelli, and F. Trifirò, *Chem. Eng. Sci.* **47**, 2641 (1992).
141. D. Collina, G. Fornasari, A. Rinaldo, F. Trifirò, G. Leofanti, G. Paparatto, and G. Petrini, in: *Preparation of Catalysts VI* (G. Poncelet, J. Martens, B. Delmon, P.A. Jacobs, and P. Grange, eds.), Studies in Surface Science and Catalysis Vol. 91, Elsevier Science: Amsterdam (1995), p. 401.
142. A. Bendandi, G. Fornasari, M. Guidoreni, L. Kubelkova, M. Lucarini, and F. Trifirò, *Topics Catal.* **3**, 337 (1996).
143. M.A. Mantegazza, G. Petrini, G. Fornasari, A. Rinaldo, and F. Trifirò, *Catal. Today* **32**, 297 (1996).
144. Y. Barbaux, D. Bouqueniaux, G. Fornasari, and F. Trifirò, *Appl. Catal. A* **125**, 303 (1995).
145. D. Pinelli, F. Trifirò, A. Vaccari, E. Giamello, and G. Pedulli, *Catal. Lett.* **13**, 21 (1992).

146. F. Bourdin, M. Costantini, M. Jouffret, and G. Lartigan, *DE Patent* 2,064,497 (1971).

147. J. Varagnat, *Ind. Eng. Chem. Prod. Res. Dev.* **15**, 212 (1976).

148. P. Maggioni, *US Patent* 3,914,323 (1975).

149. F. Minisci and P. Maggioni, *Chim. Ind. (Milan)* **61**, 834 (1979).

150. A. Esposito, M. Taramasso, and C. Neri, *US Patent* 4,396,783 (1983).

151. G. Bellussi, M. Clerici, F. Buonomo, U. Romano, A. Esposito, and B. Notari, *EP Patent* 200,260 B1 (1986).

152. B. Notari, in: *Innovation in Zeolite Material Science* (P.J. Grobet, W. Mortier, E.F. Vansant, and G. Schulz-Ekloff, eds.), Studies in Surface Science and Catalysis Vol. 37, Elsevier Science: Amsterdam (1987), p. 413.

153. T. Hamamoto, N. Kuroda, N. Takamitu, and S. Umemura, *Nippon Kagaku Kaishi*, 1850 (1980).

154. S.W. Brown, A. Hackett, A. Johnstone, A.M. King, K.M. Reeve, W.R. Sanderson, and M. Service, in: *Proceedings DGMK Conference on Selective Oxidation in Petrochemistry*, Goslar (Germany), DGMK: Hamburg (1992), p. 339.

155. B. Notari, in: *Structure–Activity and Selectivity Relationships in Heterogeneous Catalysis* (R.K. Grasselli and A.W. Sleight, eds.), Studies in Surface Science and Catalysis Vol. 67, Elsevier Science: Amsterdam (1987), p. 243.

156. A.J.H.P. van der Pol, A.J. Verduyn, and J.H. van Hooff, *Appl. Catal., A* **92**, 113 (1992).

157. A. Thangaraj, R. Kumar, and P. Ratnasamy, *J. Catal.* **131**, 294 (1991).

158. R. Millini, E. Previde Massara, G. Perego, and G. Bellussi, *J. Catal.* **137**, 497 (1992).

159. A. Esposito, C. Neri, and F. Buonomo, *DE Patent* 3,309,669 (1983).

160. A. Tuel, S. Moussa-Khouzami, Y.B. Taarit, and C. Naccache, *J. Mol. Catal.* **68**, 45 (1991).

161. J.A. Martens, Ph. Buskens, P.A. Jacobs, A. van der Pol, J.H.C. van Hooff, C. Ferrini, H.W. Kouwenhove, P.J. Kooyman, and H. van Bekkum, *Appl. Catal. A* **99**, 71 (1993).

162. G. Perego, G. Bellussi, C. Corno, M. Taramasso, F. Buonomo, and A. Esposito, in: *New Developments in Zeolite Science and Technology* (Y. Murakami, A. Iijima, and J.W. Ward, eds.), Studies in Surface Science and Catalysis Vol. 28, Elsevier Science: Amsterdam (1986), p. 129.

163. T. Katsuki and K.B. Sharpless, *J. Am. Chem. Soc.* **102**, 5976 (1980).

164. J.E. Lyons, P.E. Ellis, and V.A. Durante, in: *Structure–Activity and Selectivity Relationships in Heterogeneous Catalysis* (R.K. Grasselli and A.W. Sleight, eds.), Studies in Surface Science and Catalysis Vol. 67, Elsevier Science: Amsterdam (1987), p. 99.

165. J.E. Lyons, P.E. Ellis, R.W. Wagner, P.B. Thompson, H.B. Gray, M.E. Hughes, and J.A. Hodge, in: *Proceedings Symposium on Natural Gas Upgrading*, San Francisco (1992), p. 307.

166. T.G. Traylor and S. Tsuchiya, *Inorg. Chem.* **26**, 1338 (1987).

167. C.L. Hill, A.M. Khenkin, M.S. Weeks, Y. Hou, in: *Catalytic Selective Oxidation* (S.T. Oyama, and J.W. Hightower, eds.), ACS Symposium Series 523, American Chemical Society: Washington DC (1993), p. 67.

168. C.M. Tourné and G.F. Tourné, *Bull. Soc. Chim. Fr.* **4**, 1124 (1969).

169. C.L. Hill and R.B. Brown, *J. Am. Chem. Soc.* **108**, 536 (1986).

170. R. Neumann and C. Abu-Gnim, *J. Chem. Soc. Chem. Comm.*, 1324 (1989).

171. D.E. Katsoulis and M.T. Pope, *J. Chem. Soc. Dalton Trans.*, 1483 (1989).
172. D. Mansuy, J.F. Bartoli, P. Battioni, D.K. Lyon, and R.G. Finke, *J. Am. Chem. Soc.* **113**, 7222 (1991).
173. M. Leyrie, M. Fournier, and R. Massart, *C.R. Acad. Sci. Ser. C* **273**, 1569 (1971).
174. C.M. Tourné, G.F. Tourné, S.A. Malik, and T.J.R. Weakly, *J. Inorg. Nucl. Chem.* **32**, 3875 (1970).
175. R. Neumann and M. Levin, in: *Dioxygen Activation and Homogeneous Catalytic Oxidation* (L.I. Simandi, ed.), Studies in Surface Science and Catalysis Vol. 66, Elsevier Science: Amsterdam (1991), p. 121.
176. R.F. Renneke and C.L. Hill, *J. Am. Chem. Soc.* **110**, 5461 (1988).
177. J.A. Cusumano, *CHEMTECH*, 482 (1992).
178. N. Mizuno, D.K. Lyon, and R.G. Finke, *J. Catal.* **128**, 84 (1991).
179. R.G. Finke, D.K. Lyon, K. Nomiya, S. Sur, and N. Mizuno, *Inorg. Chem.* **29**, 1784 (1990).
180. R.A. Sheldon, in: *Heterogeneous Catalysis and Fine Chemicals II* (M. Guisnet, J. Barrault, C. Bouchoule, D. Duprez, G. Perot, R. Maurel, and C. Montassier, eds.), Studies in Surface Science and Catalysis Vol. 59, Elsevier Science: Amsterdam (1991), p. 33.
181. P.J.M. Dijkgraaf, H.A. Duisters, B.F.M. Kuster, and K. van der Wiele, *J. Catal.* **112**, 337 (1988).
182. H.E. van Dam, P. Duijverman, A.P.G. Kieboom, and H. van Bekkum, *Appl. Catal.* **33**, 373 (1987).
183. R.A. Sheldon, J.D. Chen, J. Dakka, and E. Neeleman, in: *New Developments in Selective Oxidation II* (V. Corberan and S. Vic Bellon, eds.), Studies in Surface Science and Catalysis Vol. 82, Elsevier Science: Amsterdam (1994), p. 515.
184. R.K. Grasselli, in: *Surface Properties and Catalysis by Non-Metals* (J. Bonnelle, B. Delmon, and E. Derouane, eds.), Reidel: Dordrecht (1983), p. 273.
185. F. Figueras, *Catal. Rev.-Sci. Eng.* **30**, 457 (1988).
186. B.M. Choudary, V.L.K. Valli, and A. Durga Prasad, *J. Chem. Soc. Chem. Comm.* 721 (1990).
187. T. Kwon and T. Pinnavaia, *J. Mol. Catal.* **74**, 23 (1992).
188. T. Tatsumi, K. Yamamoto, H. Tajima, and H. Tominaga, *Chem. Lett.* 815 (1992).
189. E. Narita, P.D. Kaviratna, and T. Pinnavaia, *J. Chem. Soc. Chem. Comm.* 60 (1993).
190. M.A. Drezdson, *Inorg. Chem.* **27**, 4628 (1988).
191. R.A. Sheldon, *CHEMTECH* **21**, 566 (1991).
192. C. Venturello, E. Alneri, and M. Ricci, *J. Org. Chem.* **48**, 3831 (1983).
193. G. Centi, F. Fazzini, J.L.G. Fierro, M. Lòpez Granados, R. Sanz, and D. Serrano, in: *Preparation of Catalysts VII* (B. Delmon, P.A. Jacobs, R. Maggi, J.A. Martens, P. Grange, and G. Poncelet, eds.), Studies in Surface Science and Catalysis Vol. 118, Elsevier Science: Amsterdam (1998), p. 577.
194. G. Centi, F. Fazzini, and A. Galli, *Res. Chem. Intermed.* **24**, 541 (1998).
195. W.R. Moser (ed), *Advanced Catalysts and Nanostructured Materials*, Academic Press: San Diego (1996).

196. G. Centi, S. Perathoner, F. Trifirò, A. Aboukais, C.F. Aïssi, and M. Guelton, *J. Phys. Chem.* **96**, 2617 (1992).

197. G. Bellussi, G. Centi, S. Perathoner, and F. Trifirò, in: *Selective Oxidation Catalysts* (T. Oyama and J. Hightower, eds.), ACS Symposium Series, American Chemical Society: Washington (1992), Ch. 21, p. 281.

198. J.S. Yoo, J.A. Donohue, M.S. Kleefish, P.S. Lin, and S.D. Elfine, *Appl. Catal., A* **105**, 83 (1993).

199. J.S. Yoo, C. Choi-Feng, and J.A. Donohue, *Appl. Catal., A* **118**, 87 (1994).

200. J.S. Yoo, *Appl. Catal., A* **143**, 29 (1996).

201. G.W. Zajac, C. Choi-Feng, J. Faber, J.S. Yoo, R. Patel, and H. Hochst, *J. Catal.* **151**, 338 (1995).

202. J.S. Yoo, J.A. Donohue, and C. Choi-Feng, in: *Advanced Catalysts and Nanostructured Materials* (W.R. Moser, ed.), Academic Press: San Diego (1996), Ch. 17, p. 453.

CONTROL OF THE SURFACE REACTIVITY OF SOLID CATALYSTS
Industrial Processes of Alkane Oxidation

4.1. INTRODUCTION

The demand for innovation in the field of selective oxidation in terms of raw materials, ecological issues, and process simplification (Chapters 1–3) has stimulated industrial research toward the development of new synthetic routes starting from alkane feedstocks. Alkanes are characterized by the absence of reactive sites, such as hydrogen atoms, which can be easily abstracted, and double bonds; furthermore, they are largely less reactive than most of the possible reaction products. Thus the solid catalysts used in the selective activation and transformation of alkanes must have very special surface properties. At the same time, the problem of the control of the reactivity and the unreactive character of the alkanes (e.g., a lower rate of conversion with respect to alkenes) has also driven the search for: (i) new reactor technologies to maximize selectivity or productivity (see Chapter 3), and (ii) new integrated approaches between catalyst and process design to maximize the effectiveness of the new solutions. The development of new processes of alkane selective oxidation can thus be viewed not only as a technological breakthrough but also as a methodological one as they have opened up new areas of investigation concerning: (i) surface properties required to control the reactivity, bulk catalyst properties, and structural characteristics of the catalyst; and (ii) technological and reactor engineering solutions to improve the control of the reactivity.

Analysis of the various aspects of catalyst and process development in alkane selective oxidation is thus a good way to illustrate the relationships among chemical (reaction mechanism), material (bulk and surface characteristics, as well as textural, morphological, and mechanical properties), and engineering (reactor and process design) factors and how only an integrated approach can lead to the successful development of a new process.

The synthesis of maleic anhydride from n-butane is as yet the only industrial application of alkane selective oxidation by heterogeneous gas–solid catalytic processes, and it will be discussed in detail in this chapter. A second example is the synthesis of acrylonitrile directly from propane, a process not yet being carried out on an industrial scale, but with a good outlook for rapid commercialization. Catalysts for the first reaction (maleic anhydride from n-butane) are almost exclusively vanadium/phosphorus oxides and more specifically based on the presence of a characterizing active phase: vanadyl pyrophosphate. Different types of catalysts have been suggested as alternatives for the synthesis of acrylonitrile from propane. However, sometimes the initial step of the reaction (propane to propene) is not truly catalytic, but occurs in the gas phase (radical mechanism) after hydrocarbon activation at the reactor walls or at the catalyst surface. The discussion in this chapter regarding propane ammoxidation is thus restricted to reaction conditions and catalysts (V/Sb mixed oxides) for which there is more clear evidence of the presence of a true catalytic transformation on the surface, because there are serious problems with homogeneous–heterogeneous mixed processes in terms of scale-up and control of the reactivity in moving from laboratory models to pilot plant reactors.

Other alkane selective oxidation reactions such as oxidative dehydrogenation and ethane, propane, iso-butane, and n-pentane selective oxidation are discussed in Chapter 5, because for these reactions catalytic data obtained to date indicate the necessity for further development of the catalyst before possible exploitation can be considered. The stage of the research and priorities for future studies are thus different.

The discussion in this chapter is focused mainly on the key catalytic aspects of the vanadium/phosphorus oxide catalysts and vanadium/antimony oxide catalysts for n-butane oxidation and propane ammoxidation, respectively. However, the surface reactivity of these catalysts also depends in large measure on the feed composition and reaction conditions, which in turn depend on the choice of reactor technology. The optimal catalyst characteristics are thus strictly related to the reactor engineering aspects. This relationship, often not considered in detail, is highlighted in this chapter, whereas more general aspects of the features of industrial processes and catalysts are dealt with briefly only as a background for the more specific discussion of catalyst properties that determine the selection behavior. We published detailed discussions of industrial catalysts and processes for n-butane oxidation and propane ammoxidation some years ago.[1–3]

Several aspects of the V/P oxide catalysts for n-butane selective oxidation such as the chemistry of preparation, the nature of the active phase and active centers, the reaction mechanism, structure–activity relationships, reaction kinetics, and the role of dopants have been analyzed over the past decade in a number of reviews.[1–12] The propane ammoxidation reaction and relevant catalysts have not been studied nearly as much, but the main aspects of the nature of active catalysts, kinetics, and reaction mechanism have been reviewed.[13–17] These will not be dealt with specifically in the present work, as this discussion focuses on a detailed analysis of specific characterizing aspects of the surface catalytic chemistry of these catalysts, the open problems, and opportunities for future research.

This "nontraditional" approach to the catalytic chemistry of mixed oxides offers the advantage of presenting an alternative view of the surface reactivity of these catalysts and the factors for its control, which will lead to a better understanding of the key concepts governing the catalytic behavior. It thus offers a better methodological approach to reducing the interference between single specific aspects of a catalytic system and its general features, which will help toward a better understanding of the behavior of other catalysts and/or reactions as well as toward the design of better catalysts.

A challenge for catalysis is, in fact, the development of more comprehensive theories to explain catalytic phenomena, but the first step is to distinguish the common surface phenomena, which requires delineating the basic features that make control of surface catalytic transformations possible and identifying the specificity of a single catalyst and reaction.

4.2. MALEIC ANHYDRIDE FROM *n*-BUTANE ON VANADIUM/PHOSPHORUS OXIDES

4.2.1. Industrial Processes of Maleic Anhydride Synthesis from *n*-Butane

Maleic anhydride was originally produced almost exclusively by oxidation of benzene. Since 1974, however, benzene has been replaced by other feedstocks, first by a C4 fraction containing alkenes, and more recently by *n*-butane. The primary reasons for replacing benzene with *n*-butane are:

1. The high price of benzene (the cost of *n*-butane depends considerably on where the plant is located, but often its cost is not very different from that of fuel).
2. The loss of two carbon atoms in going from benzene to maleic anhydride (differently than in the case of *n*-butane) with a consequent lessening of the yield by weight.
3. The environmental problems associated with benzene, which is classified as a carcinogen and thus requires severe and costly process control for environment and safety monitoring.
4. Product quality: benzene gives heavy by-products such as phthalic anhydride and benzoquinone, whereas *n*-butane oxidation is a very clean reaction with minimal formation of by-products (apart from carbon oxides only acetic acid).

Plants that use benzene can be retrofitted to accommodate n-butane with relatively little effort, but productivity is lower. To operate with n-butane outside the flammability zone, its concentration in air must be reduced to about 1.8%. However, productivity can be increased by modifying the reactor, either fixed-bed or fluidized-bed, while leaving separation and product recovery devices nearly unchanged.

The demand for the intermediate maleic anhydride is expected to grow at an annual rate of around 3%. Several new plants have been constructed or announced in recent years. The main uses of maleic anhydride and its derivatives are summarized in Table 4.1.[18,19] Derivatives of growing importance are those obtained by hydrogenation (butandiol and γ-butyrolactone). The Davy McKee process for the hydrogenation of diethyl maleate is the most competitive one for the production of these intermediates.[20]

The main features of the industrial technologies currently used for maleic anhydride production are summarized in Table 4.2.[1-3] The various processes differ in terms of:

1. *Type of reactor*: Fixed-bed, fluidized-bed, or transport-bed reactors.
2. *Composition of feedstock*: In particular, the concentration of n-butane is less than 2.4% for fixed-bed reactors, in the 3.6–5% range for fluidized-bed reactors, and more than 4% for transport-bed reactors.
3. *Product recovery*: Method of recovery of maleic anhydride (aqueous or organic solvent), method of purification of maleic anhydride (azeotropic batch, continuous distillation, or thin-layer evaporator), and purity of maleic anhydride.

Although in all processes, a vanadyl pyrophosphate-base catalyst is used, its specific characteristics in terms of the nature and quantity of the promoters, method of preparation and textural/mechanical properties, shape and characteristics of the

TABLE 4.1. End Uses of Maleic Anhydride[18]

End uses	Europe, %	North America, %	Japan, %
UPR	55	51	38
Fumaric and maleic acids	6	7	11
Agricultural chemicals	6	6	—
Additives for oils	5	12	3
Alkydic resins	6	—	—
γ-Butyrolactone	—	—	4
Detergents	6	—	—
Copolymers	—	8	7
Others	16	16	17

TABLE 4.2. Industrial Technologies for Maleic Anhydride Synthesis[1]

Process, licensor	Type of reactor	Product recovery
ALMA (Lonzagroup, Lummus)	Fluidized bed	Anhydrous
Lonzagroup	Fixed bed	Anhydrous or aqueous
BP (Sohio)-UCB	Fluidized bed	Aqueous
Denka Scientific Design	Fixed bed	Aqueous
DuPont	Transport bed	Aqueous
Mitsubishi Kasei	Fluidized bed	Aqueous
Monsanto	Fixed bed	Anhydrous

pellets (and additives to obtain the required properties), procedures of catalyst activation, and regeneration are different for the various processes. As a consequence of all these variables, the various processes perform quite differently in terms of selectivity and productivity, even though all of them use the same type of catalytic active phase.

4.2.1.1. Gas Phase Composition

The composition of the feedstock is directly related to the choice of the reactor technology. In fact, an increase in the concentration of n-butane is economically favorable because it increases productivity without affecting air-compression costs and also facilitates the recovery of maleic anhydride and treatment of gaseous emissions, although owing to saturation of the active sites, the rate of n-butane depletion does not increase linearly for n-butane concentrations above about 2% (this value, of course, depends considerably on the specific V/P oxide catalyst). However, the maximum concentration of n-butane that can be used depends on: (i) the flammability limits (the lower and upper flammability limits for n-butane at room temperature are, respectively, 1.85 and 8.5% in air and 1.8 and 4.9% in oxygen), and (ii) temperature control (possibility of runaway, i.e., that the formation of carbon oxides—highly exothermic reactions favored by an increase in reaction temperature—is no longer controlled, leading to a continuous increase in reactor temperature going beyond the safety threshold).

For fixed-bed reactor processes with air, safe operation requires a concentration of n-butane of less than 1.85% upstream, but owing to a decrease in the flammability limits with increasing temperature (the flammability limit falls from 1.85% at room temperature to 1.22% at 400°C) the effective concentration of n-butane in the preheated mixture must be lower or special precautions must be adopted in the preheater (e.g., the use of an inert space-filling material such as silica).

Downstream from the catalyst, the flammability limit must be calculated for a mixture of several flammable components, including maleic anhydride, carbon monoxide, and unconverted n-butane. For a conversion of 80% and a selectivity of

70% at 400°C, the n-butane concentration at the reactor inlet (for an n-butane/air mixture) must be lower than 2.3% to avoid the possibility of a flammable mixture at the reactor outlet.[21] Note that the control of accidental explosions downstream from the reactor and before the outlet quenching is much more critical than at the inlet, because the possibility of flame propagation in the catalyst bed is minimal (it acts as a flame arrester).

Oxygen concentration in the reactor effluent is also important. For an inlet concentration of n-butane of about 3.6% in air, the oxygen outlet concentration (for 80% conversion and 70% selectivity) is about 8%, a concentration below which no flame will ignite for any concentration of organic material.

These data, in combination with considerations on stable operation (below the critical n-butane concentration/reactor temperature region to avoid possible runaway phenomena[22]) and data on the dependence of the rate of n-butane depletion on n-butane and oxygen concentrations, indicate that: (i) for fixed-bed reactor operations the upstream n-butane concentration must be in the 1.5–1.8% range; (ii) for fluidized-bed reactors the n-butane concentration must be about 4% (in fluidized-bed reactors n-butane and oxygen/air are mixed inside the catalytic bed, which acts as an inhibitor for flame propagation), but lower concentrations (in the 2.3–3.6 range) or the use of oxygen-enriched air can create problems of flammability of the outlet stream; (iii) when unreacted n-butane is recycled, the inlet oxygen concentration must be regulated using a ballast to maintain an oxygen outlet concentration below 8%; and (iv) for transport-bed reactors (CFlBR-type), where O_2 and hydrocarbon contact with the catalyst occurs in separate reactors, the concentration of n-butane is dictated only by kinetic considerations.

4.2.1.2. Reactor Technologies

The choice of the reactor determines the conversion–yield relationship, the control of the heat of reaction (and thus the feed composition), and the feasibility of catalyst regeneration. The key features of the different types of reactors used in the different processes of maleic anhydride synthesis from n-butane (Table 4.2) are summarized in Table 4.3.[23]

Fluidized-bed reactors have a number of advantages over fixed-bed reactors, including: (i) excellent heat transfer and no hot spot; (ii) high tolerance for feed impurities; (iii) safe operation with inlet gas compositions within the flammability limits; (iv) higher productivity; (v) improved turndown, turnoff, and flexibility; (v) reduced limitations for heat and mass intragranular transfer; (vi) shorter downtime for catalyst replacement; (vii) constant catalyst performance (continuous catalyst makeup is possible); (viii) production of higher-value steam; (ix) single-train operation up to 50,000 tons/year (about twice that for fixed-bed); and (x) investment cost advantage for production of higher than about 10,000 tons/year. On the

TABLE 4.3. Key Features of Different Types of Reactor Technologies Used in Maleic Anhydride Synthesis from n-Butane[23]

Feature	Fixed-bed reactor	Fluidized-bed reactor	Transport-bed reactor
Gas phase hydrodynamics	Plug flow and longitudinal dispersion	Mixed flow	Plug flow and minor longitudinal dispersion
Solid hydrodynamics	—	Mixed flow	Plug flow and longitudinal dispersion
Catalyst size	2–5 mm	0.03–0.06 mm	< 0.02 mm
Catalyst holdup	50–60%	15–25%	< 10%
Gas–solid effective contact	Fairly good	Good	Very good
Catalyst effectiveness	Fairly good	Good	Very good
Fluid residence time	Very good	Good	Fairly good
Temperature control	Fairly good	Very good	Good
Catalyst temperature control	—	Good	Very good
Catalyst regeneration	—	Very good	Very good
Inlet catalysis oxidation control	—	Fairly good	Very good
Reaction control along reactor axis	Very good	Fairly good	Very good

other hand, fluidized-bed reactors suffer from: (i) lower selectivity because of backmixing; (ii) the necessity of making hard catalyst particles (must be mechanically resistant to high attrition conditions) using additives that often lower catalyst performance; (iii) greater catalyst inventory; and (iv) problems in scale-up.

Transport-bed reactors, e.g., CFlBRs, operate in the hydrodynamic regime of transport of the catalyst with the feed (n-butane/inert virtually without oxygen, although a small quantity of oxygen is added to improve productivity) in the riser reactor with regeneration of the catalyst (with air) in a separate reactor (usually a fluidized-bed reactor). The solid catalyst circulates continuously from the riser to the regeneration reactors. In the riser, it moves upward with high fluxes (usually 50–10,000 kg/m^2) in a stream of high-velocity gas (generally 2–10 m/s).[24–26] The use of CFlBRs in catalytic oxidation allows decoupling of the two stages of the redox mechanism (reduction of the catalyst by hydrocarbon and catalyst reoxidation by oxygen) and limiting (avoiding) possible side reactions due to adsorbed oxygen species. With respect to conventional CFlBRs, the one used in maleic anhydride production from n-butane by the DuPont process (Table 4.2) is characterized by a denser bed in the riser reactor[27,28] with respect to the core–annulus setup found, e.g., in FCC reactors.

Emig et al.[29] examined the effect of operating parameters on risers of different dimensions designed to produce 20,000 tons/year of maleic anhydride and found that the economics of the process depend largely on the long residence time required for catalyst reoxidation. Large quantities of circulating catalyst are necessary to achieve acceptable productivities. The amount of maleic anhydride produced is of the order of 1 g per kg of circulating solid catalyst, there are high energy costs associated with this high recirculation rate. In addition the rate of catalyst makeup is high.

The advantages of CFlBR technology using a dense-bed transport reactor with respect to fluidized-bed and fixed-bed reactors are: (i) higher throughput, because of the higher gas velocity and the ability to independently optimize the kinetic conditions for the two stages of the redox mechanism; (ii) higher selectivity; (iii) excellent intragrain and interphase heat and mass transfer; (iv) high turndown (solid and gas retention times can be adjusted independently); (v) lower catalyst inventories (it is possible to optimize conditions for reduction and reoxidation, which occur in physically separated reactors); (vi) higher product concentration (improved recovery); and (vii) greater intrinsic safety. However, there are disadvantages: (i) the need for a catalyst with an exceptionally high attrition resistance; (ii) significant uncertainty in scale-up; (iii) the need for temperature control in the riser; (iv) the very high energy costs associated with catalyst recirculation; (v) high catalyst makeup; (vi) possible side reactions (condensation and polymerization) that occur in the absence of oxygen; (vii) low productivity per mass of circulating solid (a factor of about 10^{-3}); and (viii) production of significant quantities of carbon oxides in the regeneration unit.

As can be inferred from Table 4.2, it is thus not possible to indicate a clear reactor-type preference for the synthesis of maleic anhydride from n-butane, but different options may exist depending, e.g., on the up- and downstream integration of the process and the cost of n-butane. It may be noted, however, that the choice of reactor determines reaction conditions, which in turn affect the characteristics of the "optimal" catalyst, and thus catalyst optimization is not an independent variable with respect to the choice of the final process reactor technology.

4.2.1.3. Catalyst Formulation

In general terms, an empirical formula that represents all the catalyst formulations is $VP_aMe_bO_x$/inert. Patents describe a wide range of values of a (0.8–1.5) and a broad spectrum of promoters, including Li, B, Zn, Mg, Co, Fe, Ni, Cr, Ti, Mo, W, Bi, Zr, and rare earths. However, the preferred catalyst compositions are:

- a: 1.03–1.2
- b: 0–0.1, where Me = one or more of Zn, Li, Mo, Zr, or Fe.
- x: balances the positive charge of all the other elements.
- $inert$: colloidal silica, 0–10 wt% with respect to active component.

Older catalyst formulations were always prepared in an aqueous solvent; more recently, however, patents specify organic solvents. Hatano et al.[30] also claim the use of hydrothermal conditions for the synthesis of active material.

Of the various possible ways to improve the mechanical resistance properties of the particles, two that are among the more effective and do not cause a significant decrease in catalytic performance and (i) adding small amounts of additives such as B to the precursor, and (ii) encapsulating the active component in a thin shell of silica.[26]

Activation and regeneration procedures are also a very important part of the know-how of the different companies. Usually, the preparation method for V/P oxide catalysts leads to a precursor (a hydrated vanadyl hydrogenphosphate), which is then activated in a flow of air or nitrogen at high temperature within or outside the reactor. However, the real activation occurs inside the reactor in the flow of hydrocarbon/oxygen. Different companies have complex and long (sometimes up to 1 month) activation procedures. The usual initial conditions are a low n-butane concentration (about 0.4%) and space-velocity (about 500 h^{-1}), which are then progressively increased. The activation procedure leads to a decrease in the rate of n-butane depletion and a significant increase in selectivity. The opposite occurs during long-term deactivation.[31,32] Deactivation of the V/P oxide catalysts is not well understood, but is probably associated with loss of phosphorus and catalyst overoxidation. Catalyst regeneration, in fact, is achieved by high-temperature reduction with a flow of concentered n-butane (up to 50% in nitrogen)[33] or of n-butane/oxygen (oxygen concentration lower than the stoichiometric value; the presence of oxygen according to Blum et al.[34] is necessary for catalyst surface restructuring) and organic phosphorus compounds are added (continuously or during batch regeneration) to maintain catalyst performances.[35-38]

Selected performances of the various catalysts used in maleic anhydride production from n-butane are shown in Table 4.4, although clearly data for a homogeneous comparison are not available. The data, however, give a good idea of the catalytic behavior of the catalysts and the processes. They also provide a good comparison when analyzing the catalytic data reported in the literature and estimating the validity of assumptions concerning the nature of the active phase/sites, with reference to catalysts showing much worse performances than would be possible with better prepared samples. Owing to the difference in the type of process, data for the DuPont process are not included in Table 4.4, being not directly comparable. As a reference, a maleic anhydride selectivity of 75% was obtained at n-butane conversion up to 50% for hydrocarbon concentrations of 1–50%. At higher conversions, selectivity dropped to 60% at 90% conversion.[24] The amount of maleic anhydride produced per kilogram of catalyst is of the order of 0.5–2 g, depending on the butane concentration in the feed. Based on these data, a catalyst circulation rate of 0.5 ton/s is required for a maleic anhydride plant with a capacity of about 15,000–20,000 tons/year.

TABLE 4.4. Catalytic Behavior of VPO Catalysts in n-Butane Selective Oxidation to Maleic Anhydride Reported in Patents in Relation to the Different Types of Industrial Processes

Process	Catalytic data			n-Butane concentration, %	Note	Reference
	Temperature, °C	Conversion, %	Yield, %[a]			
ALMA	407	81.8	57.8	4.6	Fluidized-bed reactor (lab. scale, 1 kg catalyst)	39
Amoco	417	~90	51.4	1.51	Fixed-bed reactor [lab scale, 1 g catalyst, space velocity 2000 h^{-1}, presence of 3 ppm P and 1% H_2O in feed); case (1) once-through after 230 days time-on-stream; case (2) recycle mode after 282 days time-on-stream	32
	407		66.3	4.63		
BP-UCB	~410	90.1	56.1	~4	Fluidized-bed reactor (lab scale, ~1 kg PVO catalyst with P/V = 1.2 and without silica or additives), data after 116 h time-on-stream	40
Denka-Scientific Design	381	81.3	57.2	1.28	Fixed-bed reactor, data after 1600 h time-on-stream, space velocity 2500 h^{-1}, $Zn/Li/Mo$-doped VPO in the form of ~6 × 6 mm tablets with internal hole	41
Mitsui–Toatsu	430	93.0	56.5	2	Lab scale reactor (1.5 g catalyst), Mg-doped catalyst [P/(Mg+V)= 1.05, Mg/(V+Mg) = 0.05]	42
Mitsubishi–Kasei	450	90.7	54.5	4	Fluidized-bed reactor, space velocity 1000 h^{-1}, catalyst prepared under hydrothermal conditions, and containing iron (P/V + Fe) = 1.0.	30
Monsanto	435	82.0	54.0	2.4	Fixed-bed, $P_{1.2}V_{1.0}Fe_{0.0018}Li_{0.003}O_x$ diluted catalyst, space velocity 1220 h^{-1}, data after 8 months time-on-stream	43

[a] Molar to maleic anhydride

Some further general aspects of the reaction, kinetics, process options, and historical development have been discussed by Burnett et al.,[21] Contractor and Sleight,[26] Chinchen et al.,[44] Emig and Martin,[45] and Stefani et al.[46,47]

4.2.2. V/P Oxide Catalysts Synthesis and Characteristics

The catalytic behavior and structural, surface, and textural characteristics of V/P oxide catalysts depend on the following factors:

1. Method of preparation of the precursor (type of reagents, reducing agents, and solvent; temperature; use of dry or wet milling).
2. Ratio of phosphorus to vanadium.
3. Use of additives.
4. Activation and conditioning procedure for the precursor at high temperature.

4.2.2.1. Role of the Precursor Phase

The preparation of the precursor is a very critical step. There is a broad consensus[7,48–57] that the conditions necessary to obtain an optimal catalyst are: (i) the synthesis of microcrystalline $(VO)HPO_4 \cdot \frac{1}{2}H_2O$ in an organic medium, with the crystals characterized by preferential exposure of the basal plane (001); (ii) the presence of defects in stacking of the platelets; and (iii) the presence of a slight excess of phosphorus with respect to the stoichiometric ratio of 1.0 (excess phosphorus, which is not removed by washing the precipitate). All preparations result in essentially only one phase. The major differences between the various precursors relate to the morphology of the $(VO)HPO_4 \cdot \frac{1}{2}H_2O$ crystallites. Precursors prepared in an aqueous medium are more crystalline, and exposure of the basal plane (001) is less pronounced; no preferential line-broadening of the reflection corresponding to the (001) plane is observed.[48,58]

The morphology of the precursor depends on various other factors: (i) the nature of the solvent (aliphatic or aromatic alcohol)[50,53]; (ii) the ratio of phosphorus to vanadium[50]; (iii) the nature of the reducing agent for the V^{5+} starting oxide (usually V_2O_5), which, in some cases, is the solvent itself[50,53]; (iv) the properties of the V^{5+} compound, such as the morphology and particle size[59,60]; (v) reduction time, temperature, and amount of water[50,53]; (vi) use of additives such as ethyl orthosilicate[50]; and (vii) removal of excess phosphorus by washing.

The preparation chemistry of this phase is thus very complex. Indeed, with isobutanol, nonagglomerated platelets (rosette morphology) form with preferential exposure of the basal (001) plane,[50] whereas using secondary or tertiary butyl alcohol or isopropyl alcohol, well-formed nonagglomerated platelets form. With benzyl alcohol, platelets with stacking faults form with the alcohol trapped between the layers.[58] Miyake and Doi[61] have reported that the morphology of the precursor changes progressively using as the solvent isobutyl, cyclohexyl, isopropyl, 2-ethyl-1-hexyl, 2-methyl-1-butyl, and 2-butyl alcohols, although unlike other authors they indicated a direct relationship between the exposed face (and thus morphology) and selective synthesis of maleic anhydride.[54,62] They observed that only surface area (and thus catalyst activity) is related to catalyst morphology—but not the selectivity.

Different types of precursor phases have also been reported to give results comparable to those obtained from the $(VO)HPO_4 \cdot \frac{1}{2}H_2O$ precursor phase, but a more detailed inspection of the data indicate that this claim is not true. Bethke et al.[63] reported that $VO(H_2PO_4)_2$ derived catalysts can give rise to active and selective catalysts, but better results were found at 485°C with 48% selectivity to maleic anhydride at 25% n-butane conversion, a rather poor result in comparison with the behavior of well-prepared catalysts derived from the $(VO)HPO_4 \cdot \frac{1}{2}H_2O$ phase, which usually show a selectivity in the 70–85% range at the same level of n-butane conversion. Furthermore, the true difference in the catalytic behavior can be seen at high n-butane conversion (>70%) and thus catalytic data at lower conversions may be not representative, especially when speculation as to the nature of the active phase are based on these results. Hutchings et al.[64,65] also reported that ultra-selective (100% selectivity) catalysts can be obtained from the $VO(H_2PO_4)_2$ precursor phase, but the catalytic data are reported only for such a low conversion of n-butane that they are of little significance when the experimental error is taken into account. As a matter of fact, even in the work[66] that followed the preliminary communications, no attempt is made to obtain catalytic data under more representative conditions, thus leaving a big question mark as to the real validity of the reported data.

Hutchings et al.[66] also claim the reduction of $VOPO_4 \cdot 2H_2O$ as another method of preparation to obtain "improved" vanadium phosphate catalysts, but the reduction of $VOPO_4 \cdot 2H_2O$ leads to the $(VO)HPO_4 \cdot \frac{1}{2}H_2O$ precursor phase or other phases, depending on the nature of the alcohol used for the reduction. In comparison with the use of V_2O_5 as the starting compound for the synthesis of the $(VO)HPO_4 \cdot \frac{1}{2}H_2O$ precursor phase, the use of $VOPO_4 \cdot 2H_2O$ leads to more active catalysts owing to the bigger surface area, but large-surface-area samples can also be obtained from the reduction of V_2O_5, when smaller particles are used, especially when the amount of water in solution is controlled (water forms as an alcohol reduction product, but its amount can be controlled by extractive distillation during the synthesis of the precursor phase). This indicates that the nature of the V^{5+} starting sample is not the key point, but rather the details of the procedure of its reduction, which must be optimized as a function of the nature of the starting material. Much more care must be taken when a finding is generalized outside the limits of validity of the specific data with respect to all the background information, especially in the case of complex situations such as those present in the preparation of V/P oxide catalysts.

A different and more interesting type of approach has been reported by Benziger et al.,[67] who started from the observations that: (i) the vanadyl hydrogen phosphate hemihydrate $[(VO)HPO_4 \cdot \frac{1}{2}H_2O]$ precursor phase has a layer-type structure, (ii) the transformation to the active vanadyl pyrophosphate phase is a topotactic reaction,[68,69] and (iii) the nature of the organic solvent/reductant during the synthesis of $(VO)HPO_4 \cdot \frac{1}{2}H_2O$ influences the presence of stacking faults,[50,70] which play a positive role on the catalytic behavior,[53] and investigated the possibility of controlling the characteristics of the final active phase by intercalating alkyl amines between the layers of the precursor phase. The results obtained indicate that this approach is an interesting new direction of research.

4.2.2.2. Activation and Conditioning Procedure

There are conflicting and confusing data in the literature about the role of the conditions of transformation of the precursor phase to the active phase, although the patent literature clearly points out that this is a fundamental step in the preparation of V/P oxide catalysts.[56,57] Various phases can result from this transformation, depending on the temperature, time and atmosphere, precursor morphology, P/V ratio, presence of dopants, and defects in the structure. Two basic types of activation procedures have been described in the technical and patent literature:

1. Activation in an oxygen-free atmosphere at temperatures higher than 400°C, followed by introduction of the reactant mixture (n-butane in air): With this procedure, pure crystalline vanadyl pyrophosphate is formed after the first step,[50,71] and then, after addition of the reaction mixture, partial oxidation may occur.
2. Single or multistep calcination in air at temperatures lower than 400°C, followed by introduction of the reactant mixture[49,50,53]: After calcination at 280°C, the precursor is still present, even while the trapped alcohol is being released, which disrupts the structure and increases the surface area. At temperatures between 380 and 400°C, the precursor starts to decompose to an amorphous phase containing both V^{4+} and V^{5+}. After introduction of the reactant mixture, this amorphous phase can be dehydrated and transformed to vanadyl pyrophosphate and/or other oxidized phases.

The general scheme of formation and transformation of the various V/P oxide catalyst phases is summarized in Scheme 4.1. Different types of $VOPO_4$ phases, some of which can be reversibly reduced to vanadyl pyrophosphate, have been identified.[7,11,59,72–75] The oxidation of vanadyl pyrophosphate and the intermediate amorphous phase to a V^{5+} phosphate must be avoided, because it has a deleterious effect on activity and selectivity.[72–74,76–79]

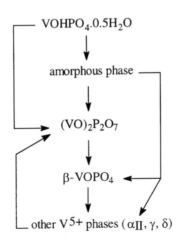

Scheme 4.1. Structural changes and interconversion of VPO phases during the activation of $(VO)HPO_4 \cdot \frac{1}{2}H_2O$.[56]

The degree of crystallinity (morphology, presence of stacking faults, preferential exposure of crystalline planes, surface area) of the precursor phase (vanadyl hydrogen phosphate hemihydrate) has a considerable effect on the structural evolution during heating and the final properties of the vanadyl pyrophosphate. The transformation of the precursor to $(VO)_2P_2O_7$ involves two water-loss steps, the first associated with water of crystallization and the second with the transformation of hydrogen phosphate to pyrophosphate. The two transformations can occur simultaneously or consecutively, depending on the nature of the precursor phase and the heat treatment.[80] When oxygen is present in the atmosphere, there is also oxidation, leading to various possible V^{5+}-phosphate phases (α_I, α_{II}, β, γ, δ) again depending on both the nature of the precursor phase and the conditions of the heat treatment. Thus there is a broad range of possibilities in terms of catalyst composition after the heat treatment (calcination) step, with identification of some of the phases (being amorphous or well dispersed on the vanadyl pyrophosphate phase) also being difficult.

Many of the studies on the structure–activity and selectivity relationships on V/P oxide catalysts have focused on identification of the different phases present in the catalyst after this heat treatment, with the aim of correlating the types and amounts of phases present to the catalytic behavior and thus identifying the active phase(s). An analysis of this question was reported in the commentary papers to a special issue of *Catalysis Today* devoted to vanadyl pyrophosphate catalysts.[9] Varying the conditions of preparation of the precursor phase and the heat treatment changes not only the phase composition, but also a series of other equally, if not more, important characteristics, such as the presence of defects, surface exposure of crystalline phases, and texture. It is thus often very difficult to definitely associate the catalytic behavior solely with the presence of a certain phase composition. Indeed, the same authors have reported contrasting indications on the nature of the active phase(s).[9]

A better approach is to correlate the change in the *in situ* surface characteristics with the parallel change in the catalytic behavior for the same catalyst, although care must still be taken to distinguish between the final catalyst transformation and a transient step in the transformation. This is made more difficult in the case of V/P oxide catalysts by the long time-on-stream required to reach the final stage of catalyst transformation during catalytic runs (sometimes more than 500 h) and by the fact that the "adaptation" of catalyst characteristics to the feed composition may lead to different final conditions. The importance of the *in situ* slow transformation process of V/P oxide was recognized in the literature only recently,[81–86] although the problem had been recognized much earlier.[7] This *in situ* change in the catalyst is often referred to as the change from "nonequilibrated" to "equilibrated" catalyst characteristics.

As a reference model to understand the final characteristics of "equilibrated" V/P oxide, it is also useful to analyze the properties of industrial V/P oxide catalysts after at least 1000 h of time-on-stream. There are very few data in the literature concerning this point, but Ebner and Thompson[87] defined the features of an "equilibrated" industrial catalyst (Monsanto process) that was kept in a flow of *n*-butane with a concentration of 1.4–2.0% in air and at least GHSV 1000 h^{-1} for approximately 200–1000 h:

- Average degree of oxidation of vanadium: 4.00–4.04
- Bulk phosphorus to vanadium ratio: 1.000–1.025
- P/V atomic ratio from XPS data: 1.5–3.0
- BET surface area: 16–25 m^2g^{-1}
- Crystalline phases from XRD: only reflections of vanadyl pyrophosphate
- TEM analysis: rectangular platelets and rodlike structures

Although no further clear data on industrial catalysts are available to verify these indications, they may be considered as being representative for all "equilibrated" V/P oxide catalysts. It should be noted, however, that more recent XPS studies estimate the surface P/V ratio have shown that there is no gross phosphorus surface enrichment, but that the value is much closer to the bulk value (14% phosphorus enrichment when the surface is assumed to terminate with pyrophosphate groups).[88]

Another important feature of the change in V/P oxide catalysts from the nonequilibrated to the equilibrated condition is the considerable change in the redox characteristics. Equilibrated catalysts cannot be reoxidized in air at 400°C, unlike freshly prepared vanadyl pyrophosphate or nonequilibrated catalyst, although equilibrated catalysts show higher selectivity than nonequilibrated ones.[89-91] This is clear evidence of the limited validity of several investigations on the relationship between catalyst redox properties and catalytic behavior,[92-97] and indicates as well the important methodological point that not only must one relationship be found, but that what happens outside of the validity of the specific relationship must be especially verified. The fact that there are such limited redox properties (in terms of dynamic surface restructuring during the catalytic reaction) of equilibrated V/P oxide catalysts indicates that there is no phase cooperation between V^{4+} and V^{5+} phosphate, as suggested from data on nonequilibrated samples, but rather that selective catalysis involves oxygen species formed at the coordinatively unsaturated vanadyl surface ions. This point will be discussed in detail in Chapter 8 in the section on the nature and catalytic role of surface oxygen species.

4.2.2.3. Role of the P/V Ratio and Catalyst Redox Properties

The optimal catalyst composition contains a slight excess of phosphorus with respect to the stoichiometric value of 1.0.[50,75,84,98] XPS analysis shows a higher P/V ratio on the surface than in the bulk,[46,87,99] although as pointed out by Coulston *et al.*[88] and Okuhara *et al.*[98] these values could be affected by incorrect calibration procedures. Secondary ion mass spectrometry (SIMS) analysis gives a surface P/V ratio greater than 5, with a bulk P/V ratio of less than 1.1,[100] values that can be explained if the surface is assumed to terminate with pyrophosphate groups.

It is generally agreed that a high P/V ratio helps to avoid the oxidation of V^{4+} in vanadyl pyrophosphate.[7,46,59,75] However, the excess of phosphorus also influences the reducibility of the catalyst. Cavani *et al.*[101-103] observed that both the reducibility of vanadium and its reoxidizability decreases with increasing P/V atomic ratio, but not in a parallel way. High oxidizability of the catalyst lowers selectivity, but a low catalyst reducibility lowers activity. Thus the optimal P/V ratio was found to be in the range 1.0–1.1.

Reducibility of V^{5+} to V^{4+} (or the inverse reaction) is not the only important parameter of catalyst behavior; the V^{4+} to V^{3+} reducibility is significant as well, although this aspect has often not been properly considered. We[7,102-106] observed the formation of V^{3+} during the catalytic reaction and the amount of V^{3+} depended on the feed composition and thus changed along the axial direction of the fixed-bed reactor. More recently Rodemerck *et al.*[107] demonstrated that the V^{4+}/V^{3+} redox couple plays a catalytic role in equilibrated $(VO)_2P_2O_7$.

The increase in the P/V ratio above 1.0 thus stabilizes the $(VO)_2P_2O_7$ phase against reoxidation to V^{5+}-phosphate phases, but the effect is more complex and factors other than the rate of this transformation also are affected. It is interesting to note that Coulston *et al.*[88] observed that the surface structure and P/V surface ratio of V/P oxide catalysts change depending on whether they are analyzed after butane interaction or reoxidation with molecular oxygen. Gai and Kourtakis[108] demonstrated structural modifications induced by a reducing environment, and using *in situ* ESR measurements Brückner *et al.*[109] showed the continuous catalytic restructuring induced by the redox changes clearly. Thus a change in the redox properties of the catalyst affects not only these properties, but also the intrinsic nature of the catalyst surface during the catalytic reaction as well as the possibilities of its dynamic *in operandi* restructuring. The presence of complex phenomena of surface restructuring during the catalytic reaction indicates that statements such as "V^{5+} species (with a low density) on the $(VO)_2P_2O_7$ matrix constitute the active sites" and "the influence of the oxidation state of a V/P oxide catalyst is demonstrated,"[11] are not true because various aspects of the surface chemistry and reactivity change at the same time by influencing the catalyst redox properties.

4.2.2.4. Role of Promoters

Dopants for V/P oxide catalysts can be classified into two main groups according to Hutchings, who reviewed data up to about 1992[110,111]:

1. Those that promote the formation of the required V/P oxide phase or avoid the formation of spurious phases.
2. Those that form solid solutions with the active phase and regulate the catalytic activity.

However, this classification does not clarify the mechanism by which these dopants interact with the V/P oxide. Some information concerning this point was reported in subsequent papers, but the general question of the real role of dopants remains largely unsolved.

Zazhigalov *et al.*[112] studied the effect of Co doping, and suggested that it leads to the formation of Co-pyrophosphate. This phase fills the micropore voids and macrodefects and hinders phosphorus transport from the bulk to the surface and its gas phase loss. Bej and Rao[113] analyzed the effect of Ce and Mo doping and indicated that these dopants block the formation of nonselective sites (possibly V^{5+}) responsible for maleic anhydride combustion at high *n*-butane conversion. Sananes *et al.*[114] studied the effect of Zn, Ti, and Zr additives and did not find changes in the V/P oxide phases, but only an enrichment of these additives on the surface, which, when added in very small amounts, promoted the rate of *n*-butane depletion. Ye *et al.*[115] studied various additives (Ge, Mo, La, Ce, Zr) and suggested that these promoters increase both the exposure of the (100) crystalline plane and the number of defects in this plane. Meisel *et al.*[116] showed that when sulfuric acid is present during crystallization of the $(VO)HPO_4 \cdot \frac{1}{2}H_2O$ phase, sulfur may partially substitute for phosphorus, leading to stacking faults. In addition, S partial substitution in V/P oxide leads to the promotion of the catalytic behavior similar to the enhanced catalytic behavior of samples obtained in an organic medium showing the presence of stacking faults. The effect of Cr was analyzed by Harrouch Batis et al.,[117] who found that it enhances the presence of amorphous V/P oxide phases, but that in large quantities on the surface it also catalyzes maleic anhydride oxidation. Cheng[118] studied the effect of In and TEOS (tetraethylorthosilicate) as promoters, and found that they reduce the thickness of the platelets and facilitate the oxidation of the precursoru, which contains disordered $(VO)HPO_4 \cdot \frac{1}{2}H_2O$. The influence of alkali and alkaline earth metal ions was analyzed by Zazhigalov et al.,[119] who found an increase in the negative charge on oxygen atoms, which was correlated with an increase in the rate of n-butane oxidation. These metals also influence the acido-base surface properties. Hutchings and Higgings[120,121] studied various promoters such as La, Ce, Cu, and Mo that mainly play a role in decreasing the crystallite size of the $(VO)HPO_4 \cdot \frac{1}{2}H_2O$ precursor phase and thus lead to final catalysts with larger surface areas, which is beneficial only under fuel-lean conditions. The same authors also studied the effect of various additives for catalysts prepared in an aqueous medium. The main effect observed was an influence on the surface area of the samples and thus on catalyst activity. With Mo, unlike with other dopants, promotion of the selectivity was also observed, indicating the possibility of formation of a phase $(MoOPO_4)$ isostructural to $VOPO_4$. The effect of Fe and Co was studied with respect to the effect in n-butane[122] and n-pentane oxidation.[123] These dopants influence the dispersion of $VOPO_4$ phases in activated samples and the morphology of the precursors. The presence of Fe hinders the crystallization of vanadyl pyrophosphate, but this is not the case for Co, which also influences catalyst microstructure. Interestingly the effect of promoters is different for n-butane and n-pentane.

The effect of dopants can be summarized as follows:

1. Addition of ions that interact with free phosphoric acid as a means of fine-tuning the optimum surface P/V ratio and acidity. Examples are basic ions such as Zn that also act to avoid the migration and loss of P.
2. Addition of ions that can substitute for P in the precursor, such as Si and S. The partial or total elimination of these ions by calcination from the vanadyl pyrophosphate structure influences the morphology and leads to defects in the $(VO)_2P_2O_7$ structure.
3. Addition of elements that substitute for vanadium and thus can act as modifiers of the reactivity, forming stable solid solutions. Ti, Zr, Ce, and Mo fall into this category.

4.2.2.5. Structure of the V/P Oxide Phases

There are different V/P oxide phases, with V in the +5, +4, and +3 oxidation state for which structures have been resolved.[68,124]

The V^{5+} phases are hydrated [$VOPO_4\tilde{A}H_2O$,[69] $VOPO_4\cdot 2H_2O$[125]] or dehydrated phosphates (α_I-, α_{II}-, β-, γ-, δ-$VOPO_4$).[68,69,126] The V^{4+} phases are hydrogen phosphates [$VOHPO_4\cdot\frac{1}{2}H_2O$,[127,128] $VOHPO_4\cdot 4H_2O$,[127,129] and $VO(H_2PO_4)_2$[68,130]], pyrophosphate [$(VO)_2P_2O_7$],[131] and metaphosphate [$VO(PO_3)_2$].[68,132] The V^{3+} phases are VPO_4[68,124] and $V(PO_3)_3$.[132]

The structure of $VOHPO_4\cdot\frac{1}{2}H_2O$ is formed by pairs of VO_6 octahedra sharing a common face. Couples of octahedra are connected together through PO_4 tetrahedra, forming the (001) planes. In one octahedron, the $V=O$ bonds are in the *cis* position. Between the (001) planes, H_2O molecules are connected through hydrogen bonds.

$(VO)_2P_2O_7$ shows a structure in which two octahedra share edges. The pairs of octahedra are connected by PO_4 tetrahedra, which gives a layer structure in the (100) plane. Double vanadyl chains [perpendicular to the (100) plane] are present (Figure 4.1). Differently from the precursor phase (vanadyl hydrogen phosphate hemihydrate), the $V=O$ bonds in the pairs of octahedra are in the *trans* position, and the layers are joined by pyrophosphate groups. Vanadium's sixth coordination site when directed inward interacts weakly with the underlying vanadyl group, and when directed toward the surface is a Lewis acid site. Pseudomorphic relations between (001) $VOHPO_4\cdot\frac{1}{2}H_2O$ and (100) $(VO)_2P_2O_7$ planes exist. The idealized surface structure of vanadyl pyrophosphate cleaved parallel to the (100) plane (Figure 4.2) is thus characterized by the presence of the following sites[7]:

1. Coupled vanadyl groups in the *trans* position. The coordinatively unsaturated vanadyl acts as a strong Lewis acid site, whereas the other vanadyl group has a pendent terminal oxygen ($V=O$).
2. Pyrophosphate or phosphate pending groups that are saturated by the creation of Brønsted acid sites of medium-high strength.

The V^{5+} phases are built with isolated tetrahedra connected by PO_4 tetrahedra, although there are some differences in the different phases. For δ- and γ-VPO_4, the structures have been proposed,[68,74] but not solved. The structural frameworks of anhydrous V/P oxide phases can be classified roughly into two groups: (i) single octahedra form single $V=O\cdots V=O\cdots$ chains in the perpendicular direction (α- and β-$VOPO_4$), and (ii) pairs of edge-sharing octahedra [δ- and γ-$VOPO_4$, $(VO)_2P_2O_7$] form double chains. The equatorial oxygens are linked to PO_4 and P_2O_7 groups, respectively. The primary difference between δ, γ-$VOPO_4$ and $(VO)_2P_2O_7$ is the presence of both unicoordinated $(V=O)^{3+}$ vanadyl bonds and double $(V=O)^{2+}\ldots(V=O)^{2+}\ldots$ chains.

There are, however, other possible phases of interest in the catalytic chemistry of V/P oxides, which for the most part have not been considered. A mixed-valence vanadium(III,III,IV) pyrophosphate has been synthesized and characterized from the structural point of view,[133] but its possible surface formation as microdomains

$(VO)_2P_2O_7$

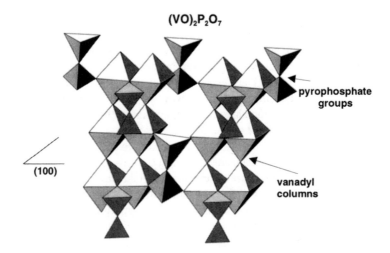

pyrophosphate groups

(100)

vanadyl columns

Figure 4.1. Structure of $(VO)_2P_2O_7$.

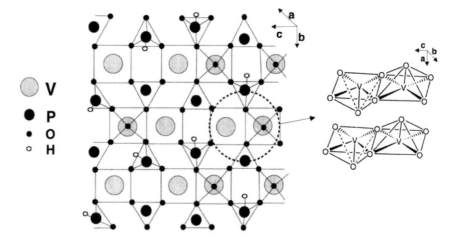

V
P
O
H

Figure 4.2. Idealized surface structure of $(VO)_2P_2O_7$.[7]

during the catalytic reaction has never been considered. Nonstoichiometric mixed valence V/P oxide phases have also been synthesized by Lii and Lee,[134] but never considered as one of the possible surface species.

The possible intercalation of transition metals into hydrated vanadyl ortho-phosphate has been shown by Datta et al.,[135] but never used as a method to introduce dopants into the catalyst in a controlled way. Organic intercalated V/P oxide catalysts are well known compounds,[136–143] but investigations into the possible use of these concepts to synthesize new V/P oxide catalysts have been started only recently.[67]

Amorphous V/P oxide phases are known to be present in V/P oxide catalysts and are known to play a significant role in the catalytic reaction and during *in situ* regeneration processes,[79,144] but only recently have studies been undertaken concerning the identification of the local structure of the amorphous phase(s). Brücker *et al.*[109] have suggested on the basis of *in situ* ESR measurements that a $(V^{3+}V^{4+})PO_4$ amorphous phase forms in a dynamic equilibrium with $(VO)_2P_2O_7$ and that there is a similarity in terms of local symmetry of vanadyl groups between these two phases.

Thus the crystal chemistry of V/P oxide catalysts, especially in relation to the real nature of the phases present and the way to control them has not yet been fully explored, despite the fact that many of the studies on structure–activity relationships in these catalysts have been centered on these aspects.

4.2.3. Advanced Aspects toward Understanding the Catalytic Chemistry of V/P Oxides

4.2.3.1. Role of Catalyst Microstructure and Topology

V/P oxides and in particular the active phase $(VO)_2P_2O_7$ show a unique and unmatched reactivity in the selective oxidation of *n*-butane to maleic anhydride (see previous sections). Furthermore, $(VO)_2P_2O_7$ shows selective behavior in the conversion of *n*-pentane to maleic and phthalic anhydrides (Chapter 5) and no other catalytic materials have been shown to have comparable behavior. On the other hand, $(VO)_2P_2O_7$ shows rather poorer properties in the selective oxidation of propane and ethane and in the oxidation of C6 linear alkanes, whereas in the case of cyclic C6 alkanes it gives rise mainly to oxidative dehydrogenation to benzene. Furthermore, when alkenes (butenes and butadiene) are oxidized on a V/P oxide catalyst selective for *n*-butane and under optimal conditions for maximizing maleic anhydride from *n*-butane, there is much poorer selectivity for maleic anhydride, notwithstanding the general consensus that these are the intermediates from *n*-butane to maleic anhydride.[2,7,9] The catalytic behavior of V/P oxide is also very dependent on the modalities of preparation and in general a structure with more defects appears to be preferable, especially in terms of stacking faults between the

(100) planes. However, the effect is different when n-butane or n-pentane are oxidized on the catalyst, i.e., the change in local structure of V/P oxide catalysts influences the catalytic performances differently depending on the type of alkane oxidized.[56,123]

Several authors have tried to explain the reactivity characteristics of $(VO)_2P_2O_7$ in maleic anhydride synthesis from n-butane in terms of the nature of the sites/phases present on its characterizing surface [the (100) plane]:

1. Cooperation between Lewis acid sites (coordinatively unsaturated V^{4+} ion) and Brønsted sites (P–OH) for the selective activation of n-butane.[145–148]
2. Formation of peroxidic-like species (η^2-type[149,150] or pseudo-ozonide[151]) by O_2 adsorption on coordinatively unsaturated V^{4+} surface cations and their role in selective alkane activation and further oxidation.
3. Presence of V–O–P bonds with the bridging oxygen playing a direct role in the reaction mechanism.[108,152,153]
4. Presence of V^{5+}-phosphate domains on the (100) $(VO)_2P_2O_7$ plane,[74,154] or V^{5+} species with a low density in the $(VO)_2P_2O_7$ matrix.[11]
5. Possibility of V^{5+}/V^{4+} and V^{4+}/V^{3+} one-electron and V^{5+}/V^{3+} two-electron redox reactions.[2,155]
6. Presence of labile lattice oxygen, but limited to the surface layer.[107,156–158]
7. Presence of a specific geometric arrangement of active, undersaturated oxygens around an adsorption site.[54,159,160]
8. Presence of surface equilibrium between V^{5+} and $V^{4+} + p^+$ (p^+ = positrons) as indicated by electrical conductivity measurements.[93] Such an electron vacancy can be filled by an electron hopping from a neighboring anion ($O^{2-} + p^+ \leftrightarrow O^-$) thus generating an oxygen species that is known to be active in alkane oxidation (see Chapter 8).

As discussed in Chapter 8, some of these conclusions are not generally valid and others must be further demonstrated, but they can explain some aspects of the reaction mechanism of n-butane oxidation to maleic anhydride, as well as some of the reasons for the unique reactivity of vanadyl pyrophosphate. However, they cannot explain the other peculiar characteristics of this catalyst outlined above and in Section 4.2.2: (i) why in the C3–C5 alkane series is propane oxidized to carbon oxides and n-butane selectively oxidized to maleic anhydride, while n-pentane forms maleic and phthalic anhydride (thus an eight-carbon-atom anhydride from the C5 alkane, which does not form from the C4 alkane), (ii) why a slight phosphorus enrichment of the surface considerably increases the selectivity but does so in a way that is not proportional to the change in catalyst redox properties, and (iii) why catalyst selectivity, and not only activity, depends on the presence of a defective $(VO)_2P_2O_7$ structure with stacking faults in the (100) planes.

One interesting hypothesis has been advanced concerning the local microstructure and topology of the active catalyst that can explain the above points. Ebner and Thompson[48,87,161] argued that:

1. The truncation of the bulk vanadyl pyrophosphate at the (100) planes may reasonably not end (due to constraints of maximum bond valence preservation for a neutral surface) with protonated orthophosphates but with protonated pyrophosphate moieties (Figure 4.3a), a result in agreement with recent improved XPS measurements.[88]

2. Single-crystal X-ray diffraction studies of vanadyl pyrophosphate indicate the presence of two polytypes, which differ with respect to the symmetry and direction of columns of vanadyl groups (Figure 4.3b).

3. The combination of these two factors (pendent surface pyrophosphate groups and structural disorder) probably results in the formation of surface cavities (Figure 4.3c) that provide steric isolation of the vanadium centers at the bottom of these clefts or cavities and may also limit the mobility of adsorbed species.

Furthermore, the pendent pyrophosphate groups that overhang the ensemble of vanadium sites in the surface clefts present a larger number of hydrogen binding sites, which can facilitate hydrogen acceptance and transport to sites of water formation and desorption and can provide sites to orient the intermediate reaction product.

The way $(VO)_2P_2O_7$ is prepared (nature of percursor and modalities for transformation) influences the degree of disorder and probably affects the orientation of the vanadyl columns, determines the symmetry and dimension of the surface cavities, and determines accessibility to the surface vanadium sites. Since syntheses of phthalic and maleic anhydrides are subject to different space constraints, it is reasonable that a change in the surface topology of the active surface influences each of the two reactions in a different way. It is also reasonable that the oxidation of propane and ethane can be less effective in impeding oxygen coordination on available sites in the surface cleft or cavity, because the products formed have smaller dimensions in the adsorbed phase and are less bound to surface sites (e.g., it is likely that acetic and acrylic acids coordinate top-end, whereas the anhydride may have a planarlike adsorption as well as a multisite coordination). This fact, in addition to the intrinsic higher rate of decarboxylation and further oxidation of the adsorbed intermediates in the case of ethane and propane (see Chapters 5 and 8), may explain why a very selective catalyst in n-butane and n-pentane oxidation can be completely unselective (under the same reaction conditions) in ethane and propane oxidation.

The presence of a slight excess of phosphorus on the surface may allow fine-tuning of the dimensions of these clefts/cavities and provide further tuning of the mechanism of orientation of adsorbed intermediates by multiple hydrogen-bond formation. It is also possible to explain where this excess phosphorus with respect to stoichiometry is located and why it does not interfere with catalyst activity by competitive coordination to the active (vanadium) sites.

It should also be noted that n-butane oxidation to maleic anhydride is a very complex transformation (a 14-electron oxidation with the insertion of 3 oxygen atoms, the abstraction of 8 hydrogen atoms and the generation of 4 water molecules) and requires very long surface residence times for complete transformation.[162–165] This necessitates a very efficient mechanism of protection of the adsorbed molecule from unselective attack by adsorbed oxygen forms reasonably present in a larger amount in the case of alkane as compared with alkene oxidation, owing to the much weaker alkane coordination to surface sites (see Chapter 8). Thus, in the case of alkane oxidation the mechanism of protection of adsorbed intermediates toward their unselective oxidation must be inherent in the catalyst characteristics, whereas in the case of alkene oxidation chemisorption of the reactant itself provides this function, at least in part (see also Section 4.2.3.4).

The concept of surface topology of vanadyl pyrophosphate as a key to interpreting and tuning its behavior can thus explain several aspects of the reactivity of V/P oxides, but present data are more on the level of hypotheses rather than

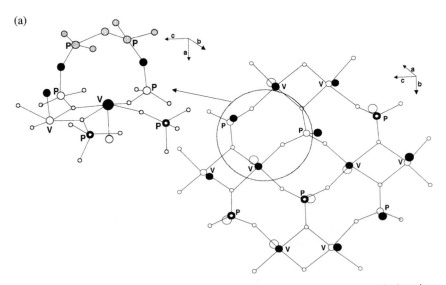

Figure 4.3. Projection of the (100) plane of vanadyl pyrophosphate illustrating the vanadyl orientations according to polytype I and example of the pyrophosphate termination on this plane (a).[175] Scheme of orientation of vanadyl columns in polytypes I and II of vanadyl pyrophosphate (b).[161] Model of the surface of the (100) plane of polytype I indicating the arrangement of surface pyrophosphate groups and the cavities created from these groups (c).[175]

(b)

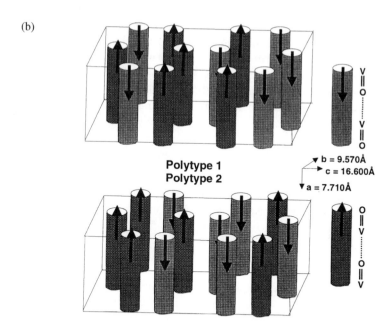

(c)

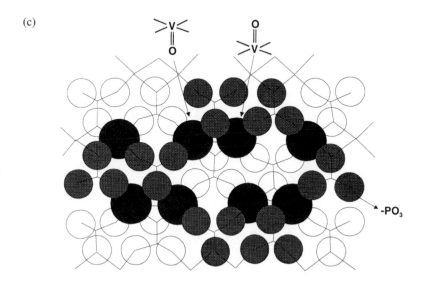

Figure 4.3. Continued

proof. It is worth noting, however, the enzyme-mimic action of vanadyl pyrophosphate in terms of this suggestion. Obtaining a better understanding of these aspects, notwithstanding the technical difficulties in the acquisition of such knowledge, is a future challenge that can establish the basis for a real breakthrough in catalyst design and understanding by no longer considering the active sites in a pseudobidimensional form but rather viewing them as three-dimensional active centers.

4.2.3.2. In Situ Surface Restructuring of VPO Catalysts

The problem of the nature of the surface restructuring induced by the catalytic reaction and by regeneration procedures is of critical importance both for understanding the real nature of the active surface so that the approach modeling the reaction mechanism will be the right one and for improvement in the performance of industrial catalysts. However, there are very few data in the literature about this central concern.

The role of the changes during *in situ* catalyst evolution up to an "equilibrated" situation was discussed in Section 4.2.2.2. During this *in situ* modification both the phase composition (the main aspect evidenced in the investigations),[82,95,123] and the degree of catalyst crystallinity change. Guliants *et al.*[81] observed that there is a surface amorphous layer present in "nonequilibrated" $(VO)_2P_2O_7$, which progressively disappears with time-on-stream with a concomitant improvement in the selectivity. This observation is consistent with the previous discussion on the role of the surface topology of $(VO)_2P_2O_7$, as the formation of a defined topology with clefts/cavities of the correct dimensions in a disordered (amorphous) phase is less effective. On the other hand, Brückner *et al.*[109] pointed out that an amorphous V/P oxide phase may form by reduction, but local structure remains nearly the same and, in particular, double vanadyl chains are still present. Thus long-range order may be lost (amorphization), but the local coordination may still be analogous, which explains why the catalytic behavior of crystalline $(VO)_2P_2O_7$ is not markedly different from that of amorphous phase.

It is also worth noting that Coulston *et al.*[88] observed a change in the surface P/V ratio when V/P oxide catalysts were cycled from reducing to reoxidizing conditions (DuPont process). This, according to Coulston *et al.*,[88] indicates a change in the intrinsic rate constants as well as a cyclic change in the surface structure. This is in agreement with the ESR data of Brückner *et al.*[109] Gleaves *et al.*[83,166] also showed that the reaction kinetics and product desorption depend considerably on the amount of active oxygen stored on the V/P oxide surface, but below the formation of structurally organized $VOPO_4$ phases.

When *n*-butane and oxygen are cofed the V/P oxide surface changes, probably in a comparable way to that observed for separate *n*-butane and oxygen interaction with the catalyst, but the process involves less structurally organized changes (i.e., only local modifications). There is thus a process of surface restructuring and possibly dynamic *in situ* modifications during the catalytic reaction.

Industrial experience indicates that *in situ* regeneration of the catalyst is sometimes necessary and in most cases the effect of regeneration is probably the same as that of surface amorphization followed by an *in situ* process of recrystallization. Partenheimer and Meyers[167] studying the regeneration of V/P oxide catalysts by CCl_4 treatment, observed that regeneration induces a solid state rearrangement leading eventually to restructuring of the vanadyl pyrophosphate surface. Stefani and Fontana[33] proposed regenerating V/P oxide by batch reduction in flowing concentrated (50% in N_2) *n*-butane at reaction temperatures for less than an hour. The main effects of treating V/P oxide catalysts this way are vanadium reduction to an average state below 4+ and surface amorphization. Blum *et al.*[34] indicated that during catalyst regeneration the microcrystalline structure of the catalyst undergoes dynamic reorientation. Ebner and Andrews[168] described a regeneration procedure in an air/steam mixture under controlled heating conditions that induces restructuring of the catalyst surface.

Steam was found to improve the selectivity to maleic anhydride.[169,170] It may have different effects on the surface reactivity, but a recent indication[171,172] is that water also induces partial dissociation of the V–O–P and P–O–P bonds with a partial opening of the pyrophosphate structure and formation of vanadyl hydrogen phosphate-like domains, which then tend to reform into the vanadyl pyrophosphate. This process leads to the formation of surface defect areas, which enables easier oxidation of subsurface vanadium sites. High-resolution electron microscopy has recently revealed that such defect areas are a result of the catalyst being exposed to steam.[173]

In conclusion, there are some indications regarding the existence of *in situ* surface restructuring processes of V/P oxide catalysts, but the data are very fragmentary and it is not possible to derive clear indications concerning the role and importance of these processes. In particular, four main aspects should be further analyzed, not only in terms of change in the surface structure but also of possible related aspects such as catalyst surface and subsurface redox processes, kinetics and energetics of surface catalytic processes, nature and amount of adsorbed species, and surface acido-base properties:

1. Role of crystallization and amorphization processes of vanadyl pyrophosphate during *in situ* activation procedures [change from nonequilibrated to equilibrated $(VO)_2P_2O_7$] and finally deactivation of the catalyst.
2. Role of surface restructuring processes during V/P oxide catalyst regeneration.
3. Presence and role of dynamic restructuring processes during the catalytic reaction, such as the creation of defect areas by the effect of local water

hydrolysis of V–O–P and P–O–P bonds followed by recrystallization of the vanadyl pyrophosphate structure.

4. Effect of continuous cycling between separate reduction and reoxidation processes (DuPont process) on the dynamics of catalyst surface structure change.

4.2.3.3. Catalyst Properties and Reactor/Process Configuration

In Section 4.2.1 it was pointed out that the choice of reactor configuration and optimal reaction conditions (e.g., feed composition, temperature, hot spot) is not independent of catalyst selection and optimization. However, this concept is not often clear in the literature, and, indeed, there have been practically no specific studies on V/P oxide catalysts dealing with these problems published.

Only Bordes and Contractor[174] have recently shown that the microscopic properties of V/P oxide catalysts must be adapted to the type of gas–solid reactor. They observed that while physical catalyst properties (e.g., size, shape, density, porosity, specific surface area) are usually studied in the industrial development of catalysts, the microscopic properties have been less commonly considered. Their discussion was focused mainly on the differences in V/P oxide catalysts required to operate under steady state conditions (fixed bed-FBR or fluidized bed-FlBR) and in cyclic reduction–reoxidation conditions (recirculating solid riser reactor—RSR). While the V/P oxide catalyst is apparently the same in both cases, its specific characteristics must be tuned for the conditions of operation.

Under steady state conditions, the reduction–reoxidation cycles occur simultaneously and thus the rates of these two reactions (which depend on several factors, such as the defective nature of the vanadyl pyrophosphate, P/V surface ratio, and the presence of domains of $VOPO_4$ phases) must be tuned to maximize the selectivity to maleic anhydride. In fact, e.g.:

1. A slight excess of P is usually needed to limit catalyst surface overoxidation, as was discussed in Section 4.2.2.
2. Fast reoxidation of the catalyst is needed, not only for a high steady state rate of reaction, but also to avoid the formation of O_2^- and O^- species, which might lower the selectivity. When oxygen is contacted with V/P oxide catalyst separately from the reduction stage, the above limitations no longer exist, and there are more degree of freedom in selecting optimal catalyst properties.

On the other hand, in RSR reactors, the more critical problem is the low catalyst productivity (of the order of 0.5–1.2 g maleic anhydride per kg of circulating V/P oxide catalyst[174]) and thus this aspect of the catalyst surface reactivity must be optimized. On the other hand, in V/P oxide catalysts for RSR application, the excessive reduction, with formation of V^{3+} microdomains[108] that lead to a fast deterioration in catalyst performance must also be avoided. As a matter of fact, the P/V ratio and catalyst redox properties for V/P oxide catalysts for FBR and RSR are different.[174]

Other catalyst characteristics also differ for the two kinds of reactor applications:

1. Owing to the presence of hot spots (usually up to 50–70°C with respect to salt bath temperature) in fixed-bed reactors, a property required of V/P oxide catalysts for this application is to avoid formation of β-$VOPO_4$ by oxidation of $(VO)_2P_2O_7$, which, being not very active/selective and poorly reducible again to $(VO)_2P_2O_7$, leads to catalyst deactivation. In RSR reactors hot-spot effects are minimal and thus this property (which creates constraints in catalyst selection especially at the stage of industrial development) is not important.

2. Intragranular mass/heat transfer effects (due both to the much smaller dimensions of the pellets used in RSR as compared with those in FBR applications and the absence of gas phase oxygen in the first case) are much less critical in RSR applications. On the other hand, in RSR-type V/P oxide catalysts, an increase in the surface area increases catalyst productivity (more correctly, oxocapacity[49]). Thus the surface areas and textural properties of RSR-type and FBR-type V/P oxide catalysts are different.

Not only do the optimal characteristics of the catalyst depend on the reactor configuration, but the sensitivity of the catalyst depends on operating parameters, such as the addition of steam to the feed. In fact, contradictory results on the effect of adding water to the feed have been observed.[169–172] The effect can be rationalized, in part, by taking into account that different catalysts for different reactor applications have been analyzed. The two effects of water on surface reactivity are: (i) to help desorption of maleic anhydride (see Chapter 8), and (ii) to limit the formation of heavier, carbonaceous-type species on the catalyst surface. When hydrocarbon and oxygen are contacted separately with the V/P oxide catalyst (RSR-type reactor), a high rate of maleic anhydride desorption is not critical for the selectivity, but under steady state conditions (FBR reactor) it is much more important, because a long surface residence time in the presence of gaseous oxygen leads to an enhanced possibility of nonselective oxidation to carbon oxides.[162] However, the formation of heavier products is much more relevant in terms of surface reactivity in the absence of gaseous oxygen, i.e., in the RSR-type reactor. Thus the same component may have different effects, depending on the reactor and the type of gas–solid contact.

4.2.3.4. Microkinetics of the Surface Transformations on V/P Oxide Catalysts

Studies of the catalytic reactivity of mixed oxides and V/P oxide catalysts, in particular, can be broadly differentiated into two classes:

1. Kinetic approach, i.e., the influence of reaction conditions on the rates of reaction is studied to derive model reaction rates (usually relatively simple) that fit experimental data on the overall transformation (the reaction from *n*-butane to maleic anhydride occurs over the entire surface without desorption of intermediate products).
2. Mechanistic approach, where using spectroscopic tools, the possible steps in the overall reaction network are identified and the structures of the possible adsorbed species and intermediates are clarified.

A further challenge is to link these and develop a microkinetic approach, in which, based on the identification of the complete surface reaction network, rates of reaction of the single (quasi-elementary) steps are determined to describe the global kinetics in terms of the individual contributions of quasi-elementary steps. The challenge is thus to describe the surface mechanism in terms of the dynamics of transformation/mobility of these surface species and not only as the presence of active sites/adsorbed species or as energetics of the transformation.[159,160]

A first step toward this objective is the development of reliable surface reaction networks, which includes identifying not only the possible reaction intermediates[7,176] but also the key factors determining the relative rates in the presence of competitive pathways and the influence of adsorbed species concentration on the relative rates of single-step reactions. This problem is discussed in detail in Chapter 8 for the case of *n*-butane and *n*-pentane oxidation on V/P oxide, and so is not gone into here. It is worth noting, however, that the development of such a microkinetic model is the basis both for a better approach to catalyst surface tuning and microstructural engineering and for improving catalyst performance through basic science. Even simple and preliminary microkinetic approaches based on these concepts[177] can be useful for understanding the surface catalytic chemistry and determining the key factors for its control and tuning.

4.2.3.5. Alkane versus Alkene Oxidation

A concept often confused in the literature on V/P oxide catalysts is that the optimal V/P oxide catalyst composition is different in alkane (*n*-butane) and in alkene (butene, butadiene) oxidation. The latter hydrocarbons are intermediate (in the adsorbed form) in the reaction pattern from *n*-butane to maleic anhydride,[7,9] and the *n*-butane activation is the rate-determining step. Thus oxidation of the intermediate (adsorbed) alkenes should be faster, and it is often hypothesized that when these alkenes are fed in directly, there should be a higher rate of reaction as well as a higher selectivity to maleic anhydride, the contrary of what is observed experimentally.[178] Alkenes and alkendiene interact very strongly with the catalyst surface, differently from the alkane, and thus the real *in situ* catalyst surface (in terms of amount of adspecies) is considerably different for alkane and alkene oxidation. Various reactivity aspects are related to this issue that offer a possible way to control and tune the surface reactivity. This aspect will be discussed in detail in Chapter 8.

4.3. PROPANE AMMOXIDATION TO ACRYLONITRILE ON VANADIUM/ANTIMONY OXIDES

4.3.1. Background on the Direct Synthesis of Acrylonitrile from Propane

Acrylonitrile is presently produced by ammoxidation of propene in fluidized-bed reactors using one of three common catalyst compositions: the Nitto catalyst (based on Fe/Sb/Te/Mo oxides), the Monsanto catalyst (based on U/Sb oxides) and the BP America catalyst (based on Bi/Mo oxides). Typical performances in propene ammoxidation can be summarized as follows:

- Feed composition: Propene, 8.3%; NH_3, 8.5%; O_2, 10%; H_2O, 35.6%; N_2, 37.6%.
- Temperature and contact time: 400–440°C and 3–6 s, respectively.
- Propene conversion and acrylonitrile yield: 97–99% and 70–80%, respectively.

The lower cost of propane relative to propene (depending on locality, the price of propane is 50 to 70% that of propene) and the worldwide propene shortage owing to its being increasingly consumed for polymerization have long provided an incentive for the development of a process for the direct conversion of propane to acrylonitrile. To justify commercial development, a process based on propane would require:

1. Productivities comparable to those achieved with propene, to enable retrofitting of existing plants, or at least of their separation units.
2. Operation at a temperature no higher than 500°C, to minimize gas phase oxidation and problems related to reactor materials.
3. The absence of gas phase chlorine promoters in the feed, to eliminate the need for expensive materials in plant equipment and avoid recovery costs.
4. Achievement of a comparable productivity of high-value by-products, such as HCN and CH_3CN, might also be necessary, if such by-products are used in downstream units.

A lower selectivity than that obtained from propene may be acceptable, because of the lower cost of the hydrocarbon, and the increased cogeneration of valuable, high-pressure steam owing to the higher release of reaction heat. Good selectivities may be achieved working at lower hydrocarbon conversion than for propene. The lower limit in selectivity, however, is determined in this case by the cost of separating by-products and recycling unreacted propane.

Several processes for the conversion of propane to acrylonitrile have been proposed by various companies, and some of them have reached the pilot-plant stage, although none is yet offered commercially. Economic estimates[179] and preliminary announcements,[180] however, indicate that a process of direct acrylonitrile synthesis from propane has a good outlook for commercialization.

The various processes for the conversion of propane to acrylonitrile that have been proposed differ in catalyst composition, operating temperature, gas phase composition, and the use of gas phase promoters.

Several years ago Monsanto researchers proposed a process for the ammoxidation of propane using circulating fluidized-bed technology.[181] The key features of the process are the separation between catalyst contact with ammonia/propane and catalyst reoxidation, similar to the DuPont process for maleic anhydride from n-butane. The catalyst, with the general composition $Sb_{1-10}U_{0.01-1}Fe_{0-1}W_{0-0.1}O_x$, is similar to that used in their propene to acrylonitrile process, but halogen compounds (haloalkanes with up to three carbon atoms) are added to the feed to promote homogeneous gas phase propane to propene oxidative dehydrogenation. The advantages of the transport-bed process over conventional fluidized-bed or fixed-bed processes are higher selectivity, greater safety, lower separation costs, and greater flexibility in catalyst selection. The necessity of adding halocompounds makes this solution unsuitable for process development, but new possibilities may open up using one of the more recently developed advanced catalysts, which are active without gas phase promoters. A specific problem with respect to n-butane, however, is the necessity of cofeeding propane and ammonia in the riser reactor. The ammonia, which strongly chemisorbs and competes for sites of propane activation (see below), can completely inhibit catalyst reactivity in the absence of oxygen.

Mitsubishi Kasei researchers[182-184] have developed a catalyst composition that gives the highest yield of acrylonitrile to date, although other researchers have not been able to reproduce their results; neither is the stability of the catalyst clear (Te is known to be a highly volatile compound). The optimal catalyst composition, $MoV_{0.3}Te_{0.23}Nb_{0.12}O_x$,[184] gives an acrylonitrile yield of about 50%, which can be further increased by the addition of Sb, B, or Ce up to nearly 60%. Key characteristics of the catalyst are not only the composition, but also the activation procedure (at high temperature under N_2) and the modality of doping (on preformed V/Mo/Nb/Te oxides).

Rhodia[185] has developed catalytic systems based on Sn/V/Sb mixed oxides. They are mainly constituted of rutile SnO_2 that contains Sb and V in solid solution, $VSbO_4$ (also with rutile structure), and Sb_2O_x. The dispersion of these components in the same matrix makes a closer interaction between components possible, thus allowing a faster transformation of the intermediate alkene to acrylonitrile.

Several years ago ICI and Power Gas proposed[186] a V/Sb based catalyst, which is selective only at high propane concentrations, but only low yields are obtained. Similar kinds of catalysts, but with improved preparation methodologies, developed by BP America led to significantly better results.[187–189] The company also patented a series of other single- and dual-catalyst compositions, with the second catalyst having the function of converting the propene intermediate to acrylonitrile. These catalyst compositions are mixed oxides such as (i) $Cr_aMo_bTe_cM_d$, where M is Mg, Ti, Sb, Fe, V, W, Cu, La, P, Ce, or Nb[190]; (ii) VSb_aM_b, where M is one or more of Sn, Ti, Fe, Mn, or Ga[188,189]; (iii) $Bi_aFe_bMo_cA_dB_e$, where A is one or more alkali or alkaline metal, B, W, Sn, or La, with B being one or more of the elements Cr, Sb, Pb, P, Cu, Ni, Co, Mn, or Mg[191]; and (iv) $Bi_aFeMo_{12}V_bD_cE_dFeG_f$, where D is one or more of the alkaline metals; E is one or more of Mn, Cr, Cu, Zn, Cd, or La; F is one or more of P, As, Sb, F, Te, W, B, Sn, Pb, or Se; and G is one or more of Co, Ni, or an alkaline earth metal.[192] The gas-feed composition was usually propane/ammonia/oxygen/water in a 5/1/2/11 molar ratio, with excess alkane and water used as a diluent. It can be observed that some of these compositions are similar to those of the Mitsubishi Kasei researchers, although the results reported are significantly poorer.

A summary of selected results from patent data is given in Table 4.5.

Apart from patent applications, few studies had been published on the direct synthesis of acrylonitrile from propane, and it is only in recent years that this topic has been of more basic scientific interest. As many of these later studies have already been reviewed,[13–17] only certain aspects of the surface reactivity will be discussed here, limited mainly to V/Sb oxides (see introduction), to provide examples of catalyst properties that offer a key for tuning the catalyst and enabling better catalyst design.

4.3.2. Role of Nonstoichiometry and Rutile Structure in V/Sb Oxide Catalysts

The discussion in the previous section and the review of catalytic data[13–17] show that optimal performance of unsupported V/Sb oxide catalysts has been found at high propane concentration, with oxygen as the limiting reactant, and at low propane conversion (in the 10–30% range), which implies the development of a process that recycles unreacted propane. Selectivity for acrylonitrile under these conditions is in the 60–70% range, which is already in the range of interest to evaluate exploitation of the catalyst, but which also clearly needs to be improved further. A central issue in the study of these catalysts is to identify the factors that enable control of the selectivity to acrylonitrile from propane, both by better design of the structural characteristics of the catalyst (the central topic of this section) and

TABLE 4.5. Catalysts and Relative Catalytic Behaviors Reported in Patents for Direct Synthesis of Acrylonitrile from Propane

Process and catalyst	Catalytic data			Feed	Note	Reference
	Temperature, °C	Conversion, %	Yield,[a] %			
Mitsubishi–Kasei						
$MoV_{0.4}Te_{0.23}Nb_{0.12}O_x$	410	86.7	55.1	$C_3H_8/NH_3/air$	Catalysts calcined in flow N_2 at 600 °C. Doping on preformed $MoV_{0.3}Te_{0.23}Nb_{0.12}O_x$, silica sol added to initial solution	184
$MoV_{0.3}Te_{0.23}Nb_{0.12}Sb_{0.016}O_x$	410	91.5	58.3	1/1.2/15		
$MoV_{0.3}Te_{0.23}Nb_{0.12}B_{0.072}O_x$	410	91.6	58.0	GHSV 1000 h^{-1}		
$MoV_{0.3}Te_{0.23}Nb_{0.12}Bi_{0.029}O_x$	410	90.0	57.3			
$MoV_{0.4}Te_{0.23}Nb_{0.12}O_x/SiO_2$ (10%)	420	88.9	53.8			
Rhodia				$C_3H_8/NH_3/O_2/H_2O/N_2$		
$VSb_5Sm_5O_x$	450	30	14.7	8/8/20/0/64	Contact time 3 s	185
$VSb_5Sm_5O_x$	470	20	10.8	25/10/20/0/45	Contact time 2.1 s	185
Power Gas–ICI				Contact time 0.5 s	ACN product only[b]	
$VSb_{2.6}O_x$	500	—	2.3	$C_3H_8/NH_3/O_2/N_2$ 80/5/10/5	280	193
$VSb_{10.6}Sn_{4.6}O_x$	480	—	0.9		120	193
$SnSb_{0.33}Ti_{0.33}O_x$	535	—	1.7	85/5/10/0	220	186
BP America				$C_3H_8/NH_3/O_2/H_2O/N_2$		
Mo/Cr/Te/O on SiO_2–Al_2O_3	480	18.9	8.7	34/11/22/11/22	2.5% C_3H_6 yield	190
$VSb_{1.4}Sn_{0.2}Ti_{0.2}Bi_{0.001}O_x$	460	15.2	9.4	51/10/29/10/0	0.3% C_3H_6 yield	187
$VSb_{1.4}Sn_{0.2}Mo_{0.001}O_x$	460	14.0	9.2	51/10/29/10/0	0.07% C_3H_6 yield	187
$VSb_{1.64}Fe_{0.015}O_x$	460	13.3	8.1	56/11/22/11/0	3.5% C_3H_6 yield	194
$Ni_{2.5}Co_{4.5}Fe_2MnBiCr_{0.5}Mo_{13}Cs_{0.05}$ $K_{0.1}O_x/33\%$ Al_2O_3	470	10.0	5.8	56/11/22/11/0	0.02% C_3H_6 yield	191
$Bi_2Mo_8Cr_3W_4Fe_2Mg_2O_x/25\%$ Al_2O_3–25% SiO_2	470	14.8	8.2	34/11/22/11/22	1.1% C_3H_6 yield	191

[a] Molar to acrylonitrile
[b] Acrylonitrile productivity in pilot plant (2 liter catalyst), $g\ h^{-1}\ L_{cat}^{-1}$

by control of the surface properties and reaction network (the central topic of Section 4.3.3).

4.3.2.1. Comparison with Other Sb-Rich Rutilelike Mixed Oxide Catalysts

V/Sb oxide catalysts show several structural and reactivity similarities to Sb-based mixed oxide catalysts[195] and in particular to iron-antimonate–based catalysts,[196–203] which are well known for propene ammoxidation to acrylonitrile, being the active phase of Nitto-type catalysts[204,205] (see Section 4.3.1):

1. A rutilelike metal–antimonate phase and α- or β-Sb_2O_4 are the two crystalline phases present, although the antimony oxide starts to become detectable by XRD only at values higher than the Sb/M stoichiometric ratio of 1.0 necessary to form the rutilelike structure.
2. Excess antimony with respect to the stoichiometric Sb/M = 1.0 is necessary to achieve complete reaction of the metal oxide with antimony oxide.
3. The selectivity to acrylonitrile markedly improves when Sb/M is above the stoichiometric value of 1.0, i.e., when there is an excess of antimony oxide.

A variety of proposals have been advanced to explain the last phenomenon of enhanced selectivity of Sb-rich samples, for iron-antimonate catalysts: (a) formation of a surface compound of stoichiometry $FeSb_2O_6$, i.e., with a trirutile-like structure[196,198,200]; (b) phase cooperation between $FeSbO_4$ and Sb_2O_4[197]; (c) formation of highly selective islands of Sb_2O_4 on top of the $FeSbO_4$ particles[199]; (d) inhibition of the formation of free Fe_2O_3 and suppression of total oxidation[206]; (e) formation of a two-dimensional antimony oxide covering the iron antimonate[201,202]; (f) synergetic interaction between Sb_2O_4 as a donor species of spillover oxygen and $FeSbO_4$ as an acceptor phase[203]; and (g) cation ordering with the formation of a trirutile-like superstructure.[207,208]

Notwithstanding the differences, these hypotheses on the nature of the active phase can be grouped into two classes: (i) those that indicate a reaction between the rutile phase and Sb_2O_4 to form a trirutile superstructure (the trirutile structure has XRD patterns virtually identical to the rutile structure and thus cannot be distinguished with this technique, nor with neutron diffraction); and (ii) those that do not consider a specific structural modification, but indicate that sites are located at the interface between the two phases or that there is cooperation between these phases and/or that the excess antimony oxide avoids nonselective side phases/sites. While there is no conclusive evidence concerning either of these two mechanisms, it may be noted that for Sb-rich rutile catalysts[195]: (i) it is possible to gently remove crystalline Sb_2O_4 (by extraction in diluted HCl solution) without affecting the catalytic performance in alkene oxidation, and (ii) the presence of Sb^{5+} sites seems to be critical. In fact, it has been observed[195] that the Sb^{5+} phase is very active/selective in alkene (amm)oxidation, but irreversibly converts to Sb_2O_4, as the latter phase is not reoxidized by gaseous oxygen. Furthermore, in fresh catalyst this phase is present over the rutile phase, even for Sb/M ratios close to 1, but quickly converts to Sb_2O_4 during catalytic tests. This change corresponds to a decrease in catalytic performance, less accentuated when the Sb/M ratio is higher than the stoichiometric value. Moreover, a change in the unit cell parameters of the rutile phase is observed when excess antimony oxide is present. All these indications suggest that a primary factor determining the surface chemistry/reactivity of Sb-rich rutile catalysts is the surface/bulk restructuring to form Sb-rich phases and thus in general the formation of nonstoichiometric rutile-type oxides.[195]

Can the same model also be used to understand the chemistry and reactivity of V/Sb oxide catalysts? This is a question of central importance and not a simple academic discussion on the nature of active species, because understanding whether this problem is relevant or not provides a key to designing better catalysts, controlling their reactivities, and, ultimately, establishing a basis for the development of a process to obtain acrylonitrile from propane.

4.3.2.2. Nature of the Phases Present

Fresh V/Sb oxide catalysts are characterized by the presence of two main crystalline phases: a rutile-type vanadium antimonate phase and α- or β-Sb_2O_4; which of the Sb_2O_4 phases is present depends on the preparation and calcination temperature.[17] In addition, there is an overlayer of well-dispersed V^{5+} and Sb^{5+} oxides on the vanadium antimonate crystallites, especially in fresh samples after calcination. With increasing Sb/V ratio, there is a change in the relative quantities of these overlying XRD amorphous phases, in addition to the change from a monophasic to biphasic system.[17] Disappearance of V^{5+} oxide is not complete, even for samples with a large excess of antimony.[17]

The nature of the vanadium antimonate rutile phase (hereinafter denoted by $\approx VSbO_4$) is a critical aspect of the V/Sb oxide catalysts and depends on the preparation conditions and especially on the oxygen partial pressure during heat treatment. Birchall and Sleight[209] demonstrated the formation of a $V_{0.92}Sb_{0.92}O_4$ phase when equimolar V_2O_5 reacts with Sb_2O_3 at 800°C in air and $V_{1.05}Sb_{0.95}O_4$ when the reaction is carried out in a closed tube. Later, Berry et al.[210,211] reported two different reduced phases: one prepared in a closed tube or in flowing commercial nitrogen, with the composition $VSb_{1-y}O_{4-2y}$ ($0 < y < 0.1$) and a second with the composition $VSb_{1-y}O_{4-1.5y}$ ($0 < y < 0.1$) prepared in flowing, oxygen-free nitrogen. The various phases differ in cell parameters; the phase obtained in the presence of oxygen has expanded $a = b$ parameters and a contracted c parameter. Teller et al.[212] also noted the possible formation of nonstoichiometric vanadium antimonate, but showed that the formation of a vanadium antimonate phase with the a and b lattice parameters (tetragonal cell) expanded and the c parameter contracted may form when vanadium antimonate grows over β-Sb_2O_4 crystallites, possibly epitaxially. Teller et al.[212] also indicate that in the epitaxially grown $V_{1-\alpha}Sb_{1-\alpha}O_4$ the metal–oxygen octahedron is somewhat squashed with respect to symmetric "normal" vanadium antimonate, possibly due to excess Sb. Therefore, their results indicate that the oxygen partial pressure during thermal treatment is not the only factor determining the nonstoichiometric and structural characteristics of the vanadium antimonate phase.

Hansen et al.[213] pointed out that the cation-deficient rutile structure prepared in air is a mixed-valence compound with the following composition: $Sb^{5+}_{0.92}V^{3+}_{0.28}V^{4+}_{0.64}\square_{0.16}O_4$. When the synthesis of vanadium antimonate is done in flowing gas with varying O_2/N_2 ratios, a continuous nonstoichiometric series of rutile-type $Sb_{0.9}V_{0.9+x}\square_{0.2-x}O_4$ ($0 < x < 0.2$) phases can be obtained[214] with a progressive increase in the V^{3+}/V^{4+} ratio from the oxidized end member $Sb^{5+}_{0.9}V^{3+}_{0.1}V^{4+}_{0.8}$ $\square_{0.2}O_4$ to the reduced end member $Sb^{5+}_{0.9}V^{3+}_{0.9}V^{4+}_{0.2}$ $\square_{0.0}O_4$ A parallel contraction of the $a = b$ parameter and expansion of the c parameter of the tetragonal rutile cell was noted. The presence of a V^{3+}/V^{4+} mixed valence in vanadium antimonate is in agreement with the results obtained by chemical analysis of V/Sb oxide samples.[215]

These results indicate that the change in cell parameters as a function of Sb/V ratio and thermal treatment in air or during catalytic tests may not be correlated only with the formation of cation ordering such as in the trirutile structure. However, all the structural investigations cited above refer to samples prepared by solid state reaction of equimolar mixtures of Sb_2O_3 and V_2O_5 at temperatures of 800°C or higher. These conditions are rather different from those for preparation of V/Sb oxide catalysts (see also the following sections regarding the role of preparation on catalyst microstructure and reactivity).[216] The question of the exact structure of the active vanadium antimonate phase and its dependence on the Sb/V ratio, method of preparation, thermal treatment, and gas phase composition is thus still open and needs to be answered. However, all data indicate the easy formation of nonstoichiometric phases in V/Sb oxide catalysts, and thus it is reasonable to consider their role in propane ammoxidation to acrylonitrile.

4.3.2.3. Nonstoichiometry of Vanadium Antimonate and Catalytic Reactivity

In the previous section it was shown that notwithstanding disagreements among various authors, there is general consensus about the formation of non-stoichiometric vanadium antimonate phases and the dependence of their charac-teristics on the method of preparation. However, when the catalytic behavior is analyzed, the role of nonstoichiometry of vanadium antimonate is usually ignored.

Andersson et al.[217–222] proposed that the selective behavior of V/Sb oxide is associated with the creation of suprasurface antimony sites over vanadium anti-monate, and that the role of surplus antimony sites is to bring about isolation of the vanadium centers at the vanadium antimonate surface. The same effect of site isolation of vanadium centers was used to explain the behavior of samples containing alumina and/or tungsten.[221,222] Zanthoff et al.[223,224] and Albonetti et al.[225,226] basically agree with this hypothesis, which explains some catalytic data but is inconsistent with: (i) the observation that Sb participates in the reaction mechanism, (ii) the effect of preparation methodology on the catalyst characteristics and reactivity, and (iii) data on the reactivity of other structurally analogous Sb-rich rutile-type catalysts. Similarly, in a recent reinvestigation of Fe/Sb oxide catalysts[227] for propene and propane ammoxidation it was shown that the presence of excess surface antimony oxides changes the distribution of nucleophilic and electrophilic oxygen species in the pool of surface oxygen species. Bowker et al.[228] observed comparable results indicating that Sb participates in the reaction mechanism and that excess Sb on the surface influences the degree of surface oxidation and the surface reactivity.

In some of our work (Centi et al.[229–231]) we focused our attention on the role of vanadium antimonate nonstoichiometry in determining the surface reactivity and characteristics of V/Sb oxide catalysts. Studying the effect of Sb/V ratio on the structural and reactivity characteristics of V/Sb oxide samples prepared by the precipitation–deposition method[229] we observed a correlation between selectivity for acrylonitrile and the amount of supported Sb^{5+} oxide, whereas the correlation was not good for α-Sb_2O_4 (Figure 4.4). This indicates that the presence of α-Sb_2O_4 and associated effects (phase cooperation, sites at the interface, isolation of active sites) are not the key parameters determining the surface reactivity, but only parameters that change in the same direction. We also observed that increasing the Sb/V ratio in the samples changes the vanadium antimonate rutile cell parameters and that seems to influence the catalytic behavior.[229] This aspect was analyzed in more detail in a subsequent paper, where the dependence of the catalytic behavior of V/Sb oxide catalysts in propane ammoxidation is studied for a series of catalysts as a function of the Sb/V ratio for samples synthesized by four different methods[230]: (i) gel to solid method (GS), (ii) deposition of vanadium on antimonic acid (DAA), (iii) solid state reaction (SR), and (iv) precipitation–deposition method (PD).

Several properties of V/Sb oxide catalysts change as a function of the method of preparation[230]: (i) quantity and dimensions of the crystallites of $\approx VSbO_4$ and Sb_2O_4; (ii) dimensions of the unit cell of $\approx VSbO_4$, which are related to its nonstoichiometric characteristics; (iii) presence and amount of amorphous phases (Sb^{5+} and V^{5+} oxides, amorphous $\approx VSbO_4$ and Sb_2O_4 oxides); and (iv) ratio between α- and β-Sb_2O_4 in the samples. Several factors come together to determine the global behavior, but a general feature of all the samples is that the selectivity for acrylonitrile increases owing to minor effects on the selectivity for carbon oxides, which depend critically on the method of preparation. Increasing the Sb/V

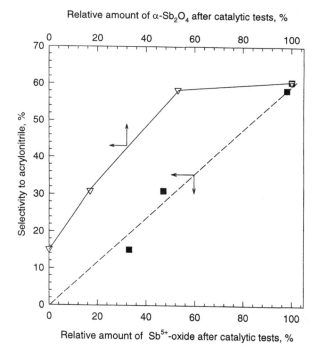

Figure 4.4. Correlation between selectivity to acrylonitrile at 480°C in VSbO samples with different Sb/V ratios (preparation by the precipitation–deposition method) and amounts of Sb^{5+} oxide and α-Sb$_2$O$_4$ in the samples.[229]

ratio also decreases the side NH$_3$ → N$_2$ reaction and increases the rate of propene intermediate conversion to acrylonitrile (thus selectivity to the latter product). It is thus possible to associate, as a dominant effect, different steps in the reaction network with different parameters in the preparation: (i) the Sb/V ratio affects primarily the rate of transformation of intermediate propene to acrylonitrile, and (ii) the chemistry of preparation is responsible for the competitive rates of propane transformation by selective and nonselective routes.

With increasing Sb/V ratio, a series of changes occurs in catalyst composition and structure: (i) the amount of crystalline and amorphous Sb oxide increases, present mainly as Sb$_2$O$_4$, but also as Sb^{5+} oxide, although the latter undergoes considerable reduction during the catalytic tests; (ii) the amount of amorphous V^{5+} oxide spread over the surface decreases, and the V^{5+} undergoes considerable reduction during the catalytic tests; and (iii) the cell characteristics of the ≈VSbO$_4$ phase change with an expansion of the cell volume and a decrease in the ratio of the c/a parameters of the tetragonal rutile unit cell.

Data in Figure 4.4 show that there is no definite correlation between selectivity for acrylonitrile and the quantity of crystalline Sb_2O_4, unlike the case of surface Sb^{5+} oxide, which tends to reduce irreversibly to Sb_2O_4 when present only as a segregated phase.[216] Correlation is observed between the quantity of V^{5+} species supported on the vanadium antimonate surface and the rate of ammonia oxidation to N_2, roughly inversely proportional to the selectivity for acrylonitrile.[229,232] When the same type of correlation is analyzed for the samples after the catalytic reaction of propane ammoxidation, the correlation is negligible (Figure 4.5), but good correlation is observed between the same reaction and the change in the c/a rutile unit cell parameter of vanadium antimonate (Figure 4.5).

This correlation between catalyst nonstoichiometry and reactivity is confirmed by the analysis of the dependence of the selectivity for acrylonitrile on the Sb/V ratio and method of preparation (Figure 4.6a) and the dependance of the rutile unit

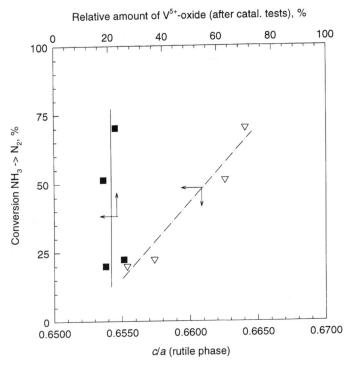

Figure 4.5. Correlation in VSbO catalysts between rate of ammonia side oxidation to nitrogen during propane ammoxidation and relative amount of V^{5+} oxide or c/a ratio in the unit cell of the rutile phase (vanadium antimonate). The amount of V^{5+} oxide is expressed as a percent with respect to the amount of V^{5+} oxide determined in the sample with the larger amount of this phase in the series of VSbO catalysts studied.[17,229]

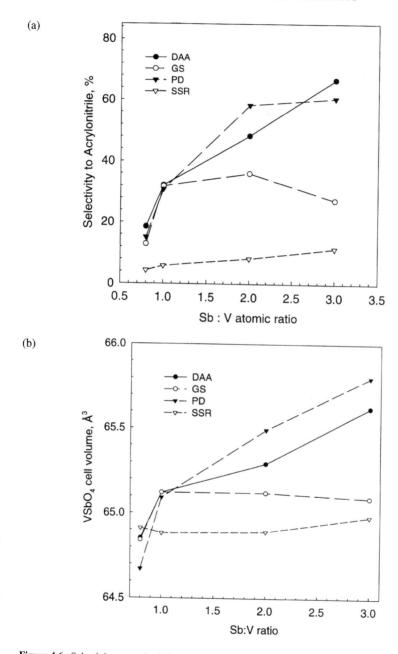

Figure 4.6. Selectivity to acrylonitrile at 480°C as a function of the Sb/V ratio in catalysts prepared by different methods (see text) (a).[230] Unit cell volume of $\approx VSbO_4$ crystallites as a function of the Sb/V ratio and method of preparation (b).[230]

cell volume on the same factors (Figure 4.6b). The good correlation in the trend as a function of the Sb/V ratio for methods of preparation strongly supports the relationship between the nonstoichiometric characteristics of vanadium antimonate owing to excess Sb and catalytic reactivity.

4.3.2.4. Nonstoichiometry of Vanadium Antimonate and Surface Characteristics

The change in the unit cell parameters with the Sb/V ratio in V/Sb oxide catalysts (Figures 4.5 and 4.6) indicates a nonstoichiometry of the rutile cell, probably due to incorporation of excess antimony. Nilsson *et al.*[217] suggested that the nonstoichiometric characteristics of vanadium antimonate affect the presence of 2-coordinated oxygen sites. Teller *et al.*[212] pointed out the presence of distortion in the metal–oxygen octahedra. Nilsson *et al.*[217] noted also that suprasurface Sb oxide reoxidizes faster. Centi and Perathoner[233] observed by IR spectroscopy that when suprasurface Sb oxide is present, the extent of surface reduction during anaerobic interaction of the catalyst with ammonia or hydrocarbon at 450°C is lower, demonstrating the role of reoxidation of suprasurface Sb oxide on the catalytic behavior. This was confirmed by transient reactivity studies,[233] which showed that the rate of acrylonitrile synthesis is controlled by the rate of catalyst reoxidation. However, transient reactivity data[233] also indicate that acrylonitrile forms only when reoxidation of the unsupported V/Sb oxide catalyst is complete.

In fact, different delays in reaching constant acrylonitrile formation after addition of gaseous oxygen to the propane and ammonia feed (during transient experiments) are observed as a function of the time of prereduction in the propane $+NH_3$ in helium feed. It should be noted, in particular, that in the case of the longer prereduction time acrylonitrile starts to form only when gaseous oxygen appears in the gas phase and reoxidation is nearly complete. The probability that there is a reversible change in the nonstoichiometric properties of the vanadium antimonate rutile phase in redox reduction–reoxidation cycles, leads to the conclusion that not only is reoxidation of suprasurface antimony oxide important (in this case the initial delay is not expected), but also that the nonstoichiometric characteristics of the vanadium antimonate rutile phase and surface restructuring phenomena associated with the change from a reduced to a reoxidized state of Sb-rich $\approx VSbO_4$ have an important role in determining the catalytic behavior and the selective synthesis of acrylonitrile.

Teller *et al.*[212] observed that the metal–oxygen octahedron is symmetric for "normal" $\approx VSbO_4$, but somewhat squashed in $\approx VSbO_4$ epitaxially grown over β-Sb_2O_4 macrocrystals and has a Sb/V ratio greater than that of "normal" $\approx VSbO_4$. The shortest O⋯O distance goes from 2.58 Å in "normal" $\approx VSbO_4$ to 2.67 Å in "epitaxially grown" $\approx VSbO_4$. The distortion of the octahedron thus does not derive from the increased metal–metal interaction as would be expected when V^{4+} ions replace V^{3+} ions in the rutile cell, but rather from the repulsive effect of oxygen anions owing to the probable formation of an Sb-rich $\approx VSbO_4$ phase.[212]

In the rutile $\approx VSbO_4$ phase an increase in the parameter a together with a decrease in the c/a value is observed, which indicates[234] an increase in d_a (apical distance) and a decrease in d_e (equatorial distance) (Figure 4.7) caused by increased electrostatic repulsion of anions due to excess localized charge on the oxygens.

The modification in the metal–oxygen octahedrons as a result of the formation of an Sb-rich $\approx VSbO_4$ phase (Figure 4.7) would be reflected by a change in surface reactivity and characteristics. In particular, an increase in the Brønsted acid character of the equatorial oxygens and an increase in the lability or redox character of the apical oxygens with respect to the normal $\approx VSbO_4$ phase is expected. By IR spectroscopy[235,236] it is possible to observe that the rate of the side reaction of $NH_3 \rightarrow N_2$ oxidation decreases with increasing Brønsted acidity of the catalysts. On the other hand, the increased redox reactivity of the apical oxygens would also favor an increase in the rate of the propene to acrylonitrile transformation. The change in Brønsted surface acidity of equatorial oxygens and that in redox reactivity of apical oxygens both act to increase the rate of transformation of intermediate propene to acrylonitrile and thus the selectivity for this product.

Furthermore, based on the proposed mechanism of propane conversion on V/Sb oxide catalysts[235,236] it is expected that these surface modifications favor a selective transformation of propane to the propene intermediate (concerted hydrogen abstraction) rather than the oxidative attack of surface oxygens to the methyl group of propane to form a propionate-like species. The latter route leads mainly to carbon oxides, whereas the former is the selective one. The expected change in

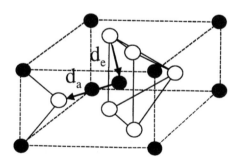

Figure 4.7. The rutile lattice with indication of the different metal–oxide distances (d_a = apical distance and d_e = equatorial distance). The arrows indicate the direction of change in bond length when nonstoichiometric antimony-rich $\approx VSbO_4$ forms.[230]

the surface properties of the rutile phase is thus consistent with the observed change in the catalytic behavior in going from a V-rich to an Sb-rich $\approx VSbO_4$ phase, both in terms of the rate of transformation of intermediate propene to acrylonitrile and the ratio between selective versus nonselective propane activation.

4.3.2.5. Role of Microstructure

Formation of the Sb-rich $\approx VSbO_4$ phase depends on the chemistry of preparation of these catalysts and their microstructure. In particular, nonstoichiometric Sb-rich $\approx VSbO_4$ is observed when $\approx VSbO_4$ microcrystals grow over Sb_2O_4 micro- or macrocrystals, explaining the considerable dependence of the catalytic behavior on the method of preparation and Sb/V ratio.[230] Whereas with preparation methods in which vanadium and/or antimony deposit over an Sb^{5+} hydroxide core, $\approx VSbO_4$ microcrystals develop over macro- or microcrystals of Sb_2O_4. In the case of solid state or gel-to-solid methods the mechanism of $\approx VSbO_4$ crystallite formation is different and does not lead to an effective formation of an Sb-rich $\approx VSbO_4$ phase with a consequent drastic plunge in catalytic performance. Thus the chemistry of growth of $\approx VSbO_4$ microcrystals influences their nonstoichiometric characteristics and thus determines their catalytic behavior.

However, it should be noted that since there is a severe type of catalyst restructuring during the catalytic reaction, there are no fixed relationships among optimal catalyst characteristics (phase composition, microstructure, nonstoichiometry). The surface reactivity depends on the feed composition[231] and the optimal catalytic behavior determined for a series of catalysts studied depends on the propane concentration in the feed, the local oxygen to propane ratio, and the reaction temperature.

In conclusion, the chemistry of V/Sb oxide catalysts in relation to their catalytic reactivity in propane ammoxidation to acrylonitrile is quite complex, owing to the presence of: (i) crystalline and amorphous phases; (ii) nonstoichiometry in the vanadium antimonate rutile phase; (iii) dependence of nonstoichiometric characteristics on composition, method of preparation, and feed composition during the catalytic tests; (iv) change in the structural and surface characteristics during the catalytic reaction; and (v) considerable dependence of catalytic performance on the microstructural characteristics of the sample.

Thus, much effort is still necessary to fully understand this structural and surface chemistry and its relationship with catalytic behavior. Present data already indicate that catalytic performance cannot be simply related to phase composition and site isolation, but that a rather more complex structural and surface chemistry, including *in situ* modifications during the catalytic reaction, determines the catalytic behavior, and this is what must be understood in order to improve and tune V/Sb oxide catalyst performance.

4.3.3. Surface Reaction Network as a Tool for Understanding and Controlling Reactivity

The possibility of designing better V/Sb oxide catalysts requires increased knowledge concerning their surface reactivity and catalytic chemistry, but the fact that different catalyst properties determine the surface reactivity in selective oxidation reactions must be taken into account when an alkane is fed in instead of an alkene. This essentially owing to the different chemisorption properties of the two hydrocarbons (see Chapter 8). A direct consequence of this fact is that in propene selective ammoxidation there is one dominant pathway of transformation from propene to acrylonitrile, whereas there are multiple pathways in propane selective ammoxidation.[235-240] A good understanding of the surface reaction network is thus necessary in order to analyze and discuss the catalytic behavior and its dependence on the reaction conditions and catalyst nature, and, in turn, establish a basis for better, more rational design for better catalysts.

Analysis of the surface reaction network in propane conversion requires identification of the surface intermediate species and their different pathways of transformation in the presence and absence of ammonia, but it is also necessary to analyze how and when the reactants themselves or some of the reaction products influence the intrinsic reactivity of the oxide surface and thus self-modify the surface reaction network. Kinetic stationary or transient reactivity studies provide useful general information on these aspects, but do not allow the level of detail required to use the surface reaction network as a tool to discuss and understand the catalytic behavior. One good approach is the use of FT-IR (Fourier transform infrared) spectroscopy,[235,236,238-240] because it gives detailed information on the nature of chemisorbed species and their surface transformations. Using the reactants and the precursors of the reactive intermediates as probe molecules (e.g., isopropyl alcohol to form the isopropylate intermediate species; the selection of the probe molecules is made on the basis of indications on the general reaction network given by reactivity studies) and studying their single-surface transformations[235,236] or their conversion in the presence of other coadsorbate species (e.g., the organic intermediates and ammonia),[238-240] one can derive an overall surface reaction network showing the different possible pathways of transformation.[241]

However, IR spectroscopy should be integrated with parallel reactivity studies, in order to distinguish between catalytically relevant surface adspecies and "spectator" species, which often are the more easily detected and more abundant ones. Particularly valuable is the combination with transient reactivity data, which in the case of propane ammoxidation to acrylonitrile gives useful indications on the type of surface accumulation processes and changes in surface reactivity.[242-244] Although the IR and transient reactivity approaches are complementary and elucidate different aspects of the surface reactivity, it is worth noting that the conclusions are in good agreement, e.g., concerning the role of "short-living" activated ammonia species in the reaction mechanism.[238–240,243,244] However, IR spectroscopy provides a more precise picture of the surface reaction network and of the factors governing the competitive pathways of transformation.

4.3.3.1. The Surface Reaction Network in Propane Ammoxidation over V/Sb Oxide Catalysts

The surface reaction network for propane (amm)oxidation is shown schematically in Scheme 4.2. It should be noted that the chemical formula for the intermediate species does not reflect the actual surface structure of the reactive intermediate, but simply indicates the chemical nature of that species. Some of the reaction steps in Scheme 4.2 are numbered, because they will be used to more clearly indicate the reaction rates of the relative steps in discussing the catalytic behavior in terms of overall surface network.

The main route of propane conversion is through the formation of propene as an intermediate, but a side reaction of acrylate formation via a propionate interme-

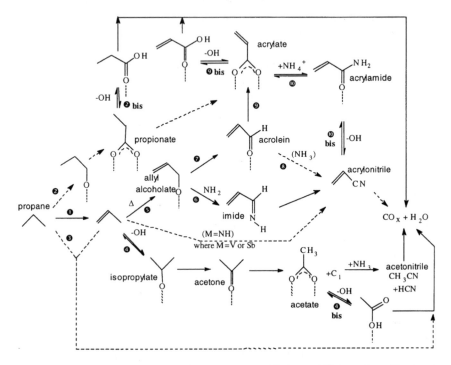

Scheme 4.2. Surface reaction network for propane ammoxidation on VSbO catalysts. The dotted arrows are the less important pathways. M is vanadium or antimony.[237]

diate is possible. The relative rate of this side reaction is higher in the case of oxidation than in ammoxidation, because chemisorption of ammonia inhibits the activity of the Lewis acid sites responsible for the alkane activation according to this side route. Direct conversion of propane to carbon oxides is also indicated, but it should be noted that the supported vanadium oxide species primarily responsible for this route transforms during the first minutes of catalytic reaction, and thus this pathway is less important under stationary conditions, especially in the presence of ammonia.

Propene can be oxidized by two routes, the first reversible and favored at low temperatures with the formation of an isopropylate species by reaction with Brønsted sites, and the second with the formation of an allyl alcoholate by abstraction of allylic hydrogen and nucleophilic lattice oxygen attack. The first adspecies easily transforms to acetone, which may undergo oxidative cleavage to an acetate species and a C1 fragment. The reaction of Brønstead acid sites with ammonia to give ammonium ions slows down the rate of the first nonselective side reaction. The second adspecies (allyl alcoholate) transforms easily to acrolein, which further converts to an acrylate species. The acrylate species interacts strongly with the surface and is thermally stable up to relatively high temperatures, but reacts faster with ammonium ions to give an acrylamide intermediate. At higher temperatures the amide transforms to acrylonitrile, but when Brønsted sites are present, the amide is hydrolyzed to reform ammonia and free, weakly bonded acrylic acid. The latter easily decarboxylates forming carbon oxides. In the presence of strong Brønsted acid sites acrylonitrile also can be hydrolyzed back to free acrylic acid and ammonium ions, and then transformed to carbon oxides. A different activated ammonia adspecies (NH_2-type species formed by heterolytic dissociation of ammonia coordinated to Lewis acid sites) reacts with the allyl alcoholate or acrolein intermediate to form the acrylimine intermediate, which transforms to acrylonitrile faster than the acrylamide intermediate.

The reaction shown in Scheme 4.2 illustrates that there are different pathways to form acrylonitrile on the surface of V/Sb oxides and that the intrinsic selectivity of each pathway is different owing to the different possible side reactions. The overall selectivity to acrylonitrile depends on the relative rates of these competitive pathways, which, in turn, depend on the reaction conditions and the nature of the catalyst.

4.3.3.2. The Surface Reaction Network as a Tool for Understanding the Surface Reactivity

There are several features of the catalytic behavior of V/Sb oxide catalysts for propane ammoxidation that can be rationalized using the surface reaction network described above.

A key aspect of the reactivity of these catalysts is that when propane and propene ammoxidation are compared on the same catalysts under the same reaction conditions, it is observed[237] that the rate constant of propene depletion is one order of magnitude higher than that of propane depletion, and thus when propene forms from propane a rapid further transformation is expected. Instead, as noted in Section 4.3.1, at maximum selectivity for acrylonitrile relatively large amounts of propene are still present. There is thus the apparent contradiction that propene alone is converted to acrylonitrile faster than propane, whereas the propene to acrylonitrile step (in propane conversion) is slower than the propane to propene step.

Another characteristic feature of the ammoxidation of C3 hydrocarbons on V/Sb oxide catalysts is that the selectivity to acrylonitrile in propene conversion is a maximum at high conversion, whereas selectivity for acrylonitrile from propane is a maximum at much lower conversions with a considerable decrease as conversion increases further, indicating that on the same catalyst and at the same reaction temperature acrylonitrile converts faster to carbon oxides when formed from propane than when formed from propene, even though propene is the reaction intermediate from propane. Clearly, understanding this contradictory aspect of the surface reactivity and how to control it may enable maintenance of good acrylonitrile selectivity up to higher propane conversion with a considerable improvement in the process economics.

An important aspect evidenced by IR studies on the coadsorption of possible reaction intermediates and ammonia[238] is the effect of ammonia chemisorption on the catalytic action of the Brønsted acid sites present on the surface of V/Sb oxide catalysts. It was, in fact, observed that they catalyze some nonselective pathways in the surface reaction network (Scheme 4.2), but that ammonia reacting with them forming ammonium ions inhibits their reactivity. For example, reaction 4 is inhibited when they react with ammonia to form ammonium ions, explaining the change in the acetonitrile/acrylonitrile ratio with the increase in the NH_3/C_3H_8 ratio.[240]

Ammonia has different effects on the surface reactivity of V/Sb oxides, apart from being involved as a reactant in the synthesis of acrylonitrile. Chemisorption on the surface Lewis acid sites owing to coordinatively unsaturated vanadium sites inhibits the rate of activation of alkane with a decrease in its rate of depletion, but especially hinders the nonselective pathway 2 → 2bis vs. pathway 1. Ammonia, interacting with Brønsted acid sites, also inhibits their effect in acid-catalyzed reactions and, in particular, inhibits reactions 4 and 2bis, as well as 4bis and 9bis, which lead to carbon chain degradation and free, weakly coordinated acids (2bis, 4bis, 9bis), which easily decarboxylate and further rapidly convert to carbon oxides. It is thus possible to rationalize why ammonia at low concentrations acts mainly as a modifier of the surface reactivity (inhibits pathways 2, 4, and others associated with Brønsted acidity), whereas at higher concentrations it also acts as a reactant for acrylonitrile synthesis.

Olefins interacting with Brønsted acid sites have an effect comparable to that of ammonia. It is useful to discuss this concept on the basis of the surface reaction network (Scheme 4.2): why there is an apparent different stability of acrylonitrile in its formation from propane or propene and why a high selectivity for acrylonitrile requires operation with a high propane concentration in the feed.[14,245]

When propene is used as a reactant, its chemisorption on the Brønsted acid sites combines with ammonia chemisorption to moderate their activity toward catalyzing side reactions. In particular, in propane ammoxidation at high conversion the rates of steps 9bis (degradation of acrylate species) and 10bis → 10 → 9bis (degradation of acrylonitrile) are quite important. It should be noted that due to both the higher rate of transformation and stronger chemisorption of the alkene with respect to alkane, the ratio of the rates of oxygen vs. nitrogen addition and thus the ratio of rates of pathway 7 → 9 vs. 6 and 8 is higher for alkane ammoxidation than for alkene ammoxidation. The formation of acrylate is thus faster in the case of propane transformation than using propene directly as a reactant, even though propene is the main intermediate in propane to acrylonitrile, because the surface population of adspecies is different and thus the surface reactivity during catalytic reaction is also different (see also Chapter 8). Although the acrylate species is rather stable, it can be transformed to weakly coordinated acrylic acid in the presence of Brønsted acid sites and water. This weakly coordinated acid is instead rather unstable and quickly decarboxylates and further oxidizes to carbon oxides. Owing to the presence of this side reaction and the higher rate of pathway 5 → 7 → 9 with respect to pathway 5 → 6 in propane ammoxidation with respect to propene ammoxidation, the selectivity to acrylonitrile from propane is lower than from propene (on the same catalyst and under the same reaction conditions). Furthermore, the propene concentration is lower when it forms from propane with respect to the case of directly feeding propene, and thus the "self-protection" effect on the nonselective catalysis by Brønsted acid sites is also reduced. At high propane conversion, where the concentrations of ammonia and propene intermediates are lower (thus their inhibiting effect on Brønsted acid activity is weaker) and the concentration of water higher (water is a main reaction product needed to shift the equilibrium toward the acids), the rates of reaction steps 4 and 9bis and of the pathway 10bis → 10 → 9 (acrylonitrile degradation) become dominant, and thus the selectivity to acrylonitrile becomes very low. On the other hand, in propene conversion, the residual propene is still enough to "shield" acrylonitrile toward its degradative pathway (10bis → 10 → 9) and thus selectivity remains high up to nearly complete conversion.

The lower concentration of surface adspecies in the case of propane ammoxidation with respect to propene ammoxidation not only favors the side reactions on organic adspecies, but also favors the side oxidation of ammonia to N_2, which reduces the effective number of surface species available for N insertion in the intermediates and thus acrylonitrile formation). This is the primary reason that the rate of propene to acrylonitrile transformation is apparently slower with respect to the same reaction, but when propene is fed in directly.

By using the surface reaction network and an analysis of the effect of chemisorption of reactants and intermediate species on the surface properties, it is possible to rationalize a series of unanswered questions on the surface reactivity of V/Sb oxide catalysts and thus enable better design.

4.3.3.3. Designing Better Catalysts

In the previous section it was shown that the reactivity of Brønsted acid sites can be inhibited by reaction with ammonia to form ammonium ions and thus the number of ammonium ions in standard test conditions should be related to the selectivity to acrylonitrile in a homogeneous series of samples. The results shown in Figure 4.8, although for only a limited number of catalysts, show that this concept is valid. On the other hand, it is not possible to simply kill Brønsted acid sites by adding, e.g., alkaline metals, because the formation of ammonium ions and the equilibrium with ammonia adsorbed as such on Lewis acid sites (first step for its heterolytic splitting)[235,236] are needed to keep a sufficient quantity of ammonia adspecies on the surface for acrylonitrile synthesis. It is thus necessary to tune the surface acido-base properties, which can be done both by a change in the Sb/V ratio and control of the nonstoichiometric characteristics of the rutilelike $\approx$VSbO$_4$ phase (see previous section).

The optimal catalytic behavior thus requires a balance between catalyst properties and effect on the surface reactivity of the chemisorption of the reactants and intermediate species. A consequence of this observation is that the optimal catalytic behavior in a homogeneous series of samples depends on the feed composition, and different catalyst compositions for V/Sb oxide samples can be optimal depending on the reaction conditions,[231] as was discussed in the previous section. A maximum in selectivity or productivity for acrylonitrile depends on the feed composition, because a change, e.g., in the rates of the $5 \rightarrow 6$ pathway vs. $5 \rightarrow 7 \rightarrow 9$ pathway (Scheme 4.2) should be reflected in different optimal reaction conditions. This shows how erroneous indications can be obtained if the behavior of catalysts is compared for only a single set of reaction conditions.

The discussion on the relationship between the surface reaction network and catalytic behavior also shows that the addition of small doses of gas dopants to the feed can be a useful way to tune the surface reactivity in a manner analogous to that previously discussed regarding the effect of ammonia and alkenes on the surface selectivity. This is a very open field of research, which although not yet substantially investigated, can be very fruitful for improving catalytic behavior.

The study of the surface reaction network is thus not only a good approach for understanding the surface catalytic chemistry in complex selective oxidation reac-

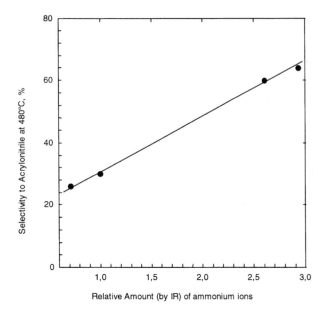

Figure 4.8. Relationship between the relative amount of ammonium ions in test conditions in a series of VSbO catalysts and selectivity to acrylonitrile at 480°C for the same samples.[238]

tions, but can also be a useful tool for determining which kinds of modifications in catalyst properties or reaction conditions can enable optimization of the catalytic behavior.

4.3.3.4. Conclusions

Analysis of the structure/composition and activity relationships suggests that the main active phase for the selective synthesis of acrylonitrile from propane is an Sb^{5+}-rich $\approx VSbO_4$ phase, although other phases participate in determining the overall catalytic behavior. Not only its presence is important, but also its non-stoichiometry characteristics, which in turn are influenced by the catalyst micro-structure determined by the preparation methodology.

The role of ammonia in the modification of surface reactivity (acid–base and redox properties) has also been shown as well as the presence of a complex surface reaction network, which explains various key features observed in the catalytic behavior.

The design of new improved catalysts for propane ammoxidation thus requires a better understanding of all the above factors and finding a way to control them, but the study of this reaction offers several clues for the development of new classes of selective catalysts in alkane functionalization.

REFERENCES

1. F. Cavani and F. Trifirò, *Appl. Catal. A* **88**, 115 (1992).

2. F. Cavani and F. Trifirò, *CHEMTECH*, **24**, 18 (1994).

3. F. Trifirò and F. Cavani, *Selective Partial Oxidation of Hydrocarbons and Related Oxidations*, Catalytica Studies Nr. 4193 SO (1994).

4. M. Malow, *Hydrocarbon Proc.* **11**, 149 (1980).

5. R.A. Varma and D.N. Saraf, *Ind. Eng. Chem. Prod. Res. Dev.* **18**, 7 (1979).

6. B.K. Hodnett, *Catal. Rev.-Sci. Eng.* **27**, 373 (1985).

7. G. Centi, F. Trifirò, J. Ebner, and V. Franchetti, *Chem. Rev.* **88**, 55 (1988).

8. G. Hutching, *Appl. Catal.* **72**, 1 (1991).

9. G. Centi (ed.), *Vanadyl Pyrophosphate Catalyst*, Vol. 16(1) of *Catal. Today* (1993).

10. F. Cavani and F. Trifirò, *Catalysis, Vol. 11*, Royal Society of Chemistry, Cambridge (1994), p. 246.

11. M. Abon and J.C. Volta, *Appl. Catal. A* **157**, 173 (1997).

12. F. Cavani and F. Trifirò, *Appl. Catal. A* **157**, 195 (1997).

13. G. Centi, R.K. Grasselli, and F. Trifirò, *Chim. Ind. (Milan)* **72**, 617 (1990).

14. G. Centi, R.K. Grasselli, and F. Trifirò, *Catal. Today* **13**, 661 (1992).

15. Y. Moro-oka and W. Ueda, *Catalysis, Vol. 11*, Royal Society of Chemistry: Cambridge (1994), p. 223.

16. V.D. Sokolovskii, A.A. Davydov, and O.Yu. Ovsitser, *Catal. Rev.-Sci. Eng.* **37**, 425 (1995).

17. G. Centi, S. Perathoner, and F. Trifirò, *Appl. Catal. A: General* **157**, 143 (1997).

18. D. Ferri, *Tecnol. Chim. (Italy)* **1**, 26 (1993).

19. H. Haase, *Chem. Ing. Tech.* **1/2**, 80 (1972).

20. N. Harris and M.W. Tuck, *Hydrocarbon Proc.* **69**, 79 (1990).

21. J.C. Burnett, R.A. Keppel, and W.D. Robinson, *Catal. Today* **1**, 537 (1987).

22. R.K. Sharma, D.L. Cresswell, and E.J. Newson, *AIChE J.* **37**, 39 (1991).

23. A. Gianetto, S. Pagliolico, G. Rovero, and B. Ruggeri, *Chem. Eng. Sci.* **45**, 2219 (1990).

24. R.M. Contractor, in: *Circulating Fluidized Bed Technology II* (P. Basu and J.F. Large, eds.), Pergamon Press: Toronto (1988), p. 467.

25. R.M. Contractor, J. Chaouki, in: *Circulating Fluidized Bed Technology III* (P. Basu, M. Horio, and M. Hasatani, eds.), Pergamon Press: Toronto (1990), p. 39.

26. R.M. Contractor and A.W. Sleight, *Catal. Today* **1**, 587 (1987).

27. R.M. Contractor, G.S. Patience, D.I. Garnett, H.S. Horowitz, G.M. Sisler, and H.E. Bergna, in: *Proceedings 4th International Conference on Circulating Fluid Beds*, Sommerset, Penn., AIChE: New York (1993).

28. R.M. Contractor, D.I. Garnett, H.S. Horowitz, H.E. Bergna, G.S. Patience, J.T. Schwartz, and G.M. Sisler, in: *New Developments in Selective Oxidation II* (V.

Corberan and S. Vic Bellon, eds.), Studies in Surface Science and Catalysis Vol. 82, Elsevier Science: Amsterdam (1994), p. 233.

29. G. Emig, K. Uihlein, and C. Häcker, in: *New Developments in Selective Oxidation II* (V. Corberan and S. Vic Bellon, eds.), Studies in Surface Science and Catalysis Vol. 82, Elsevier Science: Amsterdam (1994), p. 243.

30. M. Hatano, M. Masayoshi, K. Shima, and M. Ito, *US Patent* 5,128,299 (1992).

31. M.J. Desmond and M.A. Pepera, *US Patent* 4,801,569 (1989).

32. H. Taheri, *US Patent* 5,011,945 (1991).

33. G. Stefani and P. Fontana, *US Patent* 4,178,298 (1989).

34. P.R. Blum, E.C. Milberger, and M.L. Nicholas, *US Patent* 4,748,140 (1988).

35. R.O. Kerr, *US Patent* 3,474,041 (1969).

36. H. Taheri, *US Patent* 5,117,007 (1992).

37. T.C. Click and B.J. Barone, *US Patent* 4,515,899 (1985).

38. R.C. Edwards, *US Patent* 4,810,803 (1989).

39. G.D. Suciu, G. Stefani, and C. Fumagalli, *US Patent* 4,594,433 (1986).

40. N.J. Bremer, D.E. Dria, P.R. Blunm, E.C. Milberger, and M.L. Nicholas, *US Patent* 4,525,471 (1985).

41. B.J. Barone, *US Patent* 5,158,923 (1992).

42. J. Takashi, T. Kiyoura, Y. Kogure, and K. Kanaya, *US Patent* 5,155,235 (1992).

43. M.J. Mummey, *EP Patent* 326,536 A1 (1989).

44. G. Chinchen, P. Davies, and R.J. Sampson, in: *Catalysis: Science and Development*, Vol. 8 (J.R. Anderson and M. Boudart, eds.), Springer-Verlag: Berlin (1987), p. 1.

45. G. Emig and F. Martin, *Catal. Today* **1**, 477 (1987).

46. G. Stefani, F. Budi, C. Fumagalli, and G.D. Suciu, *Chim. Ind. (Milan)* **72**, 604 (1990).

47. G. Stefani, F. Budi, C. Fumagalli, and G.D. Suciu, in: *New Developments in Selective Oxidation* (G. Centi and F. Trifirò, eds.), Studies in Surface Science and Catalysis Vol. 55, Elsevier Science: Amsterdam (1990), p. 537.

48. J.R. Ebner and M.R. Thompson, in: *Structure-Activity and Selectivity Relationships in Heterogeneous Catalysis* (R.K. Grasselli and A.W. Sleight, eds.), Studies in Surface Science and Catalysis Vol. 67, Elsevier Science: Amsterdam (1991), p. 31.

49. R. Contractor, J.R. Ebner, and M.J. Mummey, in: *New Developments in Selective Oxidation* (G. Centi and F. Trifirò, eds.), Studies in Surface Science and Catalysis Vol. 55, Elsevier Science: Amsterdam (1990), p. 553.

50. H.S. Horowitz, C.M. Blackstone, and A.W. Sleight, *Appl. Catal.* **38**, 193 (1988).

51. I. Matsuura, *Catal. Today* **16**, 123 (1993).

52. T. Okuhara and M. Misono, *Catal. Today* **16**, 61 (1993).

53. L. Cornaglia, C.A. Sanchez, and E.A. Lombardo, *Appl. Catal. A* **95**, 117 (1993).

54. E. Bordes, *Catal. Today* **16**, 27 (1993).

55. D. Ye, A. Satsuma, A. Hattori, T. Hattori, and Y. Murakami, *Catal. Today* **16**, 113 (1993).

56. F. Cavani, A. Colombo, F. Giuntoli, F. Trifirò, P. Vazquez, and P. Venturoli, in: *Advanced Catalysis and Nanostructured Materials* (W.R. Moser, ed.), Academic Press: New York (1996), Ch. 3, p. 43.

57. F. Cavani and F. Trifirò, in: *Preparation of Catalysts VI* (G. Poncelet, J. Martens, B. Delmon, P.A. Jacobs, and P. Grange, eds.), Studies in Surface Science and Catalysis Vol. 91, Elsevier Science: Amsterdam (1995), p. 1.

58. G. Busca, F. Cavani, G. Centi, and F. Trifirò, *J. Catal.* **99**, 400 (1986).

59. M. Guilhoume, M. Roullet, G. Pajonk, B. Grzybowska, and J.C. Volta, in: *New Developments in Selective Oxidation by Heterogeneous Catalysis* (P. Ruiz and B. Delmon, eds.), Studies in Surface Science and Catalysis Vol. 72, Elsevier Science: Amsterdam (1992), p 255.

60. V.A. Zazhigalov, G.A. Komashko, A.I. Pyatnitskaya, V.M. Belousov, J. Stoch, and J. Haber, in: *Preparation of Catalysts V* (G. Poncelet, P.A. Jacobs, P. Grange, and B. Delmon, eds.), Studies in Surface Science and Catalysis, Vol. 63, Elsevier Science: Amsterdam (1991), p. 497.

61. T. Miyake and T. Doi, in: *3rd World Congress on Oxidation Catalysis* (R.K. Grasselli, S.T. Oyama, A.M. Gaffney, and J.E. Lyons, eds.), Studies in Surface Science and Catalysis Vol. 110, Elsevier Science: Amsterdam (1997), p. 835.

62. H. Igarashi, K. Tsuji, T. Okuhara, and M. Misono, *J. Phys. Chem.* **97**, 7065 (1993).

63. G.K. Bethke, D. Wang, J.M.C. Bueno, M.C. Kung, and H.H. Kung, in: *3rd World Congress on Oxidation Catalysis* (R.K. Grasselli, S.T. Oyama, A.M. Gaffney, and J.E. Lyons, eds.), Studies in Surface Science and Catalysis Vol. 110, Elsevier Science: Amsterdam (1997), p. 453.

64. I.J. Ellison, G.J. Hutchings, M.T. Sananes, and J.C. Volta, *J. Chem. Soc. Chem. Comm.* 1093 (1994).

65. M.T. Sananes, G.J. Hutchings, and J.C. Volta, *J. Chem. Soc. Chem. Comm.* 243 (1995).

66. G.J. Hutchings, M.T. Sananes, S. Sajip, C.J. Kiely, A. Burrows, I.J. Ellison, and J.C. Volta, *Catal. Today* **33**, 161 (1997).

67. J.B. Benziger, V. Guliants, and S. Sundaresam, *Catal. Today* **33**, 49 (1997).

68. E. Bordes, *Catal. Today* **1**, 499 (1987).

69. E. Bordes, J.W. Johnson, and P. Courtine, *J. Solid State Chem.* **55**, 270 (1984).

70. S. Urvin-Monshaw and A. Klein, *Chem. Eng.* **96**, 35 (1989).

71. G. Bergeret, M. David, J.P. Broyer, J.C. Volta, and G. Hecquet, *Catal. Today* **1**, 37 (1987).

72. E. Bordes, in: *Structure–Activity and Selectivity Relationships in Heterogeneous Catalysis* (R.K. Grasselli and A.W. Sleight, eds.), Studies in Surface Science and Catalysis Vol. 67, Elsevier Science: Amsterdam (1991), p. 21.

73. I. Matsuura and M. Yamazaki, in: *New Developments in Selective Oxidation* (G. Centi and F. Trifirò, eds.), Studies in Surface Science and Catalysis Vol. 55, Elsevier Science: Amsterdam (1990); p. 563.

74. F.B. Abdelouahab, R. Olier, N. Guilhaume, F. Lefevre, and J.C. Volta, *J. Catal.* **134**, 151 (1992).

75. M. O'Connor, F. Dason, and B.K. Hodnett, *Appl. Catal.* **64**, 161 (1990).

76. F. Cavani, G. Centi, F. Trifirò, and R.K. Grasselli, *Catal. Today* **3**, 185 (1988).

77. V.A. Zazhigalov, V.M. Belousov, A.I. Pyatnitskaya, G.A. Komashko, Yu.N. Merkureva, and J. Stoch, in: *New Developments in Selective Oxidation* (G. Centi and F. Trifirò, eds.), Studies in Surface Science and Catalysis Vol. 55, Elsevier Science: Amsterdam (1990); p. 617.

78. G. Koyano, T. Okuhara, and M. Misono, *Catal. Lett.* **32**, 205 (1995).

79. V.V. Guliants, J.B. Benziger, S. Sundaresan, I.E. Wachs, J.M. Jehng, and J.E. Roberst, *Catal. Today* **28**, 275 (1996).

80. F. Cavani, G. Centi, F. Trifirò, and G. Poli, *J. Thermal Anal.* **30**, 1241 (1985).

81. V.V. Guliants, J.B. Benziger, S. Sundaresan, N. Yao, and I.E. Wachs, *Catal. Lett.* **32**, 379 (1995).

82. S. Albonetti, F. Cavani, F. Trifirò, P. Venturoli, G. Calestani, M. Lopez Granados, and J.L.G. Fierro, *J. Catal.* **160**, 52 (1996).

83. Y. Schuurman and J.T. Gleaves, *Catal. Today* **33**, 25 (1977).

84. L.M. Cornaglia, C. Caspani, and E.A. Lombardo, *Appl. Catal.* **74**, 15 (1991).

85. E.A. Lombardo, C.A. Sanchez, and L.M. Cornaglia, *Catal. Today* **15**, 407 (1992).

86. G.A. Sola, B.T. Pierini, and J.O. Petunchi, *Catal. Today* **15**, 537 (1992).

87. J.R. Ebner and M.R. Thompson, *Catal. Today* **16**, 51 (1993).

88. G.W. Coulston, E.A. Thompson, and N. Herron, *J. Catal.* **163**, 122 (1996).

89. G.Busca and G.Centi, *J. Am. Chem. Soc.* **111**, 46 (1989).

90. G. Centi, G. Golinelli, and G. Busca, *J. Phys. Chem.* **94**, 6813 (1990).

91. G. Centi, F. Trifirò, G. Busca, J. Ebner, and J. Gleaves, *Faraday Discuss. Chem. Soc.* **87**, 215 (1989).

92. J.C. Volta, *Catal. Today* **32**, 29 (1996).

93. J.M. Herrmann, P. Vernoux, K.E. Béré, and M. Abon, *J. Catal.* **167**, 106 (1997).

94. M.T. Sananes-Schulz, A. Tuel, G.J. Hutchings, and J.C. Volta, *J. Catal.* **166**, 388 (1997).

95. A. Tuel, M.T. Sananes-Schulz, and J.C. Volta, *Catal. Today* **37**, 59 (1997).

96. M. Abon, K.E. Bere, A. Tuel, P. and Delichere, *J. Catal.* **156**, 28 (1995).

97. G.J. Hutchings, A. Desmartin-Chimel, R. Olier, and J.C. Volta, *Nature* **368**, 41 (1994).

98. T. Okuhara, K. Tsuji, H. Igarashi, and M. Misono, in: *Catalytic Science and Technology*, Vol. 1 (S. Yoshida, N. Takezawa, T. Ono, eds.), Kodansha: Tokyo (1990), p. 443.

99. N. Yamazoe, H. Morishige, and Y. Taraoka, in: *Successful Design of Catalysts* (T. Inui, ed.), Studies in Surface Science and Catalysis Vol. 44, Elsevier Science: Amsterdam (1988), p. 15.

100. J. Haas, C. Plog, W. Maunz, K. Mittag, K.D. Gollmer, and B. Klopries, in: *Proceedings, 9th International Congress on Catalysis*, Vol. 4 (M.J. Philips and M. Ternan, eds.), Chemical Institute of Canada: Toronto (1988), p. 1632.

101. F. Cavani, G. Centi, and F. Trifirò, *Chim. Ind. (Milan)* **74**, 182 (1992).

102. F. Cavani, G. Centi, F. Trifirò, and A. Vaccari, in: *Adsorption and Catalysis on Oxide Surfaces* (M. Che and G.C. Bond, eds.), Studies in Surface Science and Catalysis Vol. 21, Elsevier Science: Amsterdam (1985), p. 287.

103. F. Cavani, G. Centi, and F. Trifirò, *Appl. Catal.* **15**, 151 (1985).

104. G. Centi and F. Trifirò, *Chim. Ind. (Milan)* **68**, 74 (1986).

105. G. Centi, K. Dyrek, M. Labanowska, and F. Trifirò, in: *Heterogeneous Catalysis, Proceedings, 6th International Symposium, Sofia 1987*, Vol. 1, Bulgarian Academy of Sciences: Sofia (1987), p. 406.

106. G. Centi, G. Fornasari, and F. Trifirò, *J. Catal.* **89**, 44 (1984).

107. U. Rodemerck, B. Kubias, H.W. Zanthoff, G.U. Wolf, and M. Baerns, *Appl. Catal. A* **153**, 217 (1997).

108. P.L. Gai and K. Kourtakis, *Science* **267**, 661 (1995).

109. A. Brückner, B. Kubias, and B. Lücke, *Catal. Today* **32**, 215 (1996).

110. G.J. Hutchings, *Appl. Catal.* **72**, 1 (1991).

111. G.J. Hutchings, *Catal. Today* **16**, 139 (1993).

112. V. Zazhigalov, J. Haber, J. Stoch, A.I. Pyanitzkaya, G.A. Komashko, and V.M Belousov, *Appl. Catal. A* **96**, 135 (1993).

113. S.K. Bej and M.S. Rao, *Appl. Catal. A* **83**, 149 (1992).

114. M.T. Sananes, J.O. Petunchi, and E.A. Lombardo, *Catal. Today* **15**, 527 (1992).

115. D. Ye, A. Satsuma, A. Hattori, T. Hattori, and Y. Murakami, *Catal. Today* **16**, 113 (1993).

116. M. Meisel, G.U. Wolf, and A. Brückner, in: *Proceedings, DGMK Conference on Selective Oxidation in Petrochemistry*, Goslar (Germany), DGMK: Hamburg (1992), p. 27.

117. N. Harrouch Batis, H. Batis, and A. Ghorbel, *Appl. Catal. A* **147**, 347 (1996).

118. W.-H. Cheng, *Appl. Catal. A* **147**, 55 (1996).

119. V. Zazhigalov, J. Haber, J. Stoch, I.V. Bacherikova, G.A. Komashko, and A.I. Pyanitzkaya, *Appl. Catal. A* **134**, 225 (1996).

120. G.J. Hutchings and R. Higgings, *Appl. Catal. A* **154**, 103 (1997).

121. G.J. Hutchings and R. Higgings, *J. Catal.* **162**, 153 (1996).

122. M.T. Sananes-Schulz, F.B. Abdelouahab, G.J. Hutchings, and J.C. Volta, *J. Catal.* **163**, 346 (1996).

123. F. Cavani, A. Colombo, F. Trifirò, M.T. Sananes-Schultz, J.C. Volta, and G.J. Huthings, *Catal. Lett.* **43**, 241 (1997).

124. M.T. Sananes, G.J. Hutchings, and J.C. Volta, *J. Catal.* **154**, 253 (1995).
125. G.Z. Ladwig, *Z. Anorg. Chem.* **338**, 266 (1965).
126. R. Gopal and C. Calvo, *J. Solid State Chem.* **5**, 432 (1972).
127. J.W. Johnson, D.C. Johnson, A.J. Jacobson, and J.F. Brody, *J. Am. Chem. Soc.* **106**, 8123 (1984).
128. C.C. Torardi and J.C. Calabrese, *Inorg. Chem.* **23**, 1308 (1984).
129. M.E. Leonowicz, J.W. Johnson, J.F. Brody, H.F. Shannon, and J.M. Newsam, *J. Solid State Chem.* **56**, 370 (1985).
130. G. Villeneuve, *Mat. Res. Bull.* **21**, 621 (1986).
131. Y.E. Gorbunova and S.A. Linde, *Dokl. Akad. Nauk. SSSR* **245**, 5864 (1979).
132. B.C. Tofield, G.R. Crane, G.A. Aspeur, and R.C.J. Sherwood, *J. Chem. Soc. Dalton Trans.* 1806 (1975).
133. J.W. Johnson, D.C. Johnson, H.E. King, T.R. Halbert, J.F. Brody, and D.P. Goshorn, *Inorg. Chem.* **27**, 1646 (1988).
134. K.H. Lii and C.S. Lee, *Inorg. Chem.* **29**, 3298 (1990).
135. A. Datta, A.R. Saple, and R.Y. Kelkar, *J. Chem. Soc. Chem. Comm.* 1645 (1991).
136. M.R. Antonio, R.L. Barbour, and P.R. Blum, *Inorg. Chem.* **26**, 1235 (1987).
137. J.W. Jacobson, D.C. Johnston, J.F. Brody, J.C. Scanlon, and J.T. Lewandowski, *Inorg. Chem.* **24**, 1782 (1985).
138. D.C. Johnston, A.J. Jacobson, J.F. Brody, and J.T. Lewandowski, *Inorg. Chem.* **23**, 3842 (1984).
139. J.W. Johnson, A.J. Jacobson, W.M. Butler, S.E. Rosenthal, J.F. Brody, and J.T. Lewandowski, *J. Am. Chem. Soc.* **111**, 381 (1989).
140. G. Huan, A.J. Jacobson, J.W. Johnson, and E.W. Concoran Jr., *Chem. Mater.* **2**, 91 (1990).
141. J.W. Johnson, J.F. Brody, and R.M. Alexander, *Chem. Mater.* **2**, 198 (1990).
142. G. Huan, J.W. Johnson, A.J. Jacobson, and J.S. Merola, *Solid State Chem.* **89**, 220 (1990).
143. V. Guliants, J.B. Benziger, and S. Sundaresan, *Chem. Mater.* **7**, 1493 (1995).
144. G. Centi, *Catal. Today* **16**, 5 (1993).
145. G. Busca, G. Centi, and F. Trifirò, *J. Am. Chem. Soc.* **107**, 7757 (1985).
146. L.M. Cornaglia, E.A. Lombardo, J.A. Anderson, and J.L. Fierro, *Appl. Catal. A* **100**, 37 (1993).
147. S.J. Puttock and C.H. Rochester, *J. Chem. Soc. Faraday Trans. I* **82**, 2773 (1986).
148. G. Busca, G. Centi, F. Trifirò, and V. Lorenzelli, *J. Phys. Chem.* **90**, 1337 (1986).
149. B. Schiøtt, K.A. Jørgensen, and R. Hoffmann, *J. Phys. Chem.* **95**, 2298 (1991).
150. B. Schiøtt and K.A. Jørgensen, *Catal. Today* **16**, 79 (1993).
151. P.A. Agaskar, L. De Daul, and R.K. Grasselli, *Catal. Lett.* **23**, 339 (1994).
152. M.E. Lashier, T.P. Moser, and G.L. Schrader, in: *New Developments in Selective Oxidation* (G. Centi and F. Trifirò, eds.), Studies in Surface Science and Catalysis Vol. 55, Elsevier Science: Amsterdam (1990); p. 573.

153. M.E. Lashier and G.L. Schrader, *J. Catal.* **128**, 113 (1991).

154. Y. Zhang, R.P.A. Sneeden, and J.C. Volta, *Catal. Today* **16**, 39 (1993).

155. F. Cavani, S. Albonetti, and F. Trifirò, *Catal. Rev.-Sci. Eng.* **38** (1996) 413.

156. M. Abon, K.E. Béré, and P. Delichére, *Catal. Today* **33**, 15 (1997).

157. Y. Schuurman and J.T. Gleaves, *Ind. Eng. Chem. Res.* **33**, 2935 (1994).

158. M.A. Pepera, J.L. Callahan, M.J. Desmond, E.C. Milberger, P.R. Blum, and N.J. Bremer, *J. Am. Chem. Soc.* **107**, 4883 (1985).

159. J. Ziolkowski, E. Bordes, and P. Courtine, *J. Mol. Catal.* **84**, 307 (1993).

160. J. Ziolkowski, E. Bordes, and P. Courtine, *J. Catal.* **122**, 126 (1990).

161. M.R. Thompson and J.R. Ebner, in: *New Developments in Selective Oxidation by Heterogeneous Catalysis* (P. Ruiz and B. Delmon, eds.), Studies in Surface Science and Catalysis Vol. 72, Elsevier Science: Amsterdam (1992), p. 353.

162. J.T. Gleaves and G. Centi, *Catal. Today* **16**, 69 (1993).

163. J.R. Ebner and J.T. Gleaves, in: *Oxygen Complexes and Oxygen Activation by Transition Metals* (A.E. Martell and D.T. Sawyer, eds.), Plenum Press: New York (1988), p. 273.

164. G. Centi, F. Trifirò, G. Busca, J.R. Ebner, and J.T. Gleaves, *Faraday Discuss. Chem. Soc.* **87**, 215 (1989).

165. G. Gleaves, J.R. Ebner, and T.C. Kuechler, *Catal. Rev.-Sci. Eng.* **30**, 49 (1988).

166. D. Dowell and J.T. Gleaves, in: *3rd World Congress on Oxidation Catalysis* (R.K. Grasselli, S.T. Oyama, A.M. Gaffney, and J.E. Lyons, eds.), Studies in Surface Science and Catalysis Vol. 110, Elsevier Science: Amsterdam (1997), p. 199.

167. W. Partenheimer and B.L. Meyers, *Appl. Catal.* **51**, 13 (1989).

168. J.R. Ebner and W.J. Andrews, *US Patent* 5,137,860 (1991).

169. E.W. Arnold and S. Sundaresan, *Appl. Catal.* **41**, 225 (1988).

170. R.M. Contractor, H.S. Horowitz, G.M. Sisler, and E. Bordes, *Catal. Today* **37**, 51 (1997).

171. H.W. Zanthoff, M. Sananes-Schultz, S.A. Buchholz, U. Rodemerck, B. Kubias, and M. Baerns, **172**, 49 (1998).

172. B. Kubias, F. Richter, H. Papp, A. Krepel, A. Kretschmer, in: *3rd World Congress on Oxidation Catalysis* (R.K. Grasselli, S.T. Oyama, A.M. Gaffney, and J.E. Lyons, eds.), Studies in Surface Science and Catalysis Vol. 110, Elsevier Science: Amsterdam (1997), p. 461.

173. P.L. Gai, K. Kourtakis, D.R. Coulston, and G.C. Sonnichsen, *J. Phys. Chem. B* **101**, 9916 (1997).

174. E. Bordes and R.M. Contractor, *Topics Catal.* **3**, 365 (1996).

175. G. Centi, in: *Elementary Reaction Steps in Heterogeneous Catalysis* (R.W. Joyner and R.A. van Santen, eds.), Kluwer Academic: Dordrecht (1993), p. 93.

176. B. Kubias, U. Rodemerck, H.W. Zanthoff, and M. Meisel, *Catal. Today* **32**, 243 (1996).

177. G. Centi and F. Trifirò, *Chem. Eng. Science* **45**, 2589 (1990).

178. G. Centi, *Catal. Lett.* **22**, 53 (1993).
179. A. Pavone and R.H. Schwaar, *PEP Rev.* 88-2-4 (1989).
180. Newsletter, *Appl. Catal.* **67**, N5 (1990).
181. R.H. Kahney and T.D. McMinn, *US Patent* 4,000,178 (1976).
182. M. Hatano and A. Kayo, *EP Patent* 318,295 A1 (1988).
183. T. Ushikubo, K. Oshima, T. Umezawa, and K. Kiyono, *EP Patent* 512,846 A1 (1992).
184. K. Oshima, A. Kayo, T. Umezawa, K. Kiyono, and I. Sawaki, *EP Patent* 529,853 A2 (1992).
185. S. Albonetti, G. Blanchard, P. Burattin, F. Cavani, S. Masetti, and F. Trifirò, *Catal. Today* **42**, 283 (1998) and EP Patent 691, 306 (1996).
186. Power Gas and ICI, *FR Patent* 2,072,399 (1972).
187. C.S. Lynch, L.C. Glaeser, J.F. Brazdil, and M.A. Toft, *US Patent* 5,094,989 (1992).
188. J.P. Bartek, A.M. Ebner, and J.R. Brazdil, *US Patent* 5,198,580 (1993).
189. J.F. Brazdil, L.C. Glaeser, and M.A. Toft, *US Patent* 5,079,207 (1992).
190. D.D. Suresh, M.J. Seeley, J.R. Nappier, and M. Friedrich, *US Patent* 5,171,876 (1992).
191. M.J. Seeley, M.S. Friedrich, and D.D. Suresh, *US Patent* 4,978,764 (1990).
192. L.C. Glaeser, J.F. Brazdil, and M.A. Toft, *US Patent* 4,837,191 (1989).
193. Power Gas and ICI, *FR Patent* 2,072,334 (1972).
194. M.A. Toft, J.F. Brazdil, and L.C. Glaeser, *US Patent* 4,784,979 (1988).
195. G. Centi and F. Trifirò, *Catal. Rev.-Sci. Eng.* **28**, 165 (1986).
196. F. Sala and F. Trifirò, *J. Catal.* **41**, 1 (1976).
197. V. Fattore, Z.A. Fuhrman, G. Manara, and B. Notari, *J. Catal.* **37**, 223 (1975).
198. I. Aso, S. Furukawa, V. Yamazoe, and T. Seiyama, *J. Catal.* **64**, 29 (1980).
199. R.G. Teller, J.F. Brazdil, and R.K. Grasselli, *J. Chem. Soc. Faraday Trans. I* **81**, 1693 (1985).
200. M. Carbucicchio, G. Centi, and F. Trifirò, *J. Catal.* **91**, 85 (1985).
201. M.D. Allen, S. Poulston, E.G. Bithell, M.J. Goringe, and M. Bowker, *J. Catal.* **163**, 204 (1996).
202. M.D. Allen and M. Bowker, *Catal. Lett.* **33**, 269 (1995).
203. S.R.G. Carrazan, L. Cadus, Ph. Dieu, P. Ruiz, and B. Delmon, *Catal. Today* **32**, 311 (1996).
204. T. Yoshino, S. Saito, Y. Sasaki, and Y. Nakamura, *JP Patent* 3,686,138 (1972).
205. T. Yoshino, S. Saito, Y. Sasaki, and Y. Nakamura, *JP Patent* 3,657,155 (1972).
206. V.P. ShChukim, G.K. Borevskov, S.A. Ven'yaminov, and D.V. Tarasova, *Kinet. Katal.* **11**, 153 (1970).
207. F.J. Berry, J.G. Holden, and M.H. Loretto, *J. Chem. Soc. Faraday Trans. I* **83**, 615 (1987).
208. F.J. Berry, J.G. Holden, and M.H. Loretto, *Solid State Commun.* **59**, 397 (1986).
209. T. Birchall and A.E. Sleight, *Inorg. Chem.* **15**, 868 (1976).
210. F.J. Berry, M.E. Brett, and W.R. Patterson, *J. Chem. Soc. Dalton* 9 (1983).

211. F.J. Berry, M.E. Brett, and W.R. Patterson, *J. Chem. Soc. Dalton* 13 (1983).

212. R. G. Teller, M.R. Antonio, J.F. Brazdil, and R.K. Grasselli, *J. Solid State Chem.* **64**, 249 (1986).

213. S. Hansen, K. Ståhl, R. Nilsson, and A. Andersson, *J. Solid State Chem.* **102**, 340 (1993).

214. A. Landa-Canovas, J. Nilsson, S. Hansen, K. Ståhl, and A. Andersson, *J. Solid State Chem.* **116**, 369 (1995).

215. G. Centi, E. Foresti, and F. Guarneri, in: *New Developments in Selective Oxidation II* (V. Cortes Corberan and S. Vic Bellon, eds.), Studies in Surface Science and Catalysis Vol. 82, Elsevier Science: Amsterdam (1994), p. 281.

216. G. Centi and S. Perathoner, in: *Preparation of Catalysts VI* (G. Poncelet *et al.*, eds.), Studies in Surface Science and Catalysis, Vol. 63, Elsevier Science: Amsterdam (1995), p. 59.

217. R. Nilsson, T. Lindblad, and A. Andersson, *J. Catal.* **148**, 501 (1994).

218. J. Nilsson, A.R. Landa-Canovas, S. Hansen, and A. Andersson, *Catal. Today* **33**, 97 (1997).

219. R. Nilsson, T. Lindblad, A. Andersson, C. Song, and S. Hansen, in: *New Developments in Selective Oxidation II* (V. Cortes Corberan and S. Vic Bellon, eds.), Studies in Surface Science and Catalysis Vol. 82, Elsevier Science: Amsterdam (1994), p. 281.

220. R. Nilsson, T. Lindblad, and A. Andersson, *Catal. Lett.* **29**, 409 (1994).

221. J. Nilsson, A.R. Landa-Canovas, S. Hansen, and A. Andersson, *J. Catal.* **160**, 244 (1996).

222. J. Nilsson, A.R. Landa-Canovas, S. Hansen, and A. Andersson, in: *3rd World Congress on Oxidation Catalysis* (R.K. Grasselli, S.T. Oyama, A.M. Gaffney, and J.E. Lyons, eds.), Studies in Surface Science and Catalysis Vol. 110, Elsevier Science: Amsterdam (1997), p. 413.

223. H. Zanthoff, in: *Catalytic Activation and Functionalization of Light Alkanes* (E.G. Deruane, ed.), Kluwer Academic: Dordrech (1998), p. 435.

224. H.W. Zanthoff, M. Lahmer, M. Baerns, E. Klemm, M. Seitz, and G. Emig, *J. Catal.* **172**, 203 (1997).

225. S. Albonetti, G. Blanchard, P. Burattin, S. Masetti, and F. Trifirò, in: *3rd World Congress on Oxidation Catalysis* (R.K. Grasselli, S.T. Oyama, A.M. Gaffney, and J.E. Lyons, eds.), Studies in Surface Science and Catalysis Vol. 110, Elsevier Science: Amsterdam (1997), p. 403.

226. S. Albonetti, G. Blanchard, P. Burattin, T.J. Cassidy, S. Masetti, and F. Trifirò, *Catal. Lett.* **45**, 119 (1997).

227. E. van Steen, G. Kuwert, A. Naidoo, and M. Williams, in: *3rd World Congress on Oxidation Catalysis* (R.K. Grasselli, S.T. Oyama, A.M. Gaffney, and J.E. Lyons, eds.), Studies in Surface Science and Catalysis Vol. 110, Elsevier Science: Amsterdam (1997), p. 423.

228. M. Bowker, C.R. Bicknell, and P. Kerwin, *Appl. Catal. A* **136**, 205 (1996).
229. G. Centi and P. Mazzoli, *Catal. Today* **28**, 351 (1996).
230. G. Centi, P. Mazzoli, and S. Perathoner, *Appl. Catal. A* **165**, 273 (1997).
231. G. Centi, F. Guarneri, and S. Perathoner, *J. Chem. Soc. Faraday Trans.* **93**, 3391 (1997).
232. A. Andersson, S.L.T. Andersson, G. Centi, R.K. Grasselli, M. Sanati, and F. Trifirò, *Appl. Catal. A* **113**, 43 (1994).
233. G. Centi and S. Perathoner, *Appl. Catal. A* **124**, 317 (1995).
234. P.I. Sorantin and K. Schwarz, *Inorg. Chem.* **31**, 567 (1991).
235. G. Centi, F. Marchi, and S. Perathoner, *J. Chem. Soc. Faraday Trans* **92**, 5141 (1996).
236. G. Centi, F. Marchi, and S. Perathoner, *J. Chem. Soc. Faraday Trans* **92**, 5151 (1996).
237. G. Centi and S. Perathoner, *CHEMTECH* **2**, 13 (1998).
238. G. Centi and S. Perathoner, *J. Chem. Soc. Faraday Trans.* **93**, 1147 (1997).
239. G. Centi, F.Marchi, and S. Perathoner, *Appl. Catal. A: General* **149**, 225 (1997).
240. G. Centi and F. Marchi, in: *11th International Congress of Catalysis 40th Anniversary*, Baltimore (J.W. Hightower, W.N. Delgass, E. Iglesia, and A.T. Bell, eds.), Studies in Surface Science and Catalysis Vol. 101A, Elsevier Science: Amsterdam (1996), p. 277.
241. V.A. Matyshak and O.V. Krylov, *Catal. Today* **25**, 1 (1995).
242. R. Nilsson and A. Andersson, *Catal. Today* **32**, 11 (1996).
243. H.W. Zanthoff, S.A. Buchholz, and O.Y. Ovsitser, *Catal. Today* **32**, 291 (1996).
244. H.W. Zanthoff and S.A. Buchholz, *Catal. Lett.* **49**, 213 (1997).
245. R. Catani, G. Centi, F. Trifirò, and R.K. Grasselli, *Ind. Eng. Chem. Res.* **31**, 107 (1992).

CONTROL OF THE SURFACE REACTIVITY OF SOLID CATALYSTS
New Alkane Oxidation Reactions

5.1. INTRODUCTION

Since the development of the process of n-butane oxidation to maleic anhydride,[1-4] there has been considerable research interest in alternative possibilities for the functionalization of light alkanes by selective oxidation, owing to the low cost and low environmental impact of these hydrocarbons. Apart from n-butane oxidation and more recently propane ammoxidation[5,6] (see also Chapter 4), no other processes have been commercialized or have even reached a stage of near commercialization. However, intense research effort has been devoted to this area in recent years, so a brief discussion of the challenges, opportunities, and problems in the use of alkanes as feedstocks other than for maleic anhydride or acrylonitrile synthesis (Chapter 4) is worthwhile.

The discussion here is limited to heterogeneous gas phase processes with solid catalysts that use gaseous oxygen as the oxidizing agent, as these are the ones most often chosen in this area. It should be noted, however, that there are other interesting possibilities, such as oxidation of methane to methanol using N_2O as the oxidizing agent (use of an alternative oxidant) and Fe/ZSM-5-type catalysts[7] or the selective hydroxylation of alkanes in liquid by H_2O_2 and TS-1 (titanium-silicalite) as the catalyst.[8]

Two principal areas can be distinguished in the general field of new reactions of alkane oxidation: the oxidative dehydrogenation of alkanes and the selective oxidation or ammoxidation of C2–C6 alkanes other than the cases discussed in Chapter 4. A third major area is the oxyfunctionalization of methane, either the oxidative coupling to C2 hydrocarbons or its partial oxidation to methanol and formaldehyde. This third, however, is not discussed here, because: (i) after very intensive and systematic research in the last 10–15 years, interest in this area has waned owing to the impossibility of improving results to a point that would allow development of the processes on an industrial scale, (ii) the large contribution of gas phase radical-type processes (in particular in the case of oxidative coupling) makes it difficult to control the reaction and tune key features of the catalyst to adjust the surface reactivity, and (iii) the problem of the control of selectivity is largely unresolved. Furthermore, in the case of direct oxidation of methane to methanol or other oxygenated products, the commercially available two-stage process (methane conversion to syngas and then conversion of syngas to methanol, other oxygen-containing products, or hydrocarbons)[9] is so effective that, although it requires large-scale plants, there is limited incentive for the development of alternatives. Hence, the oxyfunctionalization of methane cannot be considered a good example for discussion with respect to the general approach used in this book (use of the *concept-by-example* approach to highlight open questions and trends for further research as well as to develop the basis for a new methodological approach by pointing out the limitations of current studies), although it will certainly remain a challenge for the future. For a thorough examination of the subject of methane oxyfunctionalization the reader is referred to the pioneering work of Keller and Bhasin[10] and to some recent reviews and relevant studies.[11-29]

It should be noted, however, that the most remarkable advances in this area derive from reactor and process design; e.g., continuous separation of the desired reaction products[30,31] or the use of catalytic membrane reactors.[32] This further illustrates the concept already discussed several times in these pages that a very fruitful new area of development derives from the close relationships among catalyst, reactor, and process design (more specifically, between catalyst chemistry/ engineering and reactor/process engineering), taking into account that catalyst development is not a distinct variable vis-a-vis the other two factors.

It also should be noted that various results of very selective oxidation of methane to methanol or other oxygenates such as methylformate and formic acid in the liquid phase, but using H_2O_2 (on molybdovanadophosphoric acid catalysts)[33] or hydroperoxides (on phthalocyanine complexes encapsulated in zeolites),[34] have been reported recently. However, the cost of the reactants is higher than the value of the products, making these solutions economically unattractive. Some interesting results have been obtained in the photocatalytic conversion of methane.[35-38] Recent work suggests the possibility of using visible rather than ultraviolet light for the photocatalytic reaction, which, if confirmed, would represent a significant break-through.

5.2. OXIDATIVE DEHYDROGENATION OF ALKANES

5.2.1. Dehydrogenation versus Oxidative Dehydrogenation

The recent global demand for alkenes and their shortage—as a consequence of both an increase in their primary use as monomers/comonomers and a relative decrease in their production because of changes in operating conditions in steam and catalytic cracking units—have engendered new interest in producing them from alkanes. There are several commercial or near-commercial dehydrogenation processes, including: steam active reforming, Catofin, UOP Oleflex, Linde-BASF, and Snamprogetti–Yarsintez processes, which differ from one another mainly in terms of reactor technology and type of catalyst. Two principal classes of catalysts are used: supported platinum catalysts and chromium oxide catalysts usually supported over alumina.

These catalytic dehydrogenation processes suffer from several limitations[39]:

1. Thermodynamic restrictions on conversion and selectivity.
2. Side reactions such as thermal cracking.
3. Difficulty in separating the alkene from the alkane and by-products.
4. Strong endothermic main reaction and the necessity to supply the heat at high temperature.
5. Rapid formation of coke and thus continuous necessity for catalyst regeneration as well as the loss of hydrocarbon in this side reaction.
6. Irreversible catalyst deactivation owing to the severe reaction conditions.

The severity of the reaction conditions increases with decreasing carbon chain length and thus dehydrogenation of ethane is by far the least thermodynamically favored reaction. At present there are no commercial ethane dehydrogenation processes. Propane dehydrogenation is technically feasible, but costly because of the severe reaction conditions. n-Butane dehydrogenation requires less severe conditions, but gives rise to several by-products. In the dehydrogenation of isobutane, n-butane and butenes are undesired by-products formed by skeletal isomerization that accumulate in the recycle loop and give rise to butadiene formation leading to severe deactivation by coke.[39]

There are various lines of research in place to overcome these limitations: (i) development of improved catalysts that limit some of the drawbacks especially in terms of selectivity and resistance to deactivation, (ii) coupling of dehydrogenation (endothermic reaction) with hydrogen oxidation (strong exothermic reaction; H_2 is a main by-product of dehydrogenation) to directly supply the heat of reaction required and shift the equilibrium toward the alkenes (H_2, however, is a very valuable coproduct and its combustion weakens the economics of the process), and (iii) use of a catalytic membrane to shift the equilibrium allowing operation at low temperatures (current membranes, however, are very costly so this is still a remote possibility). A fourth option being researched is the oxidative dehydrogenation of the alkanes, which overcomes the thermodynamic limitations, allows operation under relatively mild conditions, and avoids the necessity of continuous catalyst regeneration, but has a major drawback in the difficulty of controlling the consecutive oxidation to carbon oxides. Other secondary problems are the removal of the heat of reaction, the flammability of the reaction mixture, and the possibility of reaction runaway.

In contrast to alkane dehydrogenation processes, most critical problems in oxidative dehydrogenation are related to finding a catalyst and reactor technology that allow control of the consecutive oxidation to carbon oxides. Therefore, although considerable research and development remains to be done to improve the technical and economic performance of oxidative dehydrogenation before it can compete commercially with established technologies [steam cracking of either naphtha or liquified petroleum gas (LPG)], the outlook for the development on industrial-scale oxidative dehydrogenation processes still remains good, especially for light alkanes such as ethane.

The major challenge is to improve the selectivity, both in regard to reducing carbon oxide formation at high conversion and minimizing (elimination) condensable oxygenated hydrocarbons. This requires not only the better catalyst design, but also a reactor technology that can maximize catalyst effectiveness in alkene formation. These aspects will be the central topics discussed in the following sections, but without a systematic survey of relevant publications as the recent advances especially in terms of catalyst characteristics and structure–activity relationships have already been discussed in a series of recent reviews.[39–43]

The basic concept of oxidative dehydrogenation is that the reaction takes place in the presence of a hydrogen acceptor (oxygen), which gives rise to an exothermic reaction so avoiding the thermodynamic limitations of a reversible endothermic reaction (pure dehydrogenation) and at the same time avoids formation of heavier products with high C/H ratios (carbonaceous-type species) leading to fast catalyst deactivation. This concept makes it possible to distinguish between a system formed by a pure dehydrogenation catalyst coupled to a catalyst for H_2 oxidation to H_2O (only the first of the characteristics is present) and an oxidative dehydrogenation catalyst in which both operate. In the first approach a main difficulty is to find a catalyst selective only for hydrogen oxidation and not toward oxidation of the reagents and products.[39]

In the UOP process for propane dehydrogenation coupled with a second stage of hydrogen oxidation supported noble metals doped with alkali and alkaline earth metals and/or tin are used as H_2 selective oxidation catalysts, and similar types have also been proposed by other companies.[39,44–48] These are basically dehydrogenation-type catalysts modified to moderate activity and neutralize acidity (negative with respect to side reactions on the alkene products). This coupled dehydrogenation + H_2 oxidation approach makes it possible to combine the advantages of high selectivity (characteristic of pure dehydrogenation), internal heat supply necessary for the endothermic dehydrogenation reaction, and shift of the equilibrium by consuming H_2 (considering a recycling process—the H_2 + $1/2O_2 \rightarrow H_2O$ reaction—physically separated from the dehydrogenation reaction, although it takes place in the same reactor; in oxidative dehydrogenation H_2O is formed as part of the catalytic reaction cycle of alkene formation). However, the oxidative dehydrogenation approach has the potential additional advantages of avoiding catalyst deactivation and not requiring high operating temperatures, thus considerably simplifying the process and improving process economics. In addition, the effective combination of dehydrogenation and H_2 oxidation catalysts requires specific reaction conditions. In fact, while the concept has been shown to be effective in the case of styrene production from ethylbenzene (SMART process by Lummus and Monsanto[39,48]), the indications are less clear for light alkane dehydrogenation.

Membrane-assisted dehydrogenation also has the potential of avoiding some of the limitations of the dehydrogenation process. Its main advantages are that: (i) the hydrogen is removed continuously, and thus the equilibrium is shifted; and (ii) an autothermic process can be obtained when the dehydrogenation is done on one side of the membrane and an exothermic reaction (such as hydrogenation or hydrogen oxidation) occurs on the other side of the membrane. Two basic types of membranes have been proposed for use in dehydrogenation[39,49-54]: (i) nonporous membranes consisting of metals (such as Pd) or alloys (such as Pd/Ag); and (ii) porous membranes, consisting, e.g., of Vycor glass, alumina ceramics, and SiC. However, due to membrane cost, low productivity, coke formation, heat-exchange problems, low chemical and thermal resistance, and problems of irreproducibility in large-scale operation (difficulty in obtaining uniform microporosity and absence of cracks), it is not likely that membranes will be used in dehydrogenation processes of light alkanes in the near future.

In conclusion, for light alkane (ethane and propane, in particular) notwithstanding the relatively low productivity in the corresponding alkene that has been cited in the literature,[39-43] the outlook for application remains relatively good, but the problem of controlling selectivity at high conversion must be resolved. Combined catalyst design and reactor engineering is the approach necessary to address this issue.

5.2.2. Constraints in Oxidative Dehydrogenation

The flammability of the reaction mixture in light alkane oxidative dehydrogenation is a major constraint in the reaction because operation under safe conditions is not optimal for productivity. The general flammability diagram for the system alkane/oxygen/inert is shown in Figure 5.1. The diagram is a function of the type of alkane, temperature, and pressure, but although these parameters influence the numerical values, the general features remain the same so the graph in Figure 5.1 can be considered a representative model for discussing the constraints in oxidative dehydrogenation of light alkanes due to flammability and explosion limits.

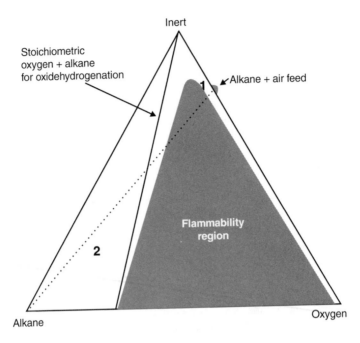

Figure 5.1. General flammability diagram for the system alkane/oxygen/inert. The dashed line shows the curve for the stoichiometric oxygen/alkane ratio for the case of alkane oxidehydrogenation.[39]

Operation inside the flammability bell is possible using special reactor configurations such as the fluidized-bed reactor (FlBR), where the continuous movement of catalyst mass efficiently inhibits radical chain propagation, but it is necessary that: (i) outside the catalyst bed the mixture composition is outside the flammability bell, and (ii) the fluidization of the catalyst is completely homogeneous to avoid gas pockets where explosions can start. Control of these parameters is very difficult, especially in the case of the occasional malfunction. Thus, for reasons of safety, it is preferable to operate the process outside the flammability bell.

Different zones of operation outside the flammability bell can be distinguished: (i) low alkane-to-oxygen ratio, i.e., the right side of the flammability bell; (ii) near the stoichiometric ratio of alkane to oxygen, i.e., the left side of the flammability bell, and (iii) high alkane-to-oxygen ratio.

In the first case, the composition must be close to the alkane/air mixture, because otherwise pure oxygen must be added to the feed, which usually renders the process unprofitable. The conversion path in the case of complete alkane conversion is indicated in Figure 5.1 by line No. 1. Under these conditions, recycling of unreacted alkane is not possible and thus a high hydrocarbon conversion per pass is required. Major disadvantages are the low productivity; high oxygen-to-alkane ratios, which favor complete combustion; and the cost of separating alkene from large volumes of gas owing to the large quantity of the inert gas (nitrogen). As high alkane conversions and high oxygen-to-alkane ratios are uneconomical for selectivity for the alkene, as will be discussed later, operation with feedstock compositions corresponding to this first option (low alkane-to-oxygen ratio) may be less suitable for alkane selective oxidation. However, when the alkene produced can be converted directly to condensable products without a preliminary separation stage (e.g., an ethane to ethene oxidative dehydrogenation unit followed directly by an ethene oxychlorination unit), operation with alkane/air mixtures can be economically advantageous, as the cost of operation with pure oxygen feedstocks (as required in the two other options discussed below) is avoided.

Therefore, the choice of feed composition for operation has constraints that derive not only from the flammability limits but also from the advantages/cost of using air versus oxygen and from the use of the reaction products. It is worthwhile noting as well that the choice of optimal catalyst and type of reactor is influenced by the choice of the feed composition, because, e.g., with low alkane-to-oxygen ratios the key catalyst parameter is selectivity at high conversion, whereas with feedstock compositions from the left side of the flammability bell the key feature must be selectivity at low conversion and productivity. The choice of the optimal reactor configuration is also determined by the type of feedstock (see below) and in turn the catalyst characteristics (e.g., thermal and mechanical resistance, porosity, particle dimensions) must be optimized with respect to the reactor characteristics. Consequently studies of catalyst performance must proceed with an awareness of the intended application and the constraints that derive from it in terms of feedstock composition, reactor design, and catalyst properties. Most of the studies that have been published leave these considerations aside and thus may not address relevant problems. It is clear that future research must involve an integrated approach taking catalyst, reactor, and process design into account.

A second possible operating zone is at the left of the flammability bell, near the stoichiometric alkane-to-oxygen ratio for oxidehydrogenation. In this zone one possibility is to operate directly with alkane and oxygen in feedstock without an inert (in Figure 5.1 operating path No. 2). The alkane in this case is in excess when oxygen is also consumed in other side reactions (i.e., in combustion) and thus, theoretically, complete oxygen conversion could be achieved. The main advantages of this operating zone are: (i) the theoretical possibility of obtaining the highest alkene productivity; (ii) the low oxygen-to-alkane ratio, which favors high selectivities; and (iii) reduced dimensions of the reactor per mass of alkene produced, but disadvantages are: (i) great amount of heat released; (ii) high alkane loss, if it is not convenient to recycle it; (iii) difficulty in achieving high conversion of alkane; and (iv) difficulty in controlling catalyst deactivation and reactor operation in general.

Most of the patent and scientific papers on light alkane oxidative dehydrogenation report catalytic results obtained using a feed composition close to the stoichiometric alkane-to-oxygen ratio, but in the presence of large quantities of an inert gas, which would correspond (in the hypothesis of translating the feed composition to a process situation) to a process in which large volumes of an inert (nitrogen) are added to the feed. This is clearly an unrealistic process option. From a practical point of view the only possibility is to add steam as the inert component, but steam considerably modifies catalyst reactivity and thus cannot be considered as equivalent to nitrogen as an inert. Catalyst performances in the case of light alkane oxidative dehydrogenation are very sensitive to feed composition and thus wrong indications can be obtained when the right feed composition is not used, even at the stage of preliminary screening of catalyst behavior.

Indeed, when the operation is carried out with recycling of the unconverted reactant, inert components are present in the reactor inlet that are either generated in the reactor (typically, carbon monoxide, which is difficult and expensive to separate from light hydrocarbons) or present as impurities in the makeup feed (typically, argon in pure oxygen, or less reactive molecules in the hydrocarbon feed). Therefore, the true operating region lies above the alkane–oxygen line (zone 2 in Figure 5.1). Sometimes an additional inert component is added, which can be easily separated from other components in plant sections downstream of the reactor, with the aim of (i) improving heat conduction properties of the fluid, (ii) diluting the components, or (iii) operating in a safer region (i.e., in the upper part of the diagram in Figure 5.1). The preferred inert is steam; in other cases a part of the off-gas is added to the hydrocarbon recycle stream. With almost all catalysts for light alkane oxidative dehydrogenation the selectivity for alkene decreases considerably with increasing alkane conversion. In addition the increase in carbon oxide formation makes control of the reaction temperature to increase conversion even more difficult (alkane combustion is a much more exothermic reaction than alkane-to-alkene conversion). Generally conversion per pass (see Section 5.2.3) must not exceed 10–20% further to be reasonable selectivities for alkenes. Although the cost of alkane feedstock is very near to that of fuel and thus unreacted hydrocarbon can be used economically to produce high-pressure steam, a hypothetical process based on state-of-the-art oxidative dehydrogenation catalysts would probably require alkane recycling and thus the use of an oxygen/steam combination to reach the required composition. This observation again illustrates the necessity of studying the reaction of light alkane oxidative dehydrogenation in the presence of steam, whereas most of the studies in the literature do not consider this aspect.

The reactor configuration also puts constraints on the catalyst requirements, not only in terms of size, shape, and thermomechanical characteristics and porosity, but also in regard to activity–selectivity properties in relation to the type of feedstock used and the optimal feedstock for the reactor type. There are three main options for reactor configurations to address the general problem of maximizing selectivity for the alkene (with respect to carbon dioxide), and minimizing formation of carbon monoxide and condensable oxygenated products. The latter products (e.g., acetic acid) are often corrosive and thus, even in traces, imply the use of expensive materials for reactor and separation units and make separation of the alkene and recycle of the alkane more expensive. Carbon monoxide formation should also be minimized, because its presence makes separation of the hydrocarbons more expensive. The three possible options are circulating-bed (CBR), multitubular fixed-bed (MFBR), and fluidized-bed (FlBR) reactors.

In the CBR reactor,[55,56] the hydrocarbon is contacted in the absence of oxygen with a reducible transition metal oxide in the riser reactor. The reduced catalyst is continuously transferred to a regeneration unit (usually a FlBR reactor), where it is reoxidized with air. The fact that there is no gaseous oxygen in the reaction (riser) unit means that high selectivities can be obtained and that there is control of the consecutive oxidation to carbon oxides, but that productivity is generally very low, of the order of a ratio 10^{-3}–10^{-4} between alkene produced and the mass of circulating catalyst. Acceptable levels of productivity require that the solid catalyst be circulated very rapidly. This implies high costs for operation (a large volume of catalyst must be moved continuously between the riser reactor and the regeneration unit). The catalyst must also have very good mechanical properties in order to minimize the makeup rate (rate of catalyst addition to maintain stable performances). The high degree of formation of fine catalyst particles as a consequence of the high rate of catalyst recirculation further increases the cost of operation.

The reactivity characteristics of the catalyst must be very different from those of catalysts selective for alkene formation when alkane and oxygen are cofed, because in the case of separate hydrocarbon and oxygen interaction with the catalyst the number of sites per catalyst mass and the turnover number of these sites are the critical determinants of catalyst performance, whereas control of the rate of consecutive and parallel oxidation to carbon oxides is the key parameter in catalyst selectivity when both alkane and oxygen are present in the feed. There are several other aspects, such as capacity to activate and adsorb oxygen, difference between rate of reduction by hydrocarbon and rate of oxidation by oxygen (i.e., steady state of the catalyst surface during catalytic reaction), which must be different in the optimal catalysts operating under separate or simultaneous hydrocarbon/oxygen redox reactions. Hence, the use of a CBR requires an *ad hoc* screening and optimization of the catalyst that is different than in the case with other types of reactors. This important feature is often not clear in the literature, and the fact that the choice of catalyst must be directly related to reactor configuration and feedstock characteristics is usually not taken into account.

The choice of the reactor is determined not only by the modality of contact between alkane and oxygen with the catalyst, but also by the possibility of controlling the heat of reaction released by the exothermic oxidative dehydrogenation of the alkane and thus avoiding the possibility of runaway phenomena. Under certain conditions, such as an accidental increase in the rate of exothermic combustion, e.g., due to an increase in the hydrocarbon feed or change in catalyst reactivity, the released heat of reaction is too great to be effectively removed in the reactor, and the increase in temperature leads to a progressive magnification of the phenomenon, since formation of carbon oxides is much more exothermic than alkene formation and is favored by the higher reaction temperature. The reactor temperature can then no longer be controlled and progressively increases (runaway) up to rupture or explosion of the reactor. Thus in the interests of safety, the reactor must have enough operating flexibility to ensure control not only of the heat of reaction released under "normal" reaction conditions but of possible runaway phenomena.

An example of the standard heat of reaction as a function of the selectivity for alkane and formation of only CO or only CO_2 as a by-product is shown in Figure 5.2. The values are estimated for the case of ethane to ethene oxidative dehydrogenation,[39] but for higher alkanes the heats of reaction are greater, so this case can be considered a basic minimal situation. The standard heat of reaction depends on the selectivity for alkene and the ratio of CO to CO_2 formed. Also shown in Figure 5.2, with dashed bars, is the possible field of application (in terms of capacity to control the amount of heat released for practical cases in terms of alkane conversion and concentration in the feed) for FBRs, MFBRs (FBRs in which the heat of reaction is used directly inside the reactor to supply the heat for an endothermic reaction such as thermal cracking of alkane to alkene[57,58]) and FlBRs. CBRs are not included because, as noted above, they are characterized by low productivities and thus control of the reaction temperature is not a problem.

The data in Figure 5.2 clearly show that FBRs cannot be used, even in the case of very high selectivities, except when alkane conversion or concentration in the feed is very low, which would make application unrealistic. MFBRs have a wider field of application, but are costly, difficult to control, and often operate under conditions closer to the practical limit of control of the heat of reaction. The use of FlBRs, which allows very good control of the reaction temperature (the catalyst fluidization allows an increase of two to three orders of magnitude in the heat transmission coefficient at the reactor walls), is thus preferred for highly exothermic reactions such as alkane oxidative dehydrogenation.[59]

As was pointed out in the case of CBRs, the use of an FlBR not only requires special catalyst characteristics (such as thermomechanical resistance), but also offers new opportunities for catalyst design. In fact, unlike with FBRs, in an FlBR it is possible to dose the amount of oxygen along the catalytic bed, in order to combine the possibility of high productivity with a high and constant alkane-to-oxygen ratio all along the catalytic bed. Although there are some technical problems especially in terms of scale-up in the construction of an FlBR with a distributed

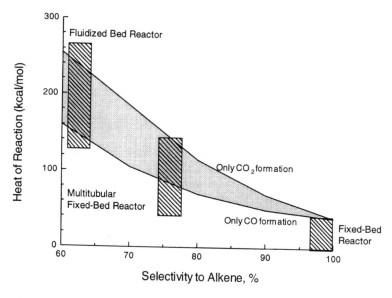

Figure 5.2. Standard heat of reaction per mole of alkene formed as a function of the selectivity to alkene formation by alkane oxidative dehydrogenation in the two hypotheses of only CO or CO_2 formation as by-products (values estimated at 650°C for the case of ethane to ethene conversion).[39] The dashed bars indicate the field of application of different reactor configurations with respect to the possibility of controlling the heat of reaction released by the alkane oxidative dehydrogenation reaction.

inlet of one reactant and back-mixing phenomena lead to a decrease in selectivity, the potential gain in selectivity (as in a CBR, but with much greater productivity) with this type of reactor configuration is very high.[60] However, characteristics of the optimal catalyst for this kind of application are different from those of optimal catalysts for an FBR, whereas most of the published data relate to the latter.[39–44] Moreover, the local feed composition in an FIBR with distributed oxygen inlet is different from that used in most of the studies with FBRs, and owing to the high sensitivity of catalyst performance to both feed composition and reactant conversion, data obtained in an FBR cannot be directly extrapolated to the case of an FIBR with distributed oxygen inlet. For example, in the latter a catalyst with the highest selectivity at very low conversion is preferred, even though the selectivity decreases considerably as conversion increases, whereas in the former a lower initial selectivity but a less pronounced decrease with increasing conversion is preferred. In fact, the dosing of small amounts of oxygen all along the catalytic bed allows consecutive oxidation to carbon oxides to be controlled (the amount of oxygen required for ethane conversion to $CO_2 + H_2O$, e.g., is seven times higher than that from ethane to ethene $+ H_2O$), whereas in an FBR control of the consecutive reaction is a more critical problem in reaching an acceptable level of productivity in alkene formation. When small doses of oxygen are continuously added to the feed, the rate of reaction must be relatively insensitive to oxygen concentration, whereas this cannot be a critical problem in an FBR. Furthermore, the catalyst must not deactivate using very high alkane-to-oxygen ratios. There are thus several different characteristics required of a catalyst to be used in FBRs or FlBRs apart from the more obvious properties such as mechanical resistance, porosity, and particle size and shape. The screening of catalytic behavior in an FBR thus cannot be directly applied for the selection of catalysts for applications in an FlBR with distributed oxygen inlet, although this concept is not usually pointed out in the literature.

In conclusion, analysis of the constraints in light alkane oxidative dehydrogenation in terms of both feed composition and reactor design indicates the need for a preliminary evaluation of process options with reference to the intended application, even at the early stage of catalyst screening, because of the close relationships among the various parameters, such as catalyst characteristics, feed composition, reactor characteristics, and constraints with regard to final reactor outlet (oxygen and hydrocarbon conversion, etc.). This problem is clearly not only related to this specific type of reaction, but is generally valid, although light alkane oxidative dehydrogenation offers a good example to illustrate the factors involved.

5.2.3. Class of Catalysts Active in Oxidative Dehydrogenation

Different types of catalysts have been reported in the literature to be selective in light alkane oxidative dehydrogenation:

1. Catalysts based on alkali and alkaline earth ions and oxides. These catalysts (e.g., the Li^+/MgO system) are basically active in methane oxidative coupling,[11-17] but also show selective behavior in ethane oxidative dehydrogenation at temperatures above about 600°C. The mechanism usually involves the formation of an ethyl radical on the surface, which then reacts further, mainly in the gas phase. The catalyst can be further promoted by adding halides and/or rare earth oxides.

2. Catalysts based on reducible transition metal oxides, which activate the paraffin at much lower temperatures than the first class of catalysts. The reaction in this case is almost completely heterogeneous. Examples of such catalysts include: (i) vanadium-based catalysts such as magnesium vanadate, supported vanadium oxide, vanadium-containing microporous materials, V/Nb oxide and V/P oxide catalysts, (ii) molybdenum-based catalysts (supported molybdenum oxides and metal molybdates), (iii) heteropolyacids (molybdophosphoric acid containing

or not vanadium, and tungstophosphoric acids with the Wells–Dawson structure), and (iv) supported chromia catalysts.

3. Other catalysts such as Sn/P oxide and B/P oxides, LaF_3/CeO_2 or LaF_3/SmO_3, Ga/zeolite, and boron oxide supported on yttria-stabilized zirconia (this last catalyst operates in an electrochemical apparatus).

We will not review the characteristics and performances in light alkane oxidative dehydrogenation of these catalysts here, because these aspects have already been thoroughly analyzed in the literature.[39–44] This discussion is instead oriented toward the analysis of problems and opportunities for the use of these classes of catalysts in order to highlight the outlook in terms of catalyst design.

It should be pointed out that the nature of the alkane influences not only the reactivity, but also the choice of catalyst. The best catalytic systems for ethane oxidehydrogenation are not the optimal ones for higher paraffins, and vice versa. This is related to the differences in the nature of the mechanism of alkane activation and reactivity of the alkene formed with a consequent difference in the key catalyst parameters determining the catalytic behavior. Similarly, the order of activity/selectivity in a series of catalysts is determined not only by the nature of the alkane, but also by the reaction conditions (feed composition, temperature). This aspect further illustrates the necessity of linking catalyst selection to effective reaction conditions and type of alkane. Furthermore, it also proves the necessity of explaining in terms of reaction mechanism not only why a catalyst is selective, but also the mechanistic reasons determining the influence of the nature of the alkane and feed composition.

Owing to the specificity of catalytic behavior in oxidative dehydrogenation of ethane and propane or higher alkanes, we will discuss here specifically the case of ethane. Later we will extend the discussion to other alkanes to illustrate the differences in terms of reactivity and reaction mechanisms.

5.2.3.1. Alkali and Alkaline Earth–Based Catalysts

The first class of catalysts includes ions and oxides of group IA and IIA metals (e.g., Li_2CO_3/MgO), which show a highly selective behavior in ethane to ethene oxidative dehydrogenation, although at high temperatures (above about 600°C). The selectivity to ethene passes through a maximum in the 600–650°C temperature range,[61] due to the presence of different competitive pathways of transformation (influenced differently by the reaction temperature), starting from the ethyl radical formed as a first step in the reaction.[13] The ethyl radical can either form an intermediate ethoxy species or desorb. At high temperatures gas phase reactions are generally favored over surface-catalyzed reactions. Hence, at high temperatures, the desorbed ethyl radical preferentially forms ethene via reaction with molecular oxygen in the gas phase. At lower temperatures, the surface formation of ethoxy species (precursor to carbon oxides) is favored, and as expected the activation energy for ethene formation over Li^+/MgO was found to be higher than that for carbon oxide formation.[62] The selectivity to ethene thus increases with increasing temperature, but above 650°C decreases again owing to the prevalence of combustion reactions of ethane and ethene.

The highest selectivities and yields for ethene can be achieved with this catalytic system, especially when chlorine-containing compounds are also fed to the reactor or when the catalyst is doped with halides.[63–65] However, the promoter effect of halides is maintained only by continuous chlorine feed. Kolts and Guillory[66] first reported an ethene yield of about 32% and a selectivity of about 92% for Li_2CO_3/MgO catalysts at 700°C. Morales and Lunsford[61] similarly reported an ethene yield of 34%. Comparable results were subsequently found by other terms.[39]

The selective mechanism does not involve a redox-type cycle, but the catalyst is involved only in the C–H heterolytic scission with radical formation, associated with the presence of Li^+/O^- centers.[14,15,67,68] Li^+ ions, partially entering into the MgO lattice, create oxygen vacancies that react with molecular oxygen, resulting in O^{2-} ions and positive holes. An Li^+/O^- center is produced by a hole being trapped at an O^{2-} ion adjacent to Li^+ ions. O^- is then capable of abstracting a hydrogen atom from the ethane, thus forming the ethyl radical and a reduced Li^+OH^- site, which is then reoxidized to an Li^+/O^- active site (Scheme 5.1).

The effect of halides is critical for achieving ethene yields above 30%. When chlorine ions are dispersed uniformly throughout the catalyst, the conversion of ethane roughly doubles with respect to the base Li/MgO without halide additives, and the selectivity for ethene increases slightly.[63,64] This suggests that the main effect of chlorine ions is to generate ethyl radical species, but there are other effects associated with the influence of chlorine ions on the surface competitive pathways of transformation. The clear identification of these effects is difficult because there is a continuous evolution of HCl from catalyst surface to gas phase. Chlorine ions not only can change the surface reactivity of the catalyst, but gas phase reactions promoted by chlorine ions can also contribute to the overall catalytic behavior. A close correlation between catalyst performance and the rate of chlorine evolution from the catalyst has seldom been observed, but the fact that chlorine evolution from the catalyst surface (in the form of HCl) involves a change in its surface concentration and a process of bulk to surface migration of chlorine ions must be taken into account. These processes occur on a different timescale and are influenced by different factors (e.g., reaction and feed conditions), so it is very difficult to distinguish between gas phase and surface reactions promoted by chlorine ions; the fact that the catalyst properties themselves are influenced by the presence of chlorine ions adds to the difficulty.[69] The oxidizing properties of the support (e.g.,

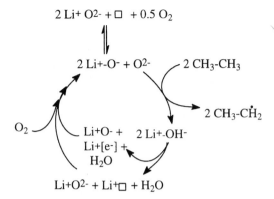

$2\ Li^+\ O^{2-} + \square + 0.5\ O_2$

$2\ Li^+\text{-}O^- + O^{2-} \quad 2\ CH_3\text{-}CH_3$

$2\ CH_3\text{-}\overset{\bullet}{C}H_2$

O_2

$Li^+O^- +\quad 2\ Li^+\text{-}OH^-$
$Li^+[e^-] +$
H_2O

$Li^+O^{2-} + Li^+\square + H_2O$

Scheme 5.1. Formation, reaction, and regeneration of active sites in Li/MgO catalyst. Elaborated from Amenomiya et al.[15]

MgO) become stronger with decreasing chlorine content (mainly due to an electronic effect), with a consequent decrease in selectivity properties of the surface.

Furthermore, the presence of chlorine ions in the gas phase also modifies the possible pathways of homogeneous reactions. Otsuka et al.[70] detected the formation of chlorinated hydrocarbons in the gas phase, indicating the possibility of formation of these molecules (on the surface or in the gas phase) and their possible role as initiators of radical-chain reactions. Burch et al.[69,71] reported that the addition of CH_3Cl to ethane in the absence of the catalyst does not promote the formation of ethene, but clearly this does not exclude the possibility that the radical-chain reaction starting from the chlorinated hydrocarbon occurs in the presence of the solid catalyst. On the other hand, Burch et al.[69,71] observed that HCl had a marked effect on ethene formation, especially in the presence of oxygen. This indicates that HCl may react with O_2 to form $HO_2^{\bullet}$ and $Cl^{\bullet}$ radicals, which also act as precursors for radical-chain reactions.

The surface chemistry of ethane oxidative dehydrogenation over Li/MgO-type catalysts and the effect of chlorine ions, in relation to homogeneous versus heterogeneous reactions, is not fully understood, although several mechanistic details are available. The very high selectivity for ethene from ethane can be accounted for by a combination of three principal effects: (i) surface-catalyzed production of ethyl radicals; (ii) gas phase dehydrogenation of ethyl radicals by molecular oxygen, by chlorine radicals, or other radical gas phase species; and (iii) a low level of combustion. The prevalence of gas phase dehydrogenation with respect to surface-catalyzed reactions involving a chlorine-modified center is probably a function of the operating conditions. The gas phase mechanism is favored when operating with a large void space in the reactor, high residence time, a catalyst with high initial loading of chlorine ions, and a high concentration of chlorine in the feed. However, chlorine ions also modify the surface properties of the oxides, inhibiting the centers responsible for total oxidation and creating special sites for the activation of ethane. The oxidative dehydrogenation of ethane over Li/MgO-type catalysts thus starts at the surface, but then depends heavily on the homogeneous–heterogeneous reactions occurring in the subsequent step.

Several other catalytic systems have been proposed operating on a comparable mechanistic basis of an Li/MgO(Cl) catalyst, such as LiCl/NiO,[70,72] which gives very high selectivity (about 98%) at 23% conversion; metal oxychlorides such as $NaCa_2Bi_3O_4Cl_6$,[73] which also show results comparable to those with LiCl/NiO, Li/TiO$_2$, and Li/Mn/TiO$_2$ catalysts[74]; and various doped Li/MgO catalysts (worth noting is the doping with sodium, boron, lanthanide oxides, and tin oxide).[66,75-77] Although some of these catalysts allow an improvement in selectivity, they do not usually lead to any remarkable improvement in ethene productivity with respect to the base Li/MgO(Cl) system. An interesting new system reported recently is made up of LaF_3 associated with a rare earth oxide (CeO_2[78] or SmO_2[79]) and promoted by BaF_2, which allows up to about 35% yield of ethene from ethane through a mechanism comparable to that discussed above for the Li/MgO(Cl) system. Data on catalyst stability, have not yet been reported, but it is expected that there will be problems.

In the case of doped Li/MgO catalysts a good relationship is observed when the selectivity for ethene is compared as a function of the temperature of 50% O$_2$ conversion (Figure 5.3).[75] The selectivity for ethene appears to depend on this temperature, in the sense that decreasing catalyst activity (thus higher temperatures of oxygen isoconversion) increases the selectivity for ethene. This further confirms the fundamental role of homogeneous reactions in determining overall catalytic behavior and thus that catalyst promotion is achieved favoring the gas phase desorption of the ethyl radical intermediate. A summary of the reaction network is shown in Scheme 5.2. After formation of the ethyl radical (rate-determining step), there are two possible pathways. The first is heterogeneous, whereby the ethyl radical is converted by reaction with lattice oxygen into an ethoxy species. This reaction occurs preferentially at lower temperatures (<600°C). The ethoxy species may then decompose to ethene or be further oxidized by lattice oxygen to form carboxylate species first and then carbon oxides. The ethoxy decomposition is favored at high temperatures. The second pathway from the ethyl radical intermediate occurs in the gas phase and is favored at high temperatures. The ethyl radical can then react with gas phase oxygen either by direct H atom metathesis or by first forming an ethylperoxy species ($C_2H_5O_2^\bullet$), which then decomposes to ethene and $HO_2^\bullet$ via an intermediate excited ethylhydroperoxy species.[80]

Catalysts that belong to this first class of materials (catalysts based on alkali and alkaline earth ions and oxides) show a very high selectivity and productivity

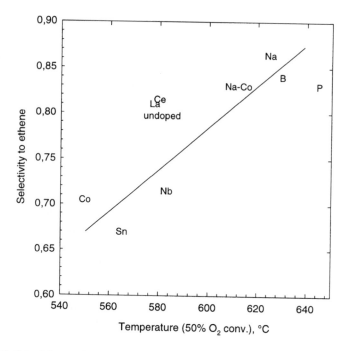

Figure 5.3. Selectivity to ethene as a function of the temperature corresponding to 50% oxygen conversion for undoped and doped Li^+/MgO catalysts. Elaborated from Swaan et al.[75]

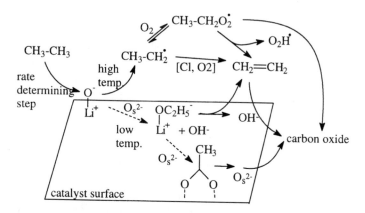

Scheme 5.2. Reaction network in ethane oxidative dehydrogenation over Li/MgO based catalysts. Elaborated from Cavani and Trifirò.[40]

for ethene, especially when promoted by halides; however, the halides themselves cause serious problems in terms of corrosion of equipment and toxicity of effluents. The conditions for the maximum in selectivity correspond to a mixed heterogeneous–homogeneous mechanism, which occurs at high temperatures (600–650°C and thus special and costly types of reactors are needed) and is very difficult to control in an industrial-size reactor. Furthermore, almost all catalysts in this class suffer from more or less serious problems of stability. Notwithstanding good catalytic performance, the outlook for application of these catalysts for a stand-alone ethane oxidative dehydrogenation process is not very good, although they may be of some interest when the presence of chlorine is not a problem, such as for a direct process of ethane dichlorination.

The existence of a large number of studies on the ethane oxidative dehydrogenation on Li/MgO-type catalysts is due not so much to intrinsic interest in this reaction, but to the fact that this ethane conversion reaction is an intermediate step in methane oxidative dimerization to ethene, and Li/MgO catalysts were the first discovered to be active in this reaction. In fact, there are only very limited data in the literature concerning the behavior of this class of catalysts in the oxidative dehydrogenation of propane; however, this is also because the presence of secondary hydrogen atoms in propane leads to significant changes in its reactivity and the reactivity of the derived radicals (propyl radical). This difference in the rate of hydrogen abstraction and the stability of the propyl radical both tend to significantly reduce occurrences of homogeneous-type reactions leading to a prevalence of heterogeneous-type reactions after the rate-determining step and a consequent lowering of the selectivity for propene. Hence, more selective class of catalysts for ethane to ethene oxidative dehydrogenation does not show comparable results in propene oxidative dehydrogenation.

It may be concluded that catalysts based on ions and oxides of group IA and IIA metals convert selectively and with the highest productivity of ethane to ethene, especially when doped with halides and/or other elements (rare earth oxides, in particular), but the drawbacks (high temperature, homogeneous–heterogeneous mixed mechanism, presence of halides, low stability) are so significant as to diminish interest in the use of this class of catalytic materials.

5.2.3.2. Catalysts Based on Transition Metal Oxides

There is a fundamental difference in this class of catalysts with respect to the Li/MgO-type oxides discussed in the previous section: these materials are based on reducible transition metal oxides, so the activity is usually significantly higher but the selectivity for alkenes is lower. Furthermore, the reaction mechanism occurs predominantly on the catalyst surface.

Oxides of molybdenum, vanadium, and niobium were shown to be active for ethane oxidation at temperatures as low as 200°C.[81] Following this pioneering work, numerous patents and papers reported studies on catalytic systems based on transition metal oxides. Attempts at improving and optimizing the selectivity have involved either preparing monophasic systems containing two or more metals or developing multiphase catalysts. Basically there are two types of catalysts in this group: vanadium- or niobium-based catalysts and molybdenum-based catalysts. In principle, rare earth–based oxides also belong to this group, but they are usually active above 600°C, with a mechanism comparable to catalysts based on ions and oxides of group IA and IIA metals. Therefore, although interesting results have been obtained with rare earth–based catalysts in ethane oxidative dehydrogenation, the same considerations hold as for Li/MgO-based catalysts.

The analysis of vanadium-based catalysts for ethane conversion, in particular, is interesting because it clarifies several general aspects of the design of catalysts for oxidative dehydrogenation of alkanes. Bulk, unsupported V_2O_5 is not an efficient catalyst for ethane oxidative dehydrogenation.[82,83] Also when it is supported over oxides with which it has a weak interaction, such as silica, the catalyst performance in oxidative dehydrogenation of alkanes remains rather poor; however, the selectivity for ethene is considerably influenced by the vanadium loading on silica which, in turn, also affects the nature of supported vanadium species. Oyama[84] reported that at 500°C the ethane conversion is influenced very little by the vanadium oxide loading on silica (up to about 10% of vanadium oxide loading the conversion of ethane remains below 1%), but the selectivity for ethene increases from about 10 to 80% passing through a maximum for a vanadium oxide loading of about 5–6%. Selectivity for bulk V_2O_5 under the same conditions is 28%, although ethane conversion is higher (5–6%).

The observed effects are attributed to a change in the structure of the supported vanadium oxide.[84,85] At low surface coverage vanadium oxide is present as a tetrahedrally coordinated vanadyl (V=O) species (linked to the surface by three V–O–Si bonds), whereas as the loading increases, small supported crystallites and finally macroscopic bulk crystallites of V_2O_5 form. At low surface coverage, ethane is believed to interact with either V=O centers (to form an intermediate ethoxy species that can be converted to ethene or acetaldehyde, the latter being a product observed with selectivities up to about 15%[83]) or a V–O–Si surface bond (to form an adsorbed ethyl species strongly interacting with the surface owing to the formation of a stable metal–carbon bond). The latter species is a precursor only of carbon oxides. For the higher loadings, V_2O_5-type species predominate, and the octahedral coordination of vanadium prevents the formation of the strongly bound ethyl species, thus making higher selectivities for ethene possible. Hence, the formation of the ethoxy intermediate is a condition for a selective reaction.

Questions regarding this model of reactivity arise from the fact that: (i) the formation of strong metal–carbon bonds for the ethyl species is not demonstrated; (ii) the vanadium coordination is not true octahedral even in well-crystallized V_2O_5

(distorted pseudopyramidal coordination)[86] and thus coordination to the metal is possible, as shown, e.g., by the presence of Lewis acidity in V_2O_5; (iii) small crystallites of V_2O_5 are in a nearly amorphous state on silica at low coverage and as expected show a large number of defects; and (iv) the model does not explain the presence of a maximum in selectivity.

Le Bars et al.[82,87] explained the difference in reactivity between unsupported and silica-supported V_2O_5 in terms of the effect of the support on the properties of vanadium oxide. At low loadings of vanadium oxide on silica, isolated vanadium ions were found to be more reducible than vanadium ions in V_2O_5, although this indication is not totally correct because it does not take into account the concepts of a matrix effect on the reducibility and delocalization of the electrons as a result of the reduction process. Thus, reducibility of an isolated vanadium species (in terms of activity per vanadium ion) may be greater than for bulk vanadium oxide, although global reducibility of the catalyst (in terms of number of lattice oxygens removed per gram of catalyst) is lower. Similarly, an isolated vanadium species seems more reducible using a reductant (e.g., H_2), but less reducible than bulk V_2O_5 in terms of spontaneous loss of lattice oxygen by thermal treatment in the absence of gaseous oxygen.

Another difference between silica-supported and unsupported V_2O_5 is that after exposure to the reactant mixture, new strong Lewis acid sites are formed on the surface of V_2O_5/SiO_2 but not on V_2O_5.[82] These sites probably derive from the creation of reduced, coordinatively unsaturated vanadium ions by reaction with the hydrocarbon. They are probably generated over both catalysts, but in unsupported V_2O_5 the oxygen bulk diffusion allows a rapid surface reconstruction that is not possible in supported catalysts. Control of the surface availability and type of oxygen species (oxygen species able to abstract hydrogen atoms from the alkane, but not so labile as to be readily inserted into the hydrocarbon molecule) is thus a key to promoting selectivity for ethene from ethane.

The support–vanadium pentoxide interaction is thus a way to control the availability and type of labile oxygen species, but their number is also a key catalyst feature which must be controlled because it determines the catalyst activity. Using alumina instead of silica under similar reaction conditions [temperature (500°C) and loading of vanadium oxide (5%)], one can obtain a conversion of ethane of 25% and a selectivity of about 60% with respect to less than 1% ethane conversion for silica-supported catalysts.[88] Le Bars et al.[89] suggested that the active sites are vanadate dimers composed of tetrahedrally coordinated, oxygen-bridging vanadium cations. These sites operate according to a redox mechanism involving transient radicals generated via one-electron transfer.

The alternative possibility is to control the properties of vanadium by forming a specific mixed oxide compound, with the same key features that promote V_2O_5 when supported in comparison to unsupported V_2O_5.

One possible approach is to start with a catalyst with good characteristics in alkane selective transformation. $(VO)_2P_2O_7$ is the active phase for butane selective oxidation to maleic anhydride (Chapter 4). This same catalyst has also been found to be active in ethane oxidative dehydrogenation to ethene, giving an ethane conversion of about 32% with selectivity for ethene of about 57% at 450°C,[90] although other authors found a comparable selectivity only at a lower conversion (about 10%).[91] Merzouki et al.[92,93] noted that at lower temperatures (300°C), the primary product is acetic acid, and ethene becomes the predominant product only at higher temperatures and low oxygen/ethane ratios. The formation of oxygenated products and especially acids, as noted in Section 5.2.2, is highly detrimental to the development of an alkane oxidative dehydrogenation process.

The peculiar activity of vanadyl pyrophosphate in alkane selective oxidation is due, in part, to the presence of unsaturated vanadyl species with vanadium in the valence-four state (see Chapter 4). Another way to create the same kind of active species is to dilute vanadium in a matrix that can stabilize V^{4+} isolated species through a structural effect. With V_2O_4, a rutile-type structure, the more reasonable approach is to dilute vanadium in an isostructural oxide having the rutile structure and compatible ionic radius of the metal, such as SnO_2. This was done by Bordoni et al.,[94–96] who studied the effect of the addition of vanadium to SnO_2 during the preparation of the mixed oxide by the sol-gel method. Two different kinds of vanadium species were found to be present, the relative amounts depending on the amount of vanadium in the V/SnO_2 sample. The species that consists of V^{4+} ions in substitutional solid solution inside the SnO_2 rutile lattice is the prevalent one at low vanadia content (i.e., V/Sn atomic ratios < 0.1) and reaches a maximum that corresponds to the composition $Sn_{0.9}V_{0.1}O_2$. A second species consisting of amorphous bulk V_2O_5 spread over the solid solution is the prevalent one at high vanadia content and has a negative effect on the selectivity, being responsible for a consecutive step of combustion. Thus, the optimal catalyst is made up of the solid solution; this material gives an ethane conversion of 13% with selectivity for ethene of 35% at 500°C.

Notwithstanding the good dispersion of vanadium and the stabilization of V^{4+} in an isostructural oxide matrix, catalytic performance in ethane oxidative dehydrogenation is not particularly good. One possibility is that the choice of SnO_2 as the oxide matrix is not optimal. However, catalysts based on SnO_2 show good behavior in ethane oxidative dehydrogenation. Conway and Lunsford[64] demonstrated a promotion effect of SnO_2 on the selectivity and stability of the Li/MgO(Cl) system in ethane oxidative dehydrogenation, similar to that observed by various authors for the case of methane oxidative coupling.[97–99] Silica-supported SnO_2 itself can activate ethane. Using IR spectroscopy, Harrison and Maunders[100] detected the formation of an ethoxy intermediate that is converted into an acetaldehyde species and further to a surface acetate species by reaction with surface hydroxyl groups. This result suggests that in V/SnO_2 catalysts, ethoxy species may form under mild conditions either via vanadium itself or via the SnO_2 and transform further to acetate species (precursor for carbon oxides) when decomposition of the ethoxy intermediate is slow as compared to attack of a second lattice oxygen to form the acetate intermediate [i.e., the reverse reaction of the addition of an hydroxyl group to ethene to form the ethoxy intermediate; this reaction usually occurs under very mild conditions, but is reversible by increasing the reaction temperature (see Chapter 8)].

The selective formation of ethene thus requires either increasing the rate of ethoxy decomposition or limiting its attack by lattice oxygen. The results with SnO_2/P_2O_5 catalysts in the oxidative dehydrogenation of ethane[101,102] are interesting in that they clarify this concept. Over SnO_2 at 550°C, the ethane conversion is fairly high (37% at 1 bar) with good selectivity for ethene (46%). When the catalyst is doped with phosphorus the catalytic behavior worsens in terms of both activity and selectivity, but with increasing operating pressure the conversion and selectivity to ethene increase considerably. Optimum selectivities higher than 90% at 37–64% conversion are obtained on a catalyst with an Sn/P ratio of about 0.5.

CH_4 forms on all the catalysts to a remarkable extent, owing to the further reaction of acetate species (formed rapidly from an ethoxy intermediate) with surface hydroxyl groups to give methane and a carbonate species, which then further decomposes to CO_2. Ethene forms by decomposition of the ethoxy intermediate. With increasing operating pressure, highly nucleophilic sites form (thought to be a result of a change in linkage of phosphate species under pressure, because the effect is not observed in SnO_2 alone[101,102]), which, catalyzing the abstraction of β-hydrogen atoms from the ethoxy intermediate, promote the formation of ethene.

These indications provide a better understanding of the behavior of V/SnO_2 catalysts.[94–96] The addition of vanadium in substitutional positions in the SnO_2 rutile matrix creates more active sites for ethane activation, but does not introduce more active sites able to catalyze the decomposition (β-hydrogen elimination) of the ethoxy intermediate. The higher activity with respect to pure SnO_2 implies lower reaction temperatures, but this then lowers the relative rate of ethoxy decomposition (favored by the higher temperatures). Therefore, the selectivity of V/SnO_2 is lower than that of pure SnO_2, although the latter requires higher reaction temperatures, which indicates that the catalytic behavior of V/SnO_2 may be promoted when other components that catalyze β-hydrogen elimination at low temperatures are introduced.

An optimal catalyst for ethane oxidative dehydrogenation should thus form the ethoxy intermediate under mild conditions, but at the same temperature should be active in β-hydrogen elimination (leading to ethene) and not in α-hydrogen elimination and subsequent nucleophilic oxygen addition (leading to acetate). While the first property (determining activity) depends on the type of M–O bonds on the surface, the second type of property (determining the selectivity) depends on: (i) the strength of the C–O–M bond in ethoxy species and the type of ethoxy species (top-on, side-on, etc.) and (ii) the nature of active sites lying near the adsorbed ethoxy species. V^{4+} ions, whose importance for ethane oxidative dehydrogenation was discussed earlier, have a radical-like character, but also a directionality in the bond formation with coordinated hydrocarbons owing to the presence of d-orbitals.

Niobium belongs to the same row of elements as vanadium, but its electronic configuration in the stable Nb^{5+} state is $4s^1$. Therefore, it has a radical-like character similar to V^{4+}, but the outer electronic shell is symmetrical (s-type orbital) and thus very directional types of bonds are not possible, avoiding the formation of stable intermediate species susceptible to nonselective conversion to carbon oxides in the case of ethane oxidative dehydrogenation. On the other hand, owing to the larger ionic radius, its reactivity is lower than that of vanadium, and the reactivity of oxygen atoms linked to niobium is different from those linked to vanadium. A combination of these two elements and a third compound with the primary function of catalyzing the decomposition of the ethoxy intermediate can thus bring about highly active and selective catalysts in ethane oxidative dehydrogenation. The very interesting results seen with catalysts based on V/Nb/Mo oxide can be understood using this perspective, although they were not discovered on the basis of this kind of reasoning.

Thorsteinson et al.[80,103] first reported the unique behavior of V/Nb/Mo oxide catalysts, indicating that a mixed oxide with the composition $Mo_{0.73}V_{0.18}Nb_{0.09}O_x$ was able to selectively oxidatively dehydrogenate ethane even at very low temperatures (10% conversion at about 285°C). As the ethane conversion increases, the selectivity for ethene decreases, although it remains high. At 25% ethane conversion, 83% selectivity was reported.[80] Burch and Swarnakar[104] found a comparable selectivity (around 80%), but at higher temperatures and for lower ethane conversion (6%). They also observed that supporting the catalyst over alumina increases the activity, but decreases the selectivity, and that eliminating niobium from the catalyst composition results in a significant decrease in both activity and selectivity. The poorer performances observed by Burch and Swarnakar[104] in comparison with those noted by Thorsteinson et al.[80,103] are probably due to differences in phase composition, as the Nb/V/Mo oxide system is quite complex. Various phases were observed such as MoO_3, $Mo_6V_9O_{40}$, $Mo_4V_6O_{25}$, and $Mo_3Nb_2O_{11}$. Merzouki et al.[92,93] reported that the method of preparation has a strong influence on the phase composition and catalytic properties. They observed, in particular, that the sample activates during the catalytic reaction and this activation is parallel to the in situ formation of a $Mo_{18}O_{52}$-like phase. They suggested that the active phase is composed of niobium ions in tetrahedral positions and connecting layers of MoO_3

and $Mo_{18}O_{52}$, whereas vanadium enters as V^{4+} ions in solid solution. However, unlike Thorsteinson et al.,[80,103] they found significant amounts of acetic acid, although the formation of this compound decreases with increasing reaction temperature and conversion.

The trend in selectivity with the reaction temperature, is highly dependent on the method of preparation of the Nb/V/Mo oxides,[93] as well as on the phase composition. Structural analysis[93] suggests the necessity of forming coherent interfaces between neighboring microdomains containing a Nb-substituted Mo_5O_{14} phase, a Nb/V-modified $Mo_{18}O_{52}$ phase, and a V-substituted MoO_2 phase, although clearly in so complex a system the roles of the various phases (the particular phase must be known) and the role if any of possible synergetic interactions between the phases is still unclear. It is also not surprising that the preparation method has a considerable influence on the catalytic behavior. Several patents have been issued for this catalytic system and possible further modifications by the addition of antimony or alkaline earth metals.[103,105–107] Claimed ethane conversions are higher than 60% at 400°C, with selectivities for ethene higher than 70%. At lower conversions, selectivities as high as 90% have been reported.

The Nb/V/Mo oxide catalysts can thus activate ethane under rather mild conditions and selectively convert it to ethene, especially when the oxygen in the feed is controlled, although the formation of small quantities of acetic acid makes the entire process more expensive. The formation of the optimized catalyst is not well understood, because the nature of the active phase is still obscure and the preparation procedure is difficult to reproduce. Worth noting, however, is the fact that the same catalysts that are highly selective from ethane to ethene show rather poor selectivity in propane and butane oxidative dehydrogenation.

High selectivities for ethene (over 90%), although for lower ethane conversions (15–20%), also can be obtained using Kegging-type heteropolymolybdates, containing antimony as a promoter of stability and tungsten as a promoter of activity.[108–110] The best catalytic performances have been obtained using $K_2P_{1.2}Mo_{10}W_1Sb_1Fe_1Cr_{0.5}Ce_{0.75}O_x$, although these catalysts suffer deactivation in long-term experiments owing to the irreversible decomposition of the oxoanion. The tungsten in this case has the analogous function of using niobium instead of vanadium to provide comparable catalytic functionality but more moderate activity to tune the sequence of catalytic transformations and optimize selectivity for ethene. On the other hand, the behavior of these catalysts shows that the high selectivity for ethene is not necessarily associated with a specific molybdenum compound, as is apparently suggested by the data on V/Nb/Mo oxides discussed earlier.

5.2.3.3. Catalysts and Reaction Mechanisms

In reviewing the principal aspects of the reactivity of the catalysts active in ethane oxidative dehydrogenation in the previous two sections, it was clear that there is a major difference between the reaction mechanisms of the two classes of catalysts. In the first group of samples, the catalyst causes the generation of ethyl radicals with the selectivity for ethene determined by the consecutive reactions occurring mainly in the gas phase, whereas in the second group the reaction occurs over the entire catalyst surface and the selectivity is determined by the competitive pathways following upon the formation of an ethoxy species as the first (stable) intermediate. Catalysts of the first class operate at high temperatures (usually > 600°C), whereas those of the second are active at lower temperatures, but the formation of oxygen-containing products (acetaldehyde or acetic acid) is a competitive reaction. High selectivities even at high conversion have been reported in both cases. Of course, this is a rough division and some of the catalysts discussed have mixed mechanisms belonging to the two classes of materials, depending on the operating temperature. On the other hand, this division into two classes is useful for the following discussion.

A summary of the catalytic data[61,63,64,76,80,82,83,92,101,104,108,111–118] in terms of selectivity and productivity for ethene as a function of the ethane conversion and the reaction temperature is shown in Figures 5.4a and b, for selectivity and productivity for ethene, respectively). It is evident from these figures that there is no correlation between maximum selectivity and productivity, nor between productivity and selectivity at high conversion or optimal catalytic behavior and operating temperature range. The data indicate that it is not possible to select a catalyst having all the necessary properties and no commercial process has yet been developed, although some of the studies have reached the pilot plant stage. In this respect, mention should be made of the results obtained by ARCO[119] and Phillips[120] using a technology of separate oxidehydrogenation of the alkane in the absence of molecular oxygen and a stage of catalyst reoxidation by gaseous oxygen. Notwithstanding the very high selectivity, the amount of catalyst recirculation per kilogram of ethene produced is quite high, making the process expensive. No outstanding catalyst and process conditions have been found yet but there are good incentives to develop an alternative technology to the thermal or steam-cracking process. There are thus good opportunities to develop new kinds of catalysts that fit the technical–economical requirements for a new process of oxidative dehydrogenation of ethane to ethene.

The ethane is usually thermally cracked into ethene at about 800°C in the presence of steam (0.5 volume ratio steam to ethane), at short residence times (≤1 s). The steam-cracking is done at 80% yield (40% per pass) and is an energetically costly and capital-intensive process. It is thus evident that a competitive new technology would have to have some necessary requisites: (i) operation at low temperatures, (ii) no requirements for special materials or equipment, and (iii) be simple in terms of separation. When the two classes of catalysts discussed above are examined from this perspective, it is clear that the first does not fit these requirements. With the second it would be necessary to lower the operating

(a)

(b)

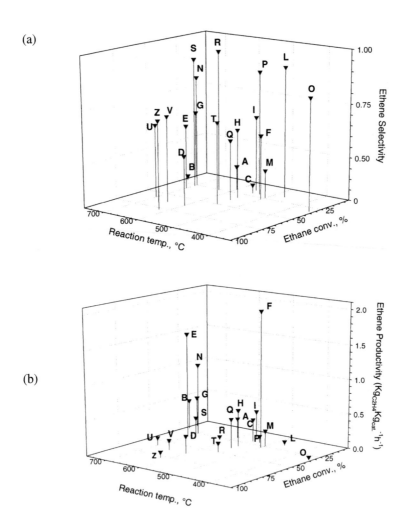

Figure 5.4. Selectivity (a) and productivity (b) to ethene as a function of the alkane conversion and reaction temperature in ethane oxidative dehydrogenation on various catalysts reported in the literature: (A) V/SiO$_2$, (B) CeO$_2$, (C) P/SnO$_2$, (D) Sr/Ce/Yb/O, (E) Sm$_2$O$_3$, (F) V/Al$_2$O$_3$, (G) Na/CeO$_2$, (H) Cr/Zr/P/O, (I) Ni/Mo/O, (L) Sb/W/PMoO, (M) Sn/V/Sb/O, (N) Li/MnO$_x$(Cl), (O) V/Nb/Mo/O, (P) B/P/O, (Q) B/Al/O, (R) Li/NiO$_x$(Cl), (S) Na/P/Mn/SiO$_2$, (T) Na/ZSM5, (U) Li/MgO, (V) Li/MgO(Cl), and (Z) Li/Sn/MgO(Cl). Elaborated and updated from Cavani and Trifirò.[39,40]

temperature, avoid the formation of oxygen containing compounds, and improve the selectivity for ethene.

The brief discussion of the characteristics and catalytic behavior of the second class of catalysts (Section 5.2.3.2) indicated that research has usually been focused primarily on finding a catalyst selective for ethane to ethene conversion, by a systematic screening procedure, and when studies of the relationship between nature of the catalyst and catalytic behavior have been done, the analysis has been directed toward identifying the nature of the active phase. Le Bars et al.,[89] e.g., studying the relationship between V/Al_2O_3 catalysts and catalytic behavior in ethane oxidative dehydrogenation, identified the active sites in vanadate dimers. Similarly, Merzouki et al.[92,93] identified the active sites in V/Nd/Mo oxides at the coherent interfaces between microdomains of Mo_5O_{14} and $Mo_{18}O_{52}$ (modified by V and Nd).

It is possible to summarize this type of approach by indicating that after a stage of catalytic screening of a series of materials and selection of one material with interesting behavior, the analysis of the structure–activity and selectivity relationships makes it possible to identify (or more often to hypothesize) the active phase responsible for the catalytic reactivity. This allows optimization of the catalytic behavior, although often the process is reversed and identification of the nature of the active phase follows the identification of optimal catalytic behavior via the screening procedure.

In the field of ethane oxidative dehydrogenation, it was pointed out in the previous section that there have been a large number of studies using this kind of approach, but in spite of the considerable effort expended no catalytic system having all of the required characteristics for process development has been found. Clearly, the investigation has not been exhaustive and new materials should be studied, but the question is whether a different kind of approach might be more useful.

In discussing the data in the previous section, an alternative way to analyze and rationalize the results was based on the concept of reaction mechanism in terms of the sequence of stages with alternative pathways for the various reaction intermediates after the rate-limiting step—the first H abstraction in the case of ethane oxidative dehydrogenation. Thus contrary to the common kinetic approach, the transformations after the rate-determining step have been considered, with the selectivity determined by the relative reaction rates for all the transformations of these intermediates. Unlike the concept of active site, this approach considers the rates of these single transformations to be functions of independent properties (functionalities) of the catalyst that must all be tuned to maximize the selective pathway from ethane to ethene. Thus, a change in one stage (e.g., activation of the ethane) and not in the other is not positive in terms of selective transformation. This concept will be discussed further in Chapter 8.

Control of selectivity thus requires that the rates of reaction of all intermediates from the start of hydrocarbon activation up to the final product, are coordinated in such a way as to avoid the surface accumulation of some intermediate that can give rise to nonselective pathways of transformation.[121,122] The development of methods for independent control and optimization of these single-step reaction rates by modifying the catalyst characteristics would be a breakthrough in the control of selectivity. The approach that involves identifying the nature of the active phase implicitly focuses attention on the problem of the first stage of hydrocarbon activation and assumes that all the necessary functionalities from the beginning to the end of the reaction are comprised within the active state as a catalytic cluster.[123,124] Therefore, the global catalytic effect is evaluated and not its subdivision into a series of single contributions that can be tuned independently. This implies that there is no consideration given to the possibility of independent monitoring (using suitable probe tests) of the single catalyst functionalities necessary for selective synthesis and how they can be changed/tuned depending on catalyst characteristics.

An innovative approach to the design of catalysts must start from the identification of the reaction mechanism in terms of the sequence of reaction stages and alternative pathways, determining how it is possible to tune the individual stages of a reaction and correlate them with the global catalytic behavior and catalyst properties. Clearly there needs to be a change in perspective for such studies so that research will focus on problems and aspects of catalytic reactivity other than those usually considered. Of course, this discussion about a different approach to catalyst design is based on a rough division between opposing approaches and in reality many of the considerations have already been related to in various studies, but not in terms of the general philosophy of research methodology, which is what is emphasized here. Schematization serves to amplify an alternative perspective on catalytic phenomena and catalyst design, and is a new avenue of research, which does not exclude the utility of both screening and structure–activity and selectivity studies, but rather integrates them in a different view.

Oxidative dehydrogenation of ethane is one reaction for which the outlook is good for the identification of new catalytic materials having the required activity/selectivity characteristics, but the research has to be undertaken from a different perspective rather than simply continued along the traditional lines.

5.2.4. Role of the Nature of the Alkane

Section 5.2.3 focused mainly on a discussion of catalysts for ethane oxidative dehydrogenation, first because propane and butane oxidative dehydrogenation and related catalysts have been discussed in detail in recent reviews,[40–43] and second because this example was more useful for pointing out new opportunities in the study of catalytic behavior and development of new improved catalysts. However, the nature of the alkane, has a significant influence on catalyst reactivity so it is necessary to clarify the relationship between nature of the alkane and catalytic behavior.

Figure 5.5 illustrates the dependence of the selectivity for corresponding alkenes in ethane, propane, and *n*-butane oxidation on unsupported V_2O_5.[83,125–128] The nature of the alkane not only has a significant effect on the selectivity at equal conversion, but also influences the rate of decrease in selectivity with increasing conversion. This effect is not specific for this catalyst. A similar effect is observed on more selective catalysts for C3–C4 alkane oxidative dehydrogenation (magnesium vanadate[42]). Table 5.1 shows a comparison of the selectivity for alkenes over magnesium orthovanadate and magnesium pyrovanadate.[42,129,130] The selectivities for oxidative dehydrogenation are similar for ethane over both magnesium orthovanadate and pyrovanadate, but the orthovanadate is much more selective for the oxidative dehydrogenation of *n*-butane and *i*-butane.

Kung *et al.*[42,129] explained this and similar results obtained over V/P oxide catalysts on the basis of the idea of a selectivity-determining step that is different from the rate-determining step. The concept is analogous to that discussed in Section 5.2.3, although it refers to a single step only. They suggested that an adsorbed alkyl forms as the first intermediate and that, depending on the number of surface VO_x units that effectively interact with the adsorbed alkyl (a function of the size of adsorbed hydrocarbon, rate of reoxidation of the vanadium active center,

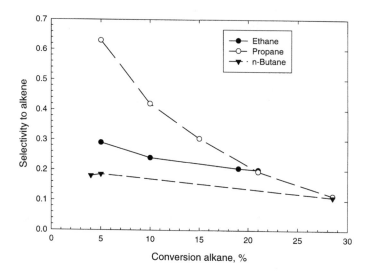

Figure 5.5. Dependence of the selectivity for the corresponding alkene in the oxidative dehydrogenation of ethane, propane, and *n*-butane on unsupported V_2O_5. Elaborated from Blasco and López Nieto.[43]

TABLE 5.1. Comparison of the Selectivity to Alkenes over Magnesium-Vanadate Catalysts [for n-butene the selectivity refers to the sum of butenes and butadiene. Elaborated from Kung and Kung.[42]]

Alkane	Mg orthovanadate			Mg pyrovanadate		
	Temperature, °C	Alkane conversion, %	Selectivity, %	Temperature, °C	Alkane conversion, %	Selectivity, %
Ethane	540	5.2	24	540	3.2	30
Propane	541	6.7	64	505	7.9	61
n-Butane	540	8.5	65.9	500	6.8	31.8
i-Butane	500	8	64	502	6.8	25

and type of catalyst), the reaction can be selective or proceed to carbon oxides. Thus, for magnesium orthovanadate, where VO_4 units are isolated from each other, adsorbed ethyl, propyl, or butyl species can only interact with one surface VO_4 unit. If one assumes that each VO_4 unit supplies a certain number of oxygen atoms to react with an adsorbed hydrocarbon molecule, then the reaction of these alkanes would show the same average oxygen stoichiometry (the average number of oxygen molecules that react with each hydrocarbon molecule). Kung et al.[42,129] indicated that this average oxygen stoichiometry is 2 on magnesium orthovanadate. In magnesium pyrovanadate this value is around 2 only for ethane and propane and doubles for n- and i-butane, because it contains V_2O_7 units (pairs of corners sharing VO_4 units), which can give twice the number of oxygen atoms for the larger hydrocarbon molecules. Similarly, in magnesium metavanadate, the presence of chains of edge-sharing VO_5 units can explain the catalytic behavior.[131] For larger molecules such as butane the interaction with more than one VO_5 unit results in very poor performance (selectivity is only 14% at 13% butane conversion[131]). For smaller molecules such as propane, lowering of the selectivity with respect to ortho and pyro magnesium vanadate occurs, but is less pronounced than with the larger molecules.[132]

A limit in this geometrical interpretation of the selectivity (the basis of the explanation of Kung et al.)[42,129] is its qualitative approach. A geometrical model should determine the nature of the adsorbed intermediate by common quantum-mechanical methods and then estimate the number of lattice oxygens that could really interact with the adsorbed intermediate, which is not proportional to the dimensions of the molecule. In fact, the surface potential of a catalyst is clearly not uniform, and there are no energetic reasons why an intermediate is simply adsorbed flat on the surface. In the absence of interactions with the surface sites, the top-on configuration is the energetically favorable, but interactions usually occur, so the configuration of the adsorbed intermediate may change considerably as a function of: (i) the local structure of the neighboring active site, (ii) the nature of the adsorbed molecule, and (iii) the nature of the surface–hydrocarbon bond. From this last point it is clear that an alkyl radical intermediate (such as that suggested as the first and key intermediate by Kung *et al.*[42,129,131]) unlike an alkoxy species, would have a very weak interaction with the surface. As discussed in more detail in Chapter 8, but also seen in Section 5.2.3, the chemistry of catalytic transformation of an alkyl radical is significantly different from that of an alkoxy species, and the later is the more reasonable intermediate over reducible catalysts such as magnesium vanadate, although clearly the general problem of the mechanism of alkane activation is still under debate and no unequivocal conclusions can be reached.

There are two critical aspects in the geometrical model discussed above: (i) the effect of electronic surface modification (and surface relaxation) by extraction of a second (or more) lattice oxygen near a first oxygen vacancy (created by oxygen incorporation in the hydrocarbon), and (ii) the role of stabilization and reactivity of intermediate species. The geometrical model discussed above has within it the implicit assumption that there is no difference when one lattice oxygen is released from the active site (due to generation of water after extraction of two hydrogen atoms) or when two or more lattice oxygens are incorporated into a hydrocarbon as a first step leading finally to carbon oxides. Clearly, the local creation of more than one oxygen vacancy is an energetically unfavorable process, unless surface reconstruction occurs rapidly. This process is even less favorable if the electrons given by the hydrocarbon to surface sites are considered. In a material that is not highly conductive, such as magnesium vanadate, such electrons are not transported away from the active site quickly. In fact, although the mechanism of oxidative degradation of hydrocarbons is not well understood (see Chapter 8), it is quite well accepted that the first step is the formation of a carboxylate adsorbed species (thus without extraction of lattice oxygens) and then, depending on the reactivity of the carboxylate [in particular, if β-hydrogens are present or if decomposition or disproportionation occurs (see Chapter 8)], carbon oxides may form rather than other oxygen-containing species. It also must be taken into account that carboxylate species are often quite stable and unreactive (especially on basic catalysts such as magnesium vanadate) and can cause self-inhibition of the reactivity with respect to this pathway, thus blocking nonselective sites. Therefore, the formation of carbon oxides clearly cannot be associated simply with the presence of two or more labile oxygens near an adsorption site.

Furthermore, after abstraction of the first hydrogen from alkane and generation of a surface OH group there should be an electronic relaxation of the surface site (its geometry changes), but also the nature of the repulsive interaction between active site and hydrocarbon changes. It is also important to consider the nature of the possible interaction between the catalyst and the products after intermediate transformation. Clearly, this type of stabilization of the consecutive products of transformation varies, again depending on the nature of the hydrocarbon, the presence of reactive hydrogens, and the possibility of multiple attack. Other basic differences are also important, such as the fact that there is twice the probability of abstracting a second hydrogen in propane than in ethane.

Kung and Kung[42] also noted that their model based on a fixed number of oxygen atoms supplied from each VO_4 unit is simplistic and that it was also necessary to determine how easily oxygen anion vacancies can be generated at this site and how fast vacancies can be replenished by migration of oxygen ions from nearby positions in the lattice (rate relative to the surface residence time of the hydrocarbon species). This indicates that catalyst reducibility is another important parameter. In fact, a correlation between selectivity for the oxidative dehydrogenation of alkanes and reducibility (reduction potential) has been observed in a series of vanadates of different cations,[133,134] although Kung and Kung[42] noted that not all cations that are difficult to reduce form selective catalysts. Calcium and barium vanadate, e.g., form such strong carbonate species that during catalytic reaction the catalyst may decompose to V_2O_5 and the corresponding carbonate.[135,136] However, the reducibility should not be correlated with the effect of the nature of the alkane on the selectivity. Although no systematic data are available, there are indications that the behavior in the ethane to n-butane series of these catalysts does not depend on catalyst reducibility, but rather on intrinsic characteristics of the catalyst itself.

It is also worth noting that although magnesium vanadate does not form such strong carbonate species (it decomposes at lower temperatures), it is expected to form rather stable carboxylate species, which in $situ$ could influence the reactivity. The stability of these carboxylate species is thought to be inversely proportional to catalyst reducibility, although there is no direct evidence for that. Thus oxygen lability and the formation of species that might self-inhibit surface reactivity both depend on catalyst reducibility, but the stability of carboxylate species is a function of both the nature of the alkane and the catalyst characteristics. The literature is devoid of any attempt to consider this aspect, but it is clearly a complementary explanation for the surface reactivity of oxidative dehydrogenation catalysts and the effect of the nature of alkane upon it.

The acido-base properties of the catalyst also influence the selectivity in alkane oxidative dehydrogenation,[137–140] while influencing the selectivity in various ways depending on the nature of the alkane. Illustrated in Figure 5.6a is the effect on the dependence of the selectivity to alkene (at isoconversion of the alkane) of the length of the linear carbon chain in undoped and doped VO_x/Al_2O_3 catalysts,[140] while Figure 5.6b shows the selectivity for alkene at isoconversion (30% alkane conversion) and temperature (550°C) feeding ethane, propane, or n-butane for a series of vanadium-based catalysts.[139] Other catalysts such as vanadium oxide supported on Mg/Al hydrotalcite and sepiolite have also been observed to follow the same trend.[43] Vanadium oxide supported on alumina is active and selective in the oxidative dehydrogenation of ethane, but shows low selectivity in n-butane conversion. However, the opposite trend is observed when the VO_x/Al_2O_3 sample is doped with potassium (Figure 5.6a). The addition of potassium does not substantially change the type of VO_x species (although it influences other surface properties), but rather affects the surface acidity. Also the nature of the vanadium-containing catalyst has an opposing influence on the selectivity to ethane and n-butane (Figure 5.6b) and the behavior can be roughly correlated with the increase in surface acidity of the catalyst in going from vanadium supported on MgO to VAPO-5 catalysts.

The acido-base characteristics of both the alkane and its corresponding alkene depend on the carbon chain length. Dadyburjor et al.[141] showed that the acid character of a hydrocarbon decreases as the number of carbon atoms and/or its degree of saturation decrease. Thus C4 alkenes have the lower acidic character and higher basic character, while ethane has the higher acidic character. This idea is not completely correct because in these hydrocarbons important aspects determining acido-base behavior are the polarizability of the bonds in the molecule and the charge delocalization, but it is indicative of the trend. Thus, if the acido-base character of the catalyst is important, the trend in the selectivity for ethene in ethane oxidative dehydrogenation should be different from that observed for n-butane. The data in Figure 5.6 show this relationship, although clearly when the acido-base characteristics of the catalyst are changed, other properties also change at the same time.

These data could be interpreted in terms of an acido-base interaction between the reactant and particularly the products with the catalyst and the effect of this parameter on the selectivity. It is thus reasonable that the alkene intermediate formed from short-chain alkanes (ethane) and having the higher acidity character will be weakly adsorbed on acid sites (its desorption on acid catalysts such as VO_x/Al_2O_3 is thus still possible), whereas less acidic alkenes (such as those from n-butane) require stronger basic catalysts to limit the interaction of the alkene with the catalyst surface. It is thus reasonable that in the analysis of a series of catalysts in n-butane oxidative dehydrogenation an inverse relationship is observed between catalyst acidity and selectivity for alkenes (Figure 5.7).[135,140,142,143] However, this does not explain why selectivity for ethene should go in the opposite direction from that for n-butane, because in terms of acido-base properties the differences among the various alkanes (and their corresponding alkenes) as a function of the carbon atom chain are much less pronounced than the catalytic effect.

Several authors have pointed to the role of acido-base characteristics of the catalyst in analyzing the catalytic behavior of oxidative dehydrogenation catalysts. Pantazidis et al.[144] observed that the most active and selective catalysts (magnesium vanadate for propane conversion) combine both a strong Lewis acidity and mild

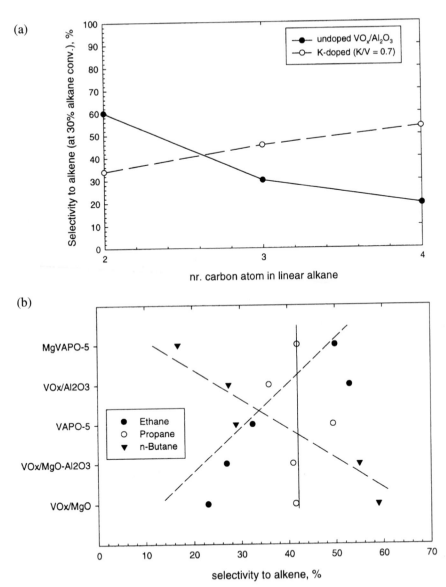

Figure 5.6. Selectivity to oxidative dehydrogenation products in ethane, propane, and n-butane conversion (30% isoconversion) on undoped and K-doped VO_x/Al_2O_3 catalysts (vanadium content: 4wt%, K/V ratio in doped catalyst = 0.7). Elaborated from Galli *et al.*[140] (a). Selectivity to alkene in ethane, propane, and n-butane conversion (data at 550°C and isoconversion of 30%) on VO_x/MgO (11 wt% of V atoms), $VO_x/MgO-Al_2O_3$ (15 wt% of V atoms), VO_x/Al_2O_3 (4 wt% of V atoms), VAPO-5 (0.6 wt% of V atoms), and MgVAPO-5 (Mg-containing aluminovanadophosphate, 2.4 wt% of V atoms). Elaborated from Concepcion *et al.*[139] (b).

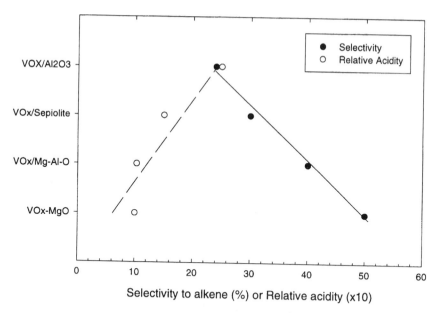

Figure 5.7. Selectivity to C4-alkenes (at 40% isoconversion) in *n*-butane oxidative dehydrogenation on a series of catalysts with respect to their relative acidity estimated by IR spectra of the pyridine adsorption. Catalyst: VO_x/MgO (20 wt% V_2O_5), VO_x/Mg-Al-O (27 wt% V_2O_5), $VO_x/Sepiolite$ (6.7 wt% V_2O_5), and VO_x/Al_2O_3 (7 wt% V_2O_5). Elaborated from Blasco and López Nieto.[43]

basicity. Using electrical conductivity measurements, they identify the presence of anionic vacancies during the catalytic reaction and attribute the Lewis acid behavior to these sites to those, while lattice oxygen is responsible for the mild basic behavior. This acido/base site is responsible for the alkane selective activation, but the oxygen vacancy may also react with gaseous oxygen in the presence of lattice oxygen to generate O^- sites responsible for the combustion. Thus a good balance between acido-base characteristics is necessary. Apart from the fact that the idea of the generation of O^- species by reaction of an oxygen vacancy with gaseous oxygen and lattice oxygen (O^{2-}) is not very reasonable from the chemical and thermodynamic points of view, it is also clear that this concept does not provide an explanation of the dependence of selectivity in the C2 → C4 alkane series on catalyst and acido-base characteristics.

Grabowski *et al.*,[145] in examining the effect of alkali metals in V_2O_5/TiO_2 catalysts for propane oxidative dehydrogenation, suggested a contrary model of the surface sites with respect to that presented by Pantazidis *et al.*[144] Grabowski *et al.*[145] indicated that the increase in the yield and selectivity for propene from propane in alkali-promoted catalysts can be correlated with the increase in catalyst basicity and decrease in acidity, because this favors alkene desorption. Courcot *et al.*,[146] came to the same conclusion about potassium-doped V_2O_5/TiO_2 catalysts, indicating a relationship between selectivity for propene from propane and selectivity in isopropanol dehydrogenation to acetone. They consider the latter reaction indicative of catalyst basicity, but this is not completely true because isopropanol of dehydrogenation requires bifunctional catalytic behavior. Furthermore, doping an oxide catalyst with potassium decreases the Brønsted acidity of the catalyst but introduces medium-strong Lewis acidity owing to the K^+ surface ions.[147] Thus doping with potassium has two opposing effects on the surface acidity, but Courcot et al.[146] did not consider its influence on the Lewis acidity. Watling et al.[148] studying the oxidative dehydrogenation of propane over niobia-supported vanadium oxide also concluded that the selectivity for propene requires the presence of weak acid sites to avoid strong alkene adsorption, but focused primarily on Lewis acid sites and thus even if apparently analogous, their conclusions were opposite to those of Grabowski et al.[145] and Courcot et al.[146] Madeira et al.[149] also observed that addition of an alkali (cesium) promoted oxidative dehydrogenation activity (n-butane over α-NiMoO$_4$), although they also noted that the formation of coke on the catalyst (which, as would be expected, involves strong acid sites) leads to an enhancement of both activity and selectivity.[150] Savary et al.[151] analyzing the propane oxidative dehydrogenation on a VPO/TiO$_2$ catalyst and in particular using a dynamic in situ IR method to study the effect of water or pyridine additions on the nature of acid sites and the relationship to catalytic behavior concluded that under catalytic conditions the increase in the number of Brønsted surface acid sites enhances the selectivity for propene, while inhibition of Lewis acid sites reduces CO_x formation.

In conclusion, the analysis of the effect of the carbon atom chain on selectivity is indicative of the fact that without an in-depth study of the reaction mechanism and the relationship of each step to the properties of the catalyst (especially in stages where alternative pathways of reaction determine the selectivity), the identification of a single factor as the determinant of catalytic behavior is uncertain. Moreover it leads to several contradictions, as shown in the case of the role of surface acidity, and does not explain particular characteristics such as the dependence of the selectivity on the alkane carbon chain length. It is thus necessary to link surface properties to the surface reaction network and not limit analysis to the catalytic activity as a global parameter, e.g., correlating behavior with cation reducibility,[134] number of labile lattice anions,[152] or structural isolation of active sites,[153] because these relationships do not allow generalizations beyond the limited field for which they were derived. It is reasonable that the suggested factors are important and influence the reactivity, but without considering the individual stage at which they play their role and the fact that the overall catalytic behavior results from a combination of multiple tuned effects, it is not possible to establish the basis for a new design of oxidative dehydrogenation catalysts. The role of the nature of the alkane is an important factor in the analysis of catalyst reactivity and surface properties, but its explanation has been bypassed by most authors because it requires this further step of linking catalyst characteristics to the surface reaction network.

5.2.5. Conclusions

The oxidative dehydrogenation of light alkanes has been an active area of research for several research groups in recent years, owing to industrial interest in finding an alternative technology to pure alkane dehydrogenation. Several types of catalysts have been developed, although none of them meets the requirements for industrial exploitation. Moreover, the reaction mechanism and the nature of the selective sites for the oxidative dehydrogenation of short-chain alkanes are still under discussion. Here, however, our concern is not with reviewing the types of catalysts used and the problem of identification of the nature of the active sites, because these aspects have been discussed in detail in some recent reviews[40–43] and are not instructive in terms of possible new developments in this field. Rather, our focus is on two aspects that are often ignored, but which constitute the basis for a reconsideration of the entire field from a different perspective: the necessity of skillful matching between catalyst design and both reactor engineering[154] and the surface reaction network.

The development of a process of light alkane oxidative dehydrogenation is clearly difficult so the problem should not only be addressed from a very broad range of approaches but should be analyzed from a different, innovative direction. This is the reason that the emphasis herein is not on reviewing the state-of-the-art but rather on attempting to stimulate a "lateral thinking" approach.

There are several interesting possible directions of research in the field of oxidative dehydrogenation, but the two that must be singled out are oxidative dehydrogenation with electrochemical pumping of oxygen at the catalyst surface[155] and oxidative hydrogenation on catalytic membrane reactors.[156] The latter uses conventional oxidative dehydrogenation catalysts as the active elements (such as magnesium vanadate), inserting them into a microporous membrane (zeolitic type) and using separate hydrocarbon and oxygen feeds to significantly improve alkene formation (productivity, however, still remains a problem). The first approach is also innovative in terms of the type of catalyst and the reaction mechanism. Supporting boron oxide on an oxygen conductor material such as yttrium-stabilized zirconia and electrochemically pumping oxygen at the surface where the boron component is located, there is a considerable jump in the rate of alkane conversion and in the selectivity for alkene. The suggested reaction mechanism[111,155,157] is the formation of boron peroxide species that then abstract the two hydrogen atoms from the alkane (Scheme 5.3). It is worth noting that surface peroxide species are usually considered to be responsible for total oxidation (see Chapter 8), but this result illustrates how care must be taken with this kind of generalization.

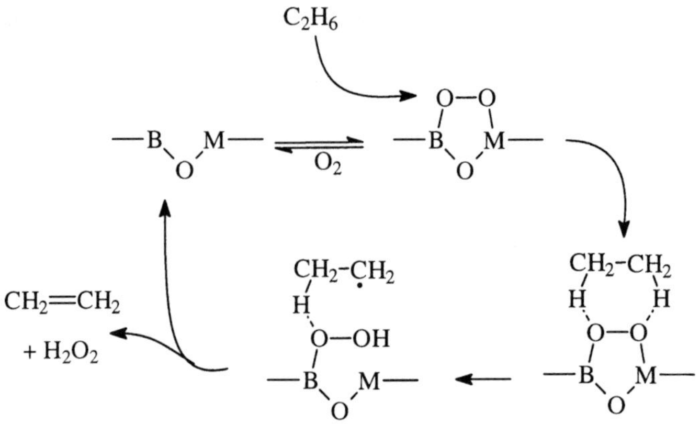

Scheme 5.3. Reaction mechanism proposed for ethane oxidative dehydrogenation on B_2O_3 on YSZ (Y-stabilized zirconia) under conditions of surface electrochemical pumping of oxygen. Elaborated from Otsuka et al.[155]

The use of different oxidizing agents must also be noted. Propane can be oxidized with SO_2 as the oxidant over oxides of aluminum, gallium, silicon, bismuth, lead, and iron.[158] CO_2 is also another possible unconventional oxidant. The selective oxidation of ethane with CO_2 yields ethene, CO and H_2O, while nonselective oxidation forms CO and H_2.[159] The catalysts used are Mn oxide and Fe/Mn oxides supported on silica, but very high temperatures are required (~800°C). With a mixture of 50% ethane and 50% CO_2, a conversion of 11% at 780°C with a selectivity for ethene of about 70% could be obtained. The use of N_2O as the oxidizing agent has been studied in much more depth over a range of different catalysts such as supported and unsupported molybdenum oxide or vanadium oxide, molybdate and vanadate, and metal-containing zeolites.[160–165] In all cases much higher selectivities for alkene than those obtained using gaseous oxygen have been observed. The use of these alternative oxidants has several drawbacks or as in the case of N_2O is too costly for process development, but it is useful to consider that gaseous oxygen is not the only possible oxidant. Ozone is another potential candidate,[166] although present results are not very promising.

There are several potential new directions that should be researched more thoroughly to address the problem of development of catalysts that are active and selective for light alkane oxidative dehydrogenation, apart from the continuous screening of new classes of materials. This is now the challenge because further optimization of known catalytic systems does not seem to be leading to industrial process development, at least for ethane and propane. For n-butane oxidation the results are more promising and mention should be made of the interesting data obtained by Dow Chemical Co. during an investigation into a new process for a styrene monomer from n-butane consisting of a first stage of n-butane oxidative dehydrogenation to butadiene followed by dimerization of butadiene to 4-vinylcyclohexene and finally oxidehydrogenation of the last product to styrene. When magnesium molybdate catalysts were used, 80% selectivity for butadiene at 40–50% conversion of n-butane could be obtained, but with separate stages of hydrocarbon and oxygen interaction with the catalyst.[167,168]

5.3. NEW TYPES OF OXIDATION OF LIGHT ALKANES

5.3.1. Introduction

Major driving forces for technological innovation in the field of selective oxidation are the use of new cheaper feedstocks and process simplification, features often found in the reactions of oxyfunctionalization (conversion to useful products by selective oxidation) of alkane discussed in this section. In Chapter 4 the only commercial example of selective functionalization of an alkane (n-butane to maleic anhydride) and a new process at a near commercialization stage (propane ammoxidation to acrylonitrile) were described. In Section 5.2 a further example (alkane oxidative dehydrogenation) was discussed, although the need for further research was clear. In addition to these three cases, there are several other interesting examples in the literature of light alkane oxyfunctionalization on solid catalysts that have, the potential to decrease feedstock costs and simplify the process technology[169]:

- Ethane oxidation to acetic acid or ammoxidation to acetonitrile.
- Propane oxidation to acrolein or acrylic acid.
- Isobutane oxidation to methacrolein or methacrylic acid.
- n-Pentane oxidation to maleic and phthalic anhydride.
- Cyclohexane oxidation to cyclohexene, cyclohexanol, and cyclohexanone or ammoxidation to adiponitrile.

No reasonably good catalysts have yet been found for any of these processes so a major research effort is needed. The state of the art on these catalytic reactions will be discussed here, but the discussion will focus on possible opportunities for future research.

Another interesting reaction of alkane functionalization is the ethane oxychlorination (oxidation in the presence of HCl) to 1,2-dichloroethane. This has been studied for many years,[170] but commercialization has been hindered by the low purity of the product obtained.[171]

Discussion here is limited to oxyfunctionalization of alkanes by gas phase processes on solid catalysts using gaseous oxygen and essentially to short-chain alkanes. For longer-chain alkanes, interesting processes in the liquid phase are possible, such as the hydroxylation of alkanes to alcohols on TS-1 using H_2O_2.[172,173] However, the potential for process development in this case is essentially dependent on the cost of H_2O_2. The challenge here, then, is the development of new processes for H_2O_2 *in situ* production (this aspect will be discussed in Chapter 7) more than the further development of catalyst properties.

It must be proved finally that interesting results are possible using an integrated (multistage) approach for some of the reactions discussed in this section. The integrated process of propane oxidation to acrolein and acrylic acid uses a first stage of propane to propene dehydrogenation.[174] The primary economic advantage of this integrated process is the elimination of the expensive intermediate unit for the separation of the alkene from the unconverted alkane. Rather, the downstream oxidation reactor is used to remove the unreacted hydrocarbon, which is easily separated from acrolein and acrylic acid. However, the mixture of hydrogen, propane, and propene in the oxidation unit presents new problems especially in connection with the larger flammability region of the mixture with respect to the case of a propene/air mixture. Moreover, conversion of the by-products of the dehydrogenation step decreases product purity and increases the cost of the separation unit. The overall process comprises several steps integrated in a single unit[174]:

1. Dehydrogenation of propane over a catalyst based on platinum supported on zinc aluminate. The conversion of propane is about 30% with selectivity for propene of over 95%.
2. Addition of air at the outlet of the dehydrogenation unit, without separation of reagents and products, and before the stream enters the oxidation reactor containing a multicomponent mixed oxide for acrolein synthesis from propene. The catalyst is similar to the commercial catalysts for this reaction,[175] but must have the particular characteristics of avoiding oxidation of the hydrogen produced in the dehydrogenation unit, in order to avoid critical hot-spot and possible runaway phenomena. Propene converts over 90% with a selectivity for acrolein of around 85%.
3. Introduction of the stream from the second reactor into a third reactor containing a V/Mo-based mixed oxide for the oxidation of acrolein to acrylic acid.[175,176] Conversion to acrolein is over 90% with selectivity for acrylic acid close to 90%.
4. Absorption of acrylic acid in a column (organic solvent) and separation of propane and propene from uncondensable gases (H_2, O_2, N_2, and light ends).
5. Recycling of the C3 fraction after desorption from the absorption column.

The overall selectivity for acrylic acid from propane is close to 75%.

An integrated process has been proposed for isobutane conversion to methacrylic acid.[177] The relevant flow diagram is shown in Figure 5.8. In the first reactor, isobutane is dehydrogenated in the presence of steam over a conventional dehydrogenation catalyst (Pt/Zn aluminate).[39] The outlet of the first reactor, after addition of O_2, is introduced into the oxidation reactor, where isobutene is converted to methacrolein on a conventional molybdate catalyst. The output from this first reactor is sent to a separation unit where methacrolein is recovered, or alternatively to a second oxidation unit to convert methacrolein to methacrylic acid. The serious problems with the process are similar to those discussed earlier for the integrated propane oxidation process. Moreover, in this case oxidation of H_2 and of light end products formed in the dehydrogenation unit must be avoided.

Although the integrated processes comprise commercial elements and catalysts and allow good global selectivities, the complexity of the multistage process, the elements of rigidity in process operation it introduces, and the problems of H_2 and light ends resulted in the absence of significant incentives for this solution using an alkane feedstock with respect to the commercial processes starting from alkenes. Thus the challenge for future research will be development of direct oxidation processes. In the following sections, discussion will be limited to catalysts for direct oxidation processes.

5.3.2. Ethane Conversion

5.3.2.1. Catalysts for Acetaldehyde and Acetic Acid Formation

Ethane can be selectively oxidized for ethene (see Section 5.2.3), acetaldehyde, and acetic acid. Of the two latter products, acetic acid is the more interesting, because its direct synthesis from ethane would not only use a cheap feedstock but

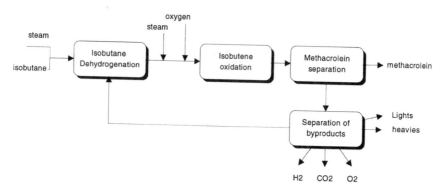

Figure 5.8. Integrated process for the three-step conversion of isobutane to methacrolein and methacrylic acid. Elaborated from Trifirò and Cavani[171] and Khoobiar.[177]

also significantly reduce the complexity of the technology with respect to the conventional naphtha oxidation, methanol synthesis, and carboxylation route.

After the pioneering work of Thorsteinson et al.[80] on V/Nb/Mo oxides, the mild oxidation of ethane was studied by several research groups on a variety of catalytic systems,[84,91–93,104,166,178–187] but acetaldehyde and acetic acid were reported to form in more than traces in only a few cases. Small amounts of acetaldehyde have been found in the oxidation of ethane over boron supported on alumina[111] or B/P oxide catalysts.[188] More recently, Otsuka et al.[179] noted that in B/P oxides the selectivity for acetaldehyde improves with the time-on-stream during the catalytic reaction, probably due to the enrichment of boron at the surface of the inert BPO_4. Based on these observations, Otsuka et al.[179] tested a catalyst prepared by grinding BPO_4 with H_3BO_3 and P_2O_5 and consecutive calcination. The acetaldehyde significantly improves to about 32% at 21% ethane conversion, although a high reaction temperature (570°C) is required. Other by-products were formaldehyde and ethene, as well as carbon oxides.

Alternative approaches to the synthesis of acetaldehyde have also been explored. Wada et al.[180] reported a selectivity for acetaldehyde of over 90%, but at very low conversion, in the photocatalytic oxidation of ethane on silica-supported MoO_3. Erdöhelyi et al.[163] found an initial selectivity of about 20% for acetaldehyde at 10% ethane conversion on Rb_2MoO_4 using N_2O instead of oxygen as the oxidant. However, the catalyst quickly deactivates. Gesser et al.[166] found comparable selectivities for acetaldehyde in the reaction of ethane with ozone, but forming a mixture of different products (formaldehyde, acetaldehyde, methanol, ethanol).

Present data for acetaldehyde synthesis from ethane, even using alternative oxidants that require milder reaction conditions than oxygen, are thus not very encouraging in terms of a possible catalytic system or catalytic technology for direct conversion of ethane to acetaldehyde. More interesting data have been obtained for the direct formation of acetic acid, although the available information still falls far short of the minimum requirements for a possible application.

The formation of significant amounts of acetic acid has recently been claimed for titania-supported catalysts[181–183] and, in particular, for titania-supported V/P oxides. The best performances are generally obtained at superatmospheric pressure (10–30 bars) in mild temperatures (250–320°C). Barthe and Blanchard[181,189] reported that for a catalyst based on a V/P mixed oxide–supported selectivity for acetic acid is 43% at 2.7% ethane conversion and 11 bars of pressure. Merzouki et al.[92,93] reported high selectivities for acetic acid at low temperatures (usually <250°C) and very low ethane conversions over ReO_3-like oxides belonging to various systems such as VO_2(B)-based oxides, $(VO)_2P_2O_7$-based compounds, and $Mo_{0.73}V_{0.18}Nb_{0.09}O_x$ catalysts. With increasing temperature and conversion, selectivity decreases considerably, mainly owing to an increase in ethene formation. They also suggested a structural model based on the presence of oxygens linked to hard acid sites, such as vanadyl groups in $(VO)_2P_2O_7$, which catalyze the direct oxidation of ethane in acetic acid at low temperatures. The structural comparison between $(VO)_2P_2O_7$ and V_4O_9 (formed after reaction of VO_2 catalysts) proposed by these authors with the indication of the position of double edge-sharing octahedra (key elements for acetic acid formation) is shown in Figure 5.9. It is worth noting that these oxides have a ReO_3-like structure.

Blum and Pepera[190] agreed with the claim that vanadyl pyrophosphate doped with rhenium oxide (empirical formula $VPRe_{0.03}O_x$) gives acetic acid and ethene from ethane, although the acetic acid selectivity was generally lower than 10%. More recently Tessier et al.[183] studied V/P oxide phases, either unsupported or supported over TiO_2. Selectivities for acetic acid up to 100% were noted, but the selectivity decreased to below 60% (at the best) at 0.5% ethane conversion, reaching values close to zero for ethane conversion above 2–3% (Figure 5.10). They also suggested that the selectivity for acetic acid could be correlated with the presence of vanadyl dimers.

Recently, the formation of acetic acid from ethane has also been reported for reduced heteropolymolybdates[187] synthesized by heat treatment of the pyridinium salt under nitrogen flow. Ethane conversion does not occur over nonreduced

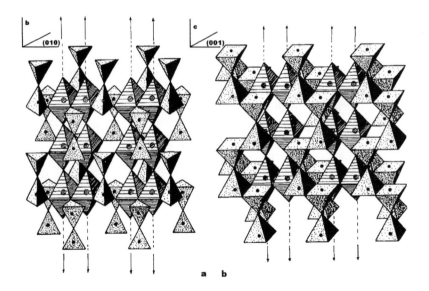

Figure 5.9. Model proposed by Merzouki et al.[93] of the structural analogy of $(VO)_2P_2O_7$ (a) and V_4O_9 (b) catalysts active in acetic acid formation from ethane. Arrows indicate the position of the double edge-sharing octahedra. Elaborated from Merzouki et al.[93]

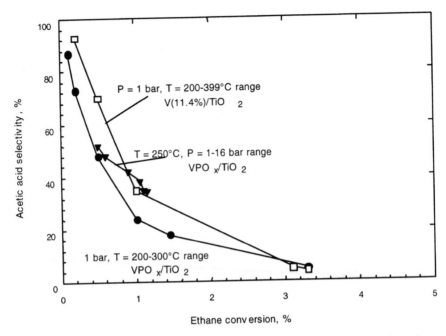

Figure 5.10. Selectivity for acetic acid from ethane on VPO_x/TiO_2 and V (11.4 mol%)/TiO_2 catalysts. Elaborated from Tessier et al.[183]

phosphomolybdate in the acid or ammonium form, because the reduction creates a defective Keggin oxoanion with oxygen vacancies, where ethane is probably activated. Logically, the introduction of vanadium in the oxoanion will increase activity and possibly selectivity, as discussed later for other types of alkane oxidation, but will also reduce the stability toward irreversible decomposition of the oxoanion, which is a critical problem of this class of catalysts for alkane selective oxidation. An acetic acid selectivity of about 10% at 1% conversion is reported for the reduced heteropolymolybdate.

5.3.2.2. Alternative Approaches for Catalyst Design

Actual data indicate that vanadium plays a key role as an active component in catalysts for acetic acid synthesis from ethane, but also point to the necessity of having a specific surface configuration of the active site to maximize the selectivity. Vanadium is often a key component in catalysts for the formation of acids or anhydrides by selective oxidation processes,[122,171,191–193] whereas other transition metals such as molybdenum favor selectivity for the formation of intermediate products such as aldehydes. Unfortunately, no attempt has yet been made to analyze the chemical reasons for this difference. However, the need for a specific element for a class of reactions (synthesis of acids or derivates), notwithstanding the differences in catalyst composition and structure,[193] suggests that the formation of the acid group is a characteristic of the element, whereas the catalyst environment (local surface structure of the active sites) determines the consecutive transformation of the adsorbed intermediate. Clearly, this is a very rough distinction, because the intrinsic reactivity of an element depends on the local environment and thus surface/bulk structure. However, it does illustrate the concept that in several cases the local catalyst (active site) structure is not the key factor determining the formation of the precursor intermediate for acid group formation, but rather governs how this intermediate is further transformed (to acid by desorption or to another species precursor of another product or carbon oxides by further oxidation). The pathways of further conversion of this precursor (acid) intermediate (logically a carboxylate species as discussed in more detail in Chapter 8) depends on: (i) the stability of the intermediate (function of the type of bond with the surface), (ii) the reactivity of other sites near the adsorbed intermediate, and (iii) the nature of the adsorbed intermediate [reactivity of α and β carbon and hydrogen atoms (see Chapter 8)]. Thus from this point of view the selectivity depends more on the mechanism of control of the consecutive surface transformations of carboxylate species than on that of the actual synthesis of this intermediate. Very few and not very systematic studies are found in the literature concerning the analysis of the factors and mechanisms that determine the surface transformation of carboxylate species,[194] whereas there is a great deal of published work on the mechanisms of selective synthesis and relationships with the catalyst structure.

In the oxidation of ethane to acetic acid (Figure 5.10), a major problem in the reactivity is the control of consecutive oxidation to carbon oxides (very strong decrease in selectivity with increasing conversion). In the literature, however, attention has generally been directed specifically toward the mechanism of synthesis of acetic acid and the role of catalyst structure, as discussed above. Hence, an innovative approach for the design of new catalysts for this reaction would be to shift the focus from the synthesis of acetic acid to the reactivity of adsorbed acetic acid (acetate-type species) and use this feature as a basis for catalyst selection.

This perspective on the problem of selectivity in acetic acid synthesis also reveals the importance of several other aspects. The reactivity of an adsorbed intermediate depends not only on the local catalyst structure, but also on the type and strength of the bond with the surface. In general terms, the desorption of an acid compound is favored over acidic-type surfaces rather than basic-type surfaces, although the problem is more complex, as discussed in Chapter 8. For this reason catalysts for acetic acid synthesis should have strong acid properties and vanadium-containing heteropoly acids (which show a strong Brønsted acidity) are good candidates for this reaction (specific data for ethane oxidation are not available, but the concept was demonstrated for other types of alkane oxidation reactions discussed in Chapter 4, Chapter 8, and further on in this chapter), although their low stability limit their use. For the same reason, doping catalysts with elements that favor a strong surface bond with carboxylate species results in a lowering of the selectivity, whereas an enhancement of the rate of acetate conversion to weakly coordinated acetic acid and its further desorption [principal steps in the mechanism of acid formation from surface carboxylate species (see Chapter 8)] increases the selectivity. The hydrolysis of acetate species to form coordinated acetic acid requires water, which, in turn, favors desorption of coordinated acetic acid by competitive adsorption. For this reason, the selectivity for acids in selective oxidation reactions is generally improved by the addition of water in the gas phase,[122] although water can also have deleterious effects [it enhances the concentration of Brønsted acid sites, which catalyze the surface transformation of adsorbed acid (see Chapter 8)].

Adding steam to the feed can enhance selectivity, but it also has drawbacks. It is possible, however, to hypothesize various mechanisms directed toward obtaining the same effect of destabilization of acetate species and enhancement of the acetic acid desorption. A negative charge at the surface, which can be achieved by changing the surface potential, electrochemically clearly destabilizes an anion and results in weaker bonding with the surface. In a semiconductor oxide the surface can become negatively charged when adsorbed species give enough electrons to the catalyst to charge it. This is probably one of the reasons why an increase in the operating pressure in ethane oxidation to acetic acid results in an increase in the selectivity, as noted in the previous section (see also Figure 5.10 for the VPO_x/TiO_2 catalyst).

The selectivity for acetic acid can be increased by enhancing its desorption from the surface, by inhibiting the further conversion of the acetate species, and, in particular, by abstracting β-hydrogen, the first step in the formation of formate species [precursor to carbon oxides (see Chapter 8)]. This control of the reactivity of acetate species could be achieved by having a suitable local structure of the catalyst or by blocking the reactivity of the sites responsible for β-hydrogen abstraction. Owing to the strong interaction of acetate species with the catalyst surface, an increase in their surface concentration can induce an effect of self-doping of the catalysts and inhibition of the rate of their surface transformation. This might be a second reason why higher selectivities for acetic acid from ethane are possible under higher operating pressures. However, it is also possible to think of alternative ways to control of the reactivity of the surface sites, e.g., adding small amounts of gas phase additives whose function is to control the surface reactivity.

Another key aspect of the reactivity in ethane oxidation to acetic acid is that catalysts are selective only at low reaction temperatures (~250°C), whereas at higher temperatures the formation of ethene is favored, which thus is not a reaction intermediate, but rather forms in a side reaction favored by the increase in temperature. This suggests that acetic acid may form through a mechanism involving a

different type of oxygen species than those active for ethane oxidative dehydrogenation, e.g., peroxidic type oxygen whose presence on oxide surfaces decreases
with increasing temperature (Chapter 8). There are no data in the literature concerning this point, but it should be studied because it can significantly influence
catalyst design. On the other hand, the selective synthesis of acetic acid requires
low temperatures, whereas desorption of acetate species (strongly bound to the
surface) requires high temperatures. As expected, the low selectivity of actual
catalysts for ethane oxidation to acetic acid is related to these two contrasting
aspects. A new catalyst design must therefore be centered on optimizing these two
functions so that they operate in the same range of reaction temperatures.

This discussion on the factors determining selectivity in ethane oxidation to
acetic acid, although largely based on hypotheses as available experimental data
are very limited, clearly illustrates that a breakthrough in catalyst performance will
only be achieved by addressing design from a different perspective than that of
simply analyzing structure–activity and selectivity relationships in known catalysts
that give poor results. There are three main aspects that require thorough analysis:

1. Mechanism of the control of surface transformations of acetate species.
2. Types of oxygen species and surface intermediates in ethane conversion to
   acetic acid.
3. How to overlap the temperature ranges of acetic acid synthesis and desorption of its intermediate precursors.

An understanding of these features can put catalyst screening on a more
scientific basis and lead to more innovative catalyst design as well as reduce the
time for catalyst development.

5.3.2.3. A New Route: Ethane Ammoxidation

Acetonitrile is currently produced by catalytic dehydration of acetamide or
dehydrogenation of ethylamine, and as a side product of the ammoxidation of
propylene to acrylonitrile,[195] but the development of an effective and economic
process for its synthesis from a low-cost hydrocarbon such as ethane can considerably expand its use. In fact, acetonitrile can be selectively hydrated over solid
Brønsted acid catalysts to give the corresponding amide, which is easily transformed to acetic acid. Thus acetonitrile can be a first step in a new process of acetic
acid synthesis (Scheme 5.4). Acetonitrile can also be selectively converted to
glycolonitrile[196] (intermediate in the manufacture of nitrilotriacetic acid or the
amino acid serine) or glycolamide[196] by oxidation on vanadium oxides in the
presence or absence of water. Furthermore, it can be converted to acrylonitrile by
reaction with methanol[197] or methane.[198] This process for acrylonitrile synthesis
starting from ethane may be a possible alternative to the commercial process from

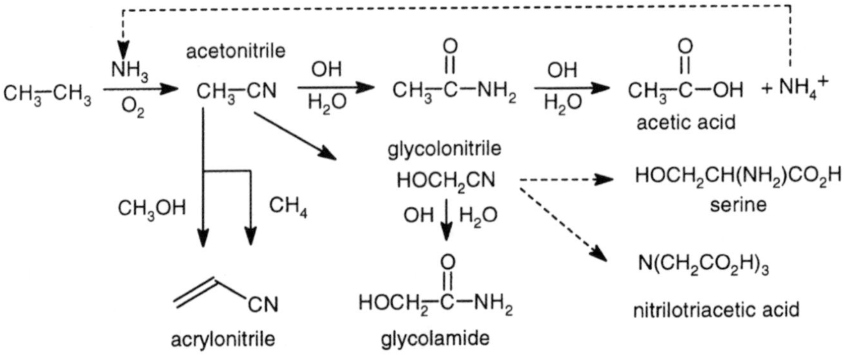

Scheme 5.4. Synthesis of acetonitrile by ethane ammoxidation and of other chemicals by catalytic conversion of acetonitrile.

propene[199] or the new one under development from propane[200] (see also Chapter 4). There is thus specific interest in the development of a new process for selective acetonitrile synthesis from ethane.

In contrast to allylic (amm)oxidation [the selective (amm)oxidation of alkenes at the allylic position such as the propene ammoxidation to acrylonitrile],[141,176,199] little is known about vinylic (amm)oxidation of alkenes such as ethene ammoxidation,[201–203] although indications suggest that different reaction mechanisms are in effect in the two cases. Similarly, it has been found that catalysts selective for propane ammoxidation (V/Sb oxide–based catalysts) are not active and are poorly selective in ethane ammoxidation, thus indicating the need to develop new classes of catalysts.[204] However, there are few studies, in the literature concerning this reaction.

Aliev et al.[205] reported that ethane can be converted to acetonitrile over Cr/Nb oxide catalysts in the 350–500°C temperature range, but a maximum yield of 10% was indicated and quite long contact times (19 s) were required. We used instead Sb/Nb oxide supported on alumina, and got significantly better results at temperatures close to 500°C and for much shorter contact times (2.6 s).[204,206] A selectivity for acetonitrile of close to 50% for up to 40% ethane conversion was found. Li and Armor[207–209] recently studied a series of metallozeolites for this reaction. Both the structure of the zeolite and the type of metal influence the catalytic behavior. Better results were found with Co-beta and Co-ZSM5, which at 450°C give a selectivity for acetonitrile close to 50% for an ethane conversion slightly lower than 40%. The catalytic behavior is thus quite comparable to that found for Sb/Nb oxides supported on alumina,[204,206] although we recorded a reaction rate some one to two orders of magnitude higher for Co-zeolite catalysts. Owing to the different reaction conditions this aspect should be further verified. Acetonitrile formation by ethane ammoxidation has also been observed over V/P oxides[129] and V/Sb oxides,[200] but with acetonitrile yields of about 1% or lower.

Owing to the dearth of studies on this reaction, information concerning the reaction mechanism is limited, although the different authors have offered various hypotheses.

Comparing $NbSbO_x/Al_2O_3$ catalysts with the analogous V/Sb oxide–based catalysts for propane ammoxidation, we observed that the most striking evidence is the following[5,204,206]: (i) Nb/Sb oxide–based catalysts, unlike those based on V/Sb oxide, lead to oxidation with carbon chain rupture and nearly equimolar formation of CO and acetonitrile from a probable common intermediate; (ii) the order of reactivity, as regards both the increase in the carbon-chain series and the difference between alkane and alkene reactivity (the case of propane and propene) is different from that observed for V/Sb oxides; and (iii) unlike with V/Sb oxides allylic type reactivity (H abstraction and O insertion) is absent in $NbSbO_x/Al_2O_3$. The detection of alkene from alkanes and the decrease in their selectivity with increasing conversion and reaction temperature indicates that the alkane oxidative dehydrogenation to the corresponding alkenes is the first step in the reaction over $NbSbO_x/Al_2O_3$ catalysts, similar to that with V/Sb oxide–based catalysts.[5] This is reasonable as niobium-based catalysts are known for their behavior in alkane oxidative dehydrogenation.[210] However, unlike V/Sb oxide–based catalysts, in $NbSbO_x/Al_2O_3$ catalysts there are no sites for allylic type oxidation.[206] This is thought to be related to the fact that antimony must be in the Sb^{5+} state to be active in allylic type oxidation and when reduced to Sb^{3+} its reoxidation is very slow, if not catalyzed by redox elements such as iron or vanadium.[211] Niobium does not catalyze this reaction, although it is in a good contact with the antimony oxide as indicated by the fact that the antimony is maintained in a highly dispersed state over the alumina.[206]

On V/Sb oxides there are two competitive routes in propane ammoxidation.[212] The propene intermediate can either: (i) react with Brønsted acid sites to give an isopropylate intermediate, which easily degrades to acetate and formate species responsible for acetonitrile and CO formation, respectively; or (ii) give allyl alcoholate-type species further transformed to acrolein and acrylonitrile. If none of the latter type of active sites is present in $NbSbO_x/Al_2O_3$, the first pathway of transformation dominates the reactivity with formation of acetonitrile and CO from propane as principal products of reaction and in nearly equimolar amounts.

In ethane ammoxidation, after the first step of ethene formation, the alkoxy intermediate may give an analogous reactivity path, but as the C–O bond forms at methylic groups rather than at methylenic positions as in isopropylate, a second oxygen attack at the same carbon atom to form an acetate species or at the neighboring carbon atom to form two formate species is possible. Both these intermediates then further rapidly transform to acetonitrile (probably via acetamide formation) and CO, respectively. The relative rate of transformation along these two pathways is about 2:1 according to the selectivity ratio (Scheme 5.5).

Li and Armor[207,208] suggested a different mechanism for the transformation on Co-zeolites. In agreement with our conclusions,[204,206] they determined that the first step in the reaction is the oxidative dehydrogenation of ethane to ethene, a fast reaction that is not rate-determining. However, they also indicated that the second step in not the oxygen addition to ethene, but rather the direct addition of adsorbed ammonia to form adsorbed ethylamine. This is a reversible reaction and the rate-limiting step, whereas the further conversion of this reactive intermediate to form acetonitrile is again a fast reaction. The proposed reaction mechanism is supported by evidence that monoethylamine is rapidly oxidatively converted to acetonitrile and that ethene is selectively (about 80% selectivity) converted to acetonitrile on Co-ZSM5.[208] The role of ammonia, besides that of reactant, is also to strongly coordinate to active sites, minimizing the number of vacant sites responsible for combustion of the ethene intermediate and thus considerably promoting selectivity (Scheme 5.5).

Sokolovskii et al.[213] in reviewing earlier information on ethane ammoxidation to acetonitrile on chromium and scandium molybdates[214-217] suggested a different mechanism. Kinetic isotope effect measurements[214] indicate that the rate-limiting

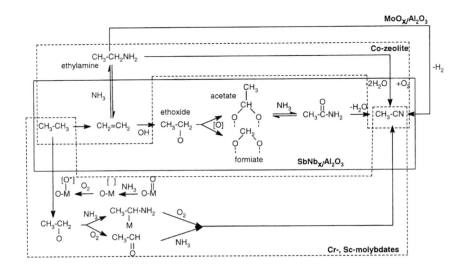

Scheme 5.5. Reaction networks in ethane or ethene ammoxidation to acetonitrile proposed for different types of catalysts.

step on these catalysts is the C–H bond cleavage in ethane, but the sites responsible for this C–H bond cleavage are created on the catalyst by the action of ammonia, first forming reduced coordinatively unsaturated molybdenum ions, which, via interaction with gaseous molecular oxygen, then form active oxygen species. The latter reacts with ethane to form an ethoxy species, which, reacting with ammonia, forms an adsorbed ethylamine species, which, in turn, is further dehydrogenated to acetonitrile. Alternatively, the ethoxy species is first oxidized to acetaldehyde, which then reacts with ammonia to give acetonitrile (Scheme 5.5). However, Peeters *et al.*[201–203] noted that on supported molybdenum oxide, two routes for acetonitrile from ethene are possible: one with the release of water as in the ammoxidation mechanism and the other with the release of hydrogen (and coproduction of ethane), resulting in oxidative ammonolysis. Under steady state conditions, acetonitrile is only formed via oxidative ammonolysis and thus without the participation of lattice oxygen (Scheme 5.5).

Thus, there are very different opinions in the literature concerning the mechanism of ethane or ethene ammoxidation to acetonitrile, and owing to the differing characteristics of the catalysts studied and limited data in support of the various hypotheses it is difficult to select any particular one as a guideline to catalyst design. This topic should be studied in more detail to determine the key factors that may be targeted to improve catalytic performance. It can be noted, however, that the data of Li and Armor[207,208] indicating the good catalytic behavior of Co-zeolites in this reaction is very interesting not only in itself, but also because few investigations of metallozeolites for selective oxidation reactions have been reported in the literature and the analysis of this new class of catalysts can open up interesting prospects (see also Chapter 7).

5.3.3. Propane Conversion

5.3.3.1. Acrolein Synthesis

The direct synthesis of acrolein from propane has not been extensively examined in the literature, although more data are available than for the case of ethane. Better catalytic results were also found in some cases, usually when there was a significant contribution of homogeneous gas phase processes in the first step of propane oxidative dehydrogenation. The catalysts employed are mainly catalysts for propene selective oxidation that also contain some element to promote alkane conversion and/or contain elements to moderate catalyst activity to shift the activity range (for propene conversion) to the higher temperatures required for a mixed heterogeneous/homogenous step of propane to propene oxidative dehydrogenation. Catalysts for a true heterogeneous mechanism of propane selective oxidation have different characteristics from those of known catalysts for propene selective oxidation.

In the oxidation of propane to acrolein, Kim et al.[218] found an acrolein yield of about 8% (64% selectivity) at 13% propane conversion (reaction temperature 500°C) on a sheelite-type Bi/V/Mo oxide doped with silver. In order to increase productivity as well as selectivity, there must be a high propane concentration (~ 30 vol%) in the feed. The first stage of the reaction is the homogeneous oxidative dehydrogenation of propane to propene.

Similar Bi/V/Mo oxide catalysts doped with various elements were described by Bartek et al.[219] in a patent for the synthesis of acrolein and acrylic acid from propane, although the second product was the principal one. The best catalyst, with the empirical composition $BiMo_{12}V_5Nb_{0.5}SbKO_x$, gave 18% propane conversion at 400°C with an acrylic acid yield of about 5% (28% selectivity) and an overall selectivity of 65% for products of partial oxidation (propene, acrolein, acrylic, and acetic acid). However, the selectivity for acrolein was about 1%.

This kind of catalyst was studied again recently by Baerns et al.,[220] who compared the behavior of supported $Ag_{0.01}Bi_{0.85}V_{0.54}Mo_{0.45}O_x$ catalysts with that of $Me_7Bi_5Mo_{12}O_x$ (Me=Mg, Ca, Zn) catalysts in propane oxidation to acrolein. They observed almost no formation of acrolein on the supported silver-doped Bi/V/Mo oxide (sheelite structure), whereas $Ca_7Bi_5Mo_{12}O_x$ and $Mg_7Bi_5Mo_{12}O_x$ gave a selectivity of 34 and 20%, respectively, to acrolein (3.3 and 2.7% yields, respectively). They suggested that two different types of lattice oxygen are involved in propane dehydrogenation to propene and consecutive reactions toward acrolein and CO_x. Under conditions of effective propane dehydrogenation to propene a competitive reaction between an allylic intermediate and propane forming propene and a propyl radical was suggested, explaining the low acrolein formation in favor of propene. In fact, they found that a two-layer fixed-bed reactor could improve the behavior, giving an acrolein yield of 7.4% (20% selectivity).

It should be noted that Bi/V/Mo oxides with a sheelite structure are highly selective to acrolein (selectivity up to 90%, depending on composition) in propene oxidation,[221–223] and Yoon et al.[224] have cited a selectivity of 73% for acrolein at 450°C on $Bi_2Mo_3O_{12}$, although for very low propane conversion (0.3%), whereas all other types of analogous molybdates (Mg, Ca, Ba, Zn, Co, Ni, Mn, Cu, Fe, Ce, etc.) were found to be selective only for propene, and not for acrolein. It is thus not clear whether or not Bi/V/Mo or Bi/Mo oxide–based catalysts can be considered selective catalysts for the propane to acrolein reaction and the reasons for the discrepancies in the types of products observed by various authors are not clearly understood (although the catalyst compositions were not exactly the same).

It is interesting to note that recent results obtained on thin MoO_3 films deposited on a gold anode of an electrochemical reactor indicate the possibility of selective synthesis of acrolein from propane depending on the crystal morphology of the MoO_3 film.[225] A porous film (obtained by sputtering at 300°C) composed of leaflike crystals with preferential orientation of the (010) plane parallel to the pore channel and perpendicular to the gold surface gave the highest activity in acrolein production. There were indications that the high activity was related to the high density of steps of this crystal morphology of MoO_3. Although data obtained under electrochemical conditions are not directly comparable with those at higher temperatures in fixed-bed reactors for heterogeneous catalytic transformations, the results show that molybdate-based catalysts are an interesting class of materials for propane conversion. Their development must now pass through a more detailed analysis of how to tune the reaction rates of each individual step of the complete reaction network, similarly to what was discussed in the previous section regarding the reaction of selective ethane oxidation.

A second class of catalysts investigated is based on metal phosphate. Takita *et al.*[226] studied the behavior of different phosphates in propane oxidation. The best acrolein yield was found for a $Mn_2P_2O_7$ catalyst at low temperatures (370°C) which gave a selectivity of about 40% at 21% propane conversion, but for a complex mixture of various products (acetaldehyde, acrolein, methanol, acetic acid, propanoic acid, acetone, and propionaldehyde). Propene selectivity under the same conditions was about 25%. Krieger and Kirch[227] claimed instead that Mo-phosphate also gives good results in propane oxidation at low temperatures (340°C), but reported a selectivity for acrolein of only about 20% at 10% propane conversion. Phosphates of the type $AV_2P_2O_{10}$ (A = Cd, Cd, Ba, Pd) with orthorhombic or monoclinic symmetry were studied recently by Savary *et al.*,[228] who found good selectivities for propene (up to over 80%, although at propane conversion below 2%), but selectivities for acrolein lower than 10%. The same catalysts were found in the same range of temperatures to be more selective for acrolein starting from propene as the reactant, and the hydrocarbon conversion was higher. Thus again a problem was found that was similar to the one discussed by Baerns *et al.*[220] of an apparent slower step of propene to acrolein transformation (starting from propane), although propene converts faster and more selectively for acrolein on the same catalyst and in the same temperature range. This is a more general problem of the reactivity of catalysts for the selective (amm)oxidation of alkanes that will be discussed in detail in Chapter 8, but indicates one of the main difficulties in catalyst design for the oxyfunctionalization of alkanes. A selective catalyst cannot be simply an assemblage of active sites for alkane to alkene oxidative dehydrogenation plus active sites selective for the second step of alkene selective oxidation.

The presence of acrolein in propane oxidation has also been observed by various other authors,[229-233] but in very small amounts. Catalytic results in propane oxidation to acrolein are shown in Figure 5.11. A maximum selectivity for acrolein of 40% even at very low propane conversion was observed, apart from the case of silver-doped Bi/V/Mo oxide (black symbols), which differ significantly from the other catalysts, but for which there is strong evidence[218,234] for the occurrence of a homogeneous gas phase propane to propene conversion.

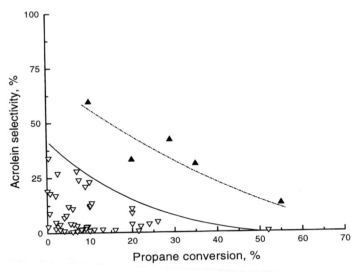

Figure 5.11. Selectivity to acrolein from propane as a function of hydrocarbon conversion for various catalysts cited in the literature (variable reaction conditions).[218,226,229–234] Data marked by solid symbols refer to reaction conditions where there is a significant contribution of gas phase dehydrogenation of propane.[234] Elaborated from Baerns et al.[220]

5.3.3.2. Acrylic Acid Synthesis

The direct synthesis of acrylic acid (as the main product) has been less extensively reported in the literature than that of acrolein, but from the industrial point of view there are more incentives to develop catalysts for the second reaction, because acrylic acid is presently produced in a two-step oxidation process from propene via acrolein,[175] and thus its direct synthesis from propane not only uses a cheaper feedstock, but also reduces process complexity.

Ai[235] studied V/P mixed oxides (catalysts for *n*-butane oxidation) doped with tellurium, a known component of the catalysts for propene conversion to acrylic acid.[236] A maximum acrylic acid yield of about 10% was obtained at 380°C, corresponding to a selectivity of 37%. However, productivity was very low owing to the long contact times (about 10 s) that were required. Another particularity was the need to use very high oxygen concentrations (76%), which are clearly unrealistic for industrial operation. In a more recent paper, Ai[237] reported a maximum yield of acrylic acid of about 8% at 45% propane conversion, but in the presence of a large coproduction of acetic acid (yield of about 5%). The catalyst had an optimal phosphorus/vanadium ratio close to 1, and no additives were used, but the oxygen concentration was still very high (65%), suggesting the occurrence of radical-type reactions leading to the coformation of acrylic and acetic acid.

In his catalytic tests, Ai[235,237] did not cofeed steam, which was instead cofed in large quantities in the experiments done by Krieger and Kirch[227] and Bartek *et al.*,[219] who cofed 40 and 65% steam, respectively. As discussed in the case of ethane, water promotes the selectivity for acids enhancing the rate of their desorption. On Bi/V/Mo oxides containing niobium, antimony, and phosphorus, as promoters, Bartek *et al.*[219] found 18% propane conversion at 400°C with an acrylic acid yield of close to 5% (selectivity 28%). They used a high propane concentration (25%) and a low oxygen-to-propane ratio (0.4), much lower than the stoichiometric value (2). Krieger and Kirch[227] reported that a P/Mo/Sb oxide ($P_{1-1.2}Mo_{12-15}Sb_{1-2}O_x$) was optimal for the reaction of propane oxidation to acrylic acid, noting a conversion of 10% at 340°C with 19% selectivity for acrylic acid. The reaction conditions were similar to those cited by Bartek *et al.*,[219] but a lower concentration of propane in the feed (10%) resulted in a higher oxygen-to-propane ratio (although still half the stoichiometric value).

The catalyst composition reported by Krieger and Kirch[227] is analogous to that expected for phospho–molybdo heteropoly acids with a Keggin structure, for which the partial introduction of antimony is known to lead to enhanced stability.[108–110] In agreement, vanadium-containing phospho–molybdo heteropoly acids with Keggin structures were found to be active and selective in propane oxidation to acrylic acid.[238] A selectivity of 22% for acrolein at 50% propane conversion and a maximum yield of acrolein close to 10% have been reported for $H_5PV_2Mo_{10}O_{40}$ heteropoly acid, but the catalyst stability was very low. The use of structural promoters of stability such as antimony and high concentrations of steam (also a promoter of stability for heteropoly compounds) can reduce the sensitivity to irreversible oxoanion decomposition, but the stability is still too low with respect to possible industrial applications. On the other hand, phospho–molybdo heteropoly acids with Keggin structures (promoted by elements such as antimony and/or phosphorus for stability and vanadium for activity/selectivity) have most of the required characteristics for selective synthesis of acrylic acid from propane (a suggested mechanism of acrylic acid formation on heteropoly acids is shown in Figure 5.12), apart from the required thermal stability. It is thus necessary to use different types of approaches to improve this feature, such as the inclusion of the oxoanion inside a rigid solid matrix such as a zeolite, which avoids the mechanism of oxoanion condensation (a reasonable first step in the mechanism of decomposition of oxoanions), but allows reactants accessibility to active sites.

Good results in the direct oxidation of propane to acrylic acid were found by Mizuno *et al.*[239,240] for iron-doped vanadophosphomolybdic heteropoly acids, with optimal behavior for $Fe_{0.08}Cs_{2.5}H_{1.26}PVMo_{11}O_{40}$, although no indications were given regarding the long-term stability. Ueda *et al.*[187,241] cited good results for reduced heteropoly acids in propane oxidation, with an optimal behavior of 50% selectivity for acrylic acid plus 24% selectivity for acetic acid at 12% propane conversion (temperature about 350°C), when 30% water is present in

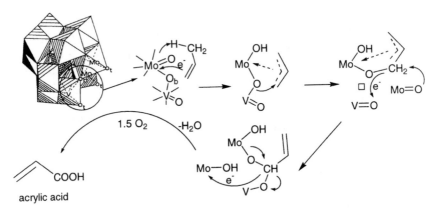

acrylic acid

Figure 5.12. Mechanism of acrylic acid formation from propane suggested by P-V-Mo heteropoly acids.[238]

the feed. However, after 30 h of time-on-stream, partial decomposition of the heteropoly acid was noted by the appearance of MoO_3 diffraction lines in the XRD patterns.

In conclusion, notwithstanding the industrial interest in a process of direct synthesis of acrylic acid from propane, the catalytic results reported up to now are still very far from a possible application. Not only are yields and productivity low, but there are other negative factors such as: (i) the coformation of various other oxygenates; (ii) the occasional presence of not true heterogeneous reactions, which make scale-up and reproducibility of the results difficult, and (iii) use in some cases of unrealistic feed compositions. On the other hand, the limited number of catalysts investigated suggests that a necessary first step is to extend the range of materials to enable a more systematic screening of the catalytic behavior of mixed oxides, also taking account of the possibility of using unconventional approaches such as placing heteropoly oxoanions inside zeolitic cages to stabilize them from deactivation.

As noted for the case of ethane partial oxidation, the analysis of the reaction mechanism should be a guideline for catalyst selection and interpretation of data, but the specific differences between propane and ethane oxidation products and the relative influence on the type of surface properties required for their synthesis must also be taken into account. In this respect, interesting data were recently reported by Barrault et al.,[242] who studied the addition of bismuth to α-$NiMoO_4$ + α-MoO_3 catalysts. The results, summarized in Table 5.2, show that the addition of bismuth to the system introduced sites of allylic type oxidation (Bi-molybdates are well-known catalysts for selective propene oxidation[199]) and as a consequence after the first step of propane oxidative dehydrogenation, oxygen insertion occurs at the 3

TABLE 5.2. Effect of Bismuth on the Propane Conversion at 375°C over α-MoO$_3$ + α-NiMoO$_4$ Catalysts [molar ratio α-MoO$_3$/(α-MoO$_3$ + α-NiMoO$_4$) = 0.83]. Elaborated from Barrault *et al.*[242]

Amount Bi	Conversion C$_3$H$_8$, %	Selectivity for C$_3$H$_6$, %	Selectivity for acrolein, %	Selectivity for acetic acid, %	Selectivity for acrylic acid, %
0	10.8	42	2.1	5.7	8.4
0.02	8.8	22	8.8	1.6	16.3
0.06	9.6	21	8.4	2.8	16.6
0.17	8.4	29	9.0	1.8	10.7

(allylic) position instead of at the 2 position as with propene. While the first type of attack leads to acrolein and acrylic acid, the second leads to acetone, which is then easily oxidatively cleaved to acetic acid and carbon oxides.[242] The introduction of bismuth thus does not change the conversion significantly, but does change the reactivity toward acrylic acid and acrolein instead of acetic acid (Table 5.2). In the presence of water, acrylic acid formation further increases, reaching a selectivity of 31% at about 15% conversion (reaction temperature 425°C).[242] The effect was interpreted as an enhanced desorption of acrylic acid in the presence of water owing to competitive chemisorption, although, as discussed earlier, the effect is probably more complex.

Stein *et al.*,[243] studying the role of adsorption in the oxidation of acrolein to acrylic acid on Mo/V/Cu oxide catalysts noted that the protonation of the acrylic anion and the following desorption of acrylic acid constitute the rate-determining step and thus there is an accumulation of the acrylic anion on the surface during the catalytic reaction (Figure 5.13) up to the point of constant activity.[244] The addition of water to the feed considerably enhances the rate of acrolein oxidation, which shows a maximum as a function of the acrolein concentration in the feed. With increasing water concentration in the feed, the concentration of acrolein at which the maximum is observed shifts progressively to higher values. This effect is in good agreement with a desorption-controlled reaction rate with the desorption of acrylic acid depending on water concentration in the feed. However, Stein *et al.*[243] also noted that the rate of catalyst reoxidation was a function of water concentration in the feed, possibly due to blockage of reduced catalyst sites by the strong chemisorbed species. Control of the rate of acrylic acid desorption and how to improve it is thus a key parameter to be considered in catalyst design.

5.3.4. Isobutane Conversion to Methacrolein and Methacrylic Acid

The most important industrial process for the synthesis of methylmethacrylate involves three consecutive steps: (i) reaction between acetone and HCN, (ii) reaction of acetone–cianohydrin with sulfuric acid, and (iii) final hydrolysis of the

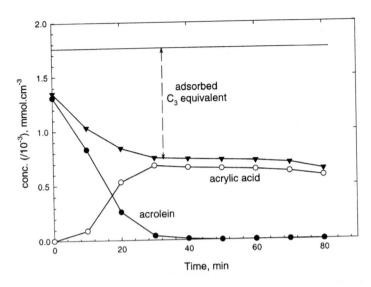

Figure 5.13. Transient catalytic behavior and surface accumulation of products on the catalyst during oxidation of acrolein in the absence of oxygen at 230°C over Mo-V-Cu-oxide catalysts. Elaborated from Stein *et al.*[243]

adduct in the presence of methanol. This process suffers from several drawbacks, among them the use of a highly toxic raw material; the coproduction of huge amounts of ammonium sulfate (which has little or no market), and the necessity for an adjacent plant for HCN production. A process to synthesize methacrylic acid directly from isobutane is thus interesting not only for the lower cost of the hydrocarbon and reduction of process complexity, but also to limit the use of toxic reactants and reduce waste.[245] The increasing interest in the development of an environmentally friendly process for methacrylic acid synthesis has stimulated research in recent years, although up to now only a limited number of catalyst classes has been studied for this reaction. Research in this area has been especially active in Japan and several patents have been issued for the direct synthesis of methacrylic acid from isobutane.[246-254]

The first such patent, assigned to Rohm & Hass,[227] claimed an active and selective catalyst based on Mo/P/Sb oxides, but the catalyst composition is close to that of heteropoly acids with Keggin structures, as noted above. Successive patents assigned to Japanese companies for this process clearly claim a heteropoly compound (in the form of a partially neutralized acid to increase the stability) with a Keggin structure,[249,253] although selectivity and the amount of methacrylic acid formed were nearly the same as that in the Rohm & Hass patent.[227] However, the productivity of the catalyst, due to a higher concentration of isobutane in the feed, was higher, as was catalyst stability, owing to a combination of improved catalyst composition and better operating conditions for the catalytic tests. The average composition for an optimal catalyst claimed in the patents is as follows: $(H_m Y_{0.2-1.5})(P_{1-1.2} Mo_{12-n} X_{0.4-1.5} O_x)$ (Y is an alkali metal ion and X an element such as V, As, and Cu) plus possibly additional elements that vary from patent to patent, as summarized in Table 5.3. No significant differences are observed in the catalytic behavior among the different compositions, but it is possible to draw some general conclusions:

1. A high concentration of isobutane and especially of steam (up to about 65%) was claimed in all patents. Steam is not only ballast to decrease the partial pressure of isobutane in the recycle loop to avoid operating inside the flammability zone, but also improves the activity and stability of the catalyst.
2. The addition of elements such as vanadium and copper plays a minor role in determining the catalyst activity, but significantly promotes selectivity for methacrylic acid, whereas the effect on methacrolein selectivity is small.
3. Other elements such as osmium and arsenic do not seem to have a specific key role in determining the reactivity, whereas alkaline metals such as antimony, promote catalyst stability but depress catalyst activity.
4. Methacrolein with selectivity in the 10–20% range was always found together with the methacrylic acid.
5. The reaction temperature was always kept lower than about 350°C to avoid irreversible decomposition of the heteropoly oxoanion.

Under these conditions, stable catalytic behavior up to over 1000 h of time-on-stream has been reported,[246] but isobutane conversion does not exceed about 15%. With increasing conversion, the selectivity to methacrylic acid decreases considerably and thus a commercial process would require an economical method for recycling unconverted isobutane and methacrolein. Kawakami et al.[252] described the use of an organic solvent (a mixture of decane, undecane, and dodecane) that efficiently adsorbs the isobutane and methacrolein from the off-gas. When oxygen is used as the oxidant, CO must first be oxidized to CO_2 at low temperature and followed by the successive adsorption of the latter in conventional solvents.

More critical is the separation of methacrylic acid from by-products, such as acetic acid, that remain in the aqueous solution in which methacrylic acid is adsorbed. This separation considerably affects the cost of the process and thus catalyst development must be directed not only toward improving selectivity to methacrylic acid, but especially toward reducing or even eliminating other by-products such as acetic acid, acetone, and maleic anhydride.

TABLE 5.3. Catalytic Activity of Heteropoly Compounds in Isobutane Oxidation to Methacrylic Acid. Elaborated from Trifirò and Cavani[171]

Catalyst	Reference	Temperature, °C	Contact time, s	Molar ratio, % $C_4/O_2/H_2O/N_2$	i-Butane conversion, %	Methacrolein selectivity, %	Methacrylic acid selectivity, %
$H_3PMo_{12}O_{40}$	242	340	3.6	60/20/20/0	10.5	21.1	18.3
$H_xP_{1.1}Mo_{12}V_{0.2}Cu_{0.1}As_{0.5}O_y$	242	330	3.6	25/11.5/20/43.5	9.0	18.2	55.2
$H_3KPMo_{11}VCu_{0.2}O_y$	243	300	3.6	25/11.5/20/43.5	7.8	21.8	53.4
$H_xCs_{1.1}P_{1.1}Mo_{12}V_{1.1}Cu_{0.11}O_y$	242	320	3.6	30/15/20/35	10.3	16.3	55.7
$H_xPMo_{12}SbO_y$	227	340	6.1	10/13/30/47	10.0	20.0	50.0
$H_xK_{0.8}PMo_{12}V_{0.5}O_y$	246	320	2.4	10/16.8/10/63.8	9.4	14.8	54.2
$H_xK_{0.8}PMo_{12}V_{0.5}B_{0.2}O_y$	246	320	2.4	10/16.8/10/63.8	10.2	14.3	54.1
$H_xK_{0.3}PMo_{12}V_2Ir_{0.02}Cu_{0.1}Co_{0.2}O_y$	246	320	2.4	10/16.8/10/63.8	17.0	10.8	46.3
$H_xK_{0.5}Cs_{0.5}Ba_{0.2}P_{1.5}Mo_{12}VO_{S0.04}Cu_{0.2}O_y$	246	320	2.4	10/16.8/10/63.8	16.3	10.0	50.1
$H_xCs_{1.5}P_{1.2}Mo_{12}VCu_{0.2}O_y$	239	320	2.4	10/16.8/10/63.8	8.8	15.2	51.2
$H_xCs_{1.5}PMo_{12}V_{0.5}As_{0.4}O_y$	239	320	2.4	10/16.8/10/63.8	8.6	22.5	42.1
$H_xCs_2PMo_{12}Cu_{0.1}As_{0.4}O_y$	239	320	2.4	10/16.8/10/63.8	12.0	10.2	51.0

More recently, the reaction of direct isobutane oxidation to methacrylic acid has been studied from a more fundamental point of view, although all authors used heteropoly compounds.[187,255-263] Li and Ueda[187] studied phosphomolybdic acid in the acid, ammonium, and reduced form, the last prepared by heat treatment of the pyridinium salt. The reduced form of the heteropoly acid shows both higher activity and selectivity for methacrylic acid (at 300°C a selectivity close to 50% for acrylic acid plus about 20% for acetic acid and 5% for acrylic acid for an isobutane conversion of 21% have been reported[187]) with minimal formation of methacrolein, formed instead with selectivity up to 30–40% over the other two parent phosphomolybdic heteropoly compounds. However, temperatures even slightly higher than 300° C are very detrimental to catalyst performance. At 340°C the methacrylic acid selectivity fall to less than 20%.[187] This is a very critical aspect for a possible industrial application and illustrates the major problem of this class of catalysts: the low thermal stability. Li and Ueda[187] suggested that the thermal treatment of the pyridinium salt of the heteropoly compound creates an oxygen-defective structure that provides Lewis acid sites as well as allowing the generation of active oxygen forms, although the latter are ill defined.

The particular activity/selectivity characteristics of heteropoly compounds for the synthesis of acids (acrylic and methacrylic acids) from light alkanes are related to the presence of these sites in combination with high proton acidity and fast electron transfer and delocalization ability.[187] Jalowiecki-Duhamel et al.[258] basically agree with this concept, although they emphasize the idea that in the reduced state of heteropoly compounds (they studied a catalyst with the following composition: $Cs_{1.6}H_{2.4}P_{1.7}Mo_{11}V_{1.1}O_{40}$) the anionic vacancies are able to store reactive hydrogen species, which diffuse through the solid. Thus a lacunar heteropoly compound should have greater ability to selectively activate the alkane, in agreement with the results of Li and Ueda.[187] This is also consistent with the observations made for heteropoly acids in n-butane selective oxidation,[264] for which it was also proposed that the active species is a partially decomposed, lacunar form, and that in Wells–Dawson type heteropolyoxotungstates for isobutane oxidation the selective form is generated in situ when partial reduction occurs because of a high hydrocarbon concentration in the feed.[265] The differences in the proposed mechanism of isobutane conversion on the oxidized and partially reduced surface of these catalysts are shown in Scheme 5.6.

The partial neutralization of the heteropoly acid leads to enhancement of the catalytic behavior in isobutane conversion. An optimal behavior in terms of selectivity to methacrylic acid is observed in $Cs_xH_{3-x}PMo_{12}O_{40}$ compounds for an x value of 2 (at 340°C a selectivity for methacrylic acid of 34% for an isobutane conversion of 11%),[259] which may be further enhanced by the addition of transition metal ions in secondary positions and vanadium in primary (oxoanion) positions.[260] A selectivity of 36% at 15% isobutane conversion ($T = 320$°C) has been reported for $Cs_{2.5}Ni_{0.08}H_{01.34}PMo_{11}V_1O_{40}$.[260,261] The maximum reported methacrylic acid

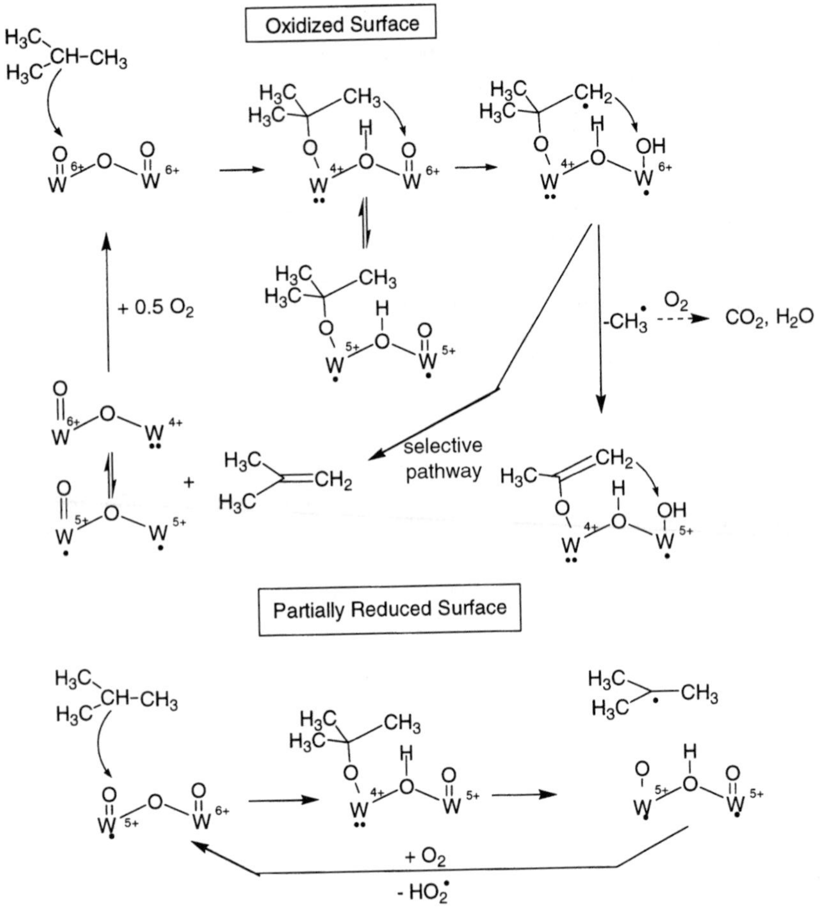

Scheme 5.6. Proposed mechanisms of the isobutane oxidation to isobutene over oxidized and partially reduced surfaces of Wells–Dawson heteropolyoxotungstate ($K_6P_2W_{18}O_{62}$). Elaborated from Cavani et al.[265]

yield was 9%, but with formation of methacrolein and acetic acid as the principal by-products. Good catalytic results have also been reported using iron as the dopant,[255–257] but maximum yields of methacrylic acid were lower than 5%. Methacrolein, acetic acid, and maleic anhydride were the main by-products, apart from carbon oxides.

The reaction mechanism of methacrylic acid formation from isobutane has been studied by Busca et al.[255] mainly by FT-IR analysis of iron-doped $K_1(NH_4)_2PMo_{12}O_{40}$. Scheme 5.7 summarizes the proposed reaction network.

Isobutane can be activated by the catalyst surface at relatively low temperatures at the tertiary carbon by oxidative breaking of the weakest C–H bond. This produces an alkoxide, a hydroxy group and two electrons transferred to the catalyst, which is consequently reduced. The alkoxy evolves toward the formation of adsorbed isobutene, which is quickly oxidized (no alkenes were detected in the gas phase during catalytic tests) in the allylic position to form methacrolein. A common intermediate was hypothesized for methacrolein and methacrylic acid: a dioxy-alkylidene species. The latter can evolve either to methacrolein, through dissociation of a C–O bond, or to a carboxylate species, via oxidation on a Mo–O bond. The carboxylate is the precursor for methacrylic acid desorption. At low temperatures, competitive pathways are the dimerization/polymerization of isobutene to heavier products and the skeletal isomerization of isobutane to n-butane from which the observed maleic anhydride is derived.

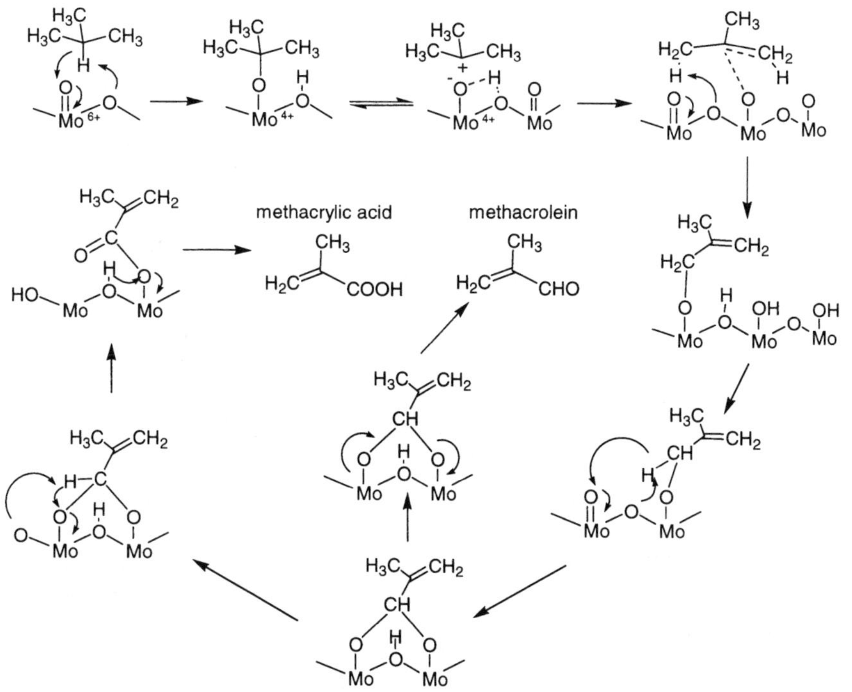

methacrylic acid

methacrolein

Scheme 5.7. Proposed reaction mechanism for the oxidation of isobutane over $K_1(NH_4)_2PMo_{12}O_{40}$ catalysts. Elaborated from Busca et al.[255]

Since the loss of selectivity for methacrylic acid is mainly due to parallel competitive reactions, objectives of the research remain the control of the competitive pathways of transformation, the identification of the reasons for the key particular reactivity characteristics of heteropoly compounds for acid formation from light alkanes, and the understanding of how to stabilize the metastable, lacunar type active forms of the heteropoly compounds.

5.3.5. n-Pentane Conversion to Maleic and Phthalic Anhydrides

C5 hydrocarbons became more available for use as a feedstock in selective oxidation processes because new restrictions on gasoline volatility effectively removed them from the gasoline pool. Although they can be used in refineries for alkylation reactions, this recent abundance has stimulated a search for processes to upgrade them to higher-value products.[266]

After the discovery of the possibility of selective synthesis of maleic and phthalic anhydride on $(VO)_2P_2O_7$ catalysts,[267] several researchers studied the selective oxidation of n-pentane over four main types of catalysts: (i) V/P oxides,[267–283] (ii) vanadophosphomolybdic heteropoly acids,[238,284,285] (iii) V/Mo mixed oxides,[274,286] and (iv) supported vanadium oxide catalysts.[287–290] While maleic anhydride formed on all these catalysts, phthalic anhydride was found only on some catalysts and in the case of V/P oxides, also depending on the preparation modalities of the catalyst. Although the molar yield of phthalic anhydride on V/P oxides can reach significant values (around 25%, plus about 20% of maleic anhydride)[272,279] the productivity is still too low for industrial exploitation, and the loss of two carbon atoms in going from n-pentane to phthalic anhydride lowers the yield by weight, a more interesting criterion for industrial purposes than molar yield. On the other hand, the synthesis of maleic anhydride from n-pentane rather than from n-butane suffers from the same drawback of lowering of yield by weight, even in the case of finding a catalyst with the same catalytic performance as V/P oxides for n-butane oxidation to maleic anhydride (Chapter 4). Moreover, present results indicate a much more drastic lowering of the selectivity with increasing hydrocarbon conversion in the case of n-pentane as compared with n-butane.

Therefore, present data do not suggest a good outlook for commercialization of a possible process of synthesis of anhydrides from n-pentane. However, from a basic science point of view it will be very interesting to obtain a better understanding of how so complex a selective synthesis is possible [n-pentane oxidation to phthalic anhydride is a 22-electron oxidation and the reaction mechanism involves the abstraction of 8 hydrogen atoms, the insertion of 3 oxygen atoms in the organic molecule, the formation of four water molecules, and the dimerization of 2 hydrocarbon molecules and transformation from a C10 molecule (dimer of n-pentane) to a C8 (phthalic anhydride) molecule; all this occurs in a single step—no intermediates have been detected—over the catalyst surface] and which catalyst properties allow this synthesis. Other particular aspects are: (i) why the same catalyst (vanadyl pyrophosphate) very selective in n-butane selective oxidation to maleic anhydride (phthalic anhydride is detected only in traces, although the C8 anhydride should in principle be favored in the case of n-butane rather than n-pentane) only gives phthalic and maleic anhydride starting from n-pentane, and (ii) why on other catalysts such as vanadophosphomolybdo heteropoly compounds, also with good behavior in n-butane oxidation, n-pentane gives rise only to maleic anhydride formation.

Two routes have been suggested for the formation of phthalic anhydride from C5 hydrocarbons. According to Hönicke et al.[286] phthalic anhydride is obtained by a Diels–Alder reaction between pentadiene and maleic anhydride, but this does not explain why butadiene (intermediate from n-butane) does not also react with maleic anhydride to form phthalic anhydride from n-butane. This is explained in the mechanism we proposed (see Scheme 8.20 and Scheme 5.8a), which is based on the fact that the pentadiene intermediate, unlike the analogous butadiene, still possesses allylic hydrogen, which can be easily further abstracted to give a cyclopentadiene intermediate that is rather unreactive toward further oxidation, but can give a templatelike surface dimerization with a second cyclopentadiene molecule.[271,272] This precursor then evolves to phthalic anhydride.

Cavani et al.[275,276] in comparing the reactivity of a series of possible intermediate-like reactants (molecules that, when adsorbed on the catalyst surface, can give rise to some of the possible reaction intermediates such as 2-pentene, pentadiene, alkylaromatics, C5–C8 linear and cycloalkanes, etc.) suggested an alternative hypothesis for the mechanism of dimerization, the key step to final formation of phthalic anhydride. They observed, in agreement with Roney,[291] that cyclopentene may be converted to decaline by acid catalysis (Scheme 5.8b) and decaline is then quickly converted to naphthalene, which might be finally converted to phthalic anhydride (see also Figure 8.9) and the related discussion). In fact, the direct oxidation of decaline yielded tetraline, naphthalene, naphthoquinone, and phthalic anhydride, although the fact that naphthalene and naphthoquinone were not detected among the by-products of n-pentane oxidation (these species are less reactive than decaline and thus form in larger amounts than phthalic anhydride in decaline oxidation)[292] indicates that this second possible mechanism of formation of the key intermediate in phthalic anhydride formation is not the dominant pathway.

Cavani et al.[275,276] also observed that while using a hydrocarbon concentration in the feed of around 1–2%, small amounts of maleic and phthalic anhydride are obtained feeding 2-pentene (much higher yields under the same conditions were obtained using n-pentane), with an increase in n-pentane concentration to about 20% the general behavior is similar to that of 2-pentene, but with a tenfold lower

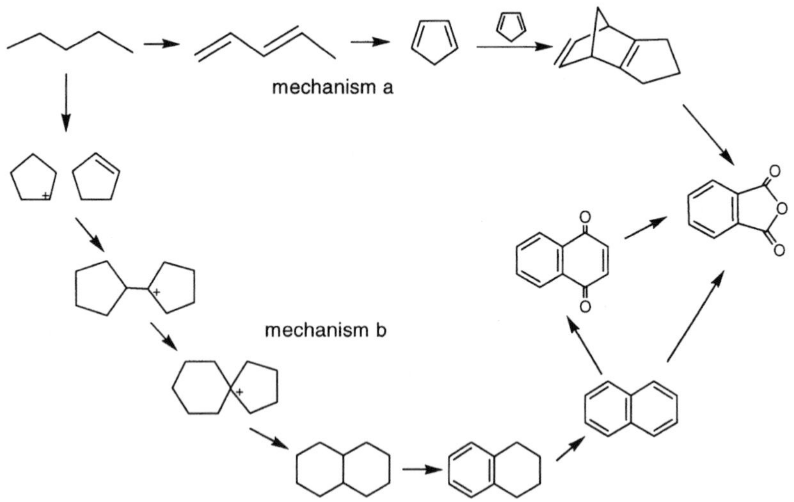

Scheme 5.8. Mechanisms proposed for the synthesis of phthalic anhydride in *n*-pentane oxidation over vanadyl pyrophosphate: (a) via "template addition" of two cyclopentadiene molecules and (b) via acid-catalyzed dimerization of two pentadiene molecules to decaline. Elaborated from Cavani and Trifirò.[275]

concentration in the feed. This could be ascribed to the different adsorption characteristics of the two reactants. Under these conditions, acrylic acid, 2-methyltetrahydrofuran, citraconic anhydride (methyl maleic anhydride), 3-penten-2-one, cinnamaldehyde (3-phenyl-propenal), and 2H-pyran-2-one were detected.[275,276]

Cavani *et al.*[275,276] observed that under anaerobic conditions butadiene forms styrene and ethylbenzene, with only traces of xylenes, on an equilibrated V/P oxide catalyst, while from 1,3-pentadiene the products obtained are mainly dimethyl ethylbenzenes. The products formed can be explained by assuming a role for surface Lewis-type acidity, which through the coordination of the unsaturated molecule addresses the reaction pathway toward the formation of specific compounds. In fact, when a classical H^+-catalyzed mechanism is operating, the formation of xylenes from butadiene would be just as likely as the formation of styrene. In the same way, starting from 1,3-pentadiene, unsaturated cyclic C10 compounds form, such as 3,4-dimethyl-5-vinyl cyclohexene or 3,5-dimethyl-4-vinyl cyclohexene. If the reaction starts from the corresponding alkanes (*n*-butane, *n*-pentane) no products are at all formed in the absence of molecular oxygen.

On the basis of this information, and also comparing the reactivity of *n*-pentane with that of *o*-xylene and of pentenes, an alternative mechanism was hypothesized that includes the following steps:

1. Oxidehydrogenation of n-pentane to pentenes and then to pentadiene.
2. Transformation of pentadiene either to maleic anhydride (possibly with a mechanism analogous to that from n-butane/butadiene, but including the loss of one carbon atom), or to the cyclic C10 compound.
3. Oxidation of the C10 intermediate to phthalic anhydride (with the loss of two carbon atoms) by oxidation of the two adjacent alkyl groups on the aromatic ring (analogously to what occurs in o-xylene oxidation; the catalyst, in agreement, was found to be selective in o-xylene oxidation).

Phthalic anhydride is the most stable product of partial oxidation that can be obtained. Other products are probably burned to carbon oxides, which explains why, under usual reaction (fuel-lean) conditions, the only products of selective oxidation are maleic and phthalic anhydrides. The same occurs from n-butane, where maleic anhydride is the only product.

Phthalic anhydride is obtained from n-butane with very low selectivity (less than 1%) even though one might expect it to be formed more easily from a C4 hydrocarbon than from n-pentane. However, the formation of phthalic anhydride is not possible because intermediate o-xylene is formed in very small amounts as compared to styrene and ethylbenzene in the acid catalyzed cyclodimerization of intermediate butadiene.

Several features of the mechanism of formation of the key intermediate in phthalic anhydride formation were clarified, but the key catalyst characteristics that determine the occurrence of this reaction and allow maximization of phthalic versus maleic anhydride formation are still not well understood. It is worth noting that the change in the vanadyl pyrophosphate catalyst from the nonequilibrated (fresh) to the equilibrated state (usually after more than 200 h of time-on-stream in n-butane oxidation) leads to a considerable increase in maleic anhydride formation from n-butane and in phthalic anhydride formation from n-pentane (Figure 5.14).[279] The change in catalyst reactivity corresponds to a change from a multiphasic system (poorly crystallized vanadyl pyrophosphate together with amorphous V^{4+}-phosphate phase and crystalline γ-V^{5+}-OPO_4) to a well-crystallized $(VO)_2P_2O_7$ catalyst only. Characterization data on this *in situ* change of the catalyst do not allow one to draw clear conclusions. However, the parallelism between the two reactions suggests that one key parameter is the catalyst's surface oxidation activity and possibly also a change in its surface topology (see also Chapter 4) with the creation of restricted surface cavities, that enable better control of the oxidation of intermediates and favor surface templatelike reactions, whereas its surface acidity does not appear to be significantly changed. This was confirmed by a study of the effect of cobalt and iron dopants on the catalytic behavior of V/P oxides in n-pentane oxidation.[276] The addition of cobalt, which favors the crystallinity of vanadyl

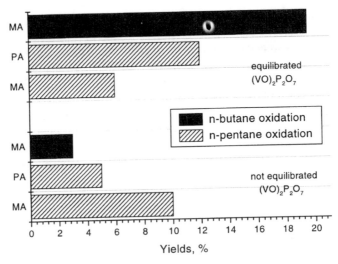

Figure 5.14. Yields to maleic anhydride (MA) and to phthalic anhydride (PA) in *n*-butane and *n*-pentane oxidation on equilibrated and nonequilibrated (fresh) vanadyl pyrophosphate catalysts at 340°C. Elaborated from Albonetti *et al.*[279]

pyrophosphate, leads to an enhancement of the selectivity for phthalic anhydride from about 30% (undoped vanadyl pyrophosphate) to about 50% (in the 5–15% *n*-pentane conversion range), whereas doping with iron (which favors nucleation of a_{II}-VOPO$_4$ and δ-VOPO$_4$) leads to lowering of the selectivity to about 20%. It therefore seems that a key parameter to maximize phthalic anhydride formation is the control of surface redox and topology characteristics of vanadyl pyrophosphate, although more detailed studies are needed.

For supported vanadium oxide catalysts, it has been suggested that the nature of the vanadia species and the oxidation state of the catalyst are both important,[290] although not specifically referred to phthalic anhydride formation from *n*-pentane. Ozkan *et al.*[289] noted that the addition of lithium or sodium to a supported V$_2$O$_5$ catalyst shifts the activity curve for phthalic anhydride formation to higher temperatures, but does not change its maximum formation. They noted instead a significant lowering of maleic anhydride formation. This indication is in contrast to the possible role of acid sites in catalyzing the formation of the key intermediate in phthalic anhydride formation, although the evidence is not strong enough to be conclusive, and some change in the reaction mechanism between V/P oxides and supported vanadium oxide is possible, as indicated by the detection of 3-methy-2,5-furandiene[289] not found on V/P oxides. It also must be noted that the formation of phthalic anhydride from *n*-pentane is much lower over supported vanadium oxide than over V/P oxides, whereas better results are obtained by feeding 1-pentene.[289]

Ozkan et al.,[288] studying the dicyclopentadiene conversion on suboxides of vanadia, also noted that the degree of vanadia reduction considerably influences the phthalic anhydride formation and suggested that maleic anhydride formation can derive from both the preoxidation ring opening and dealkylation of the dicyclopentadiene intermediate or phthalic anhydride itself, although these data are not consistent with the kinetics of phthalic and maleic anhydride formation over V/P oxides.[269] Sobalik et al.[282] suggested instead that the suppression of homogeneous reactions, e.g., by the addition of methyl tert-butylether to the feed, inhibits phthalic anhydride formation on V/P oxides, but this observation is not consistent with other published data nor with their own,[283] indicating that phthalic anhydride formation, unlike that of maleic anhydride, requires a highly ordered vanadyl pyrophosphate surface, in agreement with the previously discussed observations.

In conclusion, notwithstanding the research efforts in recent years on the study of the oxidation of n-pentane and other C5 hydrocarbons, the data do not allow clear conclusions to be drawn concerning the principal catalyst surface properties required to maximize phthalic anhydride formation, a key step for possible industrial development of the process when a C5 hydrocarbon stream may be available on site. It should also be noted that up to now only well-crystallized vanadyl pyrophosphate has been shown to form phthalic anhydride with appreciable yields, but that catalyst screening for this reaction has not been very extensive. Thus an effort in this direction may also be necessary.

5.3.6. Cyclohexane (Amm)oxidation

Very limited data are available in the literature on the selective gas phase (amm)oxidation of cyclohexane, although some interesting products can be obtained in these reactions. However, in gas phase oxidation over solid catalysts, a major problem is to limit the fast oxidative dehydrogenation, which leads to benzene at rather low temperatures.[292] It is thus probably necessary to develop new kinds of catalysts able to oxidize by ring opening the cyclohexane, without hydrogen abstraction ability.

The ammoxidation of cyclohexane to adiponitrile has been studied especially by Russian researchers,[213,293–295] who found interesting results on Mo/Sb oxide–based catalysts. Besides adiponitrile, glutaric, maleic, and fumaric acids, benzene, cyclohexene, acetonitrile, and HCN, carbon oxides were detected as by-products. The broad range of by-products and low selectivity to adiponitrile, together with a decline in activity with time-on-stream, indicates that considerable research effort is needed for industrial exploitation of the process, although the availability of cyclohexane as a side stream from refinery processes may lead to industrial interest in direct adiponitrile synthesis.

The reaction mechanism proposed is rather complex and not fully understood. One major problem, however, is the strong interaction of adiponitrile with the surface sites, which leads to its fast oligomerization on the surface, a possible reason for fast catalyst deactivation. However, the limited data available do not point to clear conclusions, and catalyst screening for this reaction has been limited to only a few samples. Preliminary investigation in this direction is therefore necessary, taking into account the strong chemisorption of both the diacid or dinitrile products as well as the fact that again in this case ring aromatization (a major problem in controlling surface selectivity) must be avoided.

REFERENCES

1. D.J. Hucknall, *Selective Oxidation of Hydrocarbons*, Academic Press: London (1974).
2. B.K. Hodnett, *Catal. Rev.-Sci. Eng.* **27**, 373 (1985).
3. G. Emig and F. Martin, *Catal. Today* **1**, 477 (1987).
4. G. Centi, F. Trifirò, J. Ebner, and V. Franchetti, *Chem. Rev.* **88**, 55 (1988).
5. G. Centi, S. Perathoner, and F. Trifirò, *Appl. Catal. A: General* **157**, 143 (1997).
6. G. Centi, R.K. Grasselli, and F. Trifirò, *Catal. Today* **13**, 661 (1992).
7. V.I. Sobolev, K.A. Dubkov, O.V. Panna, and G.I. Panov, *Catal. Today* **24**, 251 (1995).
8. M.G. Clerici and G. Bellussi, *It. Patent* 21492A (1989).
9. G.J. Hutchings (ed.), *Recent Advances in C1 Chemistry, Catal. Today* **23**, 451 (1995).
10. G.E. Keller and M.M. Bhasin, *J. Catal.* **73**, 9 (1982).
11. Catalytica, *Direct Methane Conversion to Fuels and Chemicals*. Study No. 4192 MC, Catalytica Studies Division: Mountain View, CA (1992).
12. T. Ito and J.H. Lunsford, *Nature* **314**, 721 (1985).
13. D.J. Driscoll and J.H. Lunsford, *J. Phys. Chem.* **89**, 4415 (1985).
14. T. Ito, J.X. Wang, C.H. Lin, and J.H. Lunsford, *J. Am. Chem. Soc.* **107**, 5062 (1985).
15. Y. Amemeniya, V.I. Birss, M. Goledzinowski, J. Galuszka, and A.R. Sanger, *Catal. Rev.-Sci. Eng.* **32**, 163 (1990).
16. O.V. Krylov, *Catal. Today* **18**, 209 (1993).
17. J.H. Lunsford, *Catal. Today* **6**, 235 (1990).
18. G.-M. Côme, Y. Li, P. Barbe, N. Gueritey, P.-M. Marquaire, and F. Baronnet, *Catal. Today* **30**, 215 (1996).
19. H. Wan, Z. Chao, W. Weng, X. Zhou, J. Cai, and K. Tsai, *Catal. Today* **30**, 67 (1996).
20. F. Martin-Jiménez, J.M. Blasco, L.J. Alemany, M.A. Bañares, M. Faraldos, M.A. Peña, and J.L.G. Fierro, *Catal. Lett.* **33**, 279 (1995).
21. R.G. Herman, Q. Sun, C. Shi, K. Klier, C.-B. Wang, H. Hu, I.E. Wachs, and M.M. Bhasin, *Catal. Today* **37**, 1 (1997).
22. M.A. Bañares, L.J. Alemany, M. López Granados, M. Faraldos, and J.L.G. Fierro, *Catal. Today* **33**, 73 (1997).
23. A. Parmaliana, F. Arena, F. Frusteri, D. Miceli, and V. Sokolovskii, *Catal. Today* **24**, 231 (1995).

24. T. Kobayashi, N. Guilhaume, J. Miki, N. Kitamura, and M. Haruta, *Catal. Today* **32**, 171 (1996).
25. A. Erdöhelyi, K. Fodor, and F. Solymosi, *J. Catal.* **166**, 244 (1997).
26. Q. Sun, J.-M. Jehng, H. Hu, R.G. Herman, I.E. Wachs, and K. Klier, *J. Catal.* **165**, 91 (1997).
27. F. Arena, N. Giordano, and A. Parmaliana, *J. Catal.* **167**, 66 (1997).
28. S. Pak, C.E. Smith, M.P. Rosynek, and J.H. Lunsford, *J. Catal.* **165**, 73 (1997).
29. M. Faraldos, M.A. Bañares, J.A. Anderson, H. Hu, I.E. Wachs, and J.L.G. Fierro, *J. Catal.* **160**, 214 (1996).
30. A.L. Tonkovich, R.W. Carr, and R. Aris, *Science* **262**, 221 (1993).
31. Y. Jiang, I.V. Yentekakis, and C.G. Vayenas, *Science*, **264**, 1563 (1994).
32. G Lu, S. Shen, and R. Wang, *Catal. Today* **30**, 41 (1996).
33. Y. Seki, N. Mizuno, and M. Misono, *Appl. Catal. A* **158**, L47 (1997).
34. R. Raja and P. Ratnasamy, *Appl. Catal. A* **158**, L7 (1997).
35. R.P. Noceti, C.E. Taylor, and J.R. d'Este, *Catal. Today* **33**, 199 (1997).
36. K. Ogura and M. Kataoka, *J. Mol. Catal.* **43**, 371 (1988).
37. K. Ogura, C.T. Migita, and M. Fujita, *Ind. Eng. Chem. Res.* **27**, 1387 (1988).
38. P. Maruthamuthu, M. Ashokkumar, *Int. J. Hydrogen Energy* **14**, 275 (1989).
39. F. Trifirò and F. Cavani, *Oxidative Dehydrogenation and Alternative Dehydrogenation Processes*, Catalytica Studies No. 4192 OD (1993).
40. F. Cavani and F. Trifirò, *Catal. Today* **24**, 307 (1995).
41. H.H. Kung, *Adv. Catal.* **40**, 1 (1994).
42. H.H. Kung and M.C. Kung, *Appl. Catal. A* **157**, 105 (1997).
43. T. Blasco and J.M. López Nieto, *Appl. Catal. A* **157**, 117 (1997).
44. T. Imai, *US Patent* 4,418,237 (1983).
45. J.C. Bricker, T. Imai, and D. Mackowiak, *US Patent* 4,717,779 (1988).
46. R.R. Herber and G.J. Thompson, *US Patent* 4,806,624 (1989).
47. T. Imai and R.J. Smidt, *US Patent* 4,886,928 (1989).
48. J. Romatier, M. Bentham, T., Foley, and J.A. Valentine, *Proceedings, DeWitt Petrochemical Review*, Houston, Texas (1992), K1.
49. J.N. Armor, *Appl. Catal.* **49**, 1 (1989).
50. J.N. Armor, *CHEMTECH* **22**, 557 (1992).
51. E. Kikichi, *CatTech* **1**, 67 (1997).
52. J. Shu, B.P. Grandjean, A. van Neste, and S. Kaliaguine, *Can. J. Chem. Eng.* **69**, 1036 (1991).
53. S. Ilias and R. Govind, in: *Membrane Reactor Technology* (R. Govind and N. Itoh, eds.), AIChE Symposium Series Vol. 85, No. 268, American Institute of Chemical Engineering: New York (1989), p. 18.
54. H.P. Hsieh, in: *Membrane Reactor Technology* (R. Govind and N. Itoh, eds.), AIChE Symposium Series Vol. 85, No. 268, American Institute of Chemical Engineering: New York (1989), p. 53.

55. R.G. Gastinger, A.C. Jones, and J.A. Sofranko, *EP Patent* 253,552 (1987).

56. A.D. Eastman, J.P. Guillory, C.F. Cook, and J.B. Kimble, *US Patent* 4,835,127 (1989).

57. H. Mimoun, A. Robine, S. Bonnault, and C.S. Cameron, *Appl. Catal.* **58**, 269 (1990).

58. J.H. Edwards, K.T. Do, and R.J. Tyler, in: *Natural Gas Conversion*, (A. Holman, H.J. Jens, and S. Kolboe, eds.), Studies in Surface Science and Catalysis Vol. 61, Elsevier Science: Amsterdam (1991), p. 489.

59. F.M. Dautzenberg, J.C. Schlatter, J.M. Fox, J.R. Rostrup-Nielsen, and L.J. Christiansen, *Catal. Today* **13**, 503 (1992).

60. J. Soler, J.M. Lopez Nieto, J. Herguido, M. Menendez, and J. Santamaria, *Catal. Lett.* **50**, 25 (1998).

61. E. Morales and J.H. Lunsford, *J. Catal.* **118**, 255 (1989).

62. H.M. Swaan, A. Toebes, K. Seshan, J.G. van Ommen, and J.R. Ross, *Catal. Today* **13**, 201 (1992).

63. R. Burch and S.C. Tsang, *Appl. Catal.* **65**, 259 (1990).

64. S.J. Conway and J.H. Lunsford, *J. Catal.* **131**, 513 (1991).

65. R. Burch, S. Chalker, and S.J. Hibble, *Appl. Catal. A* **96**, 289 (1993).

66. J.H. Kolts and J.P. Guillory, *EP Patent*, 205,765 (1986).

67. K. Aika and J.H. Lunsford, *J. Phys. Chem.* **81**, 1393 (1977).

68. J.X. Wang and J.H. Lunsford, *J. Phys. Chem.* **90**, 5883 (1986).

69. R. Burch, G.D. Squire, and S.C. Tsang, *Catal. Today*, **6**, 503 (1990).

70. K. Otsuka, M. Hatano, and T. Komatsu, *Catal. Today* **4**, 409 (1989).

71. R. Burch, E.M. Crabb, G.D. Squire, and S.C. Tsang, *Catal. Lett.* **2**, 249 (1989).

72. K. Otsuka, M. Hatano, and T. Komatsu, in: *Methane Conversion* (D.M. Bubby, C.D. Chang, R. Howe, and S. Yurchak, eds.), Studies in Surface Science and Catalysis Vol. 36, Elsevier Science: Amsterdam (1988), p. 383.

73. R. Burch, S. Chalker, P. Loader, J.M. Thomas, and W. Ueda, *Appl. Catal. A* **82**, 77 (1992).

74. A.D. Eastman and J.B. Kimble, *US Patent* 4,450,313 (1984).

75. H.M. Swaan, A. Toebes, K. Seshan, J.G. van Ommen, and J.R.H. Ross, *Catal. Today* **13**, 629 (1992).

76. S.J. Conway, D.J. Wang, and J.H. Lunsford, *Appl. Catal. A* **79**, L1 (1991).

77. H.M. Swaan, A. Toebes, K. Seshan, J.G. van Ommen, and J.R.H. Ross, *Catal. Today* 13, 01 (1992).

78. J.Z. Luo and H.L. Wan, *Appl. Catal. A* **158**, 137 (1997).

79. J.Z. Luo, X.P. Zhou, Z.S. Chao, and H.L. Wan, *Appl. Catal. A* **159**, 9 (1997).

80. E.M. Thorsteinson, T.P. Wilson, F.G. Young, and P.H. Kasai, *J. Catal.* **52**, 116 (1978).

81. A.F. Wagner, I.R. Slagle, D. Sarzynski, and D. Gutman, *J. Phys. Chem.* **94**, 1853 (1990).

82. J. Le Bars, A. Auroux, J.C. Vedrine, and M. Baerns, in: *New Developments in Selective Oxidation by Heterogeneous Catalysis* (P. Ruiz and B. Delmon, eds.), Studies in Surface Science and Catalysis Vol. 72, Elsevier Science: Amsterdam (1992), p. 181.

83. S.T. Oyama and G.A. Somorjai, *J. Phys. Chem.* **94**, 5022 (1990).

84. S.T. Oyama, *J. Catal.* **128**, 210 (1991).

85. S.T. Oyama, G.T. Went, K.B. Lewis, A.T. Bell, and G.A. Somorjai, *J. Phys. Chem.* **93**, 6786 (1989).

86. J. Haber, M. Witko, and R. Tokarz, *Appl. Catal. A* **157**, 3 (1977).

87. J. Le Bars, J.C. Vedrine, A. Auroux, B. Pommier, and G.M. Pajonk, *J. Phys. Chem.* **96**, 2217 (1992).

88. J. Le Bars, A. Auroux, S. Trautmenn, and M. Baerns, in: *Proceedings DGMK Conference on Selective Oxidation in Petrochemistry*, DGMK: Hamburg (1992), p. 59.

89. J. Le Bars, A. Auroux, M. Forissier, and J.C. Vedrine, *J. Catal.* **162**, 250 (1996).

90. P.R. Blum and C. Milberger, *US Patent* 4,410,752 (1983).

91. G. Centi and F. Trifirò, *Catal. Today* **3**, 151 (1988).

92. M. Merzouki, B. Taouk, L. Monceaux, E. Bordes, and P. Courtine, in: *New Developments in Selective Oxidation by Heterogeneous Catalysis* (P. Ruiz and B. Delmon, eds.), Studies in Surface Science and Catalysis Vol. 72, Elsevier Science: Amsterdam (1992), p. 165.

93. M. Merzouki, B. Taouk, L. Tessier, E. Bordes, and P. Courtine, in: *New Frontiers in Catalysis* (L. Guczi, F. Solymosi, and P. Tétényi, eds.), Studies in Surface Science and Catalysis Vol. 75, Elsevier Science: Amsterdam (1993), p. 753.

94. S. Bordoni, F. Cavani, and F. Trifirò, *Chim. Ind. (Milan)* **74**, 194 (1992).

95. S. Bordoni, F. Castellani, F. Cavani, F. Trifirò, and M.P. Kuldarni, in: *New Developments in Selective Oxidation II* (V. Corberan and S. Vic Bellon, eds.), Studies in Surface Science and Catalysis Vol. 82, Elsevier Science: Amsterdam (1994), p. 93.

96. S. Bordoni, F. Castellani, F. Cavani, F. Trifirò, and M. Gazzano, *J. Chem. Soc. Faraday Trans.* **90**, 2981 (1994).

97. C.A. Jones and J.A. Sofranko, *US Patent* 4,499,322 (1985).

98. J.A. Sofranko, J.J. Leonard, and C.A. Jones, *J. Catal.* **103**, 302 (1987).

99. H.J.F. Doval, O.A. Scelza, and A.A. Castro, *React. Kinet. Catal. Lett.* **34**, 143 (1987).

100. P.G. Harrison and B. Maunders, *J. Chem. Soc. Faraday Trans. 1* **81**, 1311 (1985).

101. A. Argent and P.G. Harrison, *J. Chem. Soc. Chem. Comm.* 1058 (1986).

102. P.G. Harrison and A. Argent, *Eur. Patent* **189**, 282 (1986).

103. F.G. Young and E.M. Thorsteinson, *US Patent* 4,250,346 (1981).

104. R. Burch and R. Swarnakar, *Appl. Catal.* **70**, 129 (1991).

105. J.H. McCain, *US Patent* 4,524,236 (1985).

106. J.H. McCain, *US Patent* 4,568,790 (1986).

107. R.M. Manyik and J.H. McCain, *US Patent* 4,596,787 (1986).

108. F. Cavani, M. Koutyrev, and F. Trifirò, *Catal. Today* **28**, 319 (1996).

109. F. Cavani, M. Koutyrev, and F. Trifirò, *Catal. Today* **24**, 365 (1995).

110. S. Albonetti, F. Cavani, F. Trifirò, and M. Koutyrev, *Catal. Lett.* **30**, 253 (1995).

111. Y. Murakami, K. Otsuka, Y. Wada, and A. Morikawa, *Bull. Chem. Soc. Jpn.* **63**, 340 (1990).

112. K. Otsuka, Y. Uragami, T. Komatsu, and M. Hatano, in: *Natural Gas Conversion* (A. Holman, H.J. Jens, and S. Kolboe, eds.), Studies in Surface Science and Catalysis Vol. 61, Elsevier Science: Amsterdam (1991), p. 15.

113. E.M. Kennedy and N.W. Cant, *Appl. Catal.* **75**, 321 (1991).

114. E.M. Kennedy and N.W. Cant, *Appl. Catal. A* **87**, 171 (1992).

115. O.J. Velle, A. Andersen, and K.J. Jens, *Catal. Today* **6**, 567 (1990).

116. M. Loukah, G. Coudurier, and J.C. Vedrine, in: *New Developments in Selective Oxidation by Heterogeneous Catalysis* (P. Ruiz and B. Delmon, eds.), Studies in Surface Science and Catalysis Vol. 72, Elsevier Science: Amsterdam (1992), p. 191.

117. E.A. Mamedov, R.M. Talyshinskii, R.G. Rizayev, J.L.G. Fierro, and V. Cortes Corberan, *Catal. Today* **32**, 177 (1996).

118. R. Juarez Lopez, N.S. Godjayeva, V. Cortes Corberan, J.L.G. Fierro, and E.A. Memedov, *Appl. Catal. A* **124**, 281 (1995).

119. C.A. Jones, J.J. Leonard, and J.A. Sofranko, *US Patent* 4,737,595 (1988).

120. A.D. Eastman and J.H. Kolts, *US Patent* 4,370,259 (1983).

121. F. Cavani and F. Trifirò, *CHEMTECH*, April, 18 (1994).

122. F. Cavani, S. Albonetti, and F. Trifirò, *Catal. Rev.-Sci. Eng.* **38** (1996) 413.

123. J.C. Vedrine, G. Coudurier, and J.-M.M. Millet, *Catal. Today* **32**, 115 (1996).

124. J.C. Vedrine, G. Coudurier, and J.-M.M. Millet, *Catal. Today* **33**, 3 (1997).

125. L. Owens and H.H. Kung, *J. Catal.* **144**, 202 (1993).

126. L. Owens and H.H. Kung, *J. Catal.* **148**, 587 (1994).

127. M. Chaar, D. Patel, M. Kung, and H.H. Kung, *J. Catal.* **105**, 483 (1987).

128. M. Chaar, D. Patel, M. Kung, and H.H. Kung, *J. Catal.* **109**, 463 (1988).

129. P.M. Michalakos, M.C. Kung, I. Jahan, and H.H. Kung, *J. Catal.* **140**, 226 (1993).

130. M.C. Kung and H.H. Kung, *J. Catal.* **134**, 668 (1992).

131. D. Patel, M.C. Kung, H.H. Kung, in: *Proceedings 9th International Congress on Catalysis* (M.J. Phillips and M. Ternan, eds.), Chemical Institute of Canada: Toronto (1988), p. 1554.

132. X. Gao, P. Ruiz, Q. Xin, X. Guo, and B. Delmon, *J. Catal.* **148**, 56 (1994).

133. K. Seshan, H.M. Swaan, R.H.H. Smits, J.G. van Ommen, and J.H.R. Ross, in: *New Developments in Selective Oxidation*, (G. Centi and F. Trifirò, eds.), Studies in Surface Science and Catalysis Vol. 55, Elsevier Science: Amsterdam (1990), p. 505.

134. O.S. Owen and H.H. Kung, *J. Mol. Catal.* **79**, 265 (1993).

135. T. Blasco, J.M. Lopez Nieto, A. Dejoz, and M.I. Vazquez, *J. Catal.* **157**, 271 (1995).

136. D. Patel, P.J. Andersen, and H.H. Kung, *J. Catal.* **125**, 130 (1990).

137. A. Corma, J.M. Lopez Nieto, N. Paredes, M. Perez, Y. Shen, H. Cao, and S.L. Suib, in: *New Developments in Selective Oxidation by Heterogeneous Catalysis* (P. Ruiz and B. Delmon, eds.), Studies in Surface Science and Catalysis Vol. 72, Elsevier Science: Amsterdam (1992), p. 213.

138. P. Concepcion, A. Dejoz, J.M. Lopez Nieto, and M.I. Vazquez, in: *Proceedings 14th Iberoamerican Symposium on Catalysis*, Chilean Chemical Society: Concepción (1994), p. 769.

139. P. Concepcion, A. Galli, J.M. Lopez Nieto, A. Dejoz, and M.I. Vazquez, *Topics Catal.* **3**, 451 (1996).

140. A. Galli, J.M. Lopez Nieto, A. Dejoz, and M.I. Vazquez, *Catal. Lett.* **34**, 51 (1995).

141. D.B. Dadyburjor, S.S. Jewur, and E. Ruckenstein, *Catal. Rev.-Sci Eng.* **19**, 293 (1979).

142. J.M. Lopez Nieto, A. Dejoz, and M.I. Vazquez, *Appl. Catal. A* **132**, 41 (1995).

143. A. Corma, J.M. Lopez Nieto, N. Paredes, and M. Perez, *Appl. Catal. A* **97**, 159 (1993).

144. A. Pantazidis, A. Auroux, J.M. Herrmann, and C. Mirodatos, *Catal. Today* **32**, 81 (1996).

145. R. Grabowski, B. Grzybowska, A. Kozlowska, J. Sloczynski, W. Wcislo, and Y. Barbaux, *Topics Catal.* **3**, 277 (1966).

146. D. Courcot, A. Ponchel, B. Grzybowska, Y. Barbaux, M. Rigole, M. Guelton, and J.P. Bonnelle, *Catal. Today* **33**, 109 (1997).

147. G. Centi, G. Golinelli, and G. Busca, *J. Phys. Chem.* **94**, 6813 (1990).

148. T.C. Watling, G. Deo, K. Seshan, I.E. Wachs, and J.A. Lercher, *Catal. Today* **28**, 139 (1996).

149. L.M. Madeira, J.M. Herrmann, F.G. Freire, M.F. Portela, and F.J. Maldonado, *Appl. Catal. A* **158**, 243 (1997).

150. F.J. Maldonado-Hodar, L.M. Palma Madeira, and M. Farinha Portela, *J. Catal.* **164**, 399 (1996).

151. L. Savary, J. Saussey, G. Costentin, M.M. Bettahar, M. Gubelmann-Bonneau, and J.C. Lavalley, *Catal. Today* **32**, 57 (1996).

152. V. Soenen, J.M. Herrmann, and J.C. Volta, *J. Catal.* **159**, 410 (1996).

153. D.L. Stern, J.N. Michaels, L. DeCaul, and R.K. Grasselli, *Appl. Catal. A* **153**, 21 (1997).

154. F. Cavani and F. Trifirò, *Chim. Ind. (Milan)* **76**, 708 (1994).

155. K. Otsuka, T. Ando, S. Suprapto, Y. Wang, K. Ebitani, and I. Yamanaka, *Catal. Today* **24**, 315 (1995).

156. A. Pantazidis, J.A. Dalmon, and C. Mirodatos, *Catal. Today* **25**, 403 (1995).

157. Y. Uragami and K. Otsuka, *J. Chem. Soc. Faraday Trans.* **88**, 3605 (1992).

158. Z.G. Osipova, S.B. Ushkov, V.B. Sokolovskii, and A.V. Kalinkin, in: *New Developments in Selective Oxidation* (G. Centi and F. Trifirò, eds.), Studies in Surface Science and Catalysis Vol. 55, Elsevier Science: Amsterdam (1990), p. 527.

159. A. Kh. Mamedov, P.A. Shiryaev, D.P. Shashkin, and O.V. Krylov, in: *New Developments in Selective Oxidation* (G. Centi, F. Trifirò, eds.), Studies in Surface Science and Catalysis Vol. 55 Elsevier Science: Amsterdam (1990), p. 477.

160. L. Mendolevici and J.H. Lunsford, *J. Catal.* **94**, 37 (1985).

161. E. Iwamatsu, K. Aika, and T. Onishi, *Bull. Chem. Soc. Jpn.* **59**, 1665 (1986).

162. A. Erdöhelyi and F. Solymosi, *J. Catal.* **129**, 497 (1991).

163. A. Erdöhelyi, F. Mate, and F. Solymosi, *J. Catal.* **135**, 63 (1992).

164. A.G. Anshits, *Catal. Today* **13**, 495 (1992).

165. L.W. Zatorski, G. Centi, J. Lopez Nieto, F. Trifirò, G. Bellussi, and V. Fattore, in: *Zeolites: Facts, Figures, Future*, Vol. 2, P.A. Jacobs and R.A. van Santen, eds.), Elsevier Science: Amsterdam (1989), p. 1243.

166. H.D. Gesser, G. Zhu, and N.R. Hunter, *Catal. Today* **24**, 321 (1995).

167. B. Khazai, G.E. Vrieland, and C.B. Murchison, *Ep. Patent* 409,355 A1 (1991).

168. G.E. Vrieland and C.B. Murchison, *Appl. Catal. A* **134**, 101 (1996).

169. F. Cavani and F. Trifirò, *Appl. Catal. A* **88**, 115 (1992).

170. M.R. Flid, I.I. Kurlyandskaya, Yu.A. Treger, and T.D. Guzhnovskaya, *3rd World Congress on Oxidation Catalysis* (R.K. Grasselli, S.T. Oyama, A.M. Gaffney, and J.E. Lyons, eds.), Studies in Surface Science and Catalysis Vol. 110, Elsevier Science: Amsterdam (1997), p. 305.

171. F. Trifirò and F. Cavani, *Selective Partial Oxidation of Hydrocarbons and Related Oxidations*, Catalytica Studies No. 4193 SO (1994).

172. M.G. Clerici, *Appl. Catal.* **68**, 249 (1991).

173. D.R.C. Huybrechts, L. de Bruijker, and P.A. Jacobs, *Nature* **345**, 240 (1990).

174. S. Khoobiar and R. Porcelli, *Eur. Patent* 117,146 A1 (1984).

175. N. Nojiri, Y. Sakai, and Y. Watanabe, *Catal. Rev.-Sci. Eng.* **37**, 145 (1995).

176. T.V. Andrushkevich, *Catal. Rev.-Sci. Eng.* **35**, 213 (1993).

177. S. Khoobiar, *US Patent*, 4,535,188 (1985).

178. V.R. Choudhary and V.H. Rane, *J. Catal.* **135**, 310 (1992).

179. K. Otsuka, Y. Uragami, and M. Hatano, *Catal. Today* **13**, 667 (1992).

180. K. Wada, K. Yoshida, Y. Watanabe, and T. Suzuki, *Appl. Catal.* **74**, L1 (1991).

181. P. Barthe and G. Blanchard, *Eur. Patent* 479,692 (1992).

182. L. Tessier, E. Bordes, F. Blaise, and M. Gubelmann-Bonneau, *Eur. Patent Appl.* 06,475 (1993).

183. L. Tessier, E. Bordes, and M. Gubelmann-Bonneau, *Catal. Today* **24**, 335 (1995).

184. M. Roy, M. Gubelmann-Bonneau, H. Ponceblanc, and J.C. Volta, *Catal. Lett.* **42**, 93 (1996).

185. R. Burch, R. Kieffer, and K. Ruth, *Topics Catal.* **3**, 355 (1996).

186. M. Banares, X. Gao, J.L.G. Fierro, and I.E. Wachs, *3rd World Congress on Oxidation Catalysis* (R.K. Grasselli, S.T. Oyama, A.M. Gaffney, and J.E. Lyons, eds.), Studies in Surface Science and Catalysis Vol. 110, Elsevier Science: Amsterdam (1997), p. 295.

187. W. Li and W. Ueda, *3rd World Congress on Oxidation Catalysis* (R.K. Grasselli, S.T. Oyama, A.M. Gaffney, and J.E. Lyons, eds.), Studies in Surface Science and Catalysis Vol. 110, Elsevier Science: Amsterdam (1997), p. 433.

188. K. Komatsu, Y. Uragami, and K. Otsuka, *Chem. Lett.* 1903 (1988).
189. P. Barthe and G. Blanchard, *Fr. Patent* 90 12,519 (1990).
190. P.R. Blum and M.A. Pepera, *Eur. Patent*, 518,548 A2 (1992).
191. F. Cavani and F. Trifirò, in: *Catalysis, Vol. 11*, (J.J. Spivey and S.K. Agarwal, eds.), Royal Society of Chemistry: Oxford (1994), p. 246.
192. G. Centi and F. Trifirò, *Appl. Catal.* 12, 1 (1984).
193. B. Grzybowska-Swierkosz, F. Trifirò, and J.C. Vedrine (eds.), Vanadia Catalysts for Selective Oxidation of Hydrocarbons and their Derivatives, *Appl. Catal. A* (Special Issue) 157 (1/2), (1997).
194. S. Rajadurai, *Catal. Rev.-Sci. Eng.* 36, 385 (1994).
195. K. Weissermel and H.J. Arpe, *Industrial Organic Chemistry*, 2nd Ed., Verlag Chemie: Weinheim, Germany (1993).
196. J.F. Brazdil, R.G. Teller, W.A. Marritt, J.C. Gleaser, and A.M. Ebner, *J. Catal.* 100, 516 (1986).
197. W. Ueda, T. Yokoyama, Y. Moro-oka, and T. Ikawa, *Ind. Eng. Chem. Prod. Res. Dev.* 24, 340 (1985).
198. W. Zhang and P.G. Smirniotis, *Natural Gas Conversion V* (A. Parmaliana, D. Sanfilippo, F. Frusteri, A. Vaccari, and F. Arena, eds.), Studies in Surface Science and Catalysis Vol. 119, Elsevier Science: Amsterdam (1998), p. 367.
199. R.K. Grasselli and J.D. Burrington, *Adv. Catal.* 30, 133 (1981).
200. A.T. Guttmann, R.K. Grasselli, and J.F. Brazdil, *US Patent* 4,746,641 (1988).
201. I. Peeters, J. Grondelle, and R.A. van Santen, in: *Heterogeneous Hydrocarbon Oxidation* (B.K. Warren and S.T. Oyama, eds.), ACS Symposium Series Vol. 683, American Chemical Society: Washington DC (1996), p. 319.
202. I. Peeters, A.W. Denier van der Gon, M.A. Rejime, P.J. Kooyman, A.M. de Jong, J. Van Grondelle, H.H. Brongersma, and R.A. van Santen, *J. Catal.* 173, 28 (1998).
203. I. Peeters, Ph.D. Thesis, Eindhoven University of Technology (The Netherlands), 1996.
204. R. Catani and G. Centi, *J. Chem. Soc. Chem. Commun.* 1081 (1991).
205. S.M. Aliev, V.D. Sokolovskii, and G.B. Borevskov, *USSR Patent* SU-738657 (1980).
206. G. Centi, S. Perathoner, *Natural Gas Conversion V* (A. Parmaliana, D. Sarfilippo, F. Frusteri, A. Vaccari, and F. Arena, eds.), Studies in Surface Science and Catalysis, Vol 119, Elsevier Science: Amsterdam (1998), p. 569.
207. Y. Li and J.N. Armor, *J. Chem. Soc. Chem. Commun.* 2013 (1997).
208. Y. Li and J.N. Armor, *J. Catal.* 173, 511 (1998).
209. Y. Li and J.N. Armor, *Eur. Patent Appl.* EP 0761,654,A2 (1997).
210. R.H. Smits, K. Seshan, and J.R. Ross, in: *New Developments in Selective Oxidation by Heterogeneous Catalysts* (P. Ruiz and B. Delmon, eds.), Studies Surface Science and Catalysis Vol, 72, Elsevier Science: Amsterdam (1992), p. 221.
211. G. Centi and F. Trifirò, *Catal. Rev.-Sci. Eng.* 28, 165 (1986).
212. G. Centi and S. Perathoner, *CHEMTECH* 2 (1998) 13.

213. V.D. Sokolovskii, A.A. Davydov, and O. Yu. Ovsitser, *Catal. Rev.-Sci. Eng.* **37**, 425 (1995).

214. S.M. Aliev and V.D. Sokolovskii, *React. Kinet. Catal. Lett.* **9**, 91 (1978).

215. S.M. Aliev and V.D. Sokolovskii, *Kinet. Katal.* **21**, 971 (1980).

216. V.D. Sokolovskii and S.M. Aliev, *Kinet. Katal.* **21**, 691 (1980).

217. S.M. Aliev, V.D. Sokolovskii, and N.S. Kotsarenko, *Kinet. Katal.* **21**, 686 (1980).

218. Y.C. Kim, W. Ueda, and Y. Moro-oka, *Catal. Today* **13**, 673 (1992).

219. J.P. Bartek, A.M. Ebner, and J.R. Brazdil, *US Patent* 5,198,580 (1993).

220. M. Baerns, O.V. Buyevskaya, M. Kubik, G. Maiti, O. Ovsitser, and O. Seel, *Catal. Today* **33**, 85 (1997).

221. W. Ueda, K. Asakawa, C.L. Chen, Y. Moro-oka, and T. Ikawa, *J. Catal.* **101**, 360 (1986).

222. W. Ueda, K. Asakawa, C.L. Chen, Y. Moro-oka, and T. Ikawa, *J. Catal.* **101**, 369 (1986).

223. S. Williams, M. Puri, A.J. Jacobson, and C.A. Mims, *Catal. Today* **37**, 43 (1997).

224. Y.S. Yoon, N. Fujikawa, W. Ueda, Y. Moro-Oka, and K.W. Lee, *Catal. Today* **24**, 327 (1995).

225. T. Tsunoda, T. Hayakawa, Y. Imai, T. Kameyama, K. Takehira, and K. Fukuda, *Catal. Today* **25**, 371 (1995).

226. Y. Takita, H. Yamashita, and K. Moritaka, *Chem. Lett.* 1733 (1989).

227. H. Krieger and L.S. Kirch, *US Patent* 4,260,822 (1981).

228. L. Savary, G. Costentin, M.M. Bettahar, A. Grandin, M. Gubelmann-Bonneau, and J.C. Lavalley, *Catal. Today* **32**, 305 (1996).

229. I. Matsuura and N. Kimura, in: *New Developments in Selective Oxidation II* (V. Corberan and S. Vic Bellon, eds.), Studies in Surface Science and Catalysis Vol. 82, Elsevier Science: Amsterdam (1994), p. 271.

230. J. Barrault and L. Magaud, in: *New Developments in Selective Oxidation II* (V. Corberan and S. Vic Bellon, eds.), Studies in Surface Science and Catalysis Vol. 82, Elsevier Science: Amsterdam (1994), p. 305.

231. C. Mazzocchia, C. Aboumrad, C. Diagne, E. Tempesti, H. Herrmann, and G. Thomas, *Catal. Lett.* **11**, 181 (1991).

232. Y.C. Kim, W. Ueda, and Y. Moro-oka, *Chem. Lett.* 531 (1989).

233. N. Giordano, J.C. Bart, P. Vitarelli, and S. Cavallaro, *Oxid. Comm.* **7**, 99 (1984).

234. Y.C. Kim, W. Ueda, and Y. Moro-oka, *Appl. Catal.* **70**, 175 (1991).

235. M. Ai, *J. Catal.* **101**, 389 (1986).

236. R.K. Grasselli, G. Centi, and F. Trifirò, *Appl. Catal.* **57**, 149 (1990).

237. M. Ai, *Catal. Today* **12**, 679 (1992).

238. G. Centi and F. Trifirò, in: *Catalysis Science and Technology*, Vol. 1, (S. Yoshida, N. Takezawa, and T. Ono, eds.), Kodansha: Tokyo, and VCH Verlag: Basel (1991), p. 225.

239. N. Mizuno, M. Tateishi, and M. Iwamoto, *Appl. Catal. A* **128**, L165 (1995).

240. N. Mizuno, D. Suh, W. Han, and T. Kudo, *J. Mol. Catal.* **114**, 309 (1996).

241. W. Ueda and Y. Suzuki, *Chem. Lett.* 541 (1995).

242. J. Barrault, C. Batiot, L. Magaud, and M. Ganne, *3rd World Congress on Oxidation Catalysis* (R.K. Grasselli, S.T. Oyama, A.M. Gaffney, and J.E. Lyons, eds.), Studies in Surface Science and Catalysis Vol. 110, Elsevier Science: Amsterdam (1997), p. 375.

243. B. Stein, C. Weimer, and J. Gaube, *3rd World Congress on Oxidation Catalysis* (R.K. Grasselli, S.T. Oyama, A.M. Gaffney, and J.E. Lyons, eds.), Studies in Surface Science and Catalysis Vol. 110, Elsevier Science: Amsterdam (1997), p. 393.

244. R. Recknagel and L. Riekert, *CHEMTECH* **46**, 324 (1994).

245. R.V. Porcelli and B. Juran, *Hydrocarbon Proc.* **65**, 37 (1986).

246. K. Nagai, Y. Nagaoka, H. Sato, and M. Ohsu, *EP Patent* 418,657 A2 (1990).

247. K. Nagai, Y. Nagaoka, H. Sato, and M. Ohsu, *EP Patent* 495,504A2 (1991).

248. S. Yamamatsu and T. Yamaguchi, *EP Patent* 425,666 A1 (1989).

249. S. Yamamatsu and T. Yamaguchi, *JP Patent* 02-0421,032 (1990).

250. S. Yamamatsu and T. Yamaguchi, *JP Patent* 02-0421,034 (1990).

251. S. Yamamatsu and T. Yamaguchi, *US Patent* 5,191,116 (1993).

252. K. Kawakami, S. Yamamatsu, and T. Yamaguchi, *JP Patent* 03-176,438 (1991).

253. T. Kuroda and M. Okita, *JP Patent*, 04-128,247 (1991).

254. H. Imai, T. Yamaguchi and M. Sugiyama, *JP Patent* 145,249 (1988).

255. G. Busca, F. Cavani, E. Etienne, E. Finocchio, A. Galli, G. Selleri, and F. Trifirò, *J. Mol. Catal. A* **114**, 343 (1996).

256. F. Cavani, E. Etienne, G. Hecquet, G. Selleri, and F. Trifirò, in: *Catalysis of Organic Reactions* (E.R. Malz, Jr., eds.), Marcel Dekker: New York (1996), p. 107.

257. F. Cavani, E. Etienne, M. Favaro, A. Galli, F. Trifirò, and G. Hecquet, *Catal. Lett.* **32**, 215 (1995).

258. L. Jalowiecki-Duhamel, A. Monnier, Y. Barbaux, and C. Hecquet, *Catal. Today* **32**, 237 (1996).

259. N. Mizuno, M. Tateishi, and M. Iwamoto, *J. Chem. Soc. Chem. Comm.*, 1411 (1994).

260. N. Mizuno, M. Tasaki, and M. Iwamoto, *J. Catal.* **163**, 87 (1996).

261. M. Misono, N. Mizuno, K. Imumaru, G. Koyano, and X.-H. Lu, in: *3rd World Congress on Oxidation Catalysis* (R.K. Grasselli, S.T. Oyama, A.M. Gaffney, and J.E. Lyons, eds.), Studies in Surface Science and Catalysis Vol. 110, Elsevier Science: Amsterdam (1997), p. 35.

262. N. Mizuno, W. Han, T. Kudo, and M. Iwamoto, *11th International Congress on Catalysis—40th Anniversary* (J.W. Hightower, W.N. Delgass, E. Iglesias, and A.T. Bell, eds.), Studies in Surface Science and Catalysis Vol. 101, Elsevier Science: Amsterdam (1996), p. 1001.

263. W. Ueda, Y. Suzuki, W. Lee, and S. Imaoka, *11th International Congress on Catalysis—40th Anniversary* (J.W. Hightower, W.N. Delgass, E. Iglesias, and A.T. Bell, eds.), Studies in Surface Science and Catalysis Vol. 101, Elsevier Science: Amsterdam (1996), p. 1061.

264. G. Centi, V. Lena, F. Trifirò, D. Ghoussoub, C.F. Aissi, M. Guelton, and J.P. Bonnelle, *J. Chem. Soc. Faraday Trans.*, **86**, 2775 (1990).

265. F. Cavani, C. Comuzzi, G. Dolcetti, R.G. Finke, A. Lucchi, F. Trifirò, and A. Trovarelli, in: *Heterogeneous Hydrocarbon Oxidation* (B.K. Warren and S.T. Oyama, eds.), ACS Symposium Series Vol. 683, American Chemical Society: Washington DC (1996), p. 141.

266. D. Braksa and D. Hönicke, in: *Proceedings DGMK Conference on Selective Oxidations in Petrochemistry*, DGMK: Hamburg (1992), p. 37.

267. M. Burattini, G. Centi, and F. Trifirò, *Appl. Catal.* **32**, 353 (1987).

268. G. Busca and G. Centi, *J. Am. Chem. Soc.* **111**, 46 (1989).

269. G. Centi, J. Nieto Lopez, D. Pinelli, and F. Trifirò, *Ind Eng. Chem. Res.* **28**, 400 (1989).

270. G. Centi, J. Lopez Nieto, D. Pinelli, F. Trifirò, and F. Ungarelli, in: *New Developments in Selective Oxidation* (G. Centi and F. Trifirò, eds.), Studies in Surface Science and Catalysis Vol. 55, Elsevier Science: Amsterdam (1990), p. 635.

271. G. Centi and F. Trifirò, *Chem. Eng. Sci.* **45**, 2589 (1990).

272. G. Centi, J.T. Gleaves, G. Golinelli, and F. Trifirò, in: *New Developments in Selective Oxidation by Heterogeneous Catalysis* (P. Ruiz and B. Delmon, eds.), Elsevier Science: Amsterdam (1992), p 231.

273. G. Golinelli and J.T. Gleaves, *J. Mol. Catal.* **73**, 353 (1992).

274. C. Fumagalli, G. Golinelli, G. Mazzoni, M. Messori, G. Stefani, and F. Trifirò, in: *New Developments in Selective Oxidation II* (V. Corberan and S. Vic Bellon, eds.), Studies in Surface Science and Catalysis Vol. 82, Elsevier Science: Amsterdam (1994), p. 221.

275. F. Cavani and F. Trifirò. *Appl. Catal. A* **157**, 195 (1997).

276. F. Cavani, A. Colombo, F. Trifirò, M.T. Sanates Schulz, J.C. Volta, and G.J. Hutchings, *Catal. Lett.* **43**, 241 (1997).

277. G. Calestani, F. Cavani, A. Duran, G. Mazzoni, G. Stefani, F. Trifirò, and P. Venturoli, in: *Science and Technology in Catalysis 1994*, Kodansha: Tokyo (1995), p. 179.

278. F. Cavani, A. Colombo, F. Giuntoli, E. Gobbi, F. Trifirò, and P. Vazquez, *Catal. Today* **32**, 125 (1996).

279. S. Albonetti, F. Cavani, F. Trifirò, P. Venturoli, G. Calestani, M. Lopez Granados, and J.L.G. Fierro, *J. Catal.* **160**, 52 (1996).

280. F. Trifirò, *Catal. Today* **16**, 91 (1993).

281. G.K. Bethke, D. Wang, J.M.C. Bueno, M.C. Kung, and H.H. Kung, in: *3rd World Congress on Oxidation Catalysis* (R.K. Grasselli, S.T. Oyama, A.M. Gaffney, and J.E. Lyons, eds.), Studies in Surface Science and Catalysis Vol. 110, Elsevier Science: Amsterdam (1997), p. 453.

282. Z. Sobalik, P. Ruiz, B. Delmon, in: *3rd World Congress on Oxidation Catalysis* (R.K. Grasselli, S.T. Oyama, A.M. Gaffney, and J.E. Lyons, eds.), Studies in Surface Science and Catalysis Vol. 110, Elsevier Science: Amsterdam (1997), p. 481.

283. Z. Sobalik, S. Gonzales, P. Ruiz, B. Delmon, in: *3rd World Congress on Oxidation Catalysis* (R.K. Grasselli, S.T. Oyama, A.M. Gaffney, and J.E. Lyons, eds.), Studies in Surface Science and Catalysis Vol. 110, Elsevier Science: Amsterdam (1997), p. 1213.

284. G. Centi, J. Lopez Nieto, C. Iapalucci, K. Bruckman, and E.M. Serwicka, *Appl. Catal.* **46**, 197 (1989).

285. T. Bergier, K. Brückman, and J. Haber, *Rev. Trev. Chim. Pays-Bas* **113**, 475 (1994).

286. D. Hönicke, K. Griesbaum, R. Augenstein, and Y. Tang, *Chem. Ing. Tech.* **59**, 222 (1987).

287. U.S. Ozkan, G. Karakas, B.T. Schilf, and S. Ang, in: *3rd World Congress on Oxidation Catalysis* (R.K. Grasselli, S.T. Oyama, A.M. Gaffney, and J.E. Lyons, eds.), Studies in Surface Science and Catalysis Vol. 110, Elsevier Science: Amsterdam (1997), p. 471.

288. U.S. Ozkan, R.E. Gooding, and B.T. Schilf, in: *Heterogeneous Hydrocarbon Oxidation* (B.K. Warren and S.T. Oyama, eds.), ACS Symposium Series Vol. 683, American Chemical Society: Washington DC (1996), p. 178.

289. U.S. Ozkan, T.A. Harris, and B.T. Schilf, *Catal. Today* **33**, 57 (1997).

290. P.M. Michalakos, K. Birkeland, and H.H. Kung, *J. Catal.* **158**, 349 (1996).

291. J.J. Roney, in: *Elementary Reaction Steps in Heterogeneous Catalysis* (R.W. Joyner and R.A. van Santen, eds.), NATO ASI Series C Vol. 398, Kluwer Academic: Dordrecht (1993); p. 51.

292. G. Centi, J.L. Nieto, F. Ungarelli, and F. Trifirò, *Catal. Lett.* **4**, 309 (1990).

293. O. Yu Ovsitser, A. Davydov, Z.G. Osipova, and V.D. Sokolovskii, *React. Kinet. Catal. Lett.* **38**, 125 (1989).

294. O. Yu Ovsitser, A. Davydov, Z.G. Osipova, and V.D. Sokolovskii, *React. Kinet. Catal. Lett.* **40**, 307 (1989).

295. O. Yu Ovsitser, Z.G. Osipova, and V.D. Sokolovskii, *React. Kinet. Catal. Lett.* **38**, 91 (1989).

NEW FIELDS OF APPLICATION FOR SOLID CATALYSTS

6.1. INTRODUCTION

Traditionally selective oxidation catalysts have been divided between catalysts for heterogeneous gas phase reactions (usually mixed oxides, apart from some relevant exceptions such as Ag/Al_2O_3 for ethene epoxidation) and catalysts for liquid phase oxidation, often complexes of transition metals. Catalysts used for one class of reactions cannot be used for the other, and even the extensive efforts made in the past to heterogenize homogeneous complexes usually failed to achieve interesting results. However, new kinds of solid catalysts have been developed, especially in the last decade, which do not fit into this broad classification.

After the discovery of titanium silicalite[1] (TS-1) and its exceptional properties in selective oxidation using hydrogen peroxide (see Chapter 3), much research activity was devoted to the study of catalytic materials having[2,3]:

1. Titanium ions in the framework positions of different zeolite structures with wider pore openings (beta, in particular) or even mesoporous materials (MCM-41), with the aim of selectively oxidizing larger molecules than is possible with TS-1.
2. Different transition metal ions, in pentasyl-type zeolites (structure similar to that of TS-1) or different types of zeolitic structures, zeo-type materials (e.g., aluminum phosphates) or mesoporous materials.

These materials opened the very active area of research on solid catalysts with isolated transition metals in a rigid structured matrix for liquid phase selective oxidation. Some of these materials also are active in gas phase reactions (e.g., zeolites containing iron and vanadium) and thus the rigid distinction between catalysts for heterogeneous and homogeneous reactions is no longer valid.

It is also worth noting that although the reaction mechanism in the liquid and gas phases is usually different (peroxo-type species in the liquid phase and redox reactions involving lattice oxygen in the gas phase), there is an increasing number of examples in which this distinction no longer exists. For example, in Au/Ti–MCM41 catalysts for the propene epoxidation using hydrogen/oxygen mixtures in gas phase conditions,[4] peroxo species generated at the gold surface migrate to isolated titanium ions of the titanium-mesoporous material to generate active species analogous to those active in TS-1 for propene epoxidation in the liquid phase.

In recent years this new class of catalysts has been the focus of extensions of a concept that may be broadly referred to as catalysis of metal ions and complexes in a constrained environment.[5] This group of new solid catalysts includes both complexes held in cages in structured matrices such as zeolites or clays, complexes anchored with short or longer ("tethered") bonds to a surface, and complexes embedded in an organic membrane.

Heteropoly compounds constitute another class of materials that cannot be strictly labeled as either heterogeneous or homogeneous catalysts.[6–9] These molecular solids are characterized by the presence of an oxoanion, which remains analogous in both the liquid and solid phases. The environment around it, however, is clearly different in the two cases with an obvious influence on its reactivity. However, here, as for the transition metal-containing micro- and mesoporous materials, it is possible to draw interesting parallels between the reactivities in the liquid and solid phases.

A third example of these catalysts at the interface between homogeneous and heterogeneous catalysis is that represented by heterogenized Wacker-type catalysts. Wacker-type reactions constitute an important class of industrial reactions for the synthesis of chemicals such as acetaldehyde from ethene and but-2-anone from but-1-ene. The synthesis is usually done in the liquid phase by Pd^{2+} complexes, in which the reoxidation by gaseous O_2 requires a cocatalyst (copper chloride) mediator. In heterogenized systems, this reoxidation step is carried out either by an oxide such as vanadium oxide or by a heteropoly compound.[10–16] As will be discussed further in Chapter 7, these materials can be considered to be at the boundary between homogeneous and heterogeneous catalysts, and during catalytic reaction a supported thin liquid film is probably present, which is the real reaction medium. The Wacker-type chemistry is not limited to the reactions described above, and others of commercial interest are possible, such as butadiene acetoxylation, an intermediate step of the Mitsubishi process to 1,4 butane-diol and tetrahydrofuran.

Also worth noting is that palladium/heteropoly acids are active in Wacker-type chemistry in the liquid as well as in the gas phase, and that they constitute the catalytic system of Catalytica, a new process of ethene conversion to acetaldehyde.[17]

A detailed analysis of these three classes of catalysts thus offers several suggestions for new areas of application of solid catalysts in selective oxidation reactions.

6.2. SELECTIVE OXIDATION IN THE LIQUID PHASE WITH SOLID MICRO- OR MESOPOROUS MATERIALS

The traditional approach for immobilizing metal catalysts usually involves addition of the transition metal to an oxide support, e.g., the heterogeneous Ti/SiO_2 catalysts used by Shell for the epoxidation of propene with ethylbenzene hydroperoxide. However, the main problem is to avoid oligomerization of active monomeric oxometal (M=O) or related species to inactive species or species that catalyze side reactions (e.g., hydrogen peroxide decomposition). Site-isolation of discrete metal centers in an environment that allows access to all the active centers can be achieved via the insertion by framework substitution of transition metal ions in a structured and rigid porous inorganic matrix {e.g., a zeolite (aluminosilicates) or zeo-type materials [similar structures, but different elemental compositions, e.g., silicalites (T = Si), aluminophosphates (AlPOs; T = Al and P), and silica aluminophosphates (SAPOs; T = Si, Al and P)]}. The isolation of metal centers in a structure, which makes them accessible to reactants, is the first key aspect of the characteristics of titanium silicalite (TS-1) and the basis of the project that led to its discovery. However, there are several other important characteristics that make its reactivity and catalytic behavior unique:

1. Stabilization of a coordination (tetrahedral) unusual for titanium ions.
2. Possibility during the reaction of partial hydrolysis of Si–O–Ti bonds to generate an octahedral peroxide titanium complex with two Si–O–Ti bonds still present to avoid leaching of the metal.
3. Rapid reformation of the tetrahedral coordination after the catalytic reaction (the characteristic that distinguishes TS-1 from other Ti-containing zeolites, because fast reformation of the tetrahedral coordination avoids leaching of the metal).
4. Presence of a hydrophobic local environment inside the zeolite channels (also an important factor in limiting deactivation by metal leaching, as well as side reactions; furthermore, the rate of reaction is adversely affected by water, but the hydrophobic local environment limits this negative effect).
5. Presence of shape-selective effects.

The possibility of extending the use of selective oxidation with hydrogen peroxide to other transition metal–containing micro- or mesoporous materials can be viewed in general terms as the ability to reproduce the characteristics of TS-1 catalysts described above, a goal not yet reached, although several of the materials developed are interesting for the synthesis of higher-value products, where the use of anhydrous media and organic peroxide lowers the stability requirements (see also Chapter 3).

In spite of the problem of stability for industrial exploitation, the considerable research in the last decade has shown clearly that confinement of the active site in channels and/or cavities can endow the resulting catalyst with unique activity over and above that observed with conventional supported catalysts. By choosing a molecular sieve with an appropriate pore size and hydrophobicity, it is possible to determine which molecules have ready access to the active site, thus creating supramolecular catalytic systems with the ability to discriminate between substrates. Hence, tailor-made catalysts can be designed that have a distinct resemblance to redox enzymes in which the protein mantle plays a similar role.[18]

The confinement of an active center in the constrained environment of a molecular sieve material influences its catalytic behavior not only because of its isolation, but also as a result of strong electrostatic interactions between acidic and basic sites on the internal surface and the substrate or reaction intermediate. This is directly analogous to substrate interactions with acidic carboxyl and basic amino groups of amino acid residues in the active sites of (redox) enzymes. Moreover, the molecular sieve can impose an unusual (high-energy) geometry at a metal site that enhances its catalytic properties. It is known in bioinorganic chemistry that this effect plays a major role in determining the activity of metalloenzymes.[19,20] The role of these secondary effects in determining the catalytic behavior of metal ions in the constrained environment of a molecular sieve has not been fully considered but these secondary effects of interactions that force the geometry of approaching molecules are fundamental to the design of the new types of supramolecular systems that can mimic the behavior of metalloenzymes.

Both the nature of the molecular sieve (pore size, hydrophobicity) and the method of confinement are important variables for the design of these materials. Their well-defined pore systems combined with their capacity for small substrate-induced structural changes enable molecular sieves to selectively discriminate among molecules based on their kinetic dimensions relative to the size of the pore openings of the zeolite cavities. The recently discovered mesoporous (alumino) silicates (MCM-41 with a regular array of one-dimensional pores with diameters in the 15–100 Å range)[21] have further extended the size of molecules that can be adsorbed in molecular sieves, thus increasing the number of potential applications.[22] Redox pillared clays[23,24] and redox aerogels[25] can in a broad sense be considered other examples of mesoporous molecular sieve materials. A further parameter for catalyst design is the method of confinement of the active element, which allows classification of these materials into three main classes, depending on the method of confinement:

- Framework substitution.
- Encapsulation of metal complexes ("ship-in-a-bottle" method).
- Grafting or tethering of metal complexes to the internal surface.

6.2.1. Framework Substitution

The synthesis of molecular sieves occurs in hydrothermal conditions (temperatures in the 150–200°C range) starting from an aqueous gel, containing a source of the framework building elements (Al, Si) and a structure-directing agent (template; usually an amine or a tetraalkylammonium salt). Instead of an aluminum source as for classical zeolites, trivalent (e.g., Fe^{3+}), or tetravalent (e.g., Ti^{4+}) salts can be added to the synthesis gel to obtain isomorphous substitution of Al or Si by the metal ion, whereby the latter occupies TO_4 tetrahedra. Isomorphous substitution is feasible when the r_{cation}/r_{oxygen} ratio is between 0.225 and 0.414. However, during calcination of the as-synthesized material to remove the template, the transition metal can change its oxidation state and be expelled from the framework. Titanium and other active metals, e.g., V, Cr, Mn, Fe, Co, Cu, Zr, and Sn have reportedly been incorporated in a variety of molecular sieve structures (silicalites, zeolites, AlPOs, SAPOs) by hydrothermal synthesis, although there is not always clear proof of their incorporation.[26,27]

All of these materials are usually referred to as "redox" molecular sieves in the literature; however, for elements such as titanium the reaction mechanism does not involve redox changes in its valence state, but only changes in its coordination environment (from tetrahedral to octahedral and back). For other transition metals, such as vanadium, the mechanism may involve changes in coordination as well as redox changes in the valence state, although this aspect is often not considered. At least in reactions with hydrogen peroxide, redox changes in the valence state of the transition metal are probably negative in terms of enhanced rate of hydrogen peroxide decomposition, a point that is also often ignored.

Apart from the presence of transition metal ions in true framework positions, several other aspects of the preparation of these isomorphically substituted molecular sieves, such as the crystal size, determine the catalytic behavior; thus except for TS-1, indications in the literature about the actual inclusion of the metal in the zeolite framework and the advantage of this isomorphic substitution are often not clear. Some features of TS-1 preparation will be discussed in the following sections in order to clarify these aspects.

Another critical problem that is also often ignored is metal leaching. Small amounts of metal leaching can have a significant effect on the observed catalytic results. In some cases, e.g., for vanadium and chromium, leached metal ions are probably responsible for all of the observed catalysis, while in other cases (e.g., cobalt in CoAPO materials) the activity is effectively related to the framework metal, but leaching of the metal causes irreversible catalyst deactivation in just a few reaction cycles. Often the experiments demonstrating that heterogeneous catalysts can be recovered and recycled without apparent loss of metal content or activity are not definitive proof of heterogeneity, because, e.g., when the filtration is not done at the reaction temperature, but rather at room temperature, there is metal adsorption on the solid.

Framework substitution can also be achieved by postsynthesis modification of molecular sieves, e.g., via direct substitution of Al in zeolites by vapor phase treatment with $TiCl_4$ or by dealumination followed by reoccupation of the vacant silanol nests.[28] The second mechanism is in essence the same as the first one; although it has two distinct steps. In fact, HCl generated by reaction of $TiCl_4$ with hydroxyl groups of the zeolite reacts with Al giving $AlCl_3$-type species, thus causing dealumination. Titanium ions then enter the zeolite framework by reacting with the zeolite defects (hydroxyl nests) generated by dealumination. Boron-containing molecular sieves are preferred for postsynthesis modification since boron is readily extracted from the framework under mild conditions.[29] Postsynthesis modification has the advantage that commercially available molecular sieves can be used as the starting material.

6.2.2. Synthesis, Characteristics, and Reactivity of Titanium Silicalite

6.2.2.1. Reactivity

Titanium silicalite is the most interesting material obtained by isomorphic substitution of trivalent metals or tetravalent metals in the framework of crystalline aluminosilicates or silicates. Titanium silicalites with MFI (TS-1) and MFI/MEL (TS-2) structures have been used in several oxidation reactions with H_2O_2 as the oxidizing agent. These reactions include the following:

- Hydroxylation of phenol to catechol and hydroquinone (see also Chapter 3)[30–37]
- Hydroxylation of benzene to phenol[38]
- Hydroxylation of toluene to cresols and of anisole to guaiacol and hydroquinone monomethylether[35,39–41]
- Oxidation of alkenes and dialkenes to epoxides (see also Chapter 3)[41–43]
- Oxidation of allylic alcohols to epoxides, followed by hydrolysis to glicydyl compounds[44,45]
- Oxidation of allylic alcohols to mixtures of ketones, aldehyde, and epoxide[46]
- Oxidation of 1,2 diols to α-hydroxyketones and dicarbonyl compounds[26]
- Oxidation of alkanes to alcohols and ketones[46–50]
- Oxidation of alcohols to ketones and carboxylic acids[51]
- Ammoximation of cyclohexanone to the oxime[52,53]
- Oxidation of vinylbenzenes to β-phenylaldehydes[51]
- Oxidehydrogenation of cyclohexanol to cyclohexanone[54]
- Synthesis of glycol monomethyl ethers from alkenes by reaction with methanol and hydrogen peroxide[55,56]

- Oxidation of ammonia to the hydroxylamine[57]
- Oxidation of ammonia to the hydroxylamine[57]
- Sulfoxidation of thioethers to the corresponding sulfoxides and sulfones[58]

Titanium silicalites have the following important features:

1. Activity in the oxidation of different organic substrates using H_2O_2 as the oxidizing agent. Reactions are usually run at less than 100°C, with aqueous H_2O_2, a substrate-to-H_2O_2 molar ratio generally greater than 2, and less than 10 wt % catalyst in the reaction medium.
2. Generally high selectivities (> 80%) to the partially oxidized product on both a substrate and a H_2O_2 base, with total conversion of the latter.
3. Easy catalyst regeneration: although the catalyst is deactivated after each reaction batch, it is easily regenerated by calcination at 500–550°C; the framework titanium is stable and is not leached by the reaction solvent.

6.2.2.2. Synthesis

It is very difficult to prepare pure TS-1 (without extraframework titanium, in the form of nanosized particles of TiO_2 anatase). The presence of extraframework titanium can lead to considerable disruption in catalytic performance, but it is very difficult to ascertain its presence. Preparation is very sensitive to such factors as the presence of alkali impurities in the tetrapropylammonium hydroxide (TPAOH, the templating agent) used in the procedure, which greatly affects the nature of the material obtained.

Titanium silicalites can be synthesized via a number of different procedures, but the method described in the first patent by Taramasso et al.[1] is generally accepted as being the best for obtaining a completely reproducible synthesis of the standard titanium silicalite catalyst EURO-TS-1.[37] In this commercial procedure, tetraethyl orthotitanate (TEOT) is the source of titanium, tetraethyl orthosilicate (TEOS) is the source of silica, and TPAOH is the base.

The Taramasso et al.[1] patent specifies the gradual addition of TEOT (15 g), followed by a 25% solution of TPAOH (800 g) to TEOS (455 g) in a glass vessel kept under a CO_2-free atmosphere. The mixture is then stirred for 1 h and carefully heated to 80–90°C to accelerate the hydrolysis and remove the ethyl alcohol formed. At this stage, it is important to avoid the formation of a gelatinous precipitate, which is a precursor of amorphous titania and monoclinic silicalite.[59] If the gelatinous precipitate does form, the procedure must be interrupted until the precipitate disappears. Alternatively, more TPAOH can be added. After about 5 h, water is added to a volume of 1.5 liters, and the solution is transferred to a titanium autoclave. The mixture is kept under hydrothermal conditions and autogenous pressure, with stirring, at 175°C. After 10 days, the autoclave is cooled and the mass of crystals recovered, and carefully washed repeatedly with hot water. The resulting product is dried and calcined at 550°C for 6 h. In commercial preparations, a crystallization time of several hours at 160°C yields crystals of an optimal size.

Other titanium sources can be used in the above preparation, including titanium tetrabutoxide dissolved in dry isopropyl alcohol, $TiCl_4$ or $TiOCl_2$, but TEOT is preferred. Silica sol can be utilized to supply silicon. In this case, hydrogen peroxide must be added to the reaction mixture to obtain titanium as the pertitanate species, because it is the only stable form in strongly basic solution.[41]

A key factor in the preparation of titanium silicalite is the purity of reagents.[31] In particular, traces of alkali must be avoided, because the effect of alkali impurities is dramatic. Increasing amounts of alkali metals lead to a TS-1 product that is characterized by a much lower diphenol yield than that obtained with TS-1 synthesized from pure TPAOH. The difference in performance derives from the fact that titanium ions do not effectively replace silicon in the zeolite framework in the presence of alkali. Rather, silicotitanates are formed, which, upon calcination, lead to the segregation of anatase. Sodium ions in excess of about 7000 ppm produce the monoclinic silicalite structure rather than the orthorombic TS-1 structure, and the diphenol yield falls to zero. Because the commercially available TPAOH is not sufficiently pure, a process for the synthesis of TPAOH has been developed.

Among the characteristics of TS-1 the dimensions of the zeolite crystals are critical (see also below). The smallest crystallite sizes are obtained using TEOS, maintaining a TPAOH/Si ratio of 0.22–0.53 (preferably, 0.35), and using highly concentrated crystallization mixtures.[33] Silicon concentrations of 5 mol/liter and above give aggregates of very small crystallites (< 0.1 μm).

TS-2 is synthesized using tetrabutylammoniumhydroxide (TBAOH) instead of TPAOH.[36,60,61] Recently, Tuel and Taarit[62] reported a synthesis of TS-2 that uses tetra-n-butylphosphonium hydroxide as the templating agent. Perego et al.[63] demonstrated that the structure of TS-2 can be represented as an intergrowth of MFI-type and MEL-type structures. This intergrowth is reflected in structural disorder that is evident from the presence of broad and sharp reflections in the X-ray diffraction pattern. The degree of disorder can be controlled by changing the reaction conditions.[60,64] During the zeolite synthesis, the proper mix of different organic templates makes it possible to control the intergrowth of the different phases, and thus their relative amounts. R_4NOH (tetraalkylammonium hydroxide), where the alkyl is ethyl, propyl, and butyl, is used as the organic template. The products are characterized by diffraction patterns that are intermediate between those of MFI and MEL zeolites.

Doubly substituted silicalites containing Ti^{4+} and Al^{3+}, Ti^{4+} and Fe^{3+}, or Ti^{4+} and Ga^{3+} have been synthesized.[65–68] The introduction of trivalent metals gives the materials an acidic character. This effect is a negative one in propene oxide synthesis, e.g., but positive when the presence of acid sites is necessary for the reaction mechanism.

6.2.2.3. Characterization

Several techniques have been used to characterize titanium silicalite, primarily to determine how titanium ions are incorporated into the zeolite framework. The powder diffraction spectrum of TS-1 is very similar to that of silicalite-1, except for some single reflections in TS-1 that are double reflections in silicalite.[41] Also, many interplanar spacings in TS-1 are significantly greater than the corresponding spacings in silicalite. The unit cell parameters of TS-1 depend linearly on the amount of titanium in the framework.[69] An accurate Rietveld refinement of the profile of the diffraction pattern demonstrates that 2.4 mol % of titanium is the limiting amount for incorporation.[69] When higher levels of incorporation are reported,[33,70] it is due to erroneous characterization.

The IR spectrum of TS-1 is characterized by a band at 960 cm^{-1} which is a measure of the level of titanium incorporation in the framework.[41,71] This band is assigned to the Si–O stretching mode of the Si–O–Ti unit.[71] The introduction of titanium reduces the local symmetry of the SiO_4 tetrahedra, splitting the degeneracy of the F_2 mode into E and A_1 modes. Moreover, using the interaction of small polar molecules with TS-1, makes it possible to show by IR spectroscopy that sixfold coordination complexes are formed by insertion of ligands in the titanium coordination sphere, and that all the titanium ions are accessible to reversible interaction with the molecules. Raman spectroscopy is a very sensitive technique for the detection of traces of TiO_2 anatase (band at 140 cm^{-1}) present as a contaminant in TS-1.[72]

Hydroxylated titanium silicalite is characterized by an electronic band at 48,000 cm^{-1} in the UV–visible diffuse reflectance spectrum owing to charge transfer from oxygen to empty d-orbitals of Ti^{4+}.[71] The band disappears as the amount of adsorbed water increases, and a new broad band is observed at 42,000 cm^{-1}. This change is due to a change from pentacoordination of hydroxylated Ti^{4+} ions to octahedral coordination for titanium with two ligands coordinated. The change in the titanium ions in titanium silicalite with increasing degree of hydration is shown in Figure 6.1.

The incorporation of titanium in silicalite leads to a progressive broadening of the characteristic multiplet in the silicalite ^{29}Si MAS-NMR spectrum owing to the distorted silicon environment. Further, a shoulder appears in the high-field side of the signal, the intensity of which increases with increasing titanium content.[41] X-ray absorption near-edge-structure and extended X-ray absorption fine-structure spectroscopic data[73,74] also confirm the coordination of framework titanium ions and the effect of ligands discussed above. Thus, the combined use of several characterization techniques has clearly and unquestionably shown that titanium ions are included in the framework and demonstrated the effect of ligands on the reversible change in the coordination of the titanium ions.

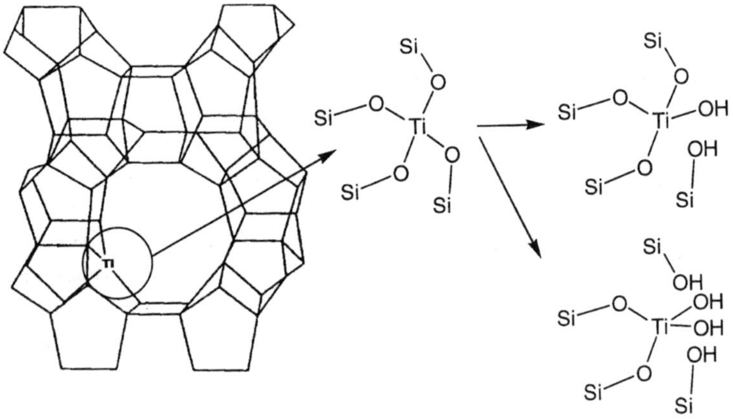

Figure 6.1. Change of titanium sites in titanium silicalite with increasing degrees of hydration.[71]

6.2.2.4. Nature of Active Species

The nature of the active species in the oxidation of different organic substrates has also been clarified in recent years, although some alternative hypotheses are still being mooted. For example, Jacobs et al.[75] recently presented evidence showing that in the reaction of Ti-containing molecular sieves (TS-1, Ti-β, Ti-MCM41) with excess H_2O_2, there is formation of O_2 in its reactive singlet state. This species is responsible for the selective oxidation behavior of these catalysts in the liquid phase, in contrast with the role of hydroperoxo titanium species discussed below.

As a result of bond polarization, Ti–O–Si bonds are cleaved by water to form silanol and titanol groups.[71,76] The hydrolysis changes the coordination of titanium from tetrahedral to square pyramidal (pentacoordinated) enabling the coordination of H_2O_2 (Scheme 6.1a). In the case of TS-1, the formation of peroxo species is clearly indicated by the yellow color of the catalyst in the presence of aqueous H_2O_2. A peroxotitanate could be formed from either a titanyl surface species or from tetrahedral lattice titanium (Scheme 6.1a), similarly to isolated titanyl groups on the surface of silica in the epoxidation catalyst developed by Shell (Ti/SiO$_2$).[77]

The mechanism of alkene epoxidation with H_2O_2 over TS-1 involves a characteristic heterolytic, peracid-like mechanism, with an electrophilic five-membered cyclic structure as the active species (Scheme 6.1b).[43] This cyclic structure is formed from a titanium hydroperoxide moiety and a protic molecule (ROH) of the solvent. In fact, in TS-1 protic solvents have a considerable positive effect on reaction rate. Thus the solvent, by virtue of its size and donor properties, plays a direct role in the epoxidation mechanism and directly affects the stability of this intermediate species.

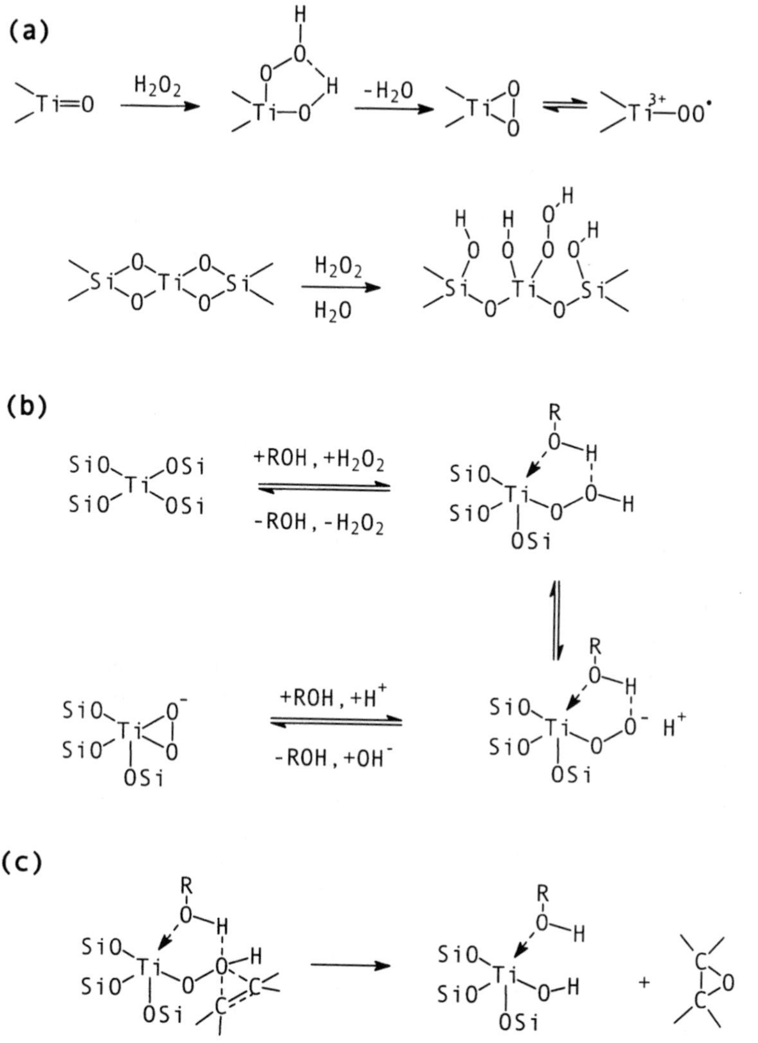

Scheme 6.1. (a) Structure of metal peroxo compounds formed by reaction of metal oxides with organic hydroperoxides or hydrogen peroxides. (b) Proposed five-membered ring species formed by interaction between titanium peroxo species and the solvent, and its change in the presence of bases.[43] (c) Proposed interaction between the five-membered titanium complex and an alkene.[43]

The epoxidation rate is highest when the solvent is methanol and lowest when it is bulky *tert*-butanol, owing to the decreased electrophilicity and increased steric constraints of the five-membered ring. The poor donor properties of water inhibit the performance of this solvent in the kinetics of epoxidation.[78] The changes in concentration of five-membered ring species also explain the observed rate enhancement in acidic solution and rate inhibition in basic solution (Scheme 6.1.b).[50,78] The oxidation mechanism of the five-membered ring species in alkene epoxidation in shown in Scheme 6.1c.

An alternative mechanism was proposed by Huybrechts *et al.*,[49,79] who suggested that titanyl groups are the active species in the oxidation of organic substrates with H_2O_2. The chemisorption of H_2O_2 leads to the formation of titanium peroxo complexes (Scheme 6.1a) in which the alkene is coordinated to the peroxo complex. Oxygen transfer occurs by intramolecular rearrangement in the alkene–titanium–peroxo complex. Clerici and Ingallina,[43] however, suggested that the mechanism was due to the presence of extraframework titania in the samples. A different hypothesis has been suggested more recently[75]: the active species is a singlet oxygen generated by the transformation of the titanium hydroperoxo complex.

The same five-membered ring species proposed as the oxidizing species in alkene epoxidation is probably also an intermediate in alkane hydroxylation. In this case, however, the five-membered ring is more likely a precursor of the true active species that homolytically abstracts hydrogen from the alkane.[50] The reaction rate in alkane hydroxylation is governed by electronic inductive effects and the diffusivity of the alkane inside the channels. The maximum rate is obtained for *n*-hexane oxidation.[50] For shorter molecules, reactivity is governed by electronic effects; for longer molecules by diffusional effects.

Methanol and secondary alcohols can be oxidized to carbonylic compounds over TS-1 with excellent selectivities. With methanol, only acetaldehyde and formaldehyde dimethyl acetal are produced, with no formic acid or methyl formate. With primary alcohols, selectivity depends on the conversion due to the presence of consecutive reactions.[31,32] Reactivity depends on chain length and branching, and, in the case of secondary alcohols, on the position of the hydroxyl groups.

The variation in reactivity can be related to a "restricted transition state" effect and hence to the accessibility of the reactive group to the catalytic site. The rate of reaction decreases with increasing chain length, and linear alcohols are generally oxidized faster than branched isomers. Because of its lower reactivity, methanol is used as the solvent for the oxidation of secondary alcohols. H_2O_2 decomposition to oxygen is less than 10% except for methanol, which promotes 20–30% decomposition. With 1,2 diols as the substrate[26] the initial product is α-hydroxyketone. Further oxidation leads to the 1,2-dicarbonyl compound, which is cleaved oxidatively in the presence of excess H_2O_2.

In the case of unsaturated alcohols, the product depends on the position of the double bond.[46] With internal bonds, epoxidation and oxidation of the hydroxy group are competitive, whereas with terminal double bonds, the epoxidation rate is slower, but the oxidation of hydroxyl groups is inhibited.

"Restricted transition state selectivity" has been invoked in the hydroxylation of substituted activated aromatic compounds such as toluene or anisole.[45] Para substitution predominates, with an isomeric distribution that differs from that observed in homogeneous catalysis.[80] Hydroxylation is selective to monosubstitu-tion.[39,40,81] Bulky substituents, such as isopropyl, retard the reaction significantly. All deactivated aromatic substrates, such as benzonitrile, clorobenzene, nitroben-zene, and benzoic acid, are nonreactive, regardless of the bulkiness of the substi-tuent. Side-chain oxidation is observed in aromatics with reactive substituents. For example, styrene is selectively oxidized to β-phenylaldehyde,[82] probably via the intermediate styrene oxide, which is rapidly isomerized by TS-1 under the same conditions. Ethylbenzene gives 30% oxidation on the aromatic ring and 70% side-chain oxidation.[32]

Tuel et al.,[83] compared the reactivity of TS-1 and an uncalcined zeolite in which internal titanium atoms are not accessible to phenol molecules. They found that catechol, the thermodynamically favored isomer, is formed initially on the external surface of the TS-1. After a short time, these external sites become inactive, because of poisoning. At this point, hydroquinone is formed preferentially on the internal titanium sites. Thus the catechol-to-hydroquinone ratio decreases with reaction time.

Two different mechanisms are possible for the ammoximation of cyclohex-anone. In the first mechanism, ammonia and hydrogen peroxide react at a titanium site to form hydroxylamine, which then reacts with cyclohexanone to yield the oxime. In the second mechanism, the ketone reacts with ammonia giving the imine. The imine then reacts with hydrogen peroxide through the agency of titanium centers to yield the oxime. This latter mechanism is similar to that proposed by Dreoni et al.[84,85] for the ammoximation of cyclohexanone in the gas phase with molecular oxygen.

Zecchina et al.[86] used UV–visible spectroscopy to show that in water tetrahe-dral titanium is transformed into a distorted, octahedrally coordinated aquohydroxo complex. In the presence of ammonia, some of the water ligands are substituted by ammonia. In the presence of H_2O_2, the hydroxo ligand is transformed into a hydroperoxo species. In the presence of both NH_3 and H_2O_2, a spectrum attributed to the formation of an octahedrally coordinated ammonia–hydroperoxo complex was observed. The latter could lead to the formation of hydroxylamine, which supports the first mechanism.

Other authors[87,88] prefer a reaction mechanism that proceeds via the imine intermediate. Based on IR observations of the formation of C=N–H groups on the surface of the TS-1, Tvaruzkova *et al.*[88] proposed that the imine path is operative. However, Zecchina *et al.*[86] suggested that it is not possible to distinguish unambiguously between the two mechanisms on the basis of spectroscopic measurements, and the hydroxylamine path is the predominant one.

6.2.3. Encapsulated Metal Complexes

The method of encapsulating metal complexes, often called the ship-in-a-bottle method, involves assembling the complex in the cavities of the microporous material, in such a way that once formed, it is too large to diffuse out.[89–92] In this way it is possible, in principle, to exploit both the characteristics of a homogeneous complex and the facility of separation and shape-selective characteristics of heterogeneous zeolitic catalysts. Three different approaches have been used to achieve encapsulation:

1. Intrazeolite complexation (flexible ligand method).
2. Intrazeolite ligand synthesis.
3. Metal complexes as templates for zeolite synthesis.

In the first method the metal complex is formed in the cavities of the microporous material (often Y-zeolite) by reaction of the metal-exchanged zeolite with a ligand that is small enough to enter the supercages of the zeolite, and, once formed, is too large to diffuse out. Examples include metal–bipyridyl and metal–salen complexes, although the latter can partially diffuse out of the supercages.

In the second method the ligand itself, constructed by intrazeolite synthesis, is too large to exit from the supercages. Examples are faujasite-encapsulated metal-lophthalocyanines.[93] The metal is first introduced into the zeolite, via ion-exchange or as a metal carbonyl or metallocene, followed by 1,2-dicyanobenzene, which reacts at elevated temperatures to form metallophthalocyanine in the supercages. An interesting observation is that in the faujasite supercages molecular models indicate that the phthalocyanine ligands are strongly deformed, but in a way analogous to the active state in enzymes.

In the template method the microporous materials are crystallized using the metal complex as the template agent. This allows the encapsulation of well-defined complexes without contamination by the free ligand or uncomplexed metal ions, but the method is limited to metal complexes such as metallophthalocyanines that are stable under the relatively severe conditions for hydrothermal synthesis of the microporous material.

6.2.4. Grafting or Tethering of Metal Complexes

This method of anchoring a metal complex to the internal surface of a micro- or mesoporous material indicates the direct formation of a bond between the metal complex and sites (often OH groups) of the micro- or mesoporous material (grafting) or the anchoring of the metal complex via a spacer ligand (tethering).[22]

Examples of the grafting method are the synthesis of surface-grafted titanium(IV) and oxomanganese by reaction of MCM-41 with titanocene dichloride $(Cp_2TiCl_2)^{22,94}$ and $Mn_2(CO)_{10}$.[95] Examples of tethering are the coordination of a manganese triazacyclononane complex to the internal surface of MCM-41[96] and of chiral molybdenum complexes (for asymmetric epoxidation) to the internal surface of mesoporous USY-zeolite.[97]

6.2.5. New "Hydrophobic" Catalytic Materials for Liquid Phase Epoxidation of Alkenes

In liquid phase heterogeneous epoxidation of alkenes the oxidizing agent of choice is almost exclusively alkylhydroperoxide, since the aqueous medium necessarily accompanying the use of H_2O_2 (the most desirable oxidizing agent for both economic and environmental reasons) interferes with selectivity for the product. In particular, hydrophilic surfaces of silica in silica-supported systems (as in Ti^{IV} or Mo^{VI} supported systems) create an aqueous environment around the metal active site, which under acid catalyzed conditions (acidity is generated by oxidation of the solvent) causes ring-opening to the corresponding glycol in the rapidly formed epoxide. In addition, the water adsorbed on the surface limits the accessibility of the alkene towards the active center, and hydrolysis of the framework metal is the primary cause of its leaching. Hydrophilicity is due to the presence of surface silanols.

The only system that is reasonably active and selective in alkene epoxidation is TS-1,[43,98] the surface of which, as is well known, exhibits highly hydrophobic properties owing to the fact that there is no aluminum properties in the framework. Partially hydrophobic materials have been obtained by anchoring TiF_4 on silica,[99] or by carrying out the synthesis of Ti-BEA zeolite in a fluoride medium[100] by trimethylsilylation of Ti-MCM-41 and Ti-MCM-48.[101]

Recently, it has been reported that it is possible to make the surface of silica gel more hydrophobic by special synthetic procedures that allow alkyl groups to be attached to the silica surface during the sol-gel synthetic procedure.[102,103] In particular the partial substitution of the tetraalkoxy-silicon precursor by a methyl-trialkoxy-silicon compound results in amorphous microporous titania–silica, which has pendant alkyl groups that make the surface hydrophobic and can epoxidize alkenes with H_2O_2.[104]

Another important factor is the procedure used for drying the gel.[105] It has been found that drying the gel by semicontinuous extraction of the solvent with supercritical CO_2 at low temperatures leads to mesoporous materials with high concentrations of Ti–O–Si connectivities, and thus to active centers for alkene epoxidation.

6.3. HETEROPOLY COMPOUNDS AS MOLECULAR-TYPE CATALYSTS

6.3.1. Introduction

Heteropolycompounds (HPC; see Figure 6.2 for an example of a transition metal (Me)–substituted heteropolycompound with Keggin structure) have been widely studied as catalytic materials. A number of reviews concerning the use of polyoxometallates as catalysts for several reactions in both homogeneous and heterogeneous media, including the oxidation of organic substrates, have been published,[106–118] and a complete review devoted specifically to the use of hetero-polycompounds as catalysts for organic oxidation appeared some years ago.[119]

The mechanism that is usually involved when molecular oxygen is the oxidizing agent in both the liquid and gas phases is a redox type, where interaction between the organic substrate and the catalyst leads to the oxidation (in terms of electron transfer) of the former and to the reduction of the polyoxometallate. The oxidation can be either a dehydrogenation or an oxygenation of the substate (O-insertion). The reduced catalyst is then reoxidized by O_2.

When a peroxygen compound is used (typically hydrogen peroxide or an alkylhydroperoxide), the mechanism involves an interaction with the catalyst, with formation of a metal–hydroperoxo or peroxo species, which acts as the O-insertion site. The oxygenation may follow either a homolytic, radical-type pathway or a heterolytic (electrophilic) one. The latter mechanism operates in the case of

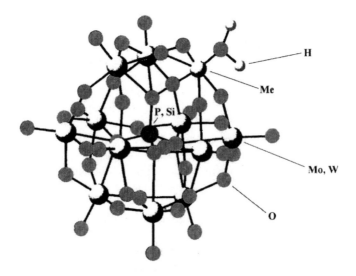

Figure 6.2. Primary structure of transition metals–substituted polyoxometalates (TMSP).

aromatic hydroxylation or alkene epoxidation. High-valence d^0 metal ions, such as V^V, Mo^{VI}, or W^{VI}, typically form these peroxo compounds, and since the oxometal in most common polyoxometallates is either Mo, V, or W, this kind of chemistry is generally observed with HPCs. Indeed, in most cases the real active species is generated by decomposition of the HPC, since the latter is often not stable in the presence of the peroxy primary oxidant, and then a radical-type mechanism takes over (with homolytic cleavage of the peroxo bond) in the reaction of alkane oxidation.

Finally, di- or trivalent transition metal ions can in part replace Mo or W ions in the framework of the heteropolycompound. In these TMSPs, where typically the guest metal ion can be divalent Fe, Ni, Mn, Cu, or Co, or trivalent Cr, the transition metal generates an active species by interaction with the oxidant, forming *in situ* a high-valence metal–oxo species, which acts as the oxidizing agent.[120-122] In practice, the polyoxometallate is a multielectron-accepting ligand for the guest transition metal, as happens in iron and manganese porphyrins, where organic ligands coordinate to the metal center (for this reason TMSP is often referred to as "inorganic porphyrin").

Monoxygen donors such as iodosobenzene, sodium periodate, or hypochlorite form this high-valence oxometal species, while in the case of alkylhydroperoxides or hydrogen peroxide interaction is possible either with the low-valence transition metal or with the high-valence d^0 oxometal (W^{VI} or Mo^{VI}). In the latter case the chemistry is similar to that described above (heterolytic peroxide chemistry), while in the former case either oxygen donation to the metal can occur, or a classic Haber–Weiss mechanism can operate (with decomposition to O_2, water, and a radical).[119] When O_2 is the oxidizing agent, a classical autoxidation mechanism develops.

6.3.2. Redox Properties of HPCs

HPCs are multielectron oxidants, and the mechanism of reduction of Keggin HPCs involves the transfer of as many as six electrons to the unit, without collapse of the structure.[123-125] The reduction increases the basicity and can be accompanied by protonation of the compound:

$$PMo_{12}O_{40}^{3-} + ae^- + bH^+ \leftrightarrow PMo_a^V Mo_{12-a}^{VI} O_{40-b}(OH)_b^{(3+b-a)-} \qquad (6.1)$$

Mixed-valence complexes are characterized by an intense blue color (hetero-polyblues), which is caused by intervalence transitions. These complexes have the same structure as the oxidized parent compounds.

The redox properties of HPCs are a function of the nature of the metal atoms in the primary structure (addenda atoms) and of both the heteroatom and the counterions. The oxidation potential decreases in the order HPC-V/Mo > HPC-Mo

> HPC-W, which means that the vanadium-containing HPCs are the strongest oxidants. This order is the same as that observed for the corresponding monomeric species, $VO^{3-} > MoO_4^{2-} > WO_4^{2-}$. The replacement of one or more molybdenum atoms in the primary structure of Keggin HPC–Mo for vanadium leads to an enhancement of the oxidation potential of the HPC, owing to the reducibility of vanadium. In fact the redox reactions

$$H_{3+n}PMo_{12-n}V_nO_{40} + ae^- + aH^+ \rightarrow H_{3+n+a}PMo_{12-n}V_{n-a}^V V_a^{IV}O_{40} \quad (6.2)$$

$$H_{3+n+a}PMo_{12-n}V_{n-a}^V V_a^{IV}O_{40} + (a/4)O_2 \rightarrow H_{3+n}PMo_{12-n}V_nO_{40} + (a/2)H_2O \quad (6.3)$$

are thermodynamically favored (redox reversibility) in aqueous media with a wide variety of organic substrates acting as electron donors (reducing agents), which allows these complexes to be used as homogeneous catalysts for oxidation with molecular oxygen. This property arises from the fact that the oxidation potential of these mixed complexes at pH 1 is approximately 0.68–0.71 V, which is higher than potentials of many organic substrates (with respect to which they can act as oxidizing agents) and lower than the potential of dioxygen (1.23 V), so that the latter can act as an oxidizing agent toward the reduced HPC. This property is exhibited by mixed-ligand Keggin HPCs (HPC-Mo/V, HPC-W/V, HPC-Mo/W), while it is not shown by HPCs with only one type of oxometal (HPC-Mo and HPC-W). However these less reactive complexes can be used efficiently as oxidation catalysts under more severe conditions, i.e., higher temperature (as in gas phase applications), under which redox reversibility is achieved.

The nature of the heteroatom affects the overall charge of the polyanion; e.g., an increase in the charge leads to a decrease in the oxidation potential for the W^{VI}/W^V couple in:

$$P^V W_{12}O_{40}^{3-} > Si^{IV}W_{12}O_{40}^{4-} = Ge^{IV}W_{12}O_{40}^{4-} > B^{III}W_{12}O_{40}^{5-}$$

$$= Fe^{III}W_{12}O_{40}^{5-} > H_2W_{12}O_{40}^{6-} = Co^{II}W_{12}O_{40}^{6-} > Cu^IW_{12}O_{40}^{7-}$$

Also the effect of the counterion can be significant. When the cation is easily reducible, the redox properties of the HPC parallel those of the cation. When the cation is not reducible (alkali metals), the reducibility of the metal in the primary structure is nevertheless affected by the nature of the cation. In particular, there is an inverse relationship between structural stability and reducibility. For instance, $K_3PMo_{12}O_{40}$ shows a lower rate of reduction than $H_3PMo_{12}O_{40}$, while the acid form of the heteropolyacid is less thermally stable than the corresponding potassium form.

6.3.3. Liquid Phase Oxidation

6.3.3.1. Oxidation with Molecular Oxygen

HPCs have been reported as catalysts for several reactions of liquid phase oxidation of organic substrates by molecular oxygen, such as:

- Oxidative dehydrogenation of alcohols to ketones.[125]
- Oxidative dehydrogenation of dienes to aromatics.[126]
- Oxidative dehydrogenation of amines.[127]
- Oxidation of aldehydes to acids.
- Oxidation of alkylbenzenes to the corresponding aldehydes.
- Oxidative bromination of arenes (use of Br^- + O_2 instead of Br_2).[128]
- Oxidative coupling of arenes.
- Acetylation of arenes.
- Oxidation of benzene to phenol.
- Oxidation of phenol or phenolic compounds to quinones.[129–131]
- Oxidation of isobutane to *tert*-butyl alcohol.
- Oxidation of propane to isopropyl alcohol and acetone.
- Oxidation of sulfur-containing compounds (mercaptans to disulfide, sulfides to sulfoxides and sulfones).[132]

In all the cases, the system being studied is the phosphomolybdate or phosphotungstate anion, where the oxometal has been in part replaced by vanadium: $PW_{12-x}V_xO_{40}^{(3+x)-}$ and $PMo_{12-x}V_xO_{40}^{(3+x)-}$.

There has at times been evidence that a two-stage redox mechanism is the one that is operating, even though, as also reported by Neumann,[119] in many cases it is likely that there are metal-centered reactions with formation of a metalloperoxo species. When a two-step redox mechanism is operating, reduction of the catalyst occurs rapidly via electron and proton transfer from the substrate (possibly through the formation of a catalyst–substrate complex). Indeed, often the HPC works in a partially reduced state, suggesting that it is the catalyst reoxidation that is the rate-determining step. Scheme 6.2 shows the generalized reaction mechanism as proposed by Neumann,[119] involving a sequence of electrons and H transfer from the organic substrate S to the V^V ions and the reoxidation of V by O_2, through the possible formation of a μ-peroxo intermediate species.

The most important homogeneous liquid phase oxidative application involving the use of an HPC-based catalyst has been studied for the Wacker process of oxidation of ethene to acetaldehyde with O_2.[133,134] A system comprising a $PMo_{12-x}V_xO_{40}^{(3+x)-}$ anion and Pd^{II} is active for this reaction, as an alternative to the industrial Wacker catalyst ($PdCl_2/CuCl_2$). The advantage lies in the fact that the HPC-based system can operate even in the absence of chloride. In this way, a less

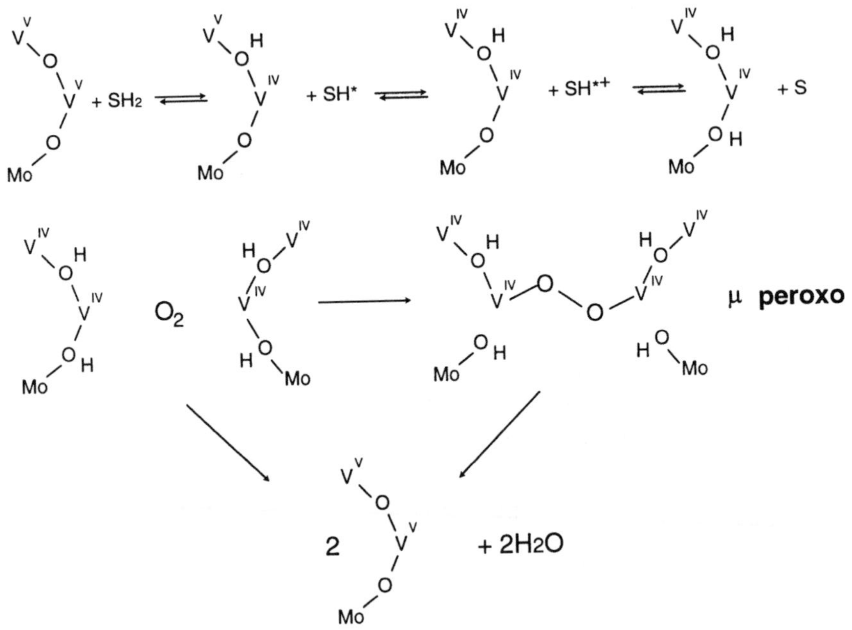

Scheme 6.2. Generalized reaction mechanism of oxidation of organic substrates catalyzed by vanadium-substituted heteropolycompounds, as proposed by Neumann.[119]

corrosive reaction medium is obtained and there is no formation of chlorinated compounds, which constitute a waste stream of the industrial process. The system operates very efficiently owing to the redox properties of vanadium in the Keggin framework, the potential of which is suitable towards the redox couple Pd^0/Pd^{II}. In fact, V acts as the reoxidizing species for Pd^0, which is reduced after the oxidation step (Scheme 6.3). The systems originally studied were characterized by very low turnover numbers and thus by the large quantities of HPC that were required. However, the system could be significantly improved by the addition of very small quantities of chloride anions to the solution, as claimed by Catalytica, which reported that this process was close to commercialization.[135]

Phillips Petroleum claimed a similar system for the oxidation of 2-butene to methylethylketone with O_2. The Phillips system uses a water/decane mixture as a two-phase diluent.[136,137] Lyons[138] also reported the use of a system consisting of a Pd salt plus the heteropolyacid $H_{11}PMo_4V_8O_{40}$ for the liquid phase oxidation of benzene with O_2 at 100°C. A ring oxidation selectivity (phenol plus phenylacetate) of 90% was achieved in the presence of excess sodium acetate. The latter inhibited biphenyl formation, probably by increasing the coordinative saturation around Pd, thus retarding the dimerization reaction.

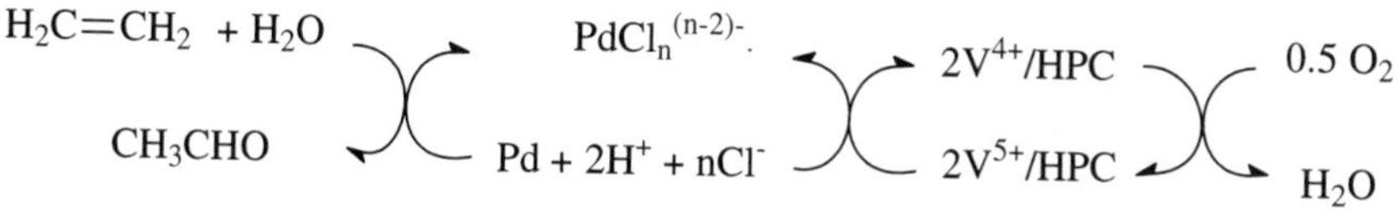

$$H_2C=CH_2 + H_2O \quad \rightarrow \quad PdCl_n^{(n-2)-} \quad \rightarrow \quad 2V^{4+}/HPC \quad \rightarrow \quad 0.5\ O_2$$

$$CH_3CHO \quad \leftarrow \quad Pd + 2H^+ + nCl^- \quad \leftarrow \quad 2V^{5+}/HPC \quad \leftarrow \quad H_2O$$

Scheme 6.3. Mechanism of Wacker-type oxidation of olefins to aldehydes catalyzed by the Pd/V-substituted heteropolycompound system.

TMSPs have been employed as catalysts for the liquid phase oxidation of light alkanes under a pressure of O_2.[139–141] Catalysts were prepared from the trilacunary P/W complexes by inserting Fe or Ni ions with three metal centers incorporated in the anion. Isopropyl alcohol and *tert*-butyl alcohol were obtained with high selectivity from propane and isobutane, respectively. The most active catalyst was the Fe_2Ni-substituted compound. With isobutane, the main product was *tert*-butyl-alcohol, with acetone and traces of *tert*-butylacetate as the by-products. Isopropyl alcohol and acetone were obtained in the oxidation of propane with a molar ratio between 0.6 and 0.8. It is possible that the mechanism involves the generation of the corresponding alkylhydroperoxide (i.e., *tert*-butylhydroperoxide in isobutane oxidation), followed by a classical radical-chain mechanism. As pointed out by Hill[108] the oxometal species originating from the interaction between the transition metal and O_2 probably constitutes a minor fraction of the reacting species, because radical-initiated autoxidation is faster than catalytic oxidation when the organic substrate is an alkane.

The use of O_2 as the primary oxidant is often limited by the fact that autoxidation pathways can be extremely poor in selectivity (except when the substrate has only one site of possible oxidative attack). This drawback can be overcome by adding a two-electron reducing agent, which is oxidized by O_2 generating a peroxo species *in situ* to the reaction medium. Typically, an aldehyde is added, which generates a peracid. In TMSPs, the substituted low-valence metal can catalyze autoxidation of the organic substrate (an aldehyde to the peracid or an alkane to the alkylhydroperoxide); the peroxy species can then further interact with the high-valence d^0 metal (W^{VI} or Mo^{VI}), generating a metal-peroxo species that is active in the epoxidation of an alkene.[142,143] The possibility of activating O_2 without reducing agents remains an important goal.[144]

6.3.3.2. Oxidation with Hydrogen Peroxide, with Organic Peroxides or Other Monoxygen Donors

Many reactions have been studied using HPCs as the catalysts with hydrogen peroxide, alkylhydroperoxides, or other monoxygen donors as the oxidizing agent, in both homogeneous system and two-phase systems (phase-transfer catalysis):

- Epoxidation of alkenes and of allyl alcohols.[145–148]
- Oxidation of cyclopentene to glutaraldehyde.
- Oxidative cleavage of alkenes and vic-diols.
- Oxidation of sec-alcohols to ketones.
- Baeyer–Villiger oxidation of cyclic ketones (with O_2 as the primary oxidant, in the presence of a reductant aldehyde).
- Oxidation of aliphatic and aromatic amines.
- Oxidation of benzene to phenol.
- Oxidation of phenols to quinones.
- Hydroxylation of arenes and phenols.
- Oxidation of cyclohexane to cyclohexanone/cyclohexanol.[149]

In oxidation reactions carried out with a peroxygen compound as the primary oxidant, the iso- and heteropolycompounds studied most intensively are those based on tunsgten, owing to its properties in the formation of metal-peroxo species. The tungsten in the primary structure has often been partially replaced with other transition metal ions in the search for more active and more selective systems, i.e., for catalysts that can oxygenate the organic substrate with specificity and do not catalyze the parallel decomposition of H_2O_2.

Epoxidation reactions are the ones that appear most often in the literature. They are usually carried out in mixed organic/water solvents, at acidic pH (2–4), moderate temperature ($< 70°C$), and with phase-transfer catalysis. Different kinds of oxidizing agents have been used, including H_2O_2 (the most common one) $NaIO_4$, $NaClO$, $PhIO$, alkylhydroperoxides, and even O_2, but in the presence of an aldehyde (typically isobutyraldehyde, which is oxidized to the corresponding peracid, thus generating the oxidizing agent in situ). Model unsaturated substrates employed range from cycloalkenes, to stilbene, to linear alkenes. Selectivity higher than 90% can be obtained, except with epoxides of lower alkenes, which are quite soluble in the aqueous phase and therefore undergo consecutive ring-opening reactions.

The HPC is generally considered to be the precursor of the true active species. It has been suggested that the latter is a peroxopolyoxometallate species, such as $PW_4O_{24}^{3-}$, which includes the $WO(\mu-O_2)(O_2)$ moiety (the so-called Venturello compound).[150] This is formed via fragmentation of the polyoxoanion in the presence of H_2O_2.[151] Therefore, the lower the stability of the HPC in solution, the easier the formation of the real active species.[152–155] The oxidant formed is catalytically very active through a heterolithic mechanism, and thus can efficiently epoxidize alkenes. This species is formed in the typical two-phase system (the organic solvent and water from aqueous H_2O_2), which is used for epoxidation and arises from the intrinsic instability of the HPC in aqueous H_2O_2. The mechanism proposed is summarized in Scheme 6.4.[109] In monophasic systems the formation of peroxo species with intact polyoxometallate has also been suggested as the oxidizing active site.[155]

Most work has been done with homogeneous transition metal–substituted Keggin-type and Wells–Dawson-type HPCs (i.e., containing Mn^{2+}, Fe^{3+}, Co^{2+}, Ni^{2+}). HPCs have been used by several groups as catalysts for double-bond epoxidation, and it has been claimed that they activate oxygen donors and catalyze oxo transfer reactions.[156–161] Organic cations are used as the counterion for the Keggin anion in order to increase the solubility of the complexes in the organic solvent. $C_6–C_{18}$ tetraalkylammonium or cetylpyridinium cation–containing complexes are employed as phase-transfer catalysts.

A few examples have also been reported that make use of supported catalytic systems; e.g., the results achieved with alumina-supported $H_3PW_{12}O_{40}$ in cycloalkene epoxidation are very good, especially in terms of selectivity for the epoxide.[162] Polyoxometallates were also intercalated in pillared layered double hydroxides and used for shape-selective epoxidation of alkenes with H_2O_2.[163] Encaged heteropoly acid catalysts have also been prepared recently by synthesizing 12-molybdophosphoric acid in the supercages of Y-type zeolites.[164]

Another reaction of interest is the hydroxylation of aromatic organic substrates in the liquid phase, i.e., of benzene to phenol and of phenol to hydroquinone and pyrocatechol, with H_2O_2. Dissolved $PMo_{12-x}V_xO_{40}^{(3+x)-}$ compounds have been used in benzene hydroxylation, at pH conditions under which the compounds were said to be structurally intact. A vanadium-peroxo species was proposed as the active site. Selectivity for phenol (on a benzene basis) was total, while yield (on an H_2O_2

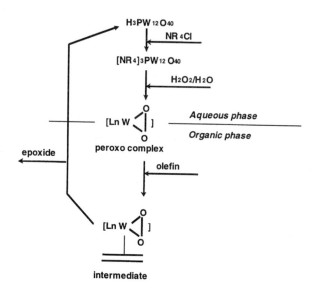

Scheme 6.4. Mechanism of two-phase epoxidation of olefins by H_2O_2 catalyzed by heteropoly compounds.[109]

basis) was higher than 90%.[116,165] In diphenol hydroxylation, one advantage of HPC-based catalysts is the possibility of affecting the diphenol ratio (ortho/pyro) by changing the catalyst composition, thus making it possible to meet market requirements. When $(Cpyr)_4PMo_{11}VO_{40}$ soluble complexes were used as the catalysts, the predominant product was hydroquinone, while with $(Cpyr)_4PW_{11}VO_{40}$ complexes, the prevailing product was catechol.[166]

TMSPs have been studied by Hill and coworkers for several oxidation reactions, i.e., alkene epoxidation and alkane oxidation.[167-170] Several different alkenes were epoxidized using hypochlorite, hydrogen peroxide, iodosylbenzene, or *tert*-butylhydroperoxide, with acetonitrile as the solvent, and with catalysts of lacunary heteropolytungstates substituted with different metals and neutralized with quaternary amines, such as $(n\text{-}Bu_4N)_4HMePW_{11}O_{39}$, where $Me = Mn^{2+}, Co^{2+}, Fe^{2+}, Cu^{2+}$, or Cr^{3+}. The type of substituent metal strongly affected the selectivity and catalyst turnover.

Alkane oxidation with TMSPs has been studied by several groups.[168,171,172] Cyclohexane was oxidized in 1,2-dichloroethane solvent, with various oxidants (iodosobenzene *tert*-butylhydroperoxide, potassium persulfate) and a $[(C_6H_{13})_4N]_4SiRu(H_2O)W_{11}O_{39}$ compound as the catalyst.[172] The latter was thus a lacunary HPC substituted with ruthenium and made soluble in an organic solvent by replacing the potassium cation with a tetra-*n*-hexylammonium cation. The yield to cyclohexylalcohol and to cyclohexanone was a function of the type of oxidizing agent. The maximum yield of the ketone was 17.7%; however higher yields were obtained by Mansuy *et al.*[170] using manganese-substituted Dawson polyoxotungstates and iodosobenzene as the oxidant ($Mn^{2+} >> Fe^{2+} > Co^{2+} >> Cu^{2+} \sim Ni^{2+} >>$ no substituent). A $Mn^V{=}O$ intermediate species, analogous to that occurring in manganese porphyrins, was postulated. Faraj and Hill[168] used *tert*-butylhydroperoxide to oxidize cyclohexane with the $Me(H_2O)PW_{11}O_{39}^{5-}$ complex as the catalyst, where Me = divalent Mn, Cu, Co, and Fe.

In view of the cost of the oxidants that have been used, the goal would be the development of systems able to operate with H_2O_2. However, with TMSP the decomposition of H_2O_2 is more rapid than with the parent unsubstituted polyoxometallates, especially when the metal is Co, Cu, or Ru, while less decomposition is observed with Mn, Cr, and Ni.[172] The compound may be unstable in the aqueous medium, with hydrolysis of the metal and its exit from the poly-oxo-metallate framework.[173,174] Good results in this direction have been obtained with "sandwich-type" polyoxometallates, such as $Fe_4(PW_9O_{34})_2^{10-}$, active for epoxidation of alkenes in a monophasic reaction environment,[175] or $(WZnMn_2)(ZnW_9O_{19})_2^{12-}$.[176] These systems minimize H_2O_2 decomposition.

Heterogenization of the polyoxometallate has been achieved by grafting the latter onto a functionalized silica with controlled surface hydrophobicity.[177] In this way it was possible to easily recover the catalyst particles from the reaction medium.

6.3.4. Gas Phase Oxidation: General Aspects

The reactivity of heteropoly compounds in gas phase selective oxidation reactions is discussed in detail in various parts of this book so only the most relevant aspects will be outlined here.

The oxidation of methacrolein to methacrylic acid is the only industrial application of HPCs as oxidation catalysts. P/Mo-based Keggin-type HPCs that also contain vanadium as a mixed addenda atom and are further doped with Cu and alkali metal ions in cationic positions have been reported as well. The composition is thus $H_{3+x-y}Cs_yPMo_{12-x}V_xO_{40}$ with $x < 2$ and $2 < y < 3$. The effect of the cations is twofold: (i) an increase in the structural stability and (ii) better control of the oxidation properties of the compound. Indeed, a balance between acidic and oxidative properties of the catalyst is fundamental. The reaction is carried out at around 280°C, and methacrylic acid is obtained with a selectivity higher than 80%.[178] The mechanism is of a redox type.

Reactions that have been reported in which HPCs can be used as gas phase oxidation catalysts are the following:

- Ammoxidation of isobutene to methacrylonitrile.
- Oxidation of butenes and n-butane to maleic anhydride.[179]
- Oxidation of acrolein to acrylic acid.[180]
- Oxidation of butadiene to furan.[181]
- Oxidative dehydrogenation of methanol to formaldehyde.[182]
- Oxidation of acetaldehyde to acetic acid.[183]
- Oxidehydrogenation of isobutyric acid to methacrylic acid.[184–189]
- Oxidation of propane to acrylic acid.[190,191]
- Oxidation of isobutane to methacrylic acid.[192–198]
- Oxidehydrogenation of light alkanes to alkenes.[199–201]

There is considerable interest in the use of HPCs as catalysts for the oxidative functionalization of light alkanes because the proper combination of acid and oxidizing properties may lead to the development of active and selective catalysts. On the other hand, the high temperatures that are usually necessary to activate light alkanes may lead to the structural decomposition of the HPCs, with a loss of the unique properties of the compound and worsening of the catalytic performance. Therefore, it is necessary to use stable salts, which are less reactive. Alternatively, it is possible to:

1. Use supported HPCs in order to obtain better spreading of the active phases (i.e., increase the specific surface area of the active compound). A problem may arise owing to the strong chemical interaction that develops between the support and the HPC, which can lead to destruction of the compound itself. Thus less reactive supports, such as silica, are needed.

2. Simultaneously improve both the structural stability and the oxidation potential of the HPC by using salts and adding transition metal ions that are known to enhance the oxidation potential of the oxometal, such as low-valence early transition metal ions in the cationic position of the HPC or vanadium as a mixed addenda oxometal.

3. Improve the thermal structural stability significantly by the addition of controlled amounts of antimony ions to ammonium salts of P/Mo Keggin-type HPCs.[202] This allows HPCs to be used as catalysts in reactions that require high temperatures, such as the oxidehydrogenation of ethane.[199] On the other hand, this problem is less dramatic when stable potassium salts of P/W Keggin-type or Wells–Dawson-type HPCs are used.[200,201]

In gas phase oxidation water is usually added to the stream. This guarantees stable catalytic performance and higher activity, probably because the presence of water can favor surface reconstruction of the heteropolyacid even under conditions at which it would usually decompose. In addition, water may favor desorption of the products, saving them from nonselective consecutive combustion.

Further key aspects of these catalytic applications of HPCs for the selective oxidation of alkanes are the following:

1. The possibility of improving the activity of P/Mo–HPCS by the addition of di/trivalent transition metal ions (Cu^{2+}, Fe^{3+}, Ni^{2+}) in the secondary framework of the compound. The partial hydrolysis of these cations (in the same way as it occurs in zeolites) and the Lewis character of some of these ions also affect the acid properties of the compounds; acidity plays a fundamental role in the activation of the C–H bond in alkanes.[112]

2. Operation must be carried out under alkane-rich conditions (large excess of isobutane with respect to molecular oxygen), since a much higher selectivity can be achieved than with operation under alkane-lean conditions. This effect is related to the redox properties of Mo^{VI} and to the oxidation state of molybdenum, which is a function of the reaction conditions. Average oxidation states lower than VI make the reaction more selective,[198] as the parallel and consecutive reactions of combustion become slower.

6.4. SOLID WACKER-TYPE CATALYSTS

Liquid phase Wacker oxidation processes were originally developed by Smidt et al.[203] for the synthesis of acetaldehyde from ethene using an aqueous solution of H_2PdCl_4 and $CuCl_2$. During the reaction Pd^{2+} is reduced to Pd^0, which is subsequently reoxidized by the Cu^{2+}/Cu^+ redox couple in the presence of excess chloride and gaseous oxygen. The process is essentially still the same, but the high corrosiveness of the solution and the formation of chlorinated by-products (especially when the carbon chain of the alkene is longer than that of the ethene) are the main drawbacks. As discussed in the previous section, a recent active line of research regards substitution of the Cu/Cl system with V-heteropoly acids, but small quantities of chloride ions are still needed. An alternative possibility is the heterogeneous gas phase version using solid catalysts,[204–211] although as is discussed in Chapter 7 the effective working system is a supported pseudoliquid phase.

After the surge of interest in the 1970s,[204–211] work in this direction was almost completely abandoned for some time. In the last 5 years, however, renewed activity in this area has led to the development of a new kind of solid Wacker-type catalyst of which there are three main classes; based on: (i) microporous samples; (ii) heteropoly compounds; (iii) supported, monolayer-type, transition metal oxides.

Palladium is the common element in the three classes, but the element that mediates the reoxidation of the reduced palladium by molecular oxygen changes. Although these catalysts perform considerably better than the first generation of solid catalysts, there are still several problems to be solved, especially regarding the stability with time, before an industrial process based on solid Wacker-type catalysts for the synthesis of ketones from alkenes can be developed. The most important reaction for this kind of application is the synthesis of butan-2-one from but-1-ene, because using conventional Wacker-type catalysts in solution the formation of chlorinated by-products occurs at an unacceptable level in view of the continuously more stringent regulations on emissions and treatment of these by-products.

6.4.1. Palladium Supported on Monolayer-Type Redox Oxides

Vanadium oxide supported on $\alpha\text{-Al}_2O_3$ and doped with Pd was the first type of monolayer redox oxide catalyst to be developed.[207,210,211] These catalysts show good activity, which can be further promoted when V_2O_5 is doped with Ti ions[207] or Mo ions.[210] A large excess of V with respect to Pd is required. V^{4+} ions are detected in the catalyst, e.g., by ESR spectroscopy, after reaction, and in general a correlation between the quantity of V^{4+} ions and the rate of reaction is noted.[210] It has been suggested that the reoxidation of V^{4+} to V^{5+} is the rate-limiting step of the reaction.

Scholten et al.[212–215] later developed an analogous catalyst, based on a vanadium pentoxide monolayer supported on a large-surface-area γ-alumina. Although better rates of reaction than the first series of samples were possible, the activity remained lower than in liquid Wacker catalysts. Another serious problem was the considerable decrease in the activity in the first hours of time-on-stream, although eventually a pseudosteady state level was reached. The authors also suggested that the activity depends on both dispersion of vanadium oxide over the support and its reducibility. As both factors are significantly better when titania is used as the support instead of alumina, owing to the formation of a pseudoamorphous homogeneous layer on titania,[216] a considerable improvement in the rate of reaction is possible using Pd supported on V_2O_5–TiO_2 catalysts.[217,218] The activity, however, is very dependent on the composition of the catalyst. For palladium acetate-based catalysts, the activity shows a first-order relationship in the palladium coverage, and the activity of the catalyst decreases with increasing vanadium oxide coverage, owing to a decrease in the latter's reducibility. It was also suggested that the difference in vanadium oxide reducibility when supported on titania and alumina is the main reason for the superior performances of titania-based catalysts. However, even in these catalysts, the activity decreases by about 60% during the first 100 h of operation, then remains stable at least up to about 700 h of time-on-stream.[218]

The deactivation was attributed to two processes[218]: the first caused by the partial reduction of the vanadium layer and the second due to rearrangement of the catalyst surface under reaction conditions. An increase in vanadium pentoxide particle size, in particular, was noted with a consequent lowering of its reducibility. The formation of carbonaceous deposits was also noted, but not considered a major factor in deactivation.

We studied the problem of the change in catalytic activity with time-on-stream for similar types of catalysts,[219,220] but arrived at different results. We noted that the rate of butan-2-one from but-1-ene initially increases in the first hour of time-on-stream and subsequently decreases. The initial change in the catalytic behavior was attributed to the formation of adsorbed species that remain chemisorbed on the catalyst surface, as the rate of reaction in the first part of the catalytic test is limited by the rate of product desorption. A progressive reduction of the vanadium oxide was also noted, which was correlated to the decrease in the rate of reaction after the maximum in butan-2-one selectivity. The progressive reduction of vanadium oxide was due not only to its slow reoxidation rate, but also to the inhibition of this rate brought about by the formation of large amounts of adsorbed species (products of partial oxidation and carboxylate species, as seen by IR characterization, and not carbonaceous-type products) on the catalyst surface. We also suggested that the same species were responsible for the crystallization of vanadium oxide particles with time-on-stream.

We observed that a similar and somewhat even better behavior in terms of selectivity could be obtained using cerium oxide instead of vanadium oxide,[220] clearly pointing out that the role of the supported transition metal oxide is only to reoxidize palladium particles and not to form a specific kind of interaction with palladium ions or particles as suggested by several authors.[207,210,213,217] Cerium oxide dispersion on alumina is very low, unlike from vanadium oxide, but can be improved using dopants and oxo-reduction treatments. This indicates that supported cerium oxide catalysts doped with palladium could be a next generation of interesting solid Wacker-type catalysts when the problem of their better dispersion can be solved. In fact, cerium oxide reoxidizes faster than vanadium pentoxide at the low reaction temperatures used for heterogeneous Wacker-type reactions (80–140°C).

An open question regarding these catalysts remains the role of the palladium salt. $PdCl_2/NaCl$ systems were observed to deactivate due to release of the chloride ion. Impregnation with $PdSO_4$ in diluted sulfuric acid was found to be a much better method of obtaining stable catalysts, but the rates of reaction and deactivation were observed to depend largely on, e.g., the method of addition and the amount of sulfuric acid.[221] The exact nature of the palladium surface complex is not known, and this contributes to the uncertainties in determining the optimal method for the preparation of these catalysts.

6.4.2. Solid Palladium-Heteropoly Compounds

As discussed earlier, the advantage of using heteropolyanions instead of chloride ions for the reoxidation of Pd^0 lies in the decreased concentration of Cl^- in solution and thus less formation of chlorinated by-products. The system was first proposed by Matveev and Koshevnikov[133] and then applied by Catalytica for liquid phase synthesis of acetaldehyde.[222] Izumi et al.[223] showed that the reoxidation of palladium is the slow step of the reaction in but-1-ene conversion. In the chloride-free medium the reaction rate depends on the acidity of the solution, decreasing as the solution acidity increases. This is due to the fact that in strongly acidic solutions palladium exists as $Pd(H_2O)_4^{2+}$ ions when coordinating anions are absent. At pH < 2 coordinated water is dissociated and replaced by two coordinated OH anions, which makes the palladium much less electrophilic and susceptible to alkene coordination, while ethene becomes less susceptible to nucleophilic attack by water.[220,221,224]

The combination of strong acidic properties and redox character is also the basis on which to explain the catalytic behavior of heterogeneous Wacker-type catalysts based on solid palladium-heteropoly compounds. Very high initial activity and selectivity were observed on these catalysts,[220,221] but fast deactivation due to reduction of molybdenum and vanadium and consequently destruction of the Keggin unit of the oxoanion occurred. Much more stable activity was observed by Stobbe-Kreemers et al.,[224] although the stability apparently remains lower than for solid Wacker-type catalysts based on supported vanadium oxide. On the other hand, heteropoly compounds constitute a very flexible and rich class of compounds where several parameters in their preparation (e.g., structure, substitution of ions in the oxoanion, or addition of elements in secondary positions) can be changed to optimize the reactivity. Therefore considerable further improvements in this class of samples can be expected.

6.4.3. Heterogenization of Wacker Catalysts in Microporous Materials

Homogeneous Wacker catalysts can be heterogenized when the anionic chloride medium can be replaced by a solid cation exchanger. This is the basic idea of zeolite-heterogenized Wacker catalysts studied by various authors and based mainly on Cu^{II} and Pd^{II} exchanged zeolites.[225–229] The early work[225–227] agrees that on an equilibrium PdCuY catalyst, prepared via a cation exchange process using the palladium amino complex, no residual ammonia is retained. More recent work by Jacobs et al.,[228,229] however, indicates that the ammonia-free catalyst initially shows good activity, but deactivates rapidly. On average, two residual ammonia ligands are required to achieve good catalyst stability.[229]

The gas phase Wacker oxidation behavior of Pd/Cu-containing zeolites depends on zeolite topology. Faujasite seems to be the only zeolite that leads to catalysts with appreciable activity, which, furthermore, is very dependent on composition. Only samples with a Cu:Pd ratio of at least 2 show catalytic activity.

The Wacker behavior of a Y-zeolite catalyst with optimized Cu/Pd composition is comparable to that of a homologous liquid phase catalyst, although the reaction kinetics are different, probably due to different rate-limiting steps. Isotope-labeling experiments using these catalysts indicate that oxygen atoms from labeled dioxygen are not incorporated into the product, suggesting that product oxygen atoms derive from water rather than from dioxygen. The reaction mechanism[229] is a combination of steps catalyzed by homogeneous transition metal catalysis and the heterogeneous zeolite. A key step is the formation of a trinuclear complex, in which the residual ammonia ligands lower the redox potential of the Pd^{II}/Pd^0 couple and inhibit the clustering of Pd^0 (main cause of deactivation) before Cu^{II} is able to reoxidize. The formation of a trinuclear complex is very dependent on zeolite topology and thus occurs only in faujasite.

Other types of analogous microporous materials should also be active. The gas phase Wacker oxidation of propene to acetone has been reported for Cu/Pd-exchanged fluorotetrasilicic mica[230] (a synthetic layer silicate), but no other types of materials have yet been developed. It is worth noting, however, that Mori et al.[231] pointed out that on proton-exchanged ultrastable Y-zeolite, butan-2-one could be synthesized from but-1-ene by direct oxidation with water in the absence of gaseous oxygen. They suggested that pentacoordinated aluminum ion have a role, but the reaction mechanism is quite unclear. However, if true, this indicates that in zeolite-based materials other types of reactivity could be responsible for Wacker-type chemistry.

REFERENCES

1. M. Taramasso, G. Perego, and N. Notari, *US Patent* 4,410,501 (1983).
2. R.A. Sheldon, *CHEMTECH*, 566 (1991).
3. R.A. Sheldon, *J. Mol. Catal. A: Chemical* **107**, 75 (1996).
4. M. Haruta, in: *3rd World Congress on Oxidation Catalysis* (R.K. Grasselli, S.T. Oyama, A.M. Gaffney, and J.E. Lyon, eds.), Studies in Surface Science and Catalysis Vol. 110, Elsevier Science: Amsterdam (1997), p. 123.
5. R.A. Sheldon, I.W.C.E. Arends, and H.E.B. Lempers, *Catal. Today* **41**, 387 (1998).
6. F. Cavani, *Catal. Today* **41**, 73 (1998).
7. G.A. Tsigdinos, *Topics Curr. Chem.* **76**, 1 (1978).
8. M.T. Pope, *Heteropoly and Isopoly Oxometallates*, Springer-Verlag: Berlin (1983).
9. M. Misono, *Catal. Rev.-Sci. Eng.* **29**, 269 (1987).
10. L. Forni and G. Gilardi, *J. Catal.* **41**, 338 (1976).
11. E. van der Heide, M. de Wind, A. Gerritsen, and J.J.F. Sholten, in: *Proceedings 9th International Congress on Catalysis*, Vol. IV (M.J. Phillips and M. Ternan, eds.), Chemical Institute of Canada: Ottawa (1988), p. 1648.
12. A.W. Stobbe-Kreemers, M. Soede, J.W. Veeman, and J.J.F. Sholten, in: *Proceedings 10th International Congress on Catalysis*, Vol. C, (L. Guczi, F. Solymosi, and P. Tetenyi, eds.), Akademie Kiado: Budapest (1993), p. 1971.
13. A.W. Stobbe-Kreemers, G. van der Lans, M. Makkee, and J.J.F. Scholten, *J. Catal.* **154**, 187 (1995).
14. G. Centi, M. Malaguti, and G. Stella, in: *New Developments in Selective Oxidation II* (V. Cortes Corberan and S. Vic Bellon, eds.), Studies in Surface Science and Catalysis Vol. 82, Elsevier Science: Amsterdam (1994), p. 461.
15. G. Centi and G. Stella, in: *Catalysis of Organic Reaction*, (M.G. Scaros and M.L. Prunier, eds.), Chemical Industries Series, Markel Dekker (1995), p. 319.
16. G. Centi, S. Perathoner, and G. Stella, in: *Catalyst Deactivation 1994* (B. Delmon and G.F. Fromant, eds.), Studies in Surface Science and Catalysis Vol. 88, Elsevier Science: Amsterdam (1994), p. 393.
17. J.A. Cusumano, *CHEMTECH*, 482 (1992).
18. R. Parton, D. De Vos, and P.A. Jacobs, in: *Zeolite Microporous Solids: Synthesis, Structure and Reactivity* (E.G. Derouane, ed.), Kluwer Academic: Dordrecht (1992), p. 555.
19. R.J.P. Williams, *J. Mol. Catal.* **1**, 31 (1976).
20. R.H. Holm, P. Kennepohl, and E.I. Solomon, *Chem. Rev.* **96**, 2239 (1996).
21. C.T. Kresge, M.E. Leonowicz, W.J. Roth, J.C. Vartuli, and J.S. Beck, *Nature* **359**, 710 (1992).
22. J.M. Thomas, *J. Mol. Catal. A: Chemical* **115**, 371 (1997).
23. T.J. Pinnavaia, *Science* **220**, 365 (1983).
24. F. Figueras, *Catal. Rev-Sci. Eng.* **30**, 457 (1988).
25. M. Schneider and A. Baiker, *Catal. Rev-Sci. Eng.* **37**, 515 (1995).

26. R.A. Sheldon, J. Dakka, *Catal. Today* **19**, 215 (1994).

27. G. Bellussi, M.S. Rigutto, in: *Advanced Zeolite Science and Applications* (J.C. Jansen, M. Stöcker, H.G. Karge, and J. Weitkamp, eds.), Studies in Surface Science and Catalysis Vol. 85, Elsevier Science: Amsterdam (1994), Ch. 6, p. 177.

28. B. Kraushaar and J.H.C. van Hooff, *Catal. Lett.* **1**, 81 (1988).

29. R. de Ruiter, K. Pamin, A.P.M. Kentgens, J.C. Jansen, and H. van Bekkum, *Zeolites* **13**, 611 (1993).

30. B. Notari, in: *Innovation in Zeolite Material Sciences* (P.J. Grobet, W.J. Mortier, E.F.Vansant, and G. Schulz-Ekloff, eds.), Studies in Surface Science and Catalysis Vol. 37, Elsevier Science: Amsterdam (1987), p. 413.

31. B. Notari, in: *Structure–Activity and Selectivity Relationships in Heterogeneous Catalysis* (R.K. Grasselli and A.W. Sleight, eds.), Studies in Surface Science and Catalysis Vol. 67, Elsevier Science: Amsterdam (1991), p. 243.

32. F. Maspero, *Chim. Ind. (Milan)* **75**, 291 (1993).

33. A.J.H.P. Van der Pol, and J.H.C. van Hooff, *Appl. Catal. A: General* **92**, 93 (1992).

34. A.J.H.P. Van der Pol, A.J. Verduyn, and J.H.C. van Hooff, *Appl. Catal. A: General* **92**, 113 (1992).

35. G. Bellussi, M.G. Clerici, F. Buonomo, U. Romano, A. Esposito, and B. Notari, *EP Patent* 200,260 B1, 1986.

36. J.S. Reddy and R. Kumar, *J. Catal.* **130**, 440, (1991).

37. J.A. Martens, Ph. Buskens, P.A. Jacobs, A. van der Pol, J.H.C. van Hooff, C. Ferrini, H.W. Kouwenhoven, P.J. Kooyman, and H. van Bekkum, *Appl. Catal. A: General* **99**, 71 (1993).

38. A. Thangaraj, R. Kumar, and P. Ratnasamy, *Appl. Catal.* **57**, L1 (1990).

39. A. Esposito, M.Taramasso, and C. Neri, *US Patent* 4,396,783 (1983).

40. A. Esposito, C. Neri, F. Buonomo, and M.Taramasso, *UK Patent* 2,116,974 B (1983).

41. G. Perego, G. Bellussi, C. Corno, M. Taramasso, F. Buonomo, and A. Esposito, in: *New Developments in Zeolite Science and Technology* (Y. Murakami, A. Iijima, and J.W. Ward, eds.), Kodansha: Tokyo/Elsevier Science: Amsterdam (1986), p. 129.

42. M.G. Clerici, and U. Romano, *EP Patent* 230,949 B1 (1987).

43. M.G. Clerici, and P. Ingallina, *J. Catal.* **140**, 71 (1993).

44. U. Romano, A. Esposito, F. Maspero, C. Neri, and M.G. Clerici, *Chim. Ind. (Milan)* **72**, 610 (1990).

45. U. Romano, A. Esposito, F. Maspero, C. Neri, and M.G. Clerici, in: *New Developments in Selective Oxidation* (G. Centi and F. Trifirò, eds.), Studies in Surface Science and Catalysis Vol. 55, Elsevier Science: Amsterdam (1990), p. 33.

46. T. Tatsumi, M.Yako, M. Nakamura, Y.Yuhara, and H.Tominaga, *J. Mol. Catal.* **78**, L41 (1993).

47. T. Tatsumi, M. Nakamura, S. Nagishi, and H. Tominaga, *J. Chem. Soc., Chem. Comm.*, 476 (1990).

48. D.R.C. Huybrechts, L. De Bruyker, and P.A. Jacobs, *Nature (London)* **345**, 240 (1990).

49. D.R.C. Huybrechts, Ph.L. Buskens, and P.A. Jacobs, *J. Mol. Catal.* **71**, 129 (1992).

50. M.G. Clerici, *Appl. Catal.* **68**, 249 (1991).

51. C. Neri and F. Buonomo, *EP Patent* 100,117 B1 (1984).

52. A. Thangaraj, S. Sivasanker, and P. Ratnasamy, *J. Catal.* **131**, 394 (1991).

53. P. Roffia, G. Leofanti, A. Cesana, M. Mantegazza, M. Padovan, G. Petrini, S. Tonti, and P. Gervasutti, in: *New Developments in Selective Oxidation* (G. Centi and F. Trifirò, eds.), Studies in Surface Science and Catalysis Vol. 55, Elsevier Science: Amsterdam (1990), p. 43.

54. B. Sulikowski and J. Klinowski, *Appl. Catal. A: General* **84**, 141 (1992).

55. C. Neri, B. Anfossi, A. Esposito, and F. Buonomo, *EP Patent* 100,119 A1 B1 (1983).

56. C. Neri, F. Buonomo, and B. Anfossi, *EP Patent* 100,118 B1 (1983).

57. M.A. Mantegazza, G. Leofanti, G. Petrini, M. Padovan, A. Zecchina, and S. Bordiga, in: *New Developments in Selective Oxidation II* (V. Corberan and S. Vic Bellon, eds.), Studies in Surface Science and Catalysis Vol. 82, Elsevier Science: Amsterdam (1994), p. 541.

58. J.S. Reddy and R. Kumar, *Zeolites* **12**, 95 (1992).

59. B. Kraushaar-Czarnetzki and J.H.C. van Hooff, *Catal. Lett.* **2**, 43 (1989).

60. G. Bellussi, A. Carati, M.G. Clerici, A. Esposito, R. Millini, and F. Buonomo, *BE Patent* 1,001,038 (1989).

61. P. Ratnasamy and R. Kumar, *Catal. Lett.* **22**, 227 (1993).

62. A. Tuel and Y.B. Taarit, *Zeolites* **13**, 357 (1993).

63. G. Perego, M. Cesari, and A. Allegra, *J. Appl. Cryst.* **17**, 403 (1984).

64. G. Perego, G. Bellussi, A. Carati, R. Millini, and V. Fattore, in: *Zeolite Synthesis* (M.L. Occelli and H.E. Robson, eds.), ACS Symposium Series 398, American Chemical Society: Washington DC (1989), p. 360.

65. G. Bellussi, M.G. Clerici, A. Giusti, and F. Buonomo, *EP Patent Appl.* 266,258 (1988).

66. G. Bellussi, M.G. Clerici, A. Carati, and A. Esposito, *EP Patent Appl.* 266,825 (1988).

67. G. Bellussi, G. Giusti, A. Esposito, and F. Buonomo, *EP Patent Appl.* 266,257 (1988).

68. G. Bellussi, A. Carati, M.G. Clerici, and A. Esposito, in: *Preparation of Catalysts V* (G. Poncelet, P.A. Jacobs, P. Grange, and B. Delmon, eds.), Studies in Surface Science and Catalysis Vol. 63, Elsevier Science: Amsterdam (1991), p. 421.

69. R. Millini, E. Previde Massara, G. Perego, and G. Bellussi, *J. Catal.* **137**, 497 (1992).

70. A. Thangaraj, S. Kumar, S.P. Mirajkar, and P. Ratnasamy, *J. Catal.* **130**, 1 (1991).

71. M.R. Boccuti, K.M. Rao, A. Zecchina, G. Leofanti, and G. Petrini, in: *Structure and Reactivity of Surfaces* (C. Morterra, A. Zecchina, and G. Costa, eds.), Studies in Surface Science and Catalysis Vol. 48, Elsevier Science: Amsterdam (1989), p. 133.

72. A. Zecchina, G. Spoto, S. Bordiga, G. Ferrero, G. Petrini, G. Leofanti, and M. Padovan, in: *Zeolite Chemistry and Catalysis* (P.A. Jacobs, N.I. Jaeger, L. Kubelkova, and B. Witcherlova, eds.), Studies in Surface Science and Catalysis Vol. 69, Elsevier Science: Amsterdam (1991), p. 251.

73. P. Behrens, J. Felshe, S. Vetter, G. Schulz-Ekloff, N. Jaeger, and W. Niemann, *J. Chem. Soc., Chem. Comm.*, 678 (1991).

74. L. Bonneviot, D. Trong On, and A. Lopez, *J. Chem. Soc. Chem. Comm.*, 685 (1993).

75. F. van Laar, D.E. de Vos, D.L. Vanoppen, and P.A. Jacobs, *Proceedings 12th International Zeolite Conference*, Baltimore, July 1998, P-241.

76. G. Bellussi, A. Carati, M.G. Clerici, G. Maddinelli, and R. Millini, *J. Catal.* **33**, 220 (1992).

77. R.A. Sheldon, *J. Mol. Catal.* **7**, 107 (1980).

78. M.G. Clerici, G. Bellussi, and U. Romano, *J. Catal.* **129**, 159 (1991).

79. D.R.C. Huybrechts, Ph.L. Buskens, and P.A. Jacobs, in: *New Developments in Selective Oxidation by Heterogeneous Catalysis* (P. Ruiz and B. Delmon, eds.), Studies in Surface Science and Catalysis Vol. 72, Elsevier Science: Amsterdam (1992), p. 21.

80. J. Varagnat, *Ind. Eng. Chem. Prod. Res. Dev.* **15**, 212 (1976).

81. G. Bellussi, M.G. Clerici, F. Buonomo, U. Romano, A. Esposito, and B. Notari, *EP Patent* 200,260 B1 (1986).

82. C. Neri and F. Buonomo, *EP Patent* 102,097 B1 (1986).

83. A. Tuel, S. Moussa-Khouzami, Y.B. Taarit, and C. Naccache, *J. Mol. Catal.* **68**, 45 (1991).

84. D.P. Dreoni, D. Pinelli, and F. Trifirò, *J. Mol. Catal.* **69**, 171 (1991).

85. D.P. Dreoni, D. Pinelli, and F. Trifirò, in: *New Developments in Selective Oxidation by Heterogeneous Catalysis* (P. Ruiz and B. Delmon, eds.), Studies in Surface Science and Catalysis Vol. 72, Elsevier Science: Amsterdam (1992), p. 109.

86. A. Zecchina, G. Spoto, S. Bordiga, F. Geobaldo, G. Petrini, G. Leofanti, M. Padovan, M. Mantegazza, and P. Roffia, in: *New Frontiers in Catalysis* (L. Guczi, F. Solymosi, and P. Tétény, eds.), Elsevier Science, Amsterdam (1993), p. 719.

87. A. Thangaraj, R. Kumar, and P. Ratnasamy, *J. Catal.* **131**, 294 (1991).

88. Z. Tvaruzkova, K. Habersberger, N. Zilkova, and P. Jiru, *Appl. Catal. A: General* **79**, 105 (1991).

89. D. E. De Vos, F. Thibault-Starzyk, P.P. Knops-Gerrits, and P.A. Jacobs, *Macromol. Symp.* **80**, 157 (1994).

90. D. E. De Vos, P.P. Knops-Gerrits, R.F. Parton, B.M. Weckhuysen, P.A. Jacobs, and R.A. Schoonheydt, *J. Incl. Phen. Mol. Recogn. Chem.* **21**, 185 (1995).

91. K.J. Balkus, A.G. Gabrielov, *J. Incl. Phen. Mol. Recogn. Chem.* **21**, 159 (1995).

92. F. Bedioui, *Coord. Chem. Rev.* **144**, 39 (1995).

93. V. Yu Zakharov, O.M. Zakharova, B.V. Romanovsky, and R.E. Mardaleishvili, *React. Kinet. Catal. Lett.* **6**, 133 (1977).

94. T. Maschmeyer, F. Rey, G. Sankar, and J.M. Thomas, *Nature* **378**, 159 (1995).

95. R. Burch, N. Cruise, D. Gleeson, and S.C. Tsang, *J. Chem. Soc. Chem. Commun.*, 951 (1996).

96. Y.V. Subba Rao, D. De Vos, T. Bein, and P.A. Jacobs, *J. Chem. Soc. Chem. Commun.*, 355 (1997).

97. A. Corma, A. Fuerte, M. Iglesias, and F. Sanchez, *J. Mol. Catal. A: Chemical* **107**, 225 (1996).

98. C.B. Dartt and M.E. Davis, *Appl. Catal. A: General* **143**, 53 (1996).

99. E. Jorda, A. Tuel, R. Teissier, and J. Kervenal, *J. Chem. Soc., Chem. Comm.*, 1775 (1995).

100. T. Blasco, M.A. Camblor, A. Corma, P. Esteve, A. Martinez, C. Prieto, and S. Valencia, *J. Chem. Soc., Chem. Comm.*, 2367 (1996).

101. T. Tatsumi, K.A. Koyano, and N. Igarashi, *J. Chem. Soc., Chem. Comm.*, 325 (1998).

102. S. Klein and W.F. Maier, *Angew. Chem. Int. Ed. Engl.* **35**, 2230 (1996).

103. F. Schwertfeger, W. Glaubitt, and U. Schubert, *J. Non-Cryst. Solids* **145**, 85 (1992).

104. H. Kochkar and F. Figueras, *J. Catal.* **171**, 420 (1997).

105. D.C.M. Dutoit, M. Schneider, and A. Baiker, *J. Catal.* **153**, 165 (1995).

106. I.V. Kozhevnikov and K.I. Matveev, *Appl. Catal.* **5**, 135 (1983)

107. M. Misono, *Catal. Rev.-Sci. Eng.* **29**, 269 (1987).

108. C.L. Hill, in: *Activation and Functionalization of Alkanes* (C.L. Hill, ed.), John Wiley & Sons: New York (1989), p. 243.

109. Y. Ono, in: *Perspectives in Catalysis* (J.M. Thomas and K.I. Zamaraev, eds.), Blackwell Scientific: Oxford (1992), p. 431.

110. M. Misono, in: *New Frontiers in Catalysis* (L. Guczi, F. Solymosi, and P. Tétény, eds.), Elsevier Science: Amsterdam (1993), p. 69.

111. N. Mizuno and M. Misono, *J. Mol. Catal.* **86**, 319 (1994).

112. F. Cavani and F. Trifirò, *Selective Partial Oxidation of Hydrocarbons*, Study No. 4193 SO, Catalytica Studies Division: Mountain View, CA (1994).

113. F. Cavani and F. Trifirò, in: *Catalysis Volume 11*, Royal Society of Chemistry: Cambridge (1994), p. 246.

114. I.V. Kozhevnikov, *Catal. Rev.-Sci. Eng.* **37**(2), 311 (1995).

115. T. Okuhara, N. Mizuno, and M. Misono, *Adv. Catal.* **41**, 113 (1996).

116. M. Misono, N. Mizuno, K. Inumaru, G. Koyano, and X.-H. Lu, in: *3rd World Congress on Oxidation Catalysis* (R.K. Grasselli, S.T. Oyama, A.M. Gaffney, and J.E. Lyons, eds.), Elsevier Science: Amsterdam (1997), p. 35.

117. C.L. Hill, A.M. Khenkin, M.S. Weeks, and Y. Heu, in: *Catalytic Selective Oxidation*, (S.T. Oyama and J.W. Hightower, eds.), ACS Symposium Series 523, American Chemical Society: Washington DC (1993), p. 67.

118. F. Cavani, *Catal. Today* **41**, 73 (1998).

119. R. Neumann, *Prog. Inorg. Chem.* **47**, 317 (1998).

120. D.E. Katsoulis and M.T. Pope, *J. Am. Chem. Soc.* **106**, 2737 (1984).

121. D.E. Katsoulis and M.T. Pope, *J. Chem. Soc., Chem. Comm.*, 1186 (1986).

122. G.A. Tsigdinos, *Topics Curr. Chem.* **76**, 1 (1978).

123. M.T. Pope, *Heteropoly and Isopoly Oxometallates*, Springer-Verlag: Berlin (1983).

124. M.T. Pope, in: *Comprehensive Coordination Chemistry*, Vol. 3 (G. Wilkinson, ed.), Pergamon Press (1987), Ch. 38, p. 1023.

125. R. Neumann and M. Levin, *J. Org. Chem.* **56**, 5707 (1991).

126. R. Neumann and M. Lissel, *J. Org. Chem.* **54**, 4607 (1989).

127. K. Nakayama, M. Hamamoto, Y. Nishiyama, and Y. Ishii, *Chem. Lett.*, 1699 (1993).

128. R. Neumann and I. Assael, *J. Chem. Soc., Chem. Comm.*, 1285 (1988).

129. K.I. Matveev, E.G. Zhizhina, and V.F. Odyakov, *React. Kinet. Catal. Lett.* **55**, 47 (1995).

130. M. Lissel, H. van de Wal, and R. Neumann, *Tetra. Lett.* **33**, 1795 (1992).

131. I.A. Weinstock, R.H. Attala, R.S. Reiner, M.A. Moen, K.E. Hammel, C.J. Houman, and C.L. Hill, *New J. Chem.* **20**, 269 (1996).

132. R.D. Gall, C.L. Hill, and J.E. Walker, *J. Catal.* **159**, 473 (1996).

133. K.I. Matveev and I.V. Kozhevnikov, *Kinet. Catal.* **21**, 855 (1980).

134. L.I. Kuznetsova and K.I. Matveev, *React. Kinet. Catal. Lett.* **3**, 305 (1975).

135. J.A. Cusumano, in: *Perspectives in Catalysis* (J.M. Thomas and K.I. Zamaraev, eds.), Blackwell: Oxford (1992), p.1.

136. T.P. Murtha, *US Patent* 4,507,507 (1985), assigned to Phillips Petroleum Co.

137. T.K. Shioyama, *US Patent* 4,550,212 (1985), assigned to Phillips Petroleum Co.

138. J.E. Lyons, *Catal. Today* **3**, 245 (1988).

139. J.E. Lyons, P.E. Ellis, H.K. Myers, G. Suld, and W.A. Lagdale, *US Patent* 4,803,187 (1989), assigned to Sun Refining & Marketing Co.

140. S.N. Shaikh, P.E. Ellis, and J.E. Lyons, *US Patent* 5,334,780 (1994).

141. P.E. Ellis and J.E. Lyons, *US Patent* 4,898,989 (1990).

142. N. Mizuno, T. Hiroshe, M. Tateishi, and M. Iwamoto, *Chem. Lett.*, 1839 (1993).

143. R. Neumann and M. Dahan, *J. Chem. Soc., Chem. Comm.*, 171 (1995).

144. R. Neumann and M. Dahan, *Nature* **388**, 353 (1997).

145. Y. Matoba, Y. Ishii, and M. Ogawa, *Synth. Commun.* **14**, 865 (1984).

146. Y. Ishii, K. Yamawaki, T. Ura, H. Yamada, T. Yoshida, and M. Ogawa, *J. Org. Chem.* **53**, 3587 (1988).

147. Y. Ishii and M. Ogawa, *J. Syn. Org. Chem.* **47**, 889 (1989).

148. M.W. Droedge and R.G. Finke, *J. Mol. Catal.* **69**, 323 (1991).

149. M.R. Cramarossa, L. Forti, M.A. Fedotov, L.G. Detusheva, V.A. Likholobov, L.I. Kuznetsova, G.L. Semin, F. Cavani, and F. Trifirò, *J. Mol. Catal. A: Chemical* **127**, 85 (1997).

150. C. Venturello, R. D'Aloiso, J.C. Bart, and M. Ricci, *J. Mol. Catal.* **32**, 107 (1985).

151. C. Aubry, G. Chottard, N. Platzer, J.M. Brégeault, R. Thouvenot, F. Chauveau, C. Huet, and H. Ledon, *Inorg. Chem.* **30**, 4409 (1991).

152. L. Salles, C. Aubry, F. Robert, G. Chottard, R. Thouvenot, H. Ledon, and J.M. Brégault, *New J. Chem.* **17**, 367 (1993).

153. A.C. Dengel, W.P. Griffith, and B.C. Parkin, *J. Chem. Soc., Dalton Trans.*, 2683 (1993).

154. D.C. Duncan, R.C. Chambers, E. Hecht, and C.L. Hill, *J. Am. Chem. Soc.* **117**, 681 (1995).

155. R. Neumann and M. de la Vega, *J. Mol. Catal.* **84**, 93 (1993).
156. C.L. Hill and X. Zhang, *Nature* **373**, 324 (1995).
157. D.K. Lyon, W.K. Miller, T. Novet, P.J. Domaille, E. Evitt, D.C. Johnson, and R.G. Finke, *J. Am. Chem. Soc.* **113**, 7209 (1991).
158. R. Neumann and M. Gara, *J. Am. Chem. Soc.* **117**, 5066 (1995).
159. N.M. Mizuno, T. Hirose, M. Tateishi, and M. Iwamoto, *J. Mol. Catal.* **88**, L125 (1994).
160. O.A. Kholdeeva, V.A. Grigoriev, G.M. Maksimov, and K.I. Zamaraev, *Topics Catal.* **3**, 313 (1996).
161. X.-Y. Zhang and M.T. Pope, *J. Mol. Catal. A: Chemical* **114**, 201 (1996).
162. S.L. Wilson and C.W. Jones, in: *3rd World Congress on Oxidation Catalysis* (R.K. Grasselli, S.T. Oyama, A.M. Gaffney, and J.E. Lyons, eds.), Elsevier Science: Amsterdam (1997), p. 603.
163. T. Tatsumi, K. Yamamoto, H. Tajima, and H. Tominaga, *Chem. Lett.*, 815 (1992).
164. S.R. Mukai, T. Masuda, I. Ogino, and K. Hashimoto, *Appl. Catal. A: General* **165**, 219 (1997).
165. K. Nomiya, H. Yanagibayashi, C. Nozaki, K. Kondoh, E. Hiramatsu, and Y. Shimizu, *J. Mol. Catal.* **114**, 181 (1996).
166. S.M. Brown, A. Hackett, A. Johnstone, A.M. King, K.M. Reeve, W.R. Sanderson, and M. Service, in: *Selective Oxidation in Petrochemistry*, DGMK: Berlin (1992), Tagungsberichte 9204, p. 339.
167. C.L. Hill and R.B. Brown, *J. Am. Chem. Soc.* **108**, 536 (1986).
168. M. Faraj and C.L. Hill, *J. Chem. Soc., Chem. Comm.*, 1487 (1987).
169. M. Faraj, C.H. Lin, and C.L. Hill, *New J. Chem.* **12**, 745 (1988).
170. D. Mansuy, J.F. Bartoli, P. Battioni, D.K. Lyon, and R.G. Finke, *J. Am. Chem. Soc.* **113**, 7209 (1991).
171. R. Neumann and C. Abu-Gnim, *J. Chem. Soc., Chem. Comm.*, 1324 (1989).
172. N.I. Kuznetsova, L.G. Detusheva, L.I. Kuznetsova, M.A. Fedotov, and V.A. Likholobov, *Kinet. Catal.* **33**, 415 (1992).
173. M. Schwegler, M. Floor, and H. van Bekkum, *Tetra. Lett.* **29**, 823 (1988).
174. D.C. Duncan, R.C. Chambers, E. Hecht, and C.L. Hill, *J. Am. Chem. Soc.* **117**, 681 (1995).
175. A.M. Khenkin and C.L. Hill, *Mendeleev Comm.* **140**, 213 (1993).
176. R. Neumann and M. Gara, *J. Am. Chem. Soc.* **116**, 5509 (1994).
177. R. Neumann and H. Miller, *J. Chem. Soc., Chem. Comm.*, 2277 (1995).
178. S. Nakamura and H. Ichihashi, in: *Proceedings 7th International Congress on Catalysis*, Tokyo (T. Seiyama and K. Tanabe, eds.), Elsevier Science: Amsterdam (1981), p. 755.
179. G. Centi, J. Lopez-Nieto, C. Iapalucci, K. Bruckman, and E.M. Serwicka, *Appl. Catal.* **46**, 197 (1989).
180. K. Bruckman, J. Haber, E. Lalik, and E.M. Serwicka, *Catal. Lett.* **1**, 35 (1988).

181.   M. Ai, *J. Catal.* **67**, 110 (1981).

182.   M.A. Banares, H. Hu, and I.E. Wachs, *J. Catal.* **155**, 249 (1995).

183.   H. Mori, N. Mizuno, and M. Misono, *J. Catal.* **131**, 133 (1991).

184.   M. Akimoto, Y. Tsuchida, K. Sato, and E. Echigoya, *J. Catal.* **72**, 83 (1981).

185.   M. Akimoto, K. Shima, H. Ikeda, and E. Echigoya, *J. Catal.* **86**, 173 (1984).

186.   O. Watzenberger, G. Emig, and D.T. Lynch, *J. Catal.* **124**, 247 (1990).

187.   S. Albonetti, F. Cavani, F. Trifirò, M. Gazzano, M. Koutyrev, F.C. Aissi, A. Aboukais, and M. Guelton, *J. Catal.* **146**, 491 (1994).

188.   Th. Ilkenhans, B. Herzog, Th. Braun, and R. Schögl, *J. Catal.* **153**, 275 (1995).

189.   L. Weismantel, J. Stoeckel, and G. Emig, *Appl. Catal. A: General* **137**, 129 (1996).

190.   W. Ueda and Y. Suzuki, *Chem. Lett.*, 541 (1995).

191.   N. Mizuno, M. Tateishi, and M. Iwamoto, *Appl. Catal. A: General* **128**, L165 (1995).

192.   F. Cavani, E. Etienne, M. Favaro, A. Galli, F. Trifirò, and G. Hecquet, *Catal. Lett.* **32**, 215 (1995).

193.   G. Busca, F. Cavani, E. Etienne, E. Finocchio, A. Galli, G. Selleri, and F. Trifirò, *J. Mol. Catal.* **114**, 343 (1996).

194.   F. Cavani, E. Etienne, G. Hecquet, G. Selleri, and F. Trifirò, in: *Catalysis of Organic Reactions* (R.E. Malz, ed.), Marcel Dekker: New York (1996), p. 107.

195.   N. Mizuno, W. Han, T. Kudo, and M. Iwamoto, in: *11th International Congress on Catalysis—40th Anniversary* (J.W. Hightower, W.N. Delgass, E. Iglesia, and A.T. Bell, eds.), Elsevier Science: Amsterdam (1996), p. 1001.

196.   N. Mizuno, M. Tateishi, and M. Iwamoto, *J. Catal.* **163**, 87 (1996).

197.   L. Jalowiecki-Duhamel, A. Monnier, Y. Barbaux, and G. Hecquet, *Catal. Today* **32**, 237 (1996).

198.   W. Ueda, Y. Suzuki, W. Lee, and S. Imaoka, in: *11th International Congress on Catalysis—40th Anniversary* (J.W. Hightower, W.N. Delgass, E. Iglesia, and A.T. Bell, eds.), Elsevier Science: Amsterdam (1996), p. 1065.

199.   F. Cavani, M. Koutyrev, and F. Trifirò, *Catal. Today* **28**, 319 (1996).

200.   F. Cavani, C. Comuzzi, G. Dolcetti, R.G. Finke, A. Lucchi, F. Trifirò, and A. Trovarelli, in: *Heterogeneous Hydrocarbon Oxidation*, ACS Symposium Series 638 (B.K. Warren and S.T. Oyama, eds.), American Chemical Society: Washington DC (1996), p. 140.

201.   C. Comuzzi, A. Primavera, A. Trovarelli, G. Bini, and F. Cavani, *Topics Catal.* **3**, 387 (1996).

202.   S. Albonetti, F. Cavani, F. Trifirò, and M. Koutyrev, *Catal. Lett.* **30**, 253 (1995).

203.   J. Smidt, W. Hafner, R. Jira, J. Sedlmeier, R. Sieber, R. Ruttinger, and H. Kopjer, *Angew. Chem.* **71**, 176 (1959).

204.   H. Komiyama and H. Inou, *J. Chem. Eng. Jpn.* **8**, 310 (1968).

205.   N.W. Cant and W.K. Hall, *J. Catal.* **16**, 220 (1970).

206.   K. Fujimoto, Y. Negami, T. Takahashi, and T. Kunugi, *Ind. Eng. Chem. Prod. Res. Dev.* **11**, 303 (1972).

207. A.N. Evnin, J.A. Rabo, and P.H. Kasai, *J. Catal.* **30**, 109 (1973).

208. Y. Izumi, Y. Fujii, and K. Urabe, *J. Catal.* **85**, 285 (1984).

209. K. Fujimoto, H. Takeda, and T. Kunugi, *Ind. Eng. Chem. Prod. Res. Dev.* **13**, 237 (1974).

210. L. Forni and G. Gilardi, *J. Catal.* **41**, 338 (1976).

211. L. Forni and G. Terzoni, *Ind. Eng. Chem. Proc. Res. Dev.* **16**, 288 (1977).

212. E. van der Heide, M. de Wind, A. Gerritsen, and J.J.F. Sholten, in: *Proceedings 9th International Congress on Catalysis*, Vol. IV (M.J. Phillips and M. Ternan, eds.), Chemical Institute of Canada: Ottawa (1988), p. 1648.

213. E. van der Heide, J.A.M. Ammerlaan, A.W. Gerritsen, and J.J.F. Scholten, *J. Mol. Catal.* **55**, 320 (1989).

214. E. van der Heide, J. Schenk, A.W. Gerritsen, and J.J.F. Scholten, *Recl. Trav. Chim. Pays-Bas.* **109**, 93 (1990).

215. E. van der Heide, M. Zwinkels, A.W. Gerritsen, and J.J.F. Scholten, *Appl. Catal. A: General* **86**, 181 (1992).

216. G. Centi, *Appl. Catal. A: General* **147** (1996) 267.

217. A.W. Stobbe-Kreemers, M. Soede, J.W. Veeman, and J.J.F. Sholten, in: *Proceedings 10th International Congress on Catalysis*, Vol. C, (L. Guczi, F. Solymosi, P. Tetenyi, eds.), Akademie Kiado: Budapest (1993), p. 1971.

218. A.W. Stobbe-Kreemers, M. Makkee, and J.J.F. Scholten, *Appl. Catal. A: General* **156**, 219 (1997).

219. G. Centi, M. Malaguti, and G. Stella, in: *New Developments in Selective Oxidation II* (V. Cortes Corberan and S. Vic Bellon, eds.), Studies in Surface Science and Catalysis Vol. 82, Elsevier Science: Amsterdam (1994), p. 461.

220. G. Centi, S. Perathoner, and G. Stella, in: *Catalyst Deactivation 1994* (B. Delmon and G.F. Fromant, eds.), Studies in Surface Science and Catalysis Vol. 88, Elsevier Science: Amsterdam (1994), p. 393.

221. G. Centi and G. Stella, in: *Catalysis of Organic Reaction* (M.G. Scaros and M.L. Prunier, eds.), Chemical Industries Series, Markel Dekker: New York (1995), p. 319.

222. J.H. Grate, D.R. Hamm, and S. Mahajan, in: *Catalysis of Organic Reactions* (J.R. Kosak and T.A. Johnson, eds.), Marcel Dekker: New York (1993), p. 213.

223. K. Urabe, F. Kimura, and U. Izumi in: *Proceedings 7th International Congress on Catalysis*, Studies in Surface Science and Catalysis Vol. 7, Kodansha: Tokyo/Elsevier Science: Amsterdam (1981), p. 1478.

224. A.W. Stobbe-Kreemers, G. van der Lans, M. Makkee, and J.J.F. Scholten, *J. Catal.* **154**, 187 (1995).

225. B. Elleuch, C. Naccache, and Y. Been Taarit, *Keynotes in Energy-Related Catalysis* (S. Kaliaguine, ed.), Studies in Surface Science and Catalysis Vol. 35, Elsevier Science: Amsterdam (1984), p. 139.

226. T. Kubota, F. Kumada, H. Tominga, and T. Kunugi, *Int. Chem. Eng.* **13**, 539 (1973).

227. H. Arai, T. Yamashiro, T. Kubo, and H. Tominga, *Bull. Jpn. Petr. Inst.* **18**, 39 (1976).

228. P.H. Espeel, M.C. Tielen, and P.A. Jacobs, *J. Chem. Soc. Chem. Comm.*, 669 (1991).

229. P.H. Espeel, G. De Peuter, M.C. Tielen, and P.A. Jacobs, *J. Phys. Chem.* **98**, 11588 (1994).

230. Y. Morikawa, S. Ishikawa, H. Tanaka, and W. Ueda, in: *Proceedings International Conference on Ion Exchange Processes*, Tokyo Inst. Techn. Press: Tokyo (1991), p. 131.

231. H. Mori, N. Mizuno, M. Tajima, S. Kagawa, and M. Iwamoto, *Catal. Lett.* **10**, 35 (1991).

NEW CONCEPTS AND NEW STRATEGIES IN SELECTIVE OXIDATION

7.1. INTRODUCTION

In the previous chapters it was shown, first, that breakthroughs in the field of selective oxidation have been driven by the need for new, more economical raw materials and the ecological issue, and, second, that progress in the field demands innovative new catalytic materials and reactor options.

Innovation requires the discovery of new catalysts, reactions, and technologies, but another fundamental component is the development of novel concepts that can open new fruitful areas of research. One example is the discovery of Ti-silicalite (TS-1) and its applications,[1,2] which stimulated widespread and more general interest in the use of microporous materials containing transition metals for liquid phase heterogeneous conversion of hydrocarbons using H_2O_2 or other peroxo compounds. The discovery of this catalyst and its application for selective oxidation with H_2O_2 is not only an industrial breakthrough, but an important scientific one as well, because it has led to the introduction of new concepts and considerably extended the horizons of selective oxidation: selective oxidation at isolated metal ions using H_2O_2 in a hydrophobic local environment, with the metal ion anchored to a quasi-rigid solid matrix (both these factors are the key parameters that enable the stability, selectivity, and nearly negligible activity in H_2O_2 decomposition of TS-1).

A survey of the literature subsequent to this discovery shows how the successive research activity was initially driven mainly by the "analogy factor," i.e., studying the reactivity of a range of analogous catalysts in which the zeolite structure, the nature of the metal, or the type of substrate to be oxidized was changed. However, the necessary step was identification of the key concept(s) determining reactivity/stability characteristics of TS-1, which is fundamental in stimulating innovation and at the same time an indicator of the dynamism of a scientific field.

An analysis of the field of selective oxidation at solid surfaces in terms of concepts developed in recent years, however, reveals a general lack of novelty with a consequent dearth of primary directions of research, and scientific activity is confined mainly to consolidating knowledge on existing catalysts and reactions, rather than being focused on exploration of new directions. Throughout this book the necessity for a different scientific approach as the key to opening new perspectives for research and the development of improved catalysts has been emphasized. The literature also shows that there are some peripheral areas under initial development that can open "fresh" new prospects and thus can be stimulii for innovations and the discovery of new applications.

This chapter will describe some examples of these new possibilities and related new concepts and point out possible (sometimes speculative) extensions of these concepts. The examples are certainly not exhaustive of all new possible directions of research in the field of selective oxidation, but will illustrate the challenges and new opportunities offered by less common approaches to the study of selective oxidation reactions.

7.2. SELECTIVE OXIDATION AT NEAR ROOM TEMPERATURE USING MOLECULAR OXYGEN

The selective oxidation processes using gaseous oxygen as the oxidizing agent can be classified as: (i) high-temperature processes usually on mixed oxide catalysts (gas–solid reaction) and based mainly on a Mars–van Krevelen type of mechanism (the active sites are lattice oxygens and gaseous oxygen simply reforms the site to close the catalytic site), and (ii) low-temperature processes in the liquid phase, usually involving a radical-type mechanism with the oxygen activated by a homogeneous catalyst such as a Co^{2+} salt or hydrocarbon autoxidation.[3,4] The latter, however, still requires relatively high temperatures/pressures and cannot be applied to a very wide range of substrates or types of oxidation reactions. A challenge in the field is thus to understand how to oxidize hydrocarbons or other organic molecules selectively under very mild conditions (room temperature, atmospheric pressure) using gaseous oxygen.

At near room temperature, oxygen and hydrocarbons are rather inert in the absence of oxygen activation, e.g., by electron transfer. Two possible approaches to solving this problem of oxygen activation have been proposed: (i) electrochemical activation of O_2, and (ii) activation of O_2 by spontaneous electron transfer from the hydrocarbon to be oxidized due to the combined action of a strong electrostatic field and molecule confinement.

7.2.1. Electrochemical Activation of Molecular Oxygen

A fuel cell device is an apparatus for converting the change in the free energy of a chemical reaction to electricity through electrochemical cell reactions. For example, for an H_2/O_2 fuel cell, when an acidic electrolyte is used, the electrochemical oxidation of H_2 to e^- and H^+ occurs at the anode ($H_2 \rightarrow 2H^+ + 2e^-$) and the reduction of O_2 with e^- and H^+ to H_2O takes place at the cathode ($\frac{1}{2}O_2 + 2H^+ + 2e^- \rightarrow H_2O$), with a net reaction of formation of water from H_2 and O_2 and electricity.

The reduction of O_2 at the cathode involves a complex sequence of steps the exact nature of which may depend on the nature of the electrocatalyst, but in general terms follows the $O_2 \rightarrow O_2^- \rightarrow O_2^{2-} \rightarrow O_2^{3-} \rightarrow H_2O$ sequence by stepwise addition of electrons. However, protons are present together with the electrons at the cathode surface and thus various other oxygenate species may form at the surface, such as HO_2, HO, H_2O_2, MO_2H, etc.[5,6] These species are known to be active in various types of selective oxidation reactions at room temperature[3] (see also Chapter 8) and thus when a hydrocarbon is present at the cathode (in the presence of a suitable catalyst) it is possible to oxidize it selectively. It has been shown that with this approach of reductive activation of O_2 by an O_2/H_2 cell it is possible to oxidize alkanes selectively and to synthesize phenol from benzene.[7–9]

The O_2/H_2 cell reactor and the principle of the method for the oxidative conversion of hydrocarbons by O_2 reductive activation is shown in Figure 7.1.

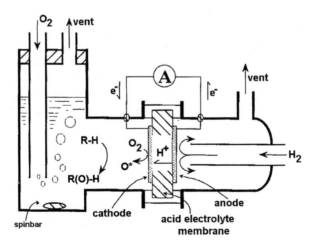

Figure 7.1. Schematic drawing of the electrochemical apparatus for the room temperature oxidation of hydrocarbons by reductive activation of a O_2 by a O_2/H_2 cell. Adapted from Otsuka.[5]

There are several potential advantages in using this approach[10–12]:

1. The reaction rate can be easily controlled by moderating the current with a variable resistor in the outer circuit; e.g., the oxidation of hydrocarbons stops when the circuit is opened.
2. The selectivity can often be controlled by the current or the anode potential, and thus a further variable to tune the behavior is available.
3. Coproduction of oxygenates and electric power at the same time is sometimes possible.

The drawbacks, on the other hand, are the cost of the reactor (electrolyte membrane, in particular) and low productivity. The latter should be increased one to two orders of magnitude for commercial application. Therefore, better electrocatalysts and appropriate electrolytes must be developed. The present target field of application is for the synthesis of fine chemicals to compensate for the low reaction rate and current efficiency. Cogeneration of electric power, when possible, is an added value.

The selective oxidation of hydrocarbons using solid oxide fuel cell systems has been investigated in detail by several groups.[13–27] Vayenas et al.[13–17] proposed the concept of non-Faradic electrochemical modification of catalytic activity (NEMCA) for catalytic oxidations, i.e., that the intrinsic catalytic activity may be increased by two to three orders of magnitude by electrochemical modification. However, the cell systems used to demonstrate the NEMCA effect are based on solid oxide electrolytes such as yttria-, calcium- or strontia-stabilized zirconia, or perovskite-type compound oxides that require high temperatures to obtain sufficient conductivity of O^{2-} or H^+. Thus the application of these solid oxide fuel cells is restricted to only a few selective oxidations of hydrocarbons at high temperatures. Such fuel cells are not suitable for room temperature selective oxidations, which require H^+-conducting electrolytes working at low temperatures ($< 100°C$).

An electrochemical cell reactor with a H^+-conducting electrolyte in a membrane was first applied by Langer et al.[26,27] at low temperatures for hydrogenation of NO into NH_2OH with cogeneration of chemicals and electricity. The system was then studied extensively and used for other types of reactions, especially by Otsuka et al.[5–12]

7.2.1.1. Benzene to Phenol

Benzene can be hydroxylated to phenol by H_2O_2 in the presence of protons and Fe(II).[28,29] When a Au-cathode is attached to a Nafion-H membrane as a protonic electrolyte, the accumulation of H_2O_2 in aqueous solution of HCl during the H_2/O_2 fuel cell reaction at room temperature (cathode: $O_2 + 2H^+ + 2e^- \rightarrow H_2O_2$) was evident.[30] When an Fe(III) salt is also present in the aqueous solution of HCl in the cathode compartment, it can be reduced to Fe(II), which then catalyzes the generation of hydroxyl radicals from H_2O_2 followed by hydroxylation of benzene to phenol:

$$H_2O_2 + Fe^{2+} + H^+ \rightarrow HO^{\bullet} + H_2O + Fe^{3+} \qquad (7.1)$$

$$Ph\text{-}H + HO^{\bullet} + Fe^{3+} \rightarrow Ph\text{-}OH + Fe^{2+} + H^+ \qquad (7.2)$$

In fact, when Fe^{3+} and benzene were added to the aqueous solution of HCl in the cathode compartment of the H_2/O_2 fuel cell, continuous formation of phenol was observed. Carbon whiskers show optimal behavior as the cathode for the hydroxylation of benzene into phenol and hydroquinone, but pretreatment with a hot aqueous solution of HNO_3 is required.[31,32] The carbon whiskers without oxidation pretreatment are not active for the reaction, owing to the creation of functional groups, such as carboxyl and quinone-hydroquinone groups, on their surfaces by the oxidation treatment with HNO_3. Moreover, they function as sites for reductive activation of O_2 even in the absence of metal cation additives in the cathode. The addition of Pd-black to the carbon whisker cathode, considerably enhances the current and the presence of Fe_2O_3 leads to a significant increase in phenol formation.[31,32] Pd-black accelerates the electrochemical reduction of O_2 (current) and formation of H_2O_2. The redox of Fe^{3+}/Fe^{2+} on Fe_2O_3 enhances the generation of $HO^{\bullet}$ from H_2O_2, which attacks benzene, producing phenol and hydroquinone via Fenton chemistry. The scheme for this reaction is shown in Scheme 7.1.

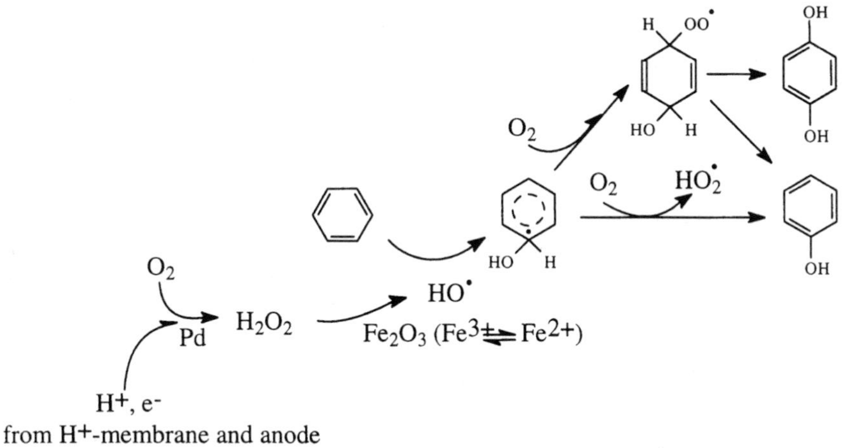

Scheme 7.1. Reaction scheme for the hydroxylation of benzene to phenol and hydroquinone at Pd-black and Fe_2O_3/carbon whisker cathode during H_2/O_2 fuel cell reactions. Adapted from Otsuka and Yamanaka.[6]

Using the H_2/O_2 fuel cell system described above it is possible not only to selectively synthesize phenol and hydroquinone from benzene (although productivity is low), but also to cogenerate electricity and oxygenates by placing a load on the outer circuit and modifying the system. Usually, when the outer circuit is loaded to obtain electric power output, the production of chemicals decreases because of the decrease in the current compared to the short-circuit conditions. However, when the carbon whisker cathode is modified with Cu(II) salts or CuO the electric power output is maximum at a cell voltage of $0.2 \sim 0.3$ V and the amount of phenol formed reaches a maximum at about 0.3 V.[6] The effect is due to the fact that under a cell voltage of about 0.30 V, Cu(II) ions reduce to Cu(I), which then catalyze the generation of H_2O_2 followed by the formation of hydroxyl radicals, active then in benzene hydroxylation, similarly to Fe(II) species:

$$O_2 + 2Cu^+ + 2H^+ \rightarrow H_2O_2 + 2Cu^{2+} \tag{7.3}$$

$$H_2O_2 + H^+ + Cu^+ \rightarrow OH^\bullet + H_2O + Cu^{2+} \tag{7.4}$$

$$C_6H_6 + HO^\bullet + Cu^{2+} \rightarrow C_6H_5OH + Cu^+ + H^+ \tag{7.5}$$

Increasing the current by loading the outer circuit decreases the rate of the reduction of Cu^+ to Cu^0 (competitive reaction), thus increasing the concentration of the Cu(I) species responsible for the formation of $HO^\bullet$, which accelerates the current as well as the oxidation of benzene into phenol. Therefore, a key factor for the hydroxylation of benzene and the reduction of O_2 (current) is the redox behavior of Cu^{2+}/Cu^+ and Cu^+/Cu^0 couples over carbon whiskers together with the influence of the external load.

7.2.1.2. Alkane Oxidation

When cyclohexane was used instead of benzene in the system described above, oxidation did not occur,[33] because the concentration of cyclohexane dissolved in an aqueous solution of HCl is very low and the reactivity of $HO^\bullet$ in an aqueous solution is lower than it would be under hydrophobic conditions.[34] Therefore, if hydrophobic conditions and a suitable electrocatalyst are selected, a powerful active oxygen species may be produced directly on the cathode, which is also able to oxygenate alkanes.

The H_2/O_2 fuel cell system for the oxidation of liquid alkanes is composed of the following sequence: $[O_2(g)$, alkane(l), cathode catalyst | H_3PO_4 aq. (1 M) in a silica wool disk | Pt-black/graphite, $H_2(g)]$. The H_3PO_4 works as an electrolyte for the H^+ conductor in the presence of hydrophobic substrates. $SmCl_3$-impregnated

graphite has been shown to have the best performance as a cathode electrocatalyst.[7-9,35-36] Although the real form of the active oxygen species on this cathode is
not known, the results of cyclic voltammetry studies suggest that the active oxygen
species could be produced from the adduct of Sm^{3+} and $HO_2^{\bullet}$, generating an
electrophilic oxygen species of $Sm^{(3-\delta)+}(HO_2^{\delta+})$.[8-9,35] The model of the active
center of this $SmCl_3$/graphite cathode is shown in Figure 7.2. The Sm^{3+} coordinated
to carboxylic groups (or other functional groups) on the graphite surface may be
the active center for the reductive activation of O_2 by e^- and H^+ and for the partial
oxidation of various hydrocarbons.

The oxidation of lighter alkanes (CH_4, C_2H_6, and C_3H_8) cannot be achieved
on the H_2/O_2 fuel cell described above for the case of cyclohexane. However, carbon
whiskers without any additives can be used successfully as cathodes for the partial
oxidation of C_3H_8 into acetone in the gas phase at room temperature.[37] The
selectivity for acetone is about 65% on the basis of C_3H_8 converted. However,
methane and ethane conversion under the same conditions is quite nonselective.
The cathodes prepared from other carbon materials such as graphite, active carbon
cloth, active carbon fiber, and carbon black were less active than the carbon
whiskers. Probably the activation of light alkanes takes place by the free $HO^{\bullet}$
evolved in the gas phase in the vicinity of the electrode/electrolyte interface.

7.2.1.3. π-Allyl and Wacker Oxidation of Alkenes

Not only the active oxygen species generated at the cathode surface (such as
in the examples discussed above) can be fruitfully applied in the synthesis of

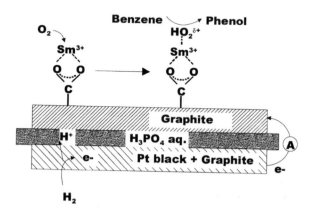

Figure 7.2. Model of the active center and active oxygen species for the partial oxidation of cyclohexane and benzene over a $SmCl_3$ embedded graphite cathode during H_2/O_2 fuel cell reactions. Adapted
from Otsuka and Yamanaka.[6]

oxygenated products at room temperature but the oxygen species generated at the anode when a hydrocarbon is oxidized instead of hydrogen can be used as well.

An interesting example is the partial oxidation of ethylene into acetaldehyde (Wacker oxidation),[38,39] one of the most important oxidation processes in today's chemical industry. It is well known that this oxidation is catalyzed through the redox couple of Pd^{2+}/Pd^0 and Cu^{2+}/Cu^+ in HCl solutions. This catalytic oxidation can be decomposed into the anode and cathode reactions:

ANODE

$$C_2H_4 + H_2O \rightarrow MeCHO + 2H^+ + 2e^- \qquad (7.6)$$

CATHODE

$$\tfrac{1}{2}O_2 + 2H^+ + 2e^- \rightarrow H_2O \qquad (7.7)$$

The selective oxidation of ethylene at the anode [Eq. (7.6)] is catalyzed by the Pd electrode and the electrochemical reduction of O_2 [Eq. (7.7)] can be catalyzed by noble metals such as a Pt electrode. The electrolyte to be used as a proton conductor held in a separator between the anode and the cathode compartments should work and be stable at temperatures above 100°C. An aqueous solution of H_3PO_4 can be used as the electrolyte.[6,40]

The selective oxidation of C_2H_4 to CH_3CHO proceeded with a selectivity greater than 97%, current efficiency better than 87%, and a yield of 7% using a Pd anode with a surface area of 3 cm^2. The reaction did not require electrical input from the outer circuit. The rate of formation of acetaldehyde can be controlled by the potential between the anode and the cathode. It is also possible using the same method, but feeding a mixture of acetic acid and ethylene in the anode compartment, to produce vinylacetate (< 10%) together with acetaldehyde (> 85%).[41]

The advantages of this fuel cell system for the synthesis of acetaldehyde compared with the current Wacker oxidation process with a mixture of C_2H_4 and O_2 in the liquid phase are the following:

1. The fuel cell used in this system is chlorine-free.
2. Since the oxidation of ethylene occurs heterogeneously at the gas–solid interface, no separation processes for products and catalysts are required.
3. It is quite easy to control the reaction rate or the current by varying the resistance.
4. The danger of explosion is reduced because C_2H_4 and O_2 are separated by a membrane holding an aqueous solution of H_3PO_4.

The drawback, however, is the cost of the apparatus coupled with its low productivity, which places the cost well beyond that of the current, large-scale, industrial production processes for acetaldehyde. However, when small-scale production units are required, the price differential becomes smaller and this technology may become commercially interesting, especially when the considerable reduction in cost that would be possible by appropriate engineering of the system is considered.

The concept of electrochemical oxidation of hydrocarbons at the anode can be extended to other alkenes such as propene.[41,42] At 92°C acrolein (selectivity 77%), acrylic acid (7%), acetone (15%), and CO_2 (6%) were observed in propene oxidation, and thus both products of a π-allyl type of oxidation [Eqs. (7.8) and (7.9); acrolein and acrylic acid] and products of Wacker-type oxidation [Eq. (7.10); acetone]:

ANODE:

$$C_3H_6 + H_2O \rightarrow CH_2CH_2CHO + 2H^+ + 2e^- \qquad (7.8)$$

$$CH_2CH_2CHO + H_2O \rightarrow CH_2CH_2CO_2H + 2H^+ + 2e^- \qquad (7.9)$$

$$C_3H_6 + H_2O \rightarrow CH_3COCH_3 + 2H^+ + 2e^- \qquad (7.10)$$

$$C_3H_6 + 6H_2O \rightarrow 3CO_2 + 18H^+ + 18e^- \qquad (7.11)$$

It is interesting that the selectivity as well as the rate of formation of these products can be changed by modifying the voltage applied to the anode (with reference to the cathode).[6,42] When a positive voltage is applied between the two electrodes, the current and the rate of formation of acetone increases considerably, that of CO_2 increases more slowly, and the rates of formation of acrolein and acrylic acid decrease. The selectivity for acetone increases to about 90% at an applied voltage higher than +0.15 V. Thus, the π-allyl oxidation is inhibited and the Wacker oxidation accelerated dramatically by a positive applied voltage. The opposite occurs with application of a negative voltage (or of an external load in the outer circuit). Therefore, the selectivity for acrolein increases when the negative voltage or the external load is applied across the cell.

The selectivity for the sum of acrolein and acrylic acid is 96% at 0.2 V. It is thus possible to control the selectivity to the Wacker oxidation and the π-allyl oxidation from the outer circuit by controlling the applied voltage. This change in product selectivities can be explained on the basis of the oxidation state of Pd. For the Wacker-type reaction mechanism Pd^{2+} cations are required, whereas the π-allyl type of oxidation takes place on Pd^0.[43] When the positive voltage is applied between the two electrodes, the potential of the Pd-black anode is more positive than that under the short-circuit conditions. This positive potential increases the concentration of Pd^{2+} and decreases that of Pd^0 in the active zone of the three-phase boundary, i.e., electrode (solid), H_3PO_4 aq. (liquid), and propene (gas). Therefore, the formation rate of acetone in Wacker oxidation is accelerated by increasing the positive voltage and that of the π-allyl oxidation is considerably decreased. When a negative voltage (or a load on the outer circuit) is applied, Pd^0 is favored, the rate of Wacker oxidation decreases sharply, and that of the π-allyl oxidation decreases gradually owing to the decrease in the current.

This example shows that in the electrochemical reactor, the cell voltage determines not only the rate of reaction, but also the selectivity and nature of active surface species.

7.2.2. Activation of Molecular Oxygen by Spontaneous Charge Transfer from a Hydrocarbon

Molecular oxygen can be activated at room temperature, inducing an electron transfer from a hydrocarbon molecule to the oxygen to form a radical cation and O_2^- (superoxide). The electron transfer occurs spontaneously when the hydrocarbon·O_2 collisional complex is in the presence of a strong electrostatic field, which stabilizes the charge transfer complex.[44–51] A zeolitic cage creates the ideal medium for confining the hydrocarbon and oxygen in a restricted environment that favors the formation of the hydrocarbon·O_2 collisional complex in the presence of a strong electrostatic field when naked alkaline or alkaline earth metals are also present.

The tail of the absorption band of the hydrocarbon·O_2 charge-transfer complex is shifted from the UV to the visible region in the presence of a strong electrostatic field. Therefore, the electron transfer can be induced at room temperature by visible photons, or even spontaneously occur thermally at temperatures slightly higher than room temperature. Alkane or alkene radical cations so produced are extremely acidic and have a strong tendency to transfer a proton to O_2^- to yield allyl (alkyl) and $HOO^\bullet$ radicals, which are the same intermediates produced in conventional liquid or gas phase autoxidation; however, their being generated at ambient temperature in a restricted environment may enable tight control and, hence, accomplish high selectivity for the final oxidation products. A solvent-free cation-exchanged zeolite such as BaY is an environment for hydrocarbon·O_2 gas phase photochemistry, ideal in terms of both: (i) enhancing the selective activation via charge transfer between the hydrocarbon and O_2 under mild conditions (it confines reactants and provides a strong electrostatic field when poorly shielded alkaline or alkaline earth extraframework cations are present), and (ii) providing a means of controlling the chemistry of the subsequently produced radicals and primary oxidation products.

The crucial functions of the supramolecular host (zeolite) are:

1. Strong adsorption/confinement of the reactants and thus a high steady state concentration of hydrocarbon·O_2 collisional pairs.
2. High electrostatic fields exerted on these pairs in the vicinity of exchangeable cations, which stabilize the energy of excited charge-transfer states determining the presence of a low-energy oxidation path accessible by visible photons.
3. Motional constraints imposed on the proposed primary radical products [alkyl (allyl, benzyl) radical, HOO radical]. This suppresses random radical coupling reactions, which otherwise would destroy product selectivity.

An example of the hydrocarbon·O_2 encounter complex aligned by the presence of the strong electrostatic field of naked Ba^{2+} extraframework cations in BaY zeolite in the case of the oxidation of toluene to benzaldehyde (see below) is shown in Figure 7.3.

These functions are not limited to cation-exchanged zeolites; any solid matrix host with microporous structure and sites of high electrostatic fields may be suitable for partial hydrocarbon oxidation. This may include organic polymeric materials with ions or ionic functional groups inside nanometer-size cavities and nonzeolite inorganic microporous frameworks featuring exchangeable cations. With use of these alternative microporous solid hosts, it may be possible to improve the desorption rates of the polar oxidation products (carbonyls, H_2O), a limiting factor in terms of productivity for current samples.

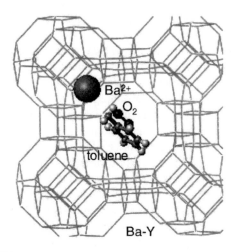

Figure 7.3. Model of the toluene·O_2 encounter complex in BaY zeolite.

Charge-transfer photochemistry of collisional complexes in zeolites need not be limited to hydrocarbon·O_2 systems, because other possible tentative examples of similar light-induced reactions in high electrostatic field environments are hydrocarbon·CO_2 or hydrocarbon·N_2 contact complexes, reactions of obvious relevance to the activation of these inert molecules. Thus this field of research pioneered by Frei et al.[44–51] opens very fruitful new areas of investigation toward the development of new kinds of room temperature processes of hydrocarbon selective oxidation, although several technical problems remain to be solved for its development, especially that of increasing productivity.

7.2.2.1. Oxidation of Alkenes

An interesting example of visible-light-induced oxidation of small alkenes by O_2 is the oxidation of 2,3-dimethyl-2-butene in alkali or alkaline earth Y zeolites. The structure of Y zeolite is that of a three-dimensional network of 13-Å spherical cages (supercages) interconnected by windows 8 Å in diameter. Loading of 0.5 Torr of the alkene and 1 atm O_2 gas at room temperature resulted in one to two hydrocarbons per supercage and one O_2 every three to four supercages on average. Exposure of the zeolite to visible or near-IR light at wavelengths as long as 750 nm induced oxidation of the alkene as detected by in situ FT-IR spectroscopy.[44,45] At –50°C it is possible to see the formation of the allyl hydroperoxide (2,3-dimethyl-3-hydroperoxy-1-butene)[47] plus a small amount of acetone (about 2%), the thermal decomposition product of this alkene hydroperoxide. The rates of hydroperoxide formation was 5×10^{-4} mol/cm^3·h for 1 W of green light, which corresponds to a space–time yield of 0.05 mol/cm^3·h for irradiation of the 1 cm^2 pellet area with 100 W visible light. At higher temperatures (room temperature) only acetone is observed.

Irradiation of propene and O_2-loaded BaY zeolite at room temperature with green- or blue-light-induced partial oxidation of the alkene.[52] Compared to 2,3-dimethyl-2-butene oxidation, shorter-wavelength photolysis light is required owing to the higher ionization potential of propene (9.7 eV vs. 8.3 eV). The products of reaction are acrolein, allyl hydroperoxide, and propylene oxide. The hydroperoxide was the main stable product (the remaining 13% was propylene oxide) when the zeolite was kept at a low temperature (–100°C). Warm-up of the zeolite after photoaccumulation of the hydroperoxide produced propylene oxide if excess propylene was kept in the matrix, but only acrolein if the olefin was removed prior to warm-up. Therefore, allyl hydroperoxide is the primary photoproduct and acrolein originates from dehydration of the hydroperoxide. Propylene oxide, on the other hand, is produced by dark O transfer from allyl hydroperoxide to excess olefin. The thermal rearrangement of the hydroperoxide to acrolein exhibits considerable temperature dependence while the epoxidation reaction does not. Hence, the aldehyde is the preferred final oxidation product of the visible-light-driven propylene oxidation at elevated zeolite temperatures.

7.2.2.2. Oxidation of Alkylaromatics

Toluene reacts with O_2 in BaY or CaY zeolite at $\lambda < 600$ nm to form benzaldehyde and H_2O without by-products,[53] although relevant data (on selectivity) were obtained only by FT-IR characterization. At $-60°C$, benzyl hydroperoxide is the main product detected[47] and warming up the matrix to room temperature resulted in spontaneous dehydration to benzaldehyde, indicating that the hydroperoxide is a reaction intermediate. Complete selectivity was sustained even upon conversion of as much as half of the toluene loaded into the zeolite. In particular, no overoxidation to benzoic acid occurred. Overoxidation of benzaldehyde in the case of the visible-light-driven reaction in the zeolite is prevented because the ionization potential of the aldehyde (9.5 eV) is higher than that of toluene (8.8 eV). Therefore, the benzaldehyde·O_2 charge-transfer absorption does not extend into the visible region, making it inaccessible to photolysis light. As a result, no further oxidation to benzoic acid can occur. In a similar study, partial oxidation of ethyl benzene to acetophenone was achieved in BaY with complete selectivity.[47]

A major problem for productivity is the strong physisorption of the polar carbonyl products inside the zeolite pores. Technical development of the reaction thus requires: (i) tuning of the zeolite characteristics (Si/Al ratio, pore structure, etc.) to find the conditions that allow product desorption from the zeolite at acceptable rates, and (ii) investigating reactor and reaction conditions needed for continuous desorption of the reaction products.

7.2.2.3. Oxidation of Alkanes

The ionization potentials of alkanes indicate that visible-light-induced oxidation through charge-transfer chemistry is also possible using these hydrocarbons.[51] Alkyl hydroperoxides are also the primary photoproducts in this case.

An interesting example is the oxidation of propane.[54] When a room temperature pellet of BaY zeolite was loaded with 150 to 300 Torr propane and 1 atm of O_2 gas and exposed to blue or green light, FT-IR data showed the formation of acetone and H_2O as final products plus isopropyl hydroperoxide. The hydroperoxide decreased in the dark with formation of acetone and H_2O.

The reaction mechanism, similar to that of alkenes, can be described as follows:

1. Alkane radical cations generated upon photoexcitation rapidly deprotonate to form alkyl radicals.[55]

2. The alkyl hydroperoxide derived from alkyl and HOO$^•$ radical recombination dehydrates spontaneously in the ionic zeolite environment via a heterolytic mechanism.

3. This heterolytic H_2O elimination leads to a carbonyl product without a side reaction, unlike homolytic peroxide bond rupture as in gas or solution phase autoxidation. The use of low-energy photons and a low-temperature environment inhibits homolysis of the peroxide bond in the zeolite.

The charge-transfer mechanism of the visible-light-induced oxidations in zeolites can be triggered in the dark if electrostatic fields and thermal energies are sufficiently high to make the hydrocarbon radical cation·O_2^- state accessible in the absence of light. Indeed, such dark thermal oxidation of cyclohexane in NaY,[51] isobutane in BaY at $T > 30°C$, and propane in BaY at $T > 50°C$ and in CaY even at room temperature has been observed.[54] The oxidation products were cyclohexanone, *tert*-butyl hydroperoxide, and acetone, respectively. As in the case of the photoinduced oxidation, selectivity was complete even at high conversion ($> 30\%$). The temperature did not exceed 80°C for cyclohexane in NaY, or 150°C in the case of propane in BaY.

It should be noted that increased concentration of residual H_2O gradually quenches the thermal reaction, probably as a result of the shielding of the Coulombic interactions by water molecules adsorbed on the cations. It is also worth noting that unsaturated hydrocarbons such as low alkenes or toluene do not exhibit thermal oxidation despite the fact that the ionization potentials and C–H bond energies are appreciably smaller than for low alkanes. This is possibly due to the strong interaction of alkenes and benzenes with exchanged cations in the supercage. As a result, the hydrocarbons reside at these cations, especially the ones that are most poorly shielded by the framework oxygens.[56,57] Hydrocarbons adsorbed on cations are unable to undergo charge transfer with O_2 since the dipole of the contact complex is antiparallel to the electrostatic field. The diminished electrostatic fields of the alkene or benzene-shielded cations experienced by other collisional pairs may be too weak to promote thermal charge transfer between these hydrocarbons and O_2 molecules, whereas the alkane·O_2 collisional pairs experience the full electrostatic field of the unshielded cations.

The pioneering work of Frei *et al.*[44–54] opens new perspectives for near-room-temperature oxidation of hydrocarbons with O_2. However, it is still necessary to: (i) investigate the behavior using more sensitive analytical methodologies than FT-IR to verify the effective high selectivity possible with the method, (ii) study the effective productivity with continuous operation, and (iii) analyze the sensitivity of the system to quenching by several components commonly present in real industrial streams, such as water and CO_2. When this information is available, it will be possible to evaluate the real possibilities for applying this reaction chemistry, after which intensive effort will be needed to improve the productivity to commercially acceptable levels.

7.2.3. Singlet Molecular Oxygen

The inertness of O_2 toward reaction with activated hydrocarbons such as alkenes largely derives from its being a triplet in the ground state. Singlet O_2 is much more reactive, but the problem is generating this reactive species. Jacobs *et al.*[58] recently demonstrated that O_2 in its reactive singlet state forms during the reaction of Ti-containing molecular sieves (TS-1, Ti-β, Ti-MCM41) with excess H_2O_2, as shown by chemical trapping of $^1O_2$ by 1-methyl-1-cyclohexene and near-IR luminescence spectroscopy (singlet to triplet relaxation).

The suggested reaction sequence for the formation of $^1O_2$ is as follows[58]:

$$Ti(OSi) + H_2O_2 \rightarrow Ti(OOH) + SiOH \qquad (7.12)$$

$$Ti(OOH) + H_2O_2 \rightarrow Ti(OH) + H_2O + {}^1O_2 \qquad (7.13)$$

It is suggested that this species is responsible for the selective oxidation behavior of these catalysts in the liquid phase, in contrast with the consolidated mechanisms of reaction suggested for TS-1 and related catalysts in the presence of H_2O_2, which involve the generation of hydroperoxo or peroxo titanium species.[59]

While the role of singlet O_2 in H_2O_2/TS-1 oxidation reactions in solution is open to question, there are other, analogous, examples in which the mechanism of generation of this $^1O_2$ reactive species is more reasonable, although they have not yet been demonstrated. Recent results have shown that using an O_2/H_2 feed it is possible to selectively hydroxylate benzene to phenol on Pt-V_2O_5/SiO_2 catalysts[60] and to obtain propene epoxidation on Au/TiO$_2$ catalysts.[61] Both reactions occur under mild conditions (around 60°C) and are very selective. The mechanism proposed in both cases involves the generation of H_2O_2 at the noble metal surface and followed by the generation of peroxo or hydroperoxo species, but the reaction conditions are quite different from those of the H_2O_2/TS-1 reaction, and the formation of singlet O_2 by a mechanism similar to that suggested by Jacobs *et al.*[58] is not unlikely. This observation, although very preliminary, suggests the possibility of generating $^1O_2$ species at solid surfaces by reaction of H_2/O_2 and then using this reactive species in selective oxidation reactions. This opens a fruitful area of research in addition to the conventional selective oxidation reactions at solid surfaces, which is very interesting especially for the synthesis of highly functionalized molecules under mild reaction conditions.

Other cases can widen the field of application of this concept. For example, it is possible to generate singlet oxygen by reaction of H_2O_2 with some catalysts $[(MoO_4^{2-},Ca(OH)_2].$[62] Although it is difficult to transpose these results to practical examples of generation and *in situ* use of singlet O_2 in selective oxidation reactions, more systematic studies of this reaction will certainly be useful. It can thus be expected that this approach (generation of H_2O_2 at gas–solid interfaces from H_2/O_2 mixtures and simultaneous use of H_2O_2 to generate singlet oxygen) will lead to the development of different kinds of selective oxidation reactions.

7.3. NEW APPROACHES TO GENERATE ACTIVE OXYGEN SPECIES

Active oxygen species may be generated under very mild conditions from gaseous O_2 as shown in the previous section, although the methods discussed generally suffer from the problems of cost of technology and/or low productivity, which must be solved before possible applications can be considered. There are, however, other interesting new ways to generate active oxygen species, which are closer to possible application but have not yet been studied systematically and thus constitute interesting alternatives for the development of new oxidation processes.

7.3.1. *In Situ* Generation of Monoxygen Donors

Dioxygen is clearly the preferred oxidant in the manufacture of bulk chemicals because of its availability, low cost, and the absence of wastes. However, the severe reaction conditions often required to activate it and/or the radical nature of the reaction involved pose problems of selectivity that limit its wider use in synthetic chemistry. As an alternative, monoxygen donors are available as milder oxidants: hydrogen peroxide, peracids, organic hydroperoxides, inorganic and metallorganic peroxides, sodium hypochlorite, iodosobenzene, and nitrous oxide. In catalytic oxidations, these are generally characterized by good activity at moderate temperatures and, often, by rather high selectivity.

Of the various monoxygen donors, hydrogen peroxide is the most attractive, because: (i) a rich and selective oxidation chemistry, based on both homogeneous and heterogeneous catalysts, is possible, and especially (ii) its oxidation product is water, which allows for the possibility of clean oxidation processes. However, in several cases the high cost of monoxygen donors precludes their use, except for the synthesis of high-value specialty products. Thus from both the industrial and scientific points of view the search for methods for the *in situ* generation of monoxygen donors from inexpensive reactants is of particular interest.

The *in situ* generation of other monoxygen donors is possible, but in general they are even more costly than for hydrogen peroxide and result in less clean oxidation reactions. However, they do have the advantage of reaction in water-free solvents, an important issue for the stability of several solid samples (e.g., transition metal–containing microporous materials)[4] that catalyze the oxidation reactions. Metallorganic peroxy and metal-oxene species are obtainable *in situ* by the reaction of the corresponding metal precursors with molecular oxygen and a reducing agent or through hydrogen peroxide. Apart from hydrogen, some reductants that are rather impractical for commercial applications have been used, such as metallic zinc and iron or metal hydrides. The oxidations carried out by Mn-porphyrin complexes and by the so-called Gif chemistry are examples of these routes. Organic peracids have been produced *in situ* in two ways: (i) the reaction of carboxylic acids with hydrogen peroxide (main method), and (ii) metal-catalyzed oxidation with a mixture of oxygen and aldehyde. Presumably, the oxidant that is formed *in situ* is a percarboxylic acid.

7.3.1.1. Methods of in Situ Generation of H_2O_2

There are various possible routes to synthesize H_2O_2.[63] The main limitation of its *in situ* generation is that the method used requires total compatibility of H_2O_2 with the oxidation catalyst and with the reagents/products involved in the overall reaction.

Hydrogen peroxide is produced commercially by the anthraquinone process (Scheme 7.2a). An alkylated anthraquinone is hydrogenated to its corresponding hydroanthraquinone, which is then oxidized with air to produce H_2O_2 and the starting anthraquinone, which is recycled after a series of purification steps to remove or chemically treat the degradation materials. If the same reaction cycle were to be used for *in situ* generation of H_2O_2, it would be expected that the hydrogenation, autoxidation, separation, and purification steps would be considerably influenced by the copresence of other reagents/products. In particular, the autoxidation step (usually carried out at moderate temperatures to prevent the degradation of the anthraquinones, with special solvent mixtures for chemical stability, good solubilizing properties for the oxidized and reduced form of the carrier, low volatility, and low water affinity) can be considerably influenced by

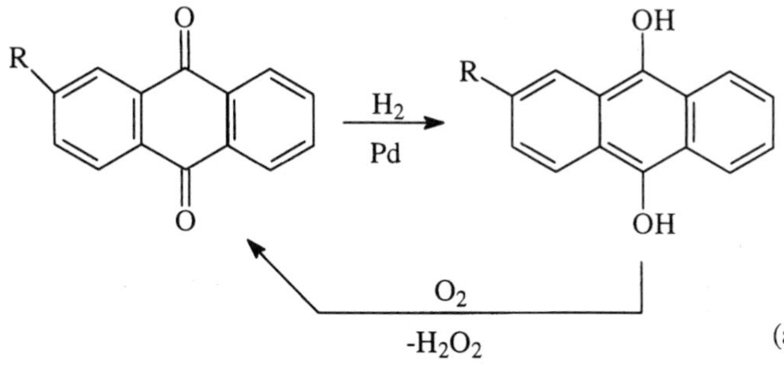

$$H_2 + O_2 \xrightarrow[\text{H}_2\text{O}/\text{H}^+/\text{Br}^-]{\text{Pd/C}} H_2O_2 \qquad \text{(b)}$$

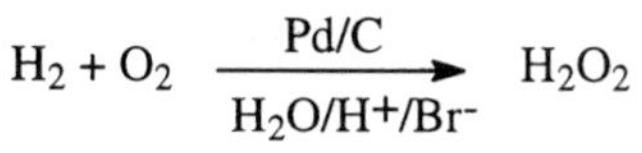

Scheme 7.2. Synthesis of H_2O_2 by the anthraquinone route (a) and by direct H_2/O_2 reaction (b).

the constraints resulting from the use of *in situ* generated H_2O_2 to catalyze selective oxidation reactions.

The direct synthesis of H_2O_2 can be done by catalytic hydrogenation of oxygen on Pd-based heterogeneous catalysts in an aqueous solution and in the presence of strong mineral acids and halide ions (Scheme 7.2).[64] Low reaction temperatures are required (0–25°C) as well as high pressures to obtain selectivities for H_2O_2 (with respect to H_2) and kinetics of practical interest. The reaction is carried out in aqueous/organic solvent mixtures (methanol, acetone and acetonitrile).

Of the two methods, the anthraquinone route is preferred for *in situ* H_2O_2 formation for various reasons[63]:

1. The moderate operating temperatures in the autoxidation step makes it suitable for oxidation reactions in which chemically unstable derivatives such as epoxides are synthesized. Also other reaction conditions are more compatible with the conditions required to use the *in situ* H_2O_2 produced.
2. The separation of low-boiling products is easier.
3. Productivity is higher.

On the other hand, in cases such as the hydroxylation of phenol and higher alkanes, the anthraquinone route both in terms of the ease of separation and the compatibility of reaction conditions is less convenient as compared to the direct synthesis from hydrogen and oxygen. The latter method, furthermore, can be more easily adapted to the creation of bifunctional catalysts (e.g., Pd/TS-1) able to generate H_2O_2 from H_2/O_2 *in situ* and then use it in selective oxidation reactions such as propene oxide synthesis.[65] Thus, the question of the optimal method for *in situ* generation of H_2O_2 depends considerably on the constraints imposed by the selective oxidation reaction of interest, in terms of the catalyst, the solvent, and the reaction conditions required. Also worth noting is the fact that the alternative approach is the *ex situ* generation of H_2O_2, in a preliminary step, although integrated into the whole process. In this case, several of the constraints and interference between conflicting conditions for the generation and the use of H_2O_2 can be minimized so that this is often the preferred solution, although plant costs are higher.

The *in situ* methods to synthesize H_2O_2 must also be modified to address the physical and chemical requirements of the selected oxidation catalyst. None of the routes to H_2O_2 is very compatible with soluble organometallic complexes (e.g., metal porphyrin complexes),[66] because there may be oxidative degradation of the organic carrier or the inhibition of the Pd catalyst, respectively. More interesting is the approach using microporous materials as catalysts.[67] The anthraquinone route is intrinsically suitable for use with titanosilicates having MFI (TS-1) and MEL (TS-2) structures.[63] The average diameter of their channel systems (~ 0.55 nm), where Ti active sites are located, prevents the diffusion of bulky molecules (alkylated anthraquinones have a cross section of about 0.6 nm) onto active sites and the resultant oxidative degradation of the organic carrier or its interference with the catalytic process. Mesoporous materials (e.g., Ti-MCM41) do not have these requirements. The coupling of the direct synthesis of H_2O_2 on heterogeneous Pd catalysts with microporous materials having selective oxidation activity with H_2O_2 (TS-1 or other transition metal–containing zeolites) thus satisfies the requirement for a minimum of interference between the two types of active sites. The noble metal can even be supported directly on titanium silicalite itself by posttreatments[65] or incorporated into it during the hydrothermal synthesis.

Another important parameter is the activity of the oxidation catalyst, which should be high to prevent H_2O_2 buildup in the medium in order to minimize side reactions. Titanium silicalites TS-1 and TS-2 are ideal catalysts in this respect, owing to their fast kinetics in dilute aqueous H_2O_2 solutions using molecules that are not bulky (e.g., propene epoxidation).[68] Worth noting, however, is that in the epoxidation of alkenes on TS-1 (potentially a major industrially interesting reaction) methanol/water mixtures should be used as solvents, because methanol acts as a cocatalyst by promoting the formation of the active species and taking part in the reaction mechanism (Scheme 7.3).

Specific effort is thus necessary especially to minimize the mismatch between the optimal conditions required by the generation of the monoxygen donor (e.g., H_2O_2, organic peroxides) and the subsequent consumption of the monoxygen donor. In this connection, more efficient and economical *ex situ* production of H_2O_2 is still an important target for future research.

Although much less studied than the generation and use of H_2O_2 in solution, H_2O_2 or analogous species may also be generated on solid catalysts by gas phase reactions. The propene epoxidation on Au/TiO$_2$ using a H_2/O_2 feed[61] is the only significant example up to now, but it is reasonable to assume that this will be one of the major research directions in the near future, although the details of the mechanism of generation of H_2O_2 on a gold surface by a H_2/O_2 reaction must first be clarified. The primary reason that Au is used in this case instead of Pd (Pd/C is the preferred catalyst for H_2O_2 synthesis from H_2/O_2 in solution) is that the latter

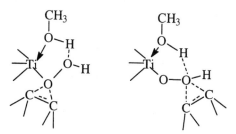

Scheme 7.3. Nature of the active complex in alkene epoxidation on TS-1. Adapted from Clerici and Ingallina.[59]

is active for the reaction only when it is in the reduced form. Unlike in the liquid phase, in the presence of gaseous O_2 oxidation of the Pd surface is fast, whereas Au is not oxidized, which illustrates the fact that notwithstanding the apparent analogy between liquid–solid and gas–solid generation of H_2O_2, the catalyst must have different properties, and thus care must be taken in extrapolating from one situation to another without having analyzed the problem in detail.

7.3.1.2.   Oxidation Reactions with in Situ Generated Hydrogen Peroxide

The *in situ* epoxidation of propene has various advantages[69]: (i) it avoids the necessity of shipping large quantities of H_2O_2 when it is not available on site, (ii) the process is much more flexible in terms of size downscaling, and (iii) there is no dependence on external suppliers. The epoxide and water are produced by the one pot reaction of an alkylated anthrahydroquinone with molecular oxygen and propene. The corresponding anthraquinone is subsequently hydrogenated to close the cycle of reactions. Hydrogen peroxide is produced in the reaction medium, then migrates into the channel system of TS-1 to form the active species at Ti sites (Scheme 7.3), which reacts with propene in the oxygen-transfer step. The organic carrier does not interfere with the catalytic process, because its cross section is larger than the diameter of the TS-1 pores. The yields of propene oxide, based on starting alkylanthraquinone, were 78 and 62%, respectively, when the ethyl- or a mixture of *tert*-butyl- and ethyl-substituted anthrahydroquinone was reacted at 30°C, with propene and air in the presence of TS-1.[63]

The hydroxylation of alkanes using a mixture of molecular hydrogen and oxygen can be carried out under mild conditions (25°C) on a 5-Å small-pore zeolite loaded with Pd(0)/Fe(II).[70] Hydrogen peroxide probably forms on Pd(0) and is consumed in the Fe(II)-promoted hydroxylation of the alkane. The turnover numbers do not exceed unity, with more than 95% of the H_2/O_2 mixture ending in water, but the system can be reasonably improved using a better catalyst than Fe(II) to catalyze the alkane hydroxylation, as well as using different types of zeolites. In fact, considerably better results are obtained on TS-1 modified by a noble metal (Pd or Pt).[71] The best turnover numbers referred to Ti are close to 10 or better using excess oxygen.[72] With *n*-hexane and *n*-octane as the substrate in methanol solutions, the maximum yield and selectivity do not exceed 21 and 57%, respectively,[63,71] and thus are significantly lower compared with the 86% yield (on H_2O_2) of the *ex situ* oxidation of *n*-hexane at 60°C.[72] The products are secondary alcohols and ketones, the latter formed by consecutive oxidation of the corresponding alcohols.

The hydroxylation of phenol can be done with an equimolar mixture of H_2/O_2 under acidic conditions on a TS-1/Pd catalyst, but a considerable drawback when acids such as HCl are used to control the pH is the halogenation of the aromatic nucleus.[73] At 90°C, with an HCl concentration of 0.5 M, H_2/O_2 pressure above 1 atm, and methanol as the solvent, the best yields were 33% on the starting phenol.[63] At room temperature Tatsumi et al.[74] reported for similar catalysts that phenol forms with a turnover number close to 13, without formation of halogenated derivatives. However, under these conditions the reaction rates are low and there is rapid deactivation from pore-plugging. An interesting new system for the hydroxylation of benzene to phenol is based on the use of precious metal catalysts supported on silica modified by various metal oxides, among which V_2O_5 is the most effective. Commercial-level productivities using a feed of H_2/O_2, a reaction temperature of about 45°C, and a $Pt/V_2O_5/SiO_2$ catalyst, have been reported, but more precise information is not available.[75]

The ammoximation of cyclohexanone using H_2O_2/TS-1 is also a reaction of considerable industrial interest, providing an environmentally cleaner route to cyclohexanonoxime.[76] Problems in developing a process of in situ generation and use of H_2O_2 include: (i) chemical incompatibility with the other reagents, (ii) complexity of separation procedures, and (iii) basicity of the ammoximation medium. The ex situ route integrated into the same process is thus a preferred solution for this reaction.

7.3.2. Generation of Active Oxygen Species by Ozone

New types of active oxygen species can also be generated from ozone at a gas–solid interface. Total oxidation of organics in solution by ozone is well known as is its role as an alternative oxidant in the oxidation of volatile organic compounds.[77,78] However, only very limited data are available on its use as a selective oxidant in the gas phase.[79] At low temperatures, ozone generates reactive peroxide species at the surface of the solid catalyst,[79] and these active oxygen species can be used for new low-temperature oxidation reactions, especially in the synthesis of fine chemicals.

This area of selective oxidation research is largely unexplored and there are not enough data available to weigh the potentialities and problems involved in the development of new types of reactions based on ozone. It should be noted, however, that the short lifetime of ozone and its surface species as well as their high reactivities limit potential applications to catalytic oxidation at very short contact times (milliseconds),[80] an area of possible industrial interest in several cases. Existing studies have focused on high temperatures and use of oxygen,[80] but lower temperatures and the use of ozone constitute a noteworthy alternative possibility.

7.3.3. Use of Nitrous Oxide as a Selective Oxidant

Nitrous oxide is another powerful precursor of active oxygen species that has been known for several years (e.g., its use for the selective dehydrogenation of alkanes).[81–83] However, it has recently been the subject of renewed interest owing to the discovery that α-type oxygen species can be generated by interaction with

ZSM5-type zeolites and used for the selective synthesis of phenol from benzene or methanol from methane.[84–92]

α-Oxygen forms by N_2O decomposition on ZSM-5-type zeolite due to the presence of either coordinatively unsaturated aluminum atoms of the zeolite (Lewis acid sites),[93] or Fe ions (impurities or added in small amounts).[84–86] At temperatures below about 300°C, N_2O decomposes on these catalysts evolving molecular nitrogen, whereas the oxygen remains stuck on the catalyst and desorbs irreversibly as molecular oxygen only at temperatures above about 300°C. The active oxygen species remaining on the catalyst is characterized by a high reactivity[94,95] and selectivity (up to 100%).

There are opposing ideas concerning the mechanism of this reaction. Some authors[89–91] consider Brønsted acid centers to be responsible for the reaction of benzene hydroxylation, whereas others[84–87] relate the catalytic activity to the ability of the zeolite to generate a special active species of surface oxygen (α-oxygen). Kinetic isotope studies[84] clearly showed that α-oxygen is the active species in benzene hydroxylation, but the question is the nature of the sites in zeolites responsible for the generation of this active species. ZSM-5 is active in this reaction, although the presence of Fe in small amounts promotes the activity. It is thus probable that the activity of Fe-free ZSM-5 is related to the presence of Fe impurities.

Iron ions entering α-sites lose their ability to activate molecular dioxygen. Their catalytic activity in the O_2 isotope exchange is four orders of magnitude lower than that of Fe atoms on the Fe_2O_3 surface.[85] In general, Fe can be present as at least four species in the ZSM-5 matrix: (i) isolated ions in tetrahedral positions of the crystal lattice, (ii) isolated ions at exchangeable cationic sites, (iii) small oxide nanocrystals inside the intracrystalline micropore space, and (iv) finely dispersed oxide phase on the outer surface of the zeolite crystals.

There are contrary indications concerning the exact role of these species in the reaction mechanism, but data suggest that the second and third types of species (isolated cation or small Fe complexes located inside the ZSM-5 micropore space) are the active ones. Panov et al.[96] suggested that α-sites are most probably associated with binuclear Fe complexes, rather than with isolated atoms, but this question should be studied more thoroughly.

7.3.3.1. Reactivity of a-Oxygen

α-Oxygen reacts with a variety of organic compounds at low temperatures, according to Scheme 7.4 for the specific case of phenol synthesis.[84] Benzene hydroxylation to phenol is the more thoroughly studied reaction, and it has recently been scaled up to pilot plant level as will be discussed later. Selectivities to phenol up to 95–99% with a productivity of around 3 mmol phenol per gram of catalyst per hour have been reported using high-silica ZSM-5 zeolites (SiO_2/Al_2O_3 about

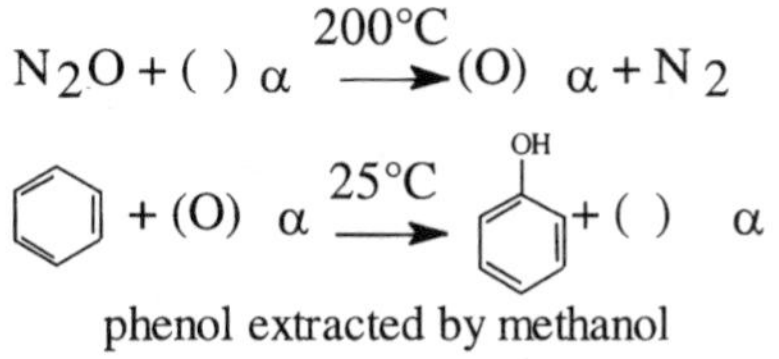

$$N_2O + (\ )\ \alpha \xrightarrow{200°C} (O)\ \alpha + N_2$$

$$\text{⬡} + (O)\ \alpha \xrightarrow{25°C} \text{⬡}^{OH} + (\ )\ \alpha$$

phenol extracted by methanol

Scheme 7.4. Reaction scheme of the formation of α-oxygen by N_2O decomposition on Fe-ZSM-5 and its reactivity with benzene at room temperature to give phenol.

100), with or without Fe (below 0.1%).[92,97,98] The catalyst deactivated with time, although new developments in catalyst preparation enable a considerable reduction in the rate of deactivation.[87] A model of the reaction of N_2O with Fe sites in ZSM-5 and the reaction of α-oxygen formed with benzene to give phenol is shown in Figure 7.4.

α-Oxygen promptly interacts with methane at room temperature. However, the product of this interaction is tightly bound to the surface and upon heating decomposes evolving CO in the gas phase.[84] However, it is possible to extract the product, in a similar way to what must be done in the stoichiometric reaction of α-oxygen with benzene at room temperature. Methanol is produced from methane with a yield of 90–100%. The reaction is very rapid and occurs readily even at $-30°C$.

Several other substrates react with α-oxygen, alkanes, alkenes, arenes, and heterocycles. In all cases, α-oxygen oxidation yields monoxygenated products. The hydroxylation of alkylaromatics is especially interesting because the competition between hydroxylation of aliphatic and aromatic carbons can be analyzed, as can the balance of electronic and steric factors influencing these transformations.

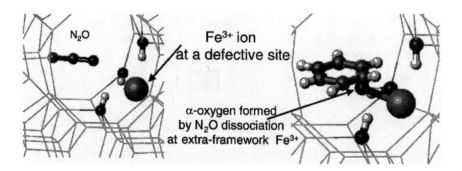

Figure 7.4. Model of the reaction of N_2O with Fe sites in Fe-ZSM-5 and subsequent reaction of the α-oxygen formed with benzene to give phenol.

Interaction of toluene with α-oxygen leads to products of both benzylic and aromatic hydroxylation (Scheme 7.5a), the latter being formed some 2.6 times more than the former. Para-cresol forms with about 50% selectivity with respect to the aromatic hydroxylation products, but meta-cresol forms in approximately the same quantity as the ortho-isomer. This is due to steric effects of the methyl group in the confined space of a zeolite channel. An increase in the size of the substituent, in fact, considerably decreases the hydroxylation of both ortho- and meta-positions (Scheme 7.5b).

Also worth noting is the fact that isopropylbenzene hydroxylation, unlike selective attack at the benzylic C–H bond in the case of radical dioxygen chemistry (cumene hydroperoxide synthesis), also leads to oxidation of primary carbon atoms of the isopropyl radical. This effect is also probably associated with shape-selectivity effects in the zeolite channels.

7.3.3.2. Use of Waste Nitrous Oxide Streams: Adipic Acid Production

Adipic acid is a large intermediate manufactured worldwide by the nitric acid oxidation of cyclohexanone and/or cyclohexanol, with yields up to 94%. However, approximately 1 mol of N_2O is produced per mol of adipic acid and thus a large outlet stream of N_2O with concentrations up to about 35% is available from this process. Furthermore, owing to the effect of N_2O has on global warming and ozone

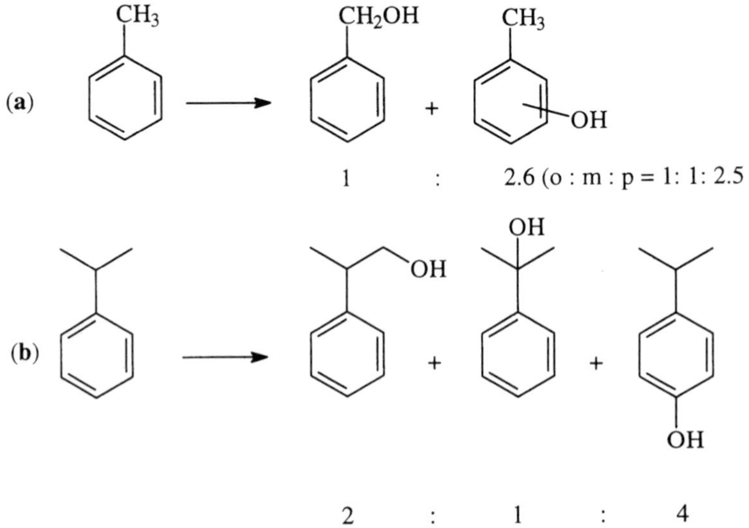

Scheme 7.5. Distribution of products in the hydroxylation of toluene (a) and isopropylbenzene (b) to corresponding alcohols by reaction with α-oxygen.[84]

depletion, a catalytic abatement unit is required to decompose N_2O into N_2 and O_2. Solutia (formerly Monsanto) has analyzed the possibility of using this waste N_2O stream to synthesize phenol from benzene and introduce it into the adipic acid process after hydrogenation of the phenol to cyclohexanone, in order to improve the efficiency of nitric acid use as well as to reduce the environmental impact of the process.[99] The overall adipic acid process scheme is shown in Scheme 7.6. Typical performances obtained in pilot plant experimentation (reaction temperature in the 400–450°C range, contact time 1–2 s) are: (i) selectivity of benzene to phenol 97–98%, (ii) selectivity of N_2O to phenol 85%, and (iii) productivity, 4 mmol phenol per gram of catalyst per hour.[87]

The selectivity of the catalyst for conversion of benzene to phenol is quite insensitive to the reaction temperature and so it is possible to use a simple adiabatic reactor, which considerably increases the economic advantages of the process. The high selectivity and easy separation of unreacted benzene by simple distillation also result in very little production of organic waste (< 2% of the phenol manufactured), thus limiting the environmental impact of the process.

It is worth noting that due to the high selectivity, cost estimates[87] have shown that the production of phenol may be competitive with the current commercial process (taking into consideration the worldwide surplus of the acetone coproduced in the classical cumene process of phenol synthesis), even when N_2O is synthesized for the purpose rather than being available as a cheap by-product as in adipic acid production. A key to success in this case could be the direct *in situ* generation of the oxidizing intermediate, which reduces the costs for two separate processes. However, it is necessary to develop selective and efficient catalysts for the *in situ* synthesis of N_2O from ammonia. The use of two ammonia molecules to generate one monoxygen donor molecule makes the synthesis of N_2O relatively expensive

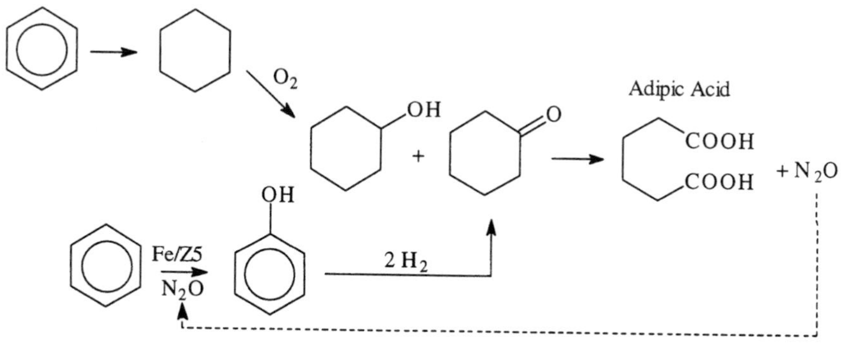

Scheme 7.6. Overall reaction scheme in a new adipic acid process from Solutia (formerly Monsanto) which includes reuse of the waste N_2O stream to hydroxylate benzene to phenol.[87]

and thus this new oxidation chemistry is recommended for the synthesis of high-value products (e.g., hydroxylated substituted aromatics)[91,97] rather than bulk products (e.g., methanol from methane) for which it cannot compete with conventional processes.

7.4. NOVEL REACTION MEDIUMS

Selective oxidation reactions can be broadly classified into liquid phase processes, either with homogeneous or heterogeneous catalysts, and gas phase processes, usually on solid surfaces. While the former reactions are usually characterized by very high selectivity, the latter type offers several advantages in terms of, e.g., ease of product separation, productivity, and eco-compatibility. Several past attempts to combine the characteristics of both and heterogenize homogeneous catalysts, have met with only limited success; however, an alternative possibility is to maintain the homogeneous chemistry, but have a thin supported liquid film, which considerably reduces problems of reactant diffusion and product and catalyst recovery. There are several technical problems and constraints in realizing this approach, but some examples will be discussed in the following section that can serve as guidelines for investigating its potential in detail.

A second novel reaction medium for oxidation reactions scarcely looked at up to now, which can offer several interesting advantages, is CO_2 under supercritical conditions. This is a completely open area of research, which requires investigation to understand both the possibilities and the limits of the technology.

7.4.1. Oxidation Reaction at Thin Supported Liquid Films

Supported liquid phase catalysis, i.e., homogeneous catalysts working in a thin liquid film supported over a solid material, which can either be inert or participate in the reaction mechanism, offers the potential advantages of combining typical features of both homogeneous and heterogeneous catalysis. There are only limited data in the literature concerning this possibility, which has never been studied specifically. In two relevant examples of industrial interest, several indications suggest that the true *in situ* catalyst operates in a thin supported liquid film: (i) heterogeneous Wacker-type catalysts for ethylene to acetaldehyde and 1-butene to butan-2-one oxidation, and (ii) Pd/Au-based supported catalysts for ethylene acetoxylation.

7.4.1.1. Heterogeneous Wacker-Type Catalysts

The Wacker reaction was first discovered by Smidt *et al.*,[100,101] who observed that ethylene is selectively oxidized to acetaldehyde using a $PdCl_2/CuCl_2$ catalyst in acetic acid solution. The oxidation of ethylene occurs in the presence of $PdCl_2$;

this reaction reduces the Pd^{2+} to Pd^0. The metal is easily oxidized via reaction with $CuCl_2$, and the cycle is completed by reoxidation of the CuCl by oxygen. At pressures of 100–150 psig and temperatures between 100–110°C, this process yields approximately 95–99% selectivities based on ethylene.

Smidt et al.[100] briefly mentioned that a Pd/Cu solution deposited on porous support particles is also active in the Wacker reaction, but the heterogeneous catalyst system was studied more in detail by Komoyama and Inoue.[102] The pore walls of porous alumina were coated with a thin film of a hydrochloric acid solution of $PdCl_2$ and $CuCl_2$. The system was active, but direct comparison with the liquid phase reaction was not possible. A major drawback was that the liquid thin film evaporates during the reaction. This problem may be minimized, but not solved, by saturating the ethylene–oxygen feed with water. However, the presence of additional water causes pore-filling, which subsequently results in catalyst deactivation.[103]

An alternative approach is to use a molten salt as the solvent for the supported liquid phase.[103] Molten salts are an interesting option here because they form mixtures over wide temperature ranges, and they also are capable of solubilizing materials such as water, metals, and oxides. Rao and Datta[103] used a $PdCl_2/CuCl_2$ + KCl/CuCl thin film (30–70 Å thick; KCl/CuCl was the molten salt) deposited on the walls of the porous silica support to selectively oxidize ethene to acetaldehyde. However, the activities obtained were not optimal.

A different approach to Wacker-type chemistry has been centered on the use of supported Pd/V_2O_5 or $Pd/H_{3+n}PV_nMo_{12-n}O_{40}$ types of catalysts.[104-112] The role of the solid component is to increase the rate of the redox cycle of palladium reoxidation ($Pd^0 \rightarrow Pd^{2+}$), as this is the slowest step. These catalysts have slightly lower to comparable activities and selectivities as compared with those of the analogous homogeneous reaction.

There are several distinct characteristics of these catalysts that point to the formation of a thin water layer over the surface and/or within micropores by capillary condensation:

1. The reaction is carried out at temperatures around 100°C and unlike all other known catalytic oxidation reactions, is negatively influenced by the reaction temperature.

2. Water is an essential reagent for the reaction, but the large quantities required (~ 20%) indicate an additional role. The water partial pressure determines selectivity, and a nonselective behavior is observed when only the water present is adsorbed on the catalyst, but in an amount large enough to sustain the reaction when it acts only as a reactant (Figure 7.5).

3. Water vapor equilibrium estimates indicate that under the conditions of best catalytic results, it is likely that a thin film of water is present on the surface or at least that micropores are filled with a water solution.

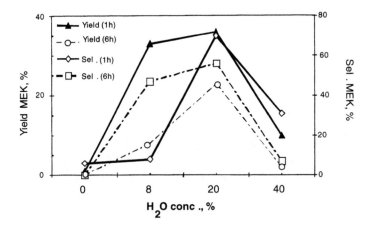

Figure 7.5. Effect of the water concentration in the feed in but-1-ene oxidation to butan-2-one over Pd/V$_2$O$_5$ on alumina catalysts.[110]

4. Under standard reaction conditions, in the case of heteropolyacids (HPA) as the reoxidizing agent for Pd, water adsorption experiments indicate that each Pd/HPA unit is surrounded by more than 20 water molecules and thus the reactants dissolve in this hydration water (pseudoliquid state). Removing this hydration water leads to much poorer catalytic behavior.

All these data thus indicate that a thin film of water forms over the surface of the catalyst during the catalytic reaction, although no direct evidence has been found. This film serves as the solvent, dissolving the anchored complex, which thus works as in the liquid phase, but with the advantage of being heterogenized so that gas phase reactants (hydrocarbon, oxygen) diffuse easily through the thin film. Figure 7.6 illustrates a model of the reaction over a thin liquid film in heterogeneous Wacker-type catalysts.

7.4.1.2. Ethene Acetoxylation to Vinyl Acetate

The vapor phase process for vinyl acetate monomer production involves the acetoxylation of gas phase ethene over a Pd/Au alloy catalyst supported on either silica or alumina. Alkali metal salts are added to promote the reaction. Zero valence palladium metal is necessary to selectively catalyze vinyl acetate formation and avoid acetaldehyde formation,[113] the latter deriving from Wacker-type chemistry catalyzed by Pd^{2+}. Other evidence, however, suggests that the reaction takes place with a Pd$^+$/OAc intermediate[114–116] or on a Pd(OAc)$_2$ acetate film on Pd crystallites.[117] In general, the mechanism can be outlined as follows: (i) ethene is dissociatively adsorbed on the palladium surface; (ii) oxygen is adsorbed on the

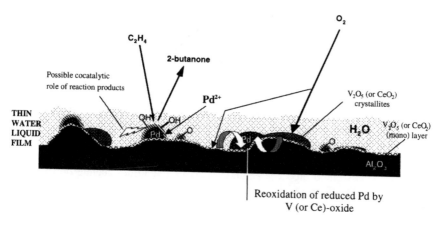

Figure 7.6. Model of the reaction over a thin liquid film in heterogeneous Wacker-type catalysts.

palladium; (iii) acetic acid is adsorbed on the catalyst, where the adsorbed oxygen then abstracts a hydrogen from the adsorbed acid; and (iv) the adsorbed ethylene and acetic acid reacts on the surface to give vinyl acetate, which ultimately desorbs. However, there is strong evidence as well that a liquid phase type of mechanism also occurs under reaction conditions, and, thus, supported liquid phase catalysis is thought to be a key factor in the vinyl acetate reaction.[118]

In heterogeneous Pd/Au-based catalysts for ethylene acetoxylation, the presence of potassium acetate (KOAc) is essential for the reaction and at typical reaction temperatures (around 150°C) in the presence of acetic acid, KOAc forms a dimeric species, probably present as a molten salt. Capillary condensation of acetic acid also occurs. Furthermore, under typical reaction conditions for industrial reactions, the adsorption of acetic acid and water on the catalyst is extensive,[119,120] even at high temperatures (at 400°C the amount of acetic acid adsorbed on the catalyst doped with potassium is around 0.8–1.0 mmol/g).[121] Under typical industrial reaction conditions, Crathorne et al.[121] indicated that the adsorbed acetic acid forms a film about three monolayers thick over the catalyst surface.

KOAc plays two very critical roles in enhancing the vinyl acetate reaction. First, its presence is essential to establish the supported liquid phase, in the absence of which the reaction rate for vinyl acetate production decreases and direct ethylene combustion occurs readily.[120,122] Potassium reacts with acetic acid, in fact, to give a dimeric form that has a melting point of 148°C and is thus present in the molten state under normal operating conditions. The second role of potassium is related to the solvation of the palladium complexes formed during reaction. The complexes, $KPd(OAc)_3$ and $K_2Pd(OAc)_4$, can dissolve in the $KH(OAc)_2$ solution to form the active centers of the catalyst. These complexes can either pyrolyze to form acetic acid or react with ethene to produce vinyl acetate. Both actions will reduce the palladium back to metal, to be reoxidized in the presence of oxygen and acetic acid. The thin liquid layer controls this reoxidation process and prevents nonselective ethene combustion.

A summary of the reaction scheme and the role of the thin liquid film is shown in Figure 7.7. It is thought that the molten salt melt acts as a protective coating on the surface of the catalyst hindering ethene combustion. The resulting isolated palladium sites are then available for preferential adsorption of ethene and oxygen, which leads to vinyl acetate formation rather than to CO_2. Some form of palladium acetate is present and can be dissolved in the liquid layer. The vinyl acetate is believed to be formed on these small clusters of palladium acetate or on dense palladium acetate surfaces.[122]

7.4.1.3. Other Cases

Some other selective oxidation reactions have been reported that probably involve supported liquid phases (molten salts, in particular), although specific studies are rare. The oxidation of naphthalene in the presence of p-xylene to form phthalic anhydride is one example. This system employs a $K_2S_2O_7/V_2O_5$ molten salt supported on silica.

Although it is not properly a selective process, there is ample evidence to indicate that the oxidation of sulfur dioxide to sulfur trioxide takes place with a catalyst (vanadium pentoxide promoted by potassium salts supported on silica)

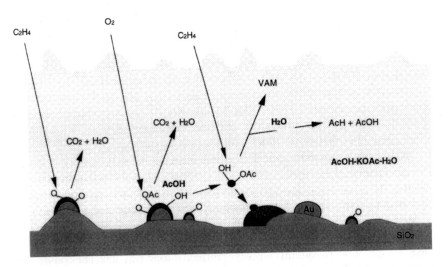

Figure 7.7. Model of the reaction in a thin supported liquid film in the case of a Pd/Au/KOAc/SiO$_2$ catalyst for ethene acetoxylation to vinyl acetate.[118]

containing a molten salt mixture on a support.[123] Under typical reaction conditions (> 420°C), a potassium pyrosulfate melt forms, which helps to distribute the vanadium oxide in the pores of the support. There also is evidence that the molten salt layer actually serves as a catalyst for this reaction.

Supported liquid films have been proposed for the gas phase oxidation of benzene to phenol.[124] In this system, Pd and $Cu_3(PO_4)$ are impregnated by a H_3PO_4 solution into silica. The evidence suggests that the phosphoric acid forms a liquid film on the surface, which helps to increase catalyst activity over catalysts prepared with more volatile solvents. As a result of the liquid film, the Cu ions become mobile within the liquid film, which enhances the opportunity for contact with immobilized Pd sites, thereby increasing activity.

7.4.2. Oxidation Reaction under Supercritical Conditions

Supercritical carbon dioxide or fluids have become increasingly important as potential processing and reaction media because they have liquidlike densities, vaporlike viscosities and diffusivities, and low surface tension, which make them effective as extraction and reaction solvents. Furthermore, conducting chemical reactions under supercritical conditions affords opportunities to manipulate the reaction environment (solvent properties) by verifying the pressure, enhance the solubilities of reactants and products, eliminate interphase transport limitations on reaction rates, and integrate reaction and separation units.[125,126]

Supercritical CO_2, in particular, has a number of advantages. It has a low critical temperature (31.1°C) and moderate critical pressure (1071 psia) and is also readily available, inexpensive, and nontoxic. Using supercritical CO_2 as a reaction medium in homogeneous catalysis may result in enhanced rates and increased selectivities, as well as ease of catalyst recovery and product purification.[127,128] A number of reactions have been shown to be advantageously conducted in super-critical CO_2,[128–130] because often in addition to the advantages cited above, chemical interaction between supercritical CO_2 and substrates and/or catalyst offer additional ways to tune activity and selectivity.[131] In particular, much study has been devoted to catalytic hydrogenation in supercritical CO_2,[132–133] especially for the enhanced hydrogen solubility in this medium and the possibility of avoiding interphase transport limitations.[132–134]

Very little attention has been focused on the possibility of using this unconventional reaction medium for selective oxidation reactions (especially for the production of fine chemicals), although the same kind of advantages may also be expected for this class of reactions.[135] In particular, there are clear advantages in the case of microporous materials containing transition metals. Worth noting is that supercritical CO_2 is an inert solvent for radical reactions and thus radical-type selective oxidation reactions can be carried out effectively in this reaction medium. Supported noble metals under mild reaction conditions are another class of potentially interesting catalysts for reactions of selective oxidation in supercritical CO_2, although to date they have been studied primarily for complete oxidation reactions after extraction of toxic wastes from water solutions by using supercritical CO_2.

In supercritical CO_2, not only O_2 but also O_3 and H_2O_2 have been used as the oxidant in order to operate under milder conditions.[136] As discussed in previous chapters and in Section 7.3, H_2O_2 is an ideal clean selective oxidizing agent, but except for the TS-1 catalyst, all other transition metal–containing microporous materials show considerable problems of stability during catalytic reactions in H_2O_2/water/solvent media due to the easy leaching of the transition metal.[4] With supercritical CO_2 it will probably be possible to prevent this leaching phenomenon, which is associated with hydrolysis of Si–O–M bonds, and have stable catalysts containing either different transition metals able to catalyze reactions not possible with Ti (e.g., oxidation of primary carbons) or containing Ti, but in larger-pore zeolites than TS-1. In addition, the solvent properties of supercritical CO_2 seem ideal for many selective oxidation reactions and can enhance substrate transport within the pore structure of the microporous material. This field of research thus appears to have considerable potential scientific and industrial interest.

Not only can CO_2 be used as a supercritical solvent, but other compounds and water itself can be used as well. In the supercritical water phase, the oxygen solubility is enhanced and many organic compounds become completely soluble. Under supercritical conditions, in fact, water acts as a fluid with a density between that of water vapor and liquid and exhibits gaslike diffusion rates along with high liquidlike collision rates so that the oxidation takes place in a homogeneous mixture with no mass-transfer limitations at the phase boundaries. Supercritical water oxidation has been developed as a technology for the complete destruction of hazardous organic species dissolved in water,[137–140] but at low temperatures, using a suitable catalyst, selective oxidation products can be synthesized. However, this possibility has not yet been investigated to any substantial degree.

Selective oxidation in supercritical fluids is thus a new open opportunity with many potential advantages, but intense research effort is necessary to determine possibilities and limitations of this new field.

7.5. CONCLUSIONS

The concepts discussed briefly above, some of which have been more thoroughly described in a special issue of the journal *Catalysis Today*,[141] show that there are new opportunities for research in selective oxidation, but a wider and more open approach must be taken rather than continuing to consolidate information on existing catalysts and reactions. Some possibilities have been discussed in this chapter, but they do not cover all possible topics systematically and thus should be considered only as stimuli for future research. The examples discussed are focused mainly on the problem of activating oxygen, clearly a central issue in the selective oxidation. However, it is emphasized that unconventional reaction media also offer several potential advantages.

At this juncture, it may be worthwhile to point out the importance of the design of new solid catalysts, a topic not discussed in this chapter, but which is of clear importance, as discussed extensively elsewhere in this volume. It should also be recalled that the concept discussed several times in this work that new reactor solutions (e.g., separate hydrocarbon and oxygen interaction with the catalyst) offer the possibility of operating with the catalyst under metastable conditions (e.g., in highly reduced states), which is not possible when the hydrocarbon and oxygen are cofed as in conventional reactor technologies. Thus not only are new catalytic materials needed, but old ones should be reanalyzed, because different catalyst characteristics are required when the reaction environment is changed.

The field of selective oxidation is by no means a mature area of research, and the development of the several new ideas and possibilities discussed here, is the challenge for future research.

REFERENCES

1. G. Bellussi, A. Carati, M.G. Clerici, G. Maddinelli, and R. Millini, *J. Catal.* **133**, 220 (1992).
2. G. Bellussi and M.S. Rigutto, in: *Advanced Zeolite Science and Applications* (J.C. Jansen, M. Stöcker, H.G. Karge, and J. Weitkamp, eds.), Studies in Surface Science and Catalysis Vol. 85, Elsevier Science: Amsterdam (1994), Ch. 6, p. 177.
3. R.A. Sheldon and J.K. Kochi, *Metal-Catalyzed Oxidations of Organic Compounds*, Academic Press: New York (1981).
4. R.A. Sheldon, in: *3rd World Congress on Oxidation Catalysis* (R.K. Grasselli, S.T. Oyama, A.M. Gaffney, and J.E. Lyon, eds.), Studies in Surface Science and Catalysis Vol. 110, Elsevier Science: Amsterdam (1997), p. 151.
5. K. Otsuka, in: *3rd World Congress on Oxidation Catalysis* (R.K. Grasselli, S.T. Oyama, A.M. Gaffney, and J.E. Lyon, eds.), Studies in Surface Science and Catalysis Vol. 110, Elsevier Science: Amsterdam (1997), p. 93.
6. K. Otsuka and I. Yamanaka, *Catal. Today* **41**, 311 (1998).
7. K. Otsuka, I. Yamanaka, and K. Hosokawa, *Nature* **345**, 697 (1990).
8. I. Yamanaka and K. Otsuka, *J. Chem. Soc. Faraday Trans.* **89**, 1791 (1993).
9. I. Yamanaka and K. Otsuka, *J. Chem. Soc. Faraday Trans.* **90**, 451 (1994).
10. K. Otsuka, S. Yokoyama, and A. Morikawa, *Chem. Lett.*, 319 (1985).
11. K. Otsuka, K. Suga, and I. Yamanaka, *Catal. Today* **6**, 587 (1990).
12. K. Otsuka, Y. Shimizu, and I. Yamanaka, *J. Chem. Soc. Chem. Commun.*, 1272 (1988).
13. C.G. Vayenas and R.D. Farr, *Science* **208**, 593 (1980).

14. C.G. Vayenas, S.I. Bebelis, and C.C. Kyriazis, *CHEMTECH* **7**, 422, (1991).

15. C.G. Vayenas, S. Bebelis, I.V. Yantekakis, and H.-G. Lintz, *Catal. Today* **11**, 303, (1992).

16. C. Pliangos, I.V. Yantekakis, and C.G. Vayenas, *J. Catal.* **154**, 124 (1995).

17. O.A. Marina, I.V. Yantekakis, C.G. Vayenas, A. Palermo, and R.M. Lambert, *J. Catal.* **166**, 218 (1997).

18. M. Stoukides, *Ind. Eng. Chem. Res.* **27**, 17451 (1988).

19. D. Eng and M. Stoukides, *Catal. Rev.-Sci. Eng.* **33**, 375 (1991).

20. M. Stoukides, *J. Appl. Electrochem.* **25**, 899 (1995).

21. B.C.H. Steeel, I. Kelly, H. Middleton, and R. Rudkin, *Solid State Ionics* **28–30**, 1547 (1988).

22. T. Hayakawa, T. Tsunoda, H. Orita, T. Kameyama, H. Takahashi, K. Takehira, and K. Fukuda, *Chem. Commun.*, 961 (1986).

23. T. Hayakawa, K. Kito, A. York, T. Tsunoda, K. Suzuki, M. Shimizu, and K. Takehira, *J. Electrochem. Soc.* **144**, 1 (1997).

24. R.W. Spillman, R.M. Spothits, and J.T. Lundguist, Jr., *CHEMTECH* **3**, 176 (1984).

25. G.A. Stafford, *Electrochem. Acta* **32**, 1137 (1987).

26. S.H. Langer and S. Yurchak, *J. Electrochem. Soc.* **116**, 1128 (1969).

27. S.H. Langer and J.A. Colucci-Rios, *CHEMTECH* **4**, 226 (1985).

28. C. Walling and R.A. Johnson, *J. Am. Chem. Soc.* **79**, 363 (1975)

29. C. Walling, *Acc. Chem. Res.* **8**, 125 (1975).

30. K. Otsuka and I. Yamanaka, *Electrochim. Acta* **35**, 319 (1990).

31. K. Otsuka, M. Kunieda, and H. Yamagata, *J. Electrochem. Soc.* **139**, 2381 (1992).

32. K. Otsuka, M. Kunieda, and I. Yamanaka, in: *New Developments in Selective Oxidation II* (V. Corberan and S. Vic Bellon, eds.), Studies in Surface Science and Catalysis Vol. 82, Elsevier Science: Amsterdam (1994), p. 703.

33. K. Otsuka, K. Hosokawa, I. Yamanaka, Y. Wada, and A. Morikawa, *Electrochim. Acta* **34**, 1485 (1989).

34. D.T. Sawyer, *Oxygen Chemistry*, Oxford University Press, New York (1991).

35. I. Yamanaka and K. Otsuka, *J. Electrochem. Soc.* **138**, 1033 (1991).

36. K. Otsuka and I. Yamanaka, *Catal. Rev. Sci. Techn.* **1**, 231 (1991).

37. Q. Zhang and K. Otsuka, *Chem. Lett.*, 363 (1997).

38. J. Smidt, W. Hafner, R. Jira, J. Sedlmerier, R. Seiber, and H. Kojer, *Angew. Chem.* **71**, 176 (1959).

39. E.W. Sterm, *Catal. Rev.* **1**, 105 (1968).

40. K. Otsuka, Y. Shimizu, and I. Yamanaka, *J. Electrochem. Soc.* **137**, 2076 (1990).

41. K. Otsuka and I. Yamanaka, *Hyoumen (Surface)* **27**, 473 (1989).

42. K. Otsuka, Y. Shimizu, I. Yamanaka, and T. Komatsu, *Catal. Lett.* **3**, 365 (1989).

43. T. Seiyama, N. Yamazoe, J. Hojo, and M. Hayakawa, *J. Catal.* **24**, 173 (1972).

44. F. Blatter and H. Frei, *J. Am. Chem. Soc.* **115**, 7501 (1993).

45. F. Blatter and H. Frei, *J. Am. Chem. Soc.* **116**, 1812 (1994).

46. H. Frei, F. Blatter, and H. Sun, *CHEMTECH* **26**, 24 (1996).
47. H. Frei and F. Blatter, *Catal. Today* **41**, 297 (1998).
48. H. Sun, F. Blatter, and H. Frei, in: *Heterogeneous Hydrocarbon Oxidation* (S. T. Oyama and B. K. Warren, eds.), ACS Symposium Series No. 638, American Chemical Society: Washington, D.C. (1996), p. 409.
49. F. Blatter, F. Moreau, and H. Frei, *J. Phys. Chem.* **98**, 13403 (1994).
50. S. Vasenkov and H. Frei, *J. Phys. Chem. B* **101**, 4539 (1997).
51. H. Sun, F. Blatter, and H. Frei, *J. Am. Chem. Soc.* **118**, 6873 (1996).
52. F. Blatter, H. Sun, and H. Frei, *Catal. Lett.* **35**, 1 (1995).
53. H. Sun, F. Blatter, and H. Frei, *J. Am. Chem. Soc.* **116**, 7951 (1994).
54. H. Sun, F. Blatter, and H. Frei, *Catal. Lett.* **44**, 247 (1997).
55. M. Iwasaki, K. Toriyama, and K. Nunome, *J. Am. Chem. Soc.* **103**, 3591 (1981).
56. J. Datka, *J. Chem. Soc. Faraday Trans. 1* **77**, 1309 (1981).
57. S.M. Auerbach, N.J. Henson, A.K. Cheetham, and H.I. Metiu, *J. Phys. Chem.* **99**, 10600 (1995).
58. F. van Laar, D.E. de Vos, D.L. Vanoppen, and P.A. Jacobs, *Proceedings 12th International Zeolite Conference*, Baltimore, July 1998, P-241.
59. M.G. Clerici and P. Ingallina, *J. Catal.* **140**, 71 (1993).
60. T. Miyake, M. Hamada, Y. Sasaki, and M. Oguri, *Appl. Catal. A* **131**, 33 (1995).
61. M. Haruta, in: *3rd World Congress on Oxidation Catalysis* (R.K. Grasselli, S.T. Oyama, A.M. Gaffney, and J.E. Lyon, eds.), Studies in Surface Science and Catalysis Vol. 110, Elsevier Science: Amsterdam (1997), p. 123.
62. J.M. Aubry and V. Nardello, in: *3rd World Congress on Oxidation Catalysis* (R.K. Grasselli, S.T. Oyama, A.M. Gaffney, and J.E. Lyon, eds.), Studies in Surface Science and Catalysis Vol. 110, Elsevier Science: Amsterdam (1997), p. 883.
63. M.G. Clerici and P. Ingallina, *Catal. Today* **41**, 351 (1998).
64. L.W. Gosser, *U.S. Patent* 4,681,751 (1987).
65. W. Laufer, R. Meiers, and W.F. Hölderich, *Proceedings 12th International Zeolite Conference*, Baltimore, July 1998, B-8.
66. I. Tabushi and A. Yazaki, *J. Am. Chem. Soc.* **103**, 7371 (1981).
67. B. Notari, *Adv. Catal.* **41**, 253 (1996).
68. M.G. Clerici, G. Bellussi, and U. Romano, *J. Catal.* **129**, 159 (1991).
69. M.G. Clerici and P. Ingallina, in: *Green Chemistry—Designing Chemistry for the Environment* (P.T. Anastas and T.C. Williamson, eds.), ACS Symposium Series 626, American Chemical Society: Washington (1996), p. 59.
70. N. Herron and C.A. Tolman, *J. Am. Chem. Soc.* **109**, 2837 (1987).
71. M.G. Clerici and G. Bellussi, *US Patent* 5,235,111 (1993).
72. M.G. Clerici, *Appl. Catal.* **68**, 249 (1991).
73. M.G. Clerici and P. Ingallina, *It. Patent* 22478 (1990).
74. T. Tatsumi, K. Yuasa, and H. Tominaga, *J. Chem. Soc., Chem. Commun.*, 1446 (1992).
75. T. Miyake, M. Hamada, Y. Sasaki, and M. Oguri, *Appl. Catal. A* **131**, 33 (1995).

76. P. Roffia, G. Leofanti, A. Cesana, M. Mantegazza, M. Padovan, G. Petrini, S. Tonti, and P. Gervasutti, in: *New Developments in Selective Oxidations* (G. Centi and F. Trifirò, eds.), Studies in Surface Science and Catalysis Vol. 55, Elsevier Science: Amsterdam (1990), p. 43.

77. A. Gervasini, G.C. Vezzoli, and V. Ragaini, *Catal. Today* **29**, 449 (1996).

78. K. Hauffe and Y. Ishikawa, *Chem. Ing. Techn.* **5**, 1035 (1974).

79. W. Li and S.T. Oyama, in: *3rd World Congress on Oxidation Catalysis* (R.K. Grasselli, S.T. Oyama, A.M. Gaffney, and J.E. Lyon, eds.), Studies in Surface Science and Catalysis Vol. 110, Elsevier Science: Amsterdam (1997), p. 873.

80. L.D. Schmidt and C.T. Goralski, in: *3rd World Congress on Oxidation Catalysis* (R.K. Grasselli, S.T. Oyama, A.M. Gaffney, and J.E. Lyon, eds.), Studies in Surface Science and Catalysis Vol. 110, Elsevier Science: Amsterdam (1997), p. 491.

81. L.W. Zatorki, G. Centi, J. Lopez Nieto, F. Trifirò, G. Bellussi, and V. Fattore, in: *Zeolites: Facts, Figures, Future* (Jacobs, P.A., van Santen R.A., eds.), Studies in Surface Science and Catalysis Vol. 49, Elsevier Science: Amsterdam (1989), p. 1243.

82. M. Che and A.J. Tench, *Adv. Catal.* **31**, 77 (1982).

83. M. Che and A.J. Tench, *Adv. Catal.* **32**, 1 (1983).

84. G.I. Panov, A.K. Uriarte, M.A. Rodkin, and V.I. Sobolev, *Catal. Today* **41**, 365 (1998).

85. V. Sobolev, G. Panov, A. Kharitonov, V. Romannikov, A. Volodin, and K. Ione, *J. Catal.* **139**, 435 (1993).

86. G. Panov, A. Kharitonov, and V. Sobolev, *Appl. Catal.* **98**, 1 (1993).

87. A.K. Uriarte, M.A. Rodkin, M.J. Gross, A.S. Kharitonov, and G. Panov, in: *3rd World Congress on Oxidation Catalysis* (R.K. Grasselli, S.T. Oyama, A.M. Gaffney, and J.E. Lyon, eds.), Studies in Surface Science and Catalysis Vol. 110, Elsevier Science: Amsterdam (1997), p. 857.

88. M. Häfele, A. Reitzmann, E. Klemm, and G. Emig, in: *3rd World Congress on Oxidation Catalysis* (R.K. Grasselli, S.T. Oyama, A.M. Gaffney, and J.E. Lyon, eds.), Studies in Surface Science and Catalysis Vol. 110, Elsevier Science: Amsterdam (1997), p. 847.

89. R. Burch and C. Howitt, *Appl. Catal. A* **103**, 135 (1993).

90. E. Suzuki, K. Nakashiro, and Y. Ono, *Chem. Lett.* **6**, 953 (1988).

91. M. Gubelmann, P. Tirel, and J. Popa, *9th International Zeolite Conference*, Montreal (July 1992). R-61.

92. M. Gubelmann, J. Popa, and P. Tirel, *EP Patent* 341,165 (1989).

93. V.I. Zholobenko, I.N. Senchenya, L.M. Kustov, and V.B. Kazansky, *Kinet. Katal.* **32**, 151 (1991).

94. A.S. Kharitonov, V.I. Sobolev, and G.I. Panov, *Kinet. Katal.* **30**, 1512 (1989).

95. G. Panov, V. Sobolev, and A. Kharitonov, *J. Mol. Catal.* **61**, 85 (1990).

96. G.I. Panov, V.I. Sobolev, K.A. Dubkov, V.N. Parmon, N.S. Ovanesyan, A.E. Shilov, and A.A. Shteinman, *React. Kinet. Catal. Lett.* **61**, 251 (1997).

97. M. Gubelmann, J. Popa, and P. Tirel, *EP Patent* 406,050 (1991).
98. A.S. Kharitonov, G.I. Panov, K.G. Ione, V.N. Romannikov, G.A. Sheveleva, L.A. Vostrikova, and V.I. Sobolev, *US Patent* 5110,995 (1992).
99. M. McCoy, *Chem. Market Reporter* **250**, 1 (1996); *Chem. Week*, Jan. 1/8, 11 (1997).
100. J. Smidt, W. Hafner, R. Jira, J. Sedlmeir, R. Sieber, R. Ruttinger, and H. Kojer, *Angew. Chem.* **71**, 176 (1959).
101. J. Smidt, W. Hafner, R. Jira, R. Sieber, J. Sedlmeir, and A. Sab, *Angew. Chemie* **74**, 92 (1962).
102. H. Komoyama and H. Inoue, *J. Chem. Eng. Jpn.* **8**, 310 (1975).
103. V. Rao and R. Datta, *J. Catal.* **114**, 377 (1988).
104. A.B. Evnin, J.A. Rabo, and P.H. Kasai, *J. Catal.* **30**, 109 (1973).
105. N.W. Cant and W.K. Hall, *J. Catal.* **16**, 220 (1970).
106. L. Forni and G. Gilardi, *J. Catal.* **41**, 338 (1976).
107. E. van der Heide, M. de Wind, A. Gerritsen, and J.J.F. Sholten, in: *Proceedings 9th International Congress on Catalysis*, Vol. IV (M.J. Phillips and M. Ternan, eds.), Chemical Institute of Canada: Ottawa (1988), p. 1648.
108. A.W. Stobbe-Kreemers, M. Soede, J.W. Veeman, and J.J.F. Sholten, in: *Proceedings 10th International Congress on Catalysis*, Vol. C (L. Guczi, F. Solymosi, and P. Tetenyi, eds.), Akademie Kiado: Budapest (1993), p. 1971.
109. A.W. Stobbe-Kreemers, G. van der Lans, M. Makkee, and J.J.F. Scholten, *J. Catal.* **154**, 187 (1995).
110. G. Centi, M. Malaguti, and G. Stella, in: *New Developments in Selective Oxidation II* (V. Cortes Corberan and S. Vic Bellon, eds.), Studies in Surface Science and Catalysis Vol. 82, Elsevier Science: Amsterdam (1994), p. 461.
111. G. Centi and G. Stella, in: *Catalysis of Organic Reaction* (Chemical Industries Series), (M.G. Scaros and M.L. Prunier, eds.), Markel Dekker: New York (1995), p. 319.
112. G. Centi, S. Perathoner, and G. Stella, in: *Catalyst Deactivation 1994* (B. Delmon and G.F. Fromant, eds.), Studies in Surface Science and Catalysis Vol. 88, Elsevier Science: Amsterdam (1994), p. 393.
113. M.I. Vargaftik, V.P Zagorodnikov, and I.I. Moiseev, *Kinet. Katal.* **22**, 951 (1981).
114. S. Nakamura and T. Yasui, *J. Catal.* **17**, 366 (1970).
115. S. Nakamura and T. Yasui, *J. Catal.* **23**, 315 (1971).
116. H. Debellefontaine and J. Besombes-Vailhe, *J. Chim. Phys.* **75**, 801 (1978).
117. S.M. Augustine and J.P. Blitz, *J. Catal.* **142**, 312 (1993).
118. C.R. Reilly and J.J. Lerou, *Catal. Today* **43**, 433 (1988).
119. B. Samamos, P. Boutry, and R. Montarnal, *J. Catal.* **23**, 19 (1971).
120. S. Tamura and T. Yasui, *Shokubail*, 21 (1979).
121. E.A. Crathorne, D. MacGowan, S.R. Morris, and A.P. Rawlinson, *J. Catal.* **149**, 254 (1994).
122. W.D. Provine, P.L. Mills, and J.J. Lerou, in: *11th International Congress of Catalysis 40th Anniversary*, Baltimore (J.W. Hightower, W.N. Delgass, E. Iglesia, and A.T.

Bell, eds.), Studies in Surface Science and Catalysis Vol. 101A, Elsevier Science: Amsterdam (1996), p.191.

123. C.N. Kenney, *Catal. Rev.-Sci. Eng.* **11**, 197 (1975).

124. K. Sasaki, T. Kitano, T. Nakai, M. Mori, S. Ito, M. Nitta, and K. Takehira, in: *New Developments in Selective Oxidation II* (V. Corberan and S. Vic Bellon, eds.), Studies in Surface Science and Catalysis Vol. 82, Elsevier Science: Amsterdam (1994), p. 451.

125. M.T. Reetz, W. Knen, and T. Strack, *Chimia* **47**, 493 (1993).

126. J.M. DeSimone, Z. Guan, and C.S. Elsbernd, *Science* **257**, 945 (1992).

127. B. Tumas, G. Brown, S. Buelow, D. Morita, D. Pesiri, and T. Walker, in: *Proceedings 1997 Annual Meeting AIChE* (Los Angeles, November 1997), p. 270j.

128. G. Kaupp, *Angew. Chem. Int. Ed. Engl.* **33**, 1452 (1994).

129. B. Subramaniam and M.A. McHugh, *Ind. Eng. Chem. Proc. Des. Dev.* **25**, 1 (1986).

130. P.E. Savage, S. Gopalan, T.I. Mizan, C.J. Martino, and E. Brock, *AIChE J.* **41**, 1723 (1995).

131. A. Fürstner, D. Kock, K. Langemann, W. Leitner, and C. Six, *Angew. Chem. Int. Ed. Engl.* **36**, 2466 (1997).

132. A. Bertucco, P. Canu, L. Devetta, and A.G. Zwahlen, *Ind. Eng. Chem. Res.* **36**, 2626 (1997).

133. S. Kainz, D. Koch, W. Baumann, and W. Leitner, *Angew. Chem. Int. Ed. Engl.* **36**, 1629 (1997).

134. P.L. Mills, P.A. Ramachandran, and R.V. Chaudari, *Rev. Chem. Eng.* **8**, 1 (1992).

135. A. Baiker, *Chem. Rev.* **99**, 453 (1999).

136. W.M. Nelson and I.K. Puri, *Ind. Eng. Chem. Res.* **36**, 3446 (1997).

137. M. Krajnc and J. Levec, *Ind. Eng. Chem. Res.* **36**, 3439 (1997).

138. M. Krajnc and J. Levec, *Appl. Catal. B* **3**, L101 (1994).

139. M. Krajnc and J. Levec, *Appl. Catal. B* **13**, 93 (1997).

140. E. Croiset and S.F. Rice, *Ind. Eng. Chem. Res.* **37**, 1755 (1998).

141. G. Centi and M. Misono (eds.), New Concepts in Selective Oxidation by Heterogeneous Catalysts, *Catal. Today* (Special issue) **41** (1998).

NEW ASPECTS OF THE MECHANISMS OF SELECTIVE OXIDATION AND STRUCTURE/ACTIVITY RELATIONSHIPS

8.1. INTRODUCTION

8.1.1. Outline and Scope of this Chapter

Obtaining a coherent and molecular-level understanding of the surface reactivity of solid catalysts active and selective in oxidation reactions is central to the research in this field, as it is a key factor for the design of new and/or improved catalysts and better tuning of their surface reactivity.

The detailed determination of the surface mechanism in selective oxidation reactions in several cases is still a challenge for the future rather than a real achievement of present research. Even though identification of the surface mechanism is usually far less precise than that achieved, for instance, in the study of the reaction mechanisms in solution by homogeneous catalysis, analysis of the surface reactivity, pathways of transformation, nature of the active sites and structure, and activity and selectivity relationships usually provides helpful indications concerning the critical factors determining the catalytic behavior and possible ways to improve it. The tendencies and directions of the research on this aspect of selective oxidation will be discussed here in order to analyze new possibilities suggested by these basic studies for the control of the catalytic behavior.

Each specific catalyst and catalytic transformation can be viewed as an individual case in terms of the mechanism of surface transformation, but a systematic discussion of the published data on the proposed mechanisms of selective oxidation is beyond the scope of this book. It should be noted, however, that the basic surface chemistry of catalytic transformation of hydrocarbons over solid catalysts is well known[1,2] so a detailed analyses of specific cases provide useful guidelines for extrapolating from the observations on a particular reaction/catalyst. However, the fact that the dominant effect governing the surface reactivity can be different depending on the nature of the hydrocarbon or solid catalyst must be taken into account.

The discussion here is limited to the selective oxidation of hydrocarbons over mixed oxides, as most of the reactions of industrial interest in the field are of this type. Following the *concept-by-example* approach used in this book, the analysis of the mechanisms of surface transformation of light hydrocarbons is used as a representative example to discuss new directions of research aimed at understanding the surface catalytic chemistry of selective oxidation reactions. Specifically, the oxidation or ammoxidation of C3–C5 alkanes and alkenes is analyzed critically, but without reviewing current knowledge on their reaction mechanism, which has recently been discussed in detail elsewhere.[1,2] The analysis is directed instead toward showing how the accepted ideas about the reaction mechanism of selective oxidation, based mainly on research done in the last two to three decades on the selective transformation of alkenes on mixed oxides, do not always fully describe the complex catalytic chemistry during these reactions. It is suggested that a new more complex model of catalytic chemistry (called here "living active surface") is necessary. This new model is presently in the early stages of development and is thus largely incomplete, but it can serve as the basis for a new perspective on catalytic chemistry at oxide surfaces.

This chapter starts with the case of propene (amm)oxidation on bismuth/molybdenum oxides as an example of the state of the art in the mechanisms of selective oxidation. The discussion is focused on the hypotheses and implicit assumptions that have been made to derive the proposed mechanism. The intent is not to comment on the reaction itself, but rather to enunciate the general "philosophy" and conclusions of the approach to see whether they constitute truly basic rules or should be considered special cases of more general phenomena. The discussion of this mode of approach and its limitations allows the development of some general guidelines for future research on the fundamental aspects of selective transformation at oxide surfaces.

The second part of this chapter describes some examples that demonstrate the importance of the new approach to studies on the surface reactivity of oxides outlined in the first part of the chapter.

8.1.2. The Established Approach to Modeling Reaction Mechanisms at Oxide Surfaces

Any brief summary of a complex topic such as that of the surface mechanism in selective oxidation reactions is usually vulnerable to criticism, but there are some common, widely accepted ideas that form the bases of the usual approaches to this topic[1,2]:

1. Selective oxidation reactions proceed through the so-called Mars–van Krevelen or redox-type mechanism.[3] The active site oxidizes the reactant and is then reoxidized by gas phase O_2 in a separate step (Scheme 8.1).
2. The reaction proceeds entirely at a single site (or cluster/ensemble of sites) (Scheme 8.2, inset). The reaction is usually schematized as a sequence of consecutive ("elementary") steps. An example of this stepwise mechanism can be a first step of abstraction of a H atom from a hydrocarbon, followed by the formation of an adsorbed complex to which a lattice oxygen atom is added, etc. (Scheme 8.2).
3. A rate-determining step controls the overall rate of reaction. Other steps [including product(s) desorption] are significantly faster and do not influence the rate-determining step.

Although the Mars–van Krevelen type of reaction mechanism is a foundation of selective oxidation catalysis, its uncritical application often fails to lead to a real understanding of the surface chemistry of oxidation catalysts. Similarly, it is useful to model the catalytic transformation in terms of the sequence of individual elementary steps but that does not explain some distinct particularities of the mechanisms of transformation over solid catalysts with respect to the transformation at isolated metal complexes in solution. An example is the possibility of a concerted (not sequential) transformation involving a transfer from substrate to surface (or vice versa) of several electrons, H and O atoms, etc. *at the same time.* Although the problem is usually not considered, the concept of a concerted mechanism opens up other aspects for discussion, such as the assumption of a rate-determining step for other than stepwise mechanisms.

It is worth noting that it is possible to carry out much more complex oxidation reactions over solid catalysts than in solution. For example, *n*-butane oxidation to

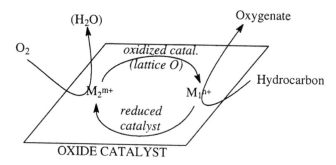

Scheme 8.1. The Mars–van Krevelen reaction scheme for selective oxidation.[3]

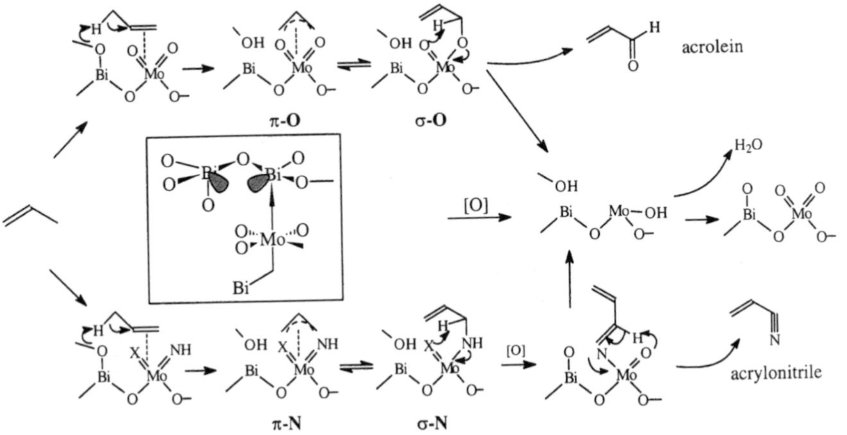

Scheme 8.2. Reaction mechanism of the selective oxidation and ammoxidation of propene over bismuth/molybdate catalysts.[12] In the lower part of the scheme (ammoxidation pathway), the symbol X indicates an imido (NH) or oxo (O) group, depending on the concentration in the feed. The inset shows a scheme of the catalytically active site of bismuth/molybdate catalysts.

maleic anhydride (a commercial process in a single step; see Chapter 4) is a 14-electron oxidation that involves the abstraction of 8 hydrogen atoms, the insertion of 3 oxygen atoms into the substrate, and use of 4 other lattice oxygens of the catalyst to form 4 water molecules from the 8 abstracted hydrogen atoms. The whole reaction proceeds on the adsorbed phase without formation of detectable intermediates in the gas phase and is much more complex than reactions in solution, thus suggesting the possibility that the reaction mechanism may also be significantly different than a stepwise single-site mode, as in solution.

In addition to these background observations, it also must be pointed out that recent preliminary evidence indicates that the reactivity of solid surfaces may be substantially different from the intrinsic reactivity of the "clean" surface without adsorbed species. Similarly, there are indications pointing to an active catalytic role of mobile surface adspecies, so an isolated single site cannot be considered as the only point of reaction.

In other words, recent developments in selective oxidation underscore the necessity of going from a localized concept of catalysis at specific active sites to a broader one in which the "collective" properties of the solid determine the catalytic behavior.[4] This is discussed in the next section, which is aimed at clarifying how the formulation of more general theories or explanations of the catalysis of selective oxidation requires a different, more general, approach than that of active (ensemble) sites.[5-9]

In addition to the common characteristics of the selective reaction mechanisms listed above, two further widely accepted ideas may be cited on the general topic of the surface reaction mechanisms of selective oxidation, although directed primarily at elucidating the reasons for the selective behavior and type of surface chemistry of different oxides:

1. *The combination of acido-base and redox properties of the oxide surface determines the mechanism of transformation.*[2,10] An example is shown in Scheme 8.3. On a basic oxide, unlike an acidic oxide where a carbocation may form by proton transfer, the alkene (π-bonded to Lewis acid sites) is susceptible to attack on the α-hydrogen from the neighboring lattice oxygen. The allyl species that forms can be side-on or end-on bonded, depending on the nature of the metal. The latter is an intermediate to a heavier product, whereas the former may undergo nucleophilic attack to give rise to a ketone or aldehyde, depending on the stereochemistry of oxygen attack, a function of the nature of the adsorbed complex. The alkene may also interact directly with surface acid Brønsted sites, which add to the π-bond system to form an alkoxide intermediate that, by further hydrogen abstraction

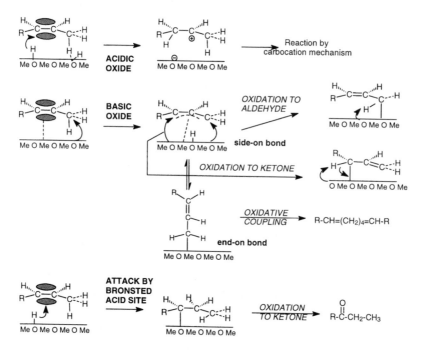

Scheme 8.3. Schematic representation of different modalities of alkene transformation depending on the acido-base and redox properties of the oxide.

transforms to a ketone, but different from those obtained by nucleophilic oxygen attack. These surface intermediate species can be in equilibrium or not, depending on the characteristics of the oxide.

2. *The nature of the oxygen species on the catalyst determines the kind of selectivity.*[7] In the process of oxygen incorporation into the oxide structure (Scheme 8.4a) various types of electrophilic activated oxygen species form before being incorporated as structural ("lattice") oxygen of the oxide. The latter has a nucleophilic character and gives rise to a different type of attack on an adsorbed alkene (Scheme 8.4b) and thus different types of products. In the process of incorporation of catalyst structural oxygen into the organic molecule, a point defect forms (Scheme 8.4c), which can be compensated for by a change in the linkage of the coordination polyhedra. If the latter process is not a rapid one, and the replenishment of surface oxygen vacancies through oxygen bulk diffusion is also slow, the population of surface electrophilic oxygen species increases with a lowering of the selectivity to partial oxidation products (Scheme 8.4b). The selectivity is thus a function of the surface geometry of active sites and the redox properties of the catalyst.

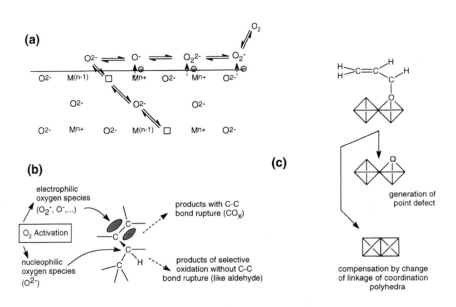

Scheme 8.4. (a) Scheme of oxygen incorporation in the oxide structure from the gas phase. (b) Model of the different types of attack on alkenes of electrophilic and nucleophilic oxygen species. (c) Model of the generation of point defects by incorporation of lattice oxygen in the organic molecule and compensation for defects by a change of linkage of coordination polyhedra.[7,27]

8.2. ACTIVE SITES OR "LIVING ACTIVE SURFACE"?

There is a basic central question in analyzing and discussing the state of the art of the mechanisms of selective oxidation of hydrocarbons: can the reaction mechanism at a solid surface be viewed as the equivalent of that occurring at metal complexes in solution, i.e., does the reaction occur at a single site or at a cluster of sites ("inorganic ensemble" of metallic oxide atoms[5,6]) in a sequence of elementary consecutive steps, or are more complex phenomena present (e.g., a concerted multistep mechanism, the movement of surface adspecies as a part of the mechanism of transformation, long-range effects on the activity of an active site, etc.) indicating that the whole solid surface (including the chemisorbed species) determines the catalytic behavior?

In other words, for heterogeneous catalysis it is correct to invoke the concept of *localized catalysis at a single site* (the equivalent of what occurs in solution at metal complexes) or do the collective properties of the solid, surface, and chemisorbed species determine the surface reactivity (a model of *living active surface* instead of catalysis at active sites)?

Most of the studies on the surface reaction mechanisms of selective oxidation implicitly assume the localized approach. However, for complex surface catalytic transformations such as C3–C5 alkane (amm)oxidation, considering the whole catalyst surface as responsible for the catalytic behavior gives a more representative picture of the real dynamics during the catalytic reaction. Obtaining a better understanding of this process is the challenge for future studies in this area and can be a great opportunity for a breakthrough in the understanding of the catalytic chemistry at oxide surfaces and for a new approach to the design of improved solid catalysts.

8.2.1. The Mechanism of Propene (Amm)oxidation

A very extensively investigated system from the mechanistic point of view is the selective oxidation or ammoxidation of propene on bismuth/molybdate catalysts.[11–25] The reaction involves a complex multistep transformation and is basically the same irrespective of the presence or absence of ammonia, apart from the transformation of molybdenum metal-dioxo species ($O{=}Mo{=}O$) to metal-imido species ($X{=}Mo{=}NH$, where $X{=}O$ or NH—see Scheme 8.2). The rate-determining step is the abstraction of an α-methyl hydrogen from a lattice oxygen linked to a bismuth ion to form a π-allyl intermediate coordinated to a molybdenum ion. The metal-oxo or metal-imido group then attacks the allyl intermediate forming a σ-bonded oxygen- or nitrogen-allyl species, which is in a rapid equilibrium with the π-bonded species. The σ-bonded species then transforms to acrolein or acrylonitrile by further abstraction(s) of α-hydrogen(s). The $-OH$ groups formed during the abstraction of the hydrogen atoms then react to form water, which desorbs from the catalyst. Finally gaseous oxygen, adsorbed dissociatively probably at the bismuth site, reoxidizes the active sites reforming the starting bismuth/molybdenum active sites.

8.2.2. Analysis of the Model of the Mechanism of Propene (Amm)oxidation

There are several key, but not explicitly evidenced, hypotheses for this mechanism, which should be pointed out:

1. *Stepwise process*: The reaction mechanism involves a cyclic sequence of steps and not simultaneous concerted steps.
2. *Unique type of site*: A unique type of active site is present on the surface (heat of adsorption independent of the surface coverage), and neither defects (already present or created during the catalytic reaction) nor defective ions (at the edges or steps of the surface) play a role in the reaction mechanism.
3. *No movement of adspecies*: The reaction occurs at a specific ensemble site (a Bi/Mo site) that possesses all of the features necessary to complete the transformation from the reactant to the product. There is no movement on the surface or desorption/readsorption of the intermediate species.
4. *No formation of long-living adspecies*: Strongly bound (long-lived) adspecies do not form. After the rate-determining step all intermediates quickly further transform or desorb; the surface coverage of the adspecies (intermediates) after the rate-determining step is negligible. The surface coverage of the species before the rate-determining step also does not play a role in determining the surface reactivity.
5. *Single pathway*: There is only a single selective pathway of reaction from the reactant to the product. Alternative pathways do not exist, and the conditions of reaction do not influence the reaction pathway.
6. *Redox-only mechanism*: The selective oxidation/ammoxidation occurs with a Mars–van Krevelen type of mechanism. There is no bulk or surface oxygen transport to reoxidize the reduced center.
7. *No role of oxygen adspecies*: Gaseous oxygen is rapidly dissociated and transformed to lattice oxygen having a nucleophilic character without the formation of adsorbed oxygen species. These species, regardless of whether they are weakly or strongly bound to the surface sites, do not play a role in the selective mechanism of hydrocarbon transformation.
8. *No role of coadsorbed species*: The reaction occurs through redox reactions only at the oxide center without the participation of other coadsorbate species in the reaction mechanism.
9. *Localized-only electron transfer*: The electrons transferred from the organic molecule to the catalyst are localized at the metal center (Mo) and

there is no electron delocalization or long range transfer. The local change in the charge does not influence neighboring sites by electronic coupling, change in valence band, etc.

10. *Model of "clean" surface*: Neither the reactivity, nor the electronic and redox properties of the active site are influenced by the chemisorption of the reactants on other surface sites. The turnover number of active sites does not depend on the surface coverage, formation of spectator adspecies, etc.

The mechanism of propene (amm)oxidation outlined in Scheme 8.2 represents an elegant effort to define in detail the surface catalytic chemistry of an important selective oxidation industrial reaction. However, with respect to the real situation during catalytic reaction, it probably suffers from the same limitations as the Langmuir-type model of an ideal surface vis-a-vis real catalysts. As with the Langmuir-type approach in its kinetic extension (Langmuir–Hinshelwood kinetic models), the formalization in terms of kinetic rate of the model reported in Scheme 8.2 provides a good description of the catalytic behavior,[12,21,22] but as has been shown in several cases is not proof of its validity, essentially owing to the insensitivity of the kinetic approach to a mechanistic verification.

Obviously, a simplification of the real complex situation is necessary in order to model the surface catalytic chemistry and some of the above remarks have already been considered, at least in part, in analyzing the surface mechanism of propene (amm)oxidation. For example, the feed composition and reaction conditions influence the catalyst reactivity, because during ammoxidation a mono- or di-imido molybdenum species may form, depending on temperature and partial pressure of the reactants. The two surface species have different turnover activities and thus, in general terms, the reactants themselves induce a change in the surface activity, but a change only in the nature of the active site. Keulks *et al.*[24,26] observed that the kinetics of propene oxidation over bismuth/molybdate are in good agreement with a Mars–van Krevelen mechanism, but not the rate of carbon oxide formation, indicating the probable role of adsorbed oxygen species in its formation. Haber[7,25,27] has pointed out in several publications that activation of gaseous oxygen at the surface of mixed oxides produces electrophilic oxygen adspecies, which, if not quickly further converted to nucleophilic lattice oxygen, attack the regions of highest electron density (the π-bond system) forming peroxo and superoxo complexes, which decompose by C–C bond cleavage to give oxygenated fragments or undergo combustion (see also Scheme 8.4 and related discussion in Section 8.1.2). On the other hand, nucleophilic (basic) lattice oxygen can attack a carbon atom after hydrogen abstraction. In this respect, it also should be noted, following observations of Bettahar *et al.*,[2] that the results leading to the reaction network reported in Scheme 8.2 may also be interpreted in terms of nucleophilic attack of O_2^- anions on the C=C bond concerted with the abstraction of a hydride ion from the CH_3 group (SN_2' type of mechanism). Other steps indicated in Scheme

8.2 have not been fully demonstrated, such as the mechanism of nitrogen insertion, but it is beyond the scope of this chapter to discuss whether the various parts of this mechanism are valid or not.

Most of the proposed mechanisms of selective oxidation at oxide surfaces[1,2] reproduce the same "philosophy" of a stepwise mechanism at a single ensemble site and usually minimal, if any, effort is devoted to analyzing whether the above ten assumptions, which in general terms underscore the importance of under-standing the surface chemistry during catalytic reaction, are valid or not. Thus, significant possibilities for using the analysis of the surface mechanism as a real investigative tool to design better catalysts are lost.[28]

8.2.3. Toward a Model of "Living Active Surface" Rather than Localized Catalysis at Active Sites

The questions of the reaction mechanism and factors determining it have been addressed by other authors from different perspectives which are closer to a dynamic view of the catalysis at the oxide surface and the hypothesis of collective properties of the surface as a key factor.

Moro-oka and Ueda[29] in reviewing the behavior of multicomponent bis-muth/molybdate catalysts concluded that bulk diffusion of oxide ions plays an important role in the enhancement of the catalytic activity and that their migration is mainly accelerated by lattice vacancies. The surface picture suggested by these studies is that the catalytic behavior is not associated with a specific localized active site, but that collaboration between different kinds of active sites proceeds quite efficiently (Scheme 8.5). One site activates molecular oxygen while another mainly consumes it for the reaction.

The dual site concept is also present in the model shown in Scheme 8.2, where different parts of the ensemble site play different roles, but the reaction occurs entirely within the ensemble site and there is no collaboration ("synergy") between spatially separated active sites.

Snyder and Hill[21] in reviewing recent results argued that oxygens associated with both bismuth and molybdenum (or bridging them) are involved in the mechanism of hydrogen abstraction or oxygen insertion into the allyl intermediate, whereas reoxidation involves adsorption and dissociation on vacancies associated with molybdenum. Reoxidation of bismuth polyhedra must involve a rapid transfer of oxygen atoms between bismuth and molybdenum polyhedra through the bulk of the catalyst. While this concept introduces a role for the bulk of the catalyst in the reaction mechanism, the concept suggested by Moro-oka and Ueda[29] is more advanced, as it introduces the possibility of cooperation between different (spatially separated) active sites in determining the global activity, with a rapid oxygen-scrambling between them. It is worth noting that this possibility of multisite cooperation and oxygen-scrambling has not been considered in analyzing the

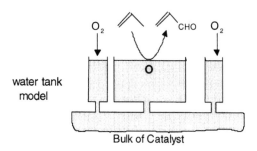

Scheme 8.5. Model of site cooperation in multicomponent bismuth molybdate.[29]

spectroscopic and transient reactivity data with ^{18}O-labeled compounds that are the evidence for the reaction mechanism shown in Scheme 8.2 and the conclusions on the role of bismuth and molybdenum oxygens in the various steps of the reaction (see above).

More advanced with respect to the concept of the whole surface as the active site is the theory of remote control proposed by Delmon,[30–32] who starts from the phenomenon of spillover first discovered for hydrogenation reactions.[33] Monatomic hydrogen species generated on a metal surface quickly diffuse over the surface of an oxide (e.g., alumina) making it active in various hydrogenation reactions after removal of the metal. In order to explain the synergy observed especially in oxidation and hydrotreating reactions from pure transition metal oxides when mixed gently together, Delmon and his coworkers proposed the theory of remote control. This model of the active surface is based on: (i) a continuous generation of active sites during the catalytic reaction, (ii) oxidation and reduction reactions occurring at different, spatially separated, sites on the surface of the catalyst (donor and acceptor species), and (iii) the presence of mobile species (O, H) which are generated at one site and quickly migrate on the oxide surface up to the acceptor site (Scheme 8.6).

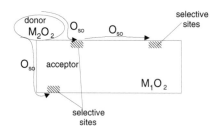

Scheme 8.6. Model of the remote control mechanism in selective oxidation reactions.[32]

Several aspects of the proposed view of the surface reactivity should be further demonstrated before wider acceptance, but it is worth noting that in this model attention is shifted from a localized reaction mechanism analogous to that occurring at complexes in the liquid phase to a true reaction mechanism on solid surfaces in which the collective properties of the surface are the key features.

The notion of active surface with respect to that of active site has also been supported by recent theoretical advances concerning catalytic concepts governing surface reactivity of solids.[34] In fact, the following have been noted[34]:

1. An influence of the coadsorbate on the nature of the bond formed between adsorbate and surface sites with a consequent dependence of the intrinsic catalyst reactivity on the surface coverage by adspecies.
2. The role of surface adlayer composition and surface phases in determining selectivity.
3. The possible role of surface species that do not participate directly in the reaction mechanism (spectator species) in influencing the surface reactivity.

These studies[34] also indicate that a "clean" catalyst surface may have a significantly different reactivity than the same surface in the presence of reactants. Furthermore, these studies show that multiple pathways of reaction are possible for which the overall rate of reaction can be influenced in various ways depending on the chemisorption phenomena.

Thus it is incorrect to speak of reaction mechanism generically; rather the specific reaction conditions for which the proposed mechanism is the dominant pathway should be indicated. This observation underscores the fact that care must be taken in interpreting the results of spectroscopic methods, where measurements are often carried out under conditions that are not representative of the surface situation during real catalysis. They make it possible to indicate a possible surface reaction mechanism, but the necessary further step, often not considered, is to determine whether this suggested mechanism corresponds to the dominant pathway *during* catalytic reaction under given reaction conditions.

The concept of modification of *in situ* reactivity of the catalyst surface when reactants are adsorbed on it has been shown to be valid not only by theoretical calculations, but also by *in situ* observation of adsorbed molecules at oxide surfaces. For Nb monomer species supported on SiO_2, Iwasawa[35-38] observed a switchover of the reaction path in ethanol conversion from dehydration to dehydrogenation when a second ethanol molecule was adsorbed on the Nb-ethoxide intermediate. Onishi *et al.*[39] noted that the chemisorption of a formic acid molecule induces a change in the reaction mechanism of decomposition of surface formate species on TiO_2 from dehydration to dehydrogenation.

Bao et al.[40] proved that the interaction of water with an oxygenated Ag surface leads to the formation of hydroxyl groups (incorporated into the subsurface region), which at higher temperatures produce strongly bound oxygen adspecies by dehydroxylation that offer an alternative to the energetically less favorable pathway for the direct formation of α-oxygen (a strongly held oxygen species responsible for partial oxidation of methanol to formaldehyde[41,42]). Bao et al.[43] also observed that two simultaneous reaction pathways operate under steady state conditions, due to the presence of different surface microstructures that are in a continuous dynamic interchange. The surface reconstruction is induced by adsorption.

These studies point out that the intrinsic reactivity of surface sites and the paths of reaction can be significantly altered by the chemisorption of the reactants themselves and by the population and nature of the adspecies. It is thus necessary to take account of these phenomena in studying the reaction mechanisms at oxide surfaces.

8.2.4. The Question of Stepwise Reaction Mechanisms

The model of a stepwise reaction mechanism in distinct steps with a rate-determining step (allyl intermediate formation in the case of Scheme 8.2) is, as observed before, a common simplification but its relevance is often not considered or fully understood.

One of the first models introduced to explain the catalytic behavior of selective (amm)oxidation catalysts was that of "site isolation" originally introduced by Callahan and Grasselli.[44] The basis for the model was the postulate that oxygen atoms must be distributed on the surface of a selective oxidation catalyst in an arrangement that provides for limiting the number of active oxygen atoms in various isolated groups. A key observation[12] was that the catalytic behavior of a CuO catalyst as a function of the degree of surface reduction can be well described using statistical methods to calculate the distribution of oxygen (site population) on the surface as a function of surface coverage (Figure 8.1). The model predicts that clusters of two to five adjacent surface lattice oxygens maximize the selectivity for acrolein from propene.

Although it has often been claimed that this model explains catalytic results in the selective oxidation of hydrocarbons, including those for bismuth/molybdate catalysts, it should be noted that it extends the model of Scheme 8.2, because the density of oxygen sites around the hydrocarbon adsorption site is a key factor for the selectivity in the "site isolation" model, while Scheme 8.2 assumes a determined "density" of oxygen sites around the hydrocarbon adsorption site. This oxygen site density is clearly a function of the collective properties of the solid. Furthermore, this model shows that selective synthesis is possible only when the rate of desorption of the product is faster than the rate of oxidation (the latter clearly dependent on the local concentration of oxygen sites around the adsorbed intermediate). The

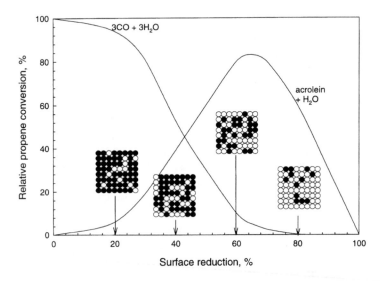

Figure 8.1. Relative propene conversion as a function of reduction of an oxidized grid. The insets show some examples of the relative degree of reduction of the oxidized grid.[12]

selectivity is thus a kinetic function of the rates of the various steps in the reaction network (including desorption of the products) and depends on the reaction conditions. Thus it is not a function only of the characteristics of the active site, another implicit assumption of the model in Scheme 8.2. Finally, the model of site isolation shows that, depending on the local concentration of oxygen sites, the modality of oxygen insertion into the intermediate and thus the pathway of transformation may change.

8.2.5. The Geometrical Approach to Oxidation Catalysis at Oxide Surfaces

The model of site isolation can be viewed basically as a geometric approach to catalysis, and the model of structure sensitivity of oxidation reactions is the natural extension. The different crystal faces of an oxide have different characteristics, such as oxygen packing and arrangement of metal-oxygen polyhedra, presence of coordinatively unsaturated metal ions (Lewis sites), metal to oxygen bond strength and polarizability, and surface OH groups. A direct consequence of the geometrical model of catalysis is thus that different faces of an oxide crystallite should also have different reactivity characteristics.

Since the first studies by Volta *et al.*[45,46] and Tatibouet *et al.*[47,48] on MoO_3 single crystals, several studies have focused on demonstrating the concept of structure sensitivity. In the case of MoO_3 single crystals the (100) side face was found to be very selective for propene oxidation to acrolein, while the (010) basal face was selective for total oxidation (Figure 8.2). It follows that the (100) face was proposed as the place for propene activation while the oxygen atom supply for insertion to form acrolein comes from the (010) face.[49] For but-1-ene oxidation, butadiene was formed on all basal (010), side (100), and apical (101, $\overline{1}$01) faces while carbon oxides formed only on the basal plane.[50] For isobutene oxidation, methacrolein was observed mainly on the side face (100) as were carbon oxides, while acetone was formed on the basal (010) and apical (101, $\overline{1}$01) planes.[50] The results are summarized in Figure 8.3.

In regard to catalytic behavior, MoO_3 crystallites having different crystalline habits were found to have effectively different reactivities, and this can be considered as proof of the correctness of the geometrical approach. It may be argued, however, that the results regarding the changing nature of the hydrocarbon (Figure 8.3) are not fully consistent with this hypothesis. Why, for instance, does the basal plane give mainly carbon oxides for propene and but-1-ene, but not for isobutene, for which the side (100) plane, the most selective plane for acrolein and butadiene, is responsible for carbon oxide formation? The geometry of the adsorbed complex is certainly relevant, but it is difficult to explain the data in Figure 8.3. Furthermore, in a purely geometrical approach to catalysis, it is difficult to explain the previous conclusion that the oxygen inserted into the hydrocarbon comes from a different crystalline face than the one where the hydrocarbon is adsorbed.[51,52]

Volta et al.[49,51] suggested that the gaseous oxygen is dissociatively adsorbed on the (010) face and diffuses to lateral faces to react with the hydrocarbon through the MoO_3 bulk by oxygen vacancy disordering. We may note, however, that the bulk oxygen transport properties of oxide crystallites depend on their habit. The absorption edge (also an index of bulk transport properties, as well as the position of the conduction band) in oxide semiconductors (TiO_2 and V_2O_5) depends on the habit of the crystallites, for instance, when grown inside zeolitic cavities.[53,54] Therefore, when the ratio of the exposed surface crystalline faces changes, the bulk electronic and ionic transport properties as well as, e.g., the conduction band and Fermi level also change. If oxygen diffusion is a key factor in determining

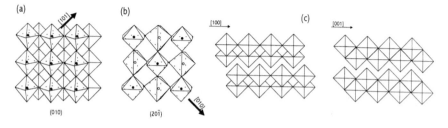

Figure 8.2. Crystal structure of MoO3: (a) (010) crystal plane; (b) (20$\overline{1}$) crystal plane; (c) view of [001] and [100] directions.[84]

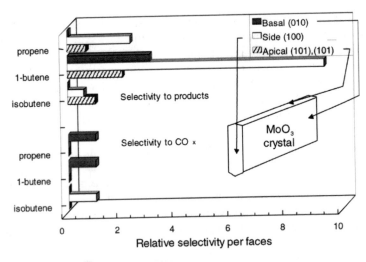

Figure 8.3. Structure sensitivity of the different faces of MoO3 crystals in alkene oxidation at 380°C.[48]

reactivity, the catalyst behavior thus depends on the habit of the oxide crystallites, even in the absence of structure sensitivity or a localized reaction at a specific crystalline face.

More recently, Volta et al.[55] integrated the earlier $^{18}O_2/Mo^{16}O_3$ studies[50] on the structure sensitivity of MoO3 with $C^{18}O_2/Mo^{16}O_3$ exchange tests. They observed[55] that at 500°C two contributions to the exchange reaction can be distinguished. The first corresponds to a fast surface oxygen exchange, and the second, which is more prominent over samples with (100) lateral faces, can be assigned to migration of oxygen atoms from lateral to basal faces and/or to a slow migration of oxygen atoms from the bulk structure of MoO_3 toward the lateral faces. A preferential migration of oxygen atoms through the MoO_3 structure along the (100) and (001) directions has also been indicated by secondary ion mass spectroscopy (SIMS) analysis of MoO_3 crystals during ^{18}O incorporation.[56]

The analysis of the structure sensitivity of MoO_3 crystallites thus suggests that:

1. The mechanism of oxygen migration is a key parameter in determining structure sensitivity;
2. The migration of oxygen is anisotropic;
3. Different mechanisms of oxygen diffusion are present characterized by different rates;
4. The reaction conditions (temperature, oxygen, and hydrocarbon partial pressures) determine which type of surface diffusion mechanism influences the rate of reaction.

These conclusions show how a careful analysis of the structure sensitivity phenomena leads to conclusions that are contrary to the starting hypothesis of a localized approach to catalysis and support instead the model of a "living active surface."

8.2.6. The Question of Reaction at a Single "Ensemble" Site

The geometrical approach to catalysis at oxide surfaces has also been presented from other perspectives, such as: (i) the crystallochemical model of active sites,[57–60] which is an extension of the concept of catalytic anisotropy[50,61–64] to include geometry and energetics of active cluster sites and reactant–surface interactions in order to determine the energetically most favorable pathway; and (ii) the idea of an "inorganic oxide cluster" (ensemble site) with a stereochemistry that fits that of the reactant molecule(s).[5,6,65]

In all cases, notwithstanding the differences in the relative theories, the common background hypotheses are that: (i) the geometry of the site is fixed and can be derived from crystallochemical considerations; and (ii) the reactant, after activation, remains stuck to the active site and does not migrate over the surface.

The latter hypothesis is based mainly on the fact that in hydrocarbon oxidation the rate-determining step is usually hydrocarbon activation by hydrogen abstraction, suggesting that (i) the consecutive steps are faster, (ii) there is no desorption of the intermediate species, and (iii) the quantity of intermediates or products adsorbed on the catalyst is negligible. However, the absence of desorption to the gas phase does not imply the absence of movement of organic adspecies on the surface.

The possibility of fast surface movement of H and O atoms produced by spillover is well demonstrated,[32,33] whereas much less study has been devoted to the rate of surface diffusion of organic adspecies. Kapoor et al.[66] reviewed this topic, and distinguished three types of surface diffusion according to the type of moving adspecies: (i) physically adsorbed species, (ii) chemisorbed species, and (iii) self-diffusion of metal atoms.

The diffusion of the second type of species is the least well known, but probably occurs through a hopping mechanism. The rate of diffusion of chemisorbed adspecies depends on the surface coverage and temperature, in addition to the type of bond formed with the surface. However, at high temperatures as during catalytic tests, the chemical bond between organic adspecies and surface sites may be significantly weaker (due to thermal lattice oxide vibrations) to allow a fast hopping movement. This problem can be studied using available theoretical simulation models, but these are usually resorted to only in the case of surface diffusion inside zeolitic cavities.

Data on the rate of the surface diffusion other than that for small molecules or species (H, O, CO, NCO) are nearly nonexistent, but the migration of different hydrocarbons on Pt zeolite has been estimated by field mass spectrometry.[67] Solymosi et al.[68,69] observed that on Rh supported over various oxides (Al_2O_3, TiO_2, MgO) during the $H_2 + CO_2$ reaction the number of formate species is about five to eight times the number of Rh atoms on the surface. Formate species form over the Rh particles and then migrate to the oxide support, indicating that the rate of surface diffusion of formate species should be of the same order of magnitude as the rate of generation and desorption of the same species.

In a study of the mechanism of deactivation by SO_2 of $V/(TiO_2\text{-}Al_2O_3)$ catalysts for the reduction of NO with ammonia, migration of sulfate species from the site, generated by oxidation of SO_2 to the support, has also been observed[70] and a similar effect was observed for Cu/Al_2O_3 catalysts in a study of the mechanism of sulfation of the surface by $SO_2 + O_2$.[71,72] In the latter case, the kinetic modeling of the reaction[72] indicates that the rate of surface migration of sulfate species is of the same order of magnitude as the rate of reaction. This indicates that even for surface species characterized by a relatively strong interaction with the surface, such as sulfate species, the surface mobility can be significant. The rate of migration depends on various factors, such as the presence of water vapor, because the equilibrium shifts between sulfate and sulfuric acid forms, the latter being less tightly bound to surface oxygens and thus more mobile.

The same observation (formation of the acid form in the presence of water vapor) is valid for several organic acids on the surface of oxides. In the absence of water vapor (as is usual in most spectroscopic studies), strongly bound and thermally stable carboxylate species form, but these can transform to their weakly bound and unstable acid counterparts in the presence of water vapor. This has been demonstrated, for instance, in the formation of acrylate species during propane oxidation on V/Sb oxides.[73] In the presence of small quantities of water vapor, the acrylate species easily transforms into a weakly interacting acrylic acid species that decarboxylates easily owing to the higher surface mobility.

This observation again demonstrates how the surface reactivity and mobility of adspecies may be significantly different when studied under conditions other than those of the working catalyst or when these differences in the test conditions (the presence of water, e.g., which is a common product of oxidation reactions, if not specifically added to the feed) are not taken into consideration.

It must be also taken into account that results from fast reactivity tests such as those possible using temporal analysis of products (TAP) reactors[74–78] show that the lifetime of adsorbed intermediate species over oxide surfaces during selective catalytic oxidation reactions can be relatively long (in the range of several milliseconds) especially for complex reactions such as n-butane oxidation to maleic anhydride over $(VO)_2P2O_7$.[78] Owing to this long surface lifetime, the movement of surface intermediates and spectator adspecies (usually more weakly adsorbed) cannot be ruled out in analyzing the surface mechanisms of selective oxidation.

It should also be pointed out that nonrandom distribution of adspecies[79] is possible as a result of adsorbate interactions or anisotropic surface diffusion, as well as being due to the fact that on an oxide surface the local heat released during catalytic oxidation cannot be rapidly and uniformly dispersed. There is thus a local increase in the surface reaction, which causes a higher local rate of reaction and thus may cause a further propagation effect. This results in local nonuniformities (on a nanometric scale) in the surface temperature and local changes in the concentration of adspecies, which causes the surface diffusion. On the other hand, as noted earlier, at the typical range of temperatures of selective oxidation reactions (300–500°C) a fast surface hopping movement is possible.

These background indications point out that the presence of surface diffusion phenomena and nonuniform distribution of adspecies (both spectator and reaction intermediates) cannot be ruled out in studying the mechanisms of selective oxidation, although the absence of specific studies of this issue precludes more precise conclusions. It is worth noting that the demonstration of a rate of diffusion of chemisorbed species of the same order of magnitude as that of their surface transformation can considerably modify our view of the surface chemistry during catalytic selective oxidation reactions.

There are several potential methodologies for determining the rate of surface diffusion of chemisorbed species. One possible experiment, which is a variation of the field mass spectrometry technique, could be as follows: The reactant is adsorbed on the oxide film supported on a metal plate and then the gas phase species are removed. Under vacuum using a low-energy pulse laser a small region of the surface is heated locally (depositing a small oxide film over the metallic surface increases the speed with which heat is dispersed), and the concentration of adspecies in that region is determined by a mass quadrupole detector. Changing the time between two consecutive pulses makes it possible to estimate the rate of diffusion of adspecies. Field emission microscopy is another powerful technique, which allows the surface diffusion of adspecies to be estimated because the surface work function depends on the adsorbate coverage.[80–83]

Thus, there are experimental techniques available to estimate the possible rate of surface diffusion of chemisorbed species at the temperatures and on the oxide surfaces typical of selective oxidation catalysts, and considerable effort should be put forth in this direction, as determination of the surface migration rate of adspecies during catalysis is of fundamental importance for modeling the surface reactivity.

8.2.7. The Role of Catalyst Reduction and Dynamics of Reaction

Oxidation catalysts are characterized by the presence of continuous redox phenomena during catalytic reaction, which may lead to a change in the surface structure and reactivity. The change depends on the reaction conditions (temperature and partial pressures of the hydrocarbon, oxygen, or other possible reactants such as NH_3), but usually the properties of oxide catalysts after the catalytic tests are different from those of the starting sample after calcination. It is thus necessary to understand whether this possible change might also be relevant to the general question under discussion in this section: Can the reactivity at oxide surfaces be correctly approached considering the model of stepwise reaction at a single (ensemble) site as a valid unquestionable rule?

Haber and Lalik[84] have recently reviewed the problem of the reactivity of the different faces of MoO_3 crystallites to analyze the effect of the degree of surface reduction. MoO_3 is a nonstoichiometric oxide, whose lattice components are in equilibrium with the gas phase. Its intrinsic defect structure consists of oxygen vacancies, the concentration of which depends on gas phase concentration and reaction temperature. The equilibrium is dynamic and thus even under steady state conditions there is a continuous process of oxygen desorption and readsorption; i.e., there is a continuous scrambling of lattice oxygens, and the surface is populated by transient oxygen species such as O^-, and O_2^- (see Scheme 8.4a). The surface coverage of MoO_3 by these species depends on the position of the Fermi level on the solid. When MoO_3 is reduced, isolated oxygen vacancies are formed at first, and the electrons released in this process become localized on Mo^{6+} ions, reducing them to Mo^{4+} ions.

Although XPS data indicate the change in the valence state of molybdenum ions,[84] the effective occurrence of this process at the temperatures of the catalytic reaction should be considered. In fact, for instance, for vanadium supported on TiO_2, conductivity measurements during the catalytic tests[85] showed that a reduced vanadium ion (V^{4+}) should be better described as a V^{5+} ion and an electron delocalized around the vanadium site. At room temperature the delocalized electron is preferentially localized at the vanadium center.[86]

This observation suggests that a process of reduction at oxide surfaces, especially those having a semiconductor nature, should not be viewed only as a change in the valence state of the metal, but rather should be considered as the production of one or more electrons which can be localized at a transition metal ion and change its valence state, or be delocalized on a surface conduction band or trapped at surface holes. The localization or delocalization of the electron at the metal center is thought to be a function of the temperature and presence of strongly bound adspecies, which can favor delocalization, such as when carbonaceous-type species form. Although this kind of species is not supposed to form in oxidation reactions, experimental evidence indicates that they do form in some cases, such as during n-butane or n-pentane selective oxidation on $(VO)_2P_2O_7$.[87]

Understanding the effective localization or nonlocalization of the electrons released during the redox reaction and their reactivity requires *in situ* techniques during the catalytic runs because, as noted above, the results obtained with *ex situ* techniques may not be representative. This issue is another critical, usually under-valued, component of the identification of the real working nature of the oxide surface during catalytic reaction and thus is the basis for a more realistic under-standing of the reaction mechanism.

When reduction of MoO_3 proceeds, oxygen vacancies aggregate, forming shear planes that may also become ordered to form the Magneli phases $[Mo_nO_{3n-1}]$. At higher temperatures the structure disintegrates into crystallites of Mo_4O_{11}, which then transform into MoO_2.[84] Alternatively, reduction starts with the formation of oxygen vacancies at the (010) plane (see Figure 8.2) by adsorption of hydrogen on terminal oxygen atoms[88] and desorption of water along the [101] series of octahedra (see Figure 8.2). This results in the formation of a layer of MoO_2 structure. This second type of mechanism predominates under mild reaction con-ditions, such as those in force during catalytic oxidation tests.[84]

When MoO_2 domains form, the interface between the MoO_3 and MoO_2 domains can have a considerable influence on the surface properties of that crystalline face, as shown by the results of Haber and Lalik.[84] In fact, the value of the work function of MoO_2 is smaller than that of MoO_3, and thus the surface of MoO_3 will be negatively charged. Haber and Lalik[84] indicated that owing to this phenomenon the population of surface electrophilic oxygen species responsible for nonselective oxidation behavior increases, but it may also be argued that: (i) the change in the surface potential of MoO_3 influences the interaction of the hydrocar-bon with the surface, and (ii) the domains of MoO_2 become positively charged, which has an influence on their reactivity characteristics. Nevertheless, the data of Haber and Lalik[84] clearly indicate that:

1. When a reduction occurs that leads to a structural rearrangement with the formation of microdomains of different phases, the surface reactivity prop-erties can be considerably influenced.
2. The mechanism of reduction depends on the crystalline face with conse-quent different modifications of the surface characteristics as a function of the degree of reduction.

When the anisotropic (structure sensitivity) reactivity properties of an oxide are studied at very low conversion it may be hypothesized that the crystallographic model of the surface based on cutting the "ideal" crystal structure can be an acceptable approximation of the active surface, but when catalytic tests are done under more severe conditions at high conversion (conditions for which major structural surface reorganization can occur), the crystallographic model can be an unrealistic approach. Hence, the data of Haber and Lalik[84] underscore the importance of understanding the dynamic processes of catalyst reconstruction driven by the reaction conditions and that:

1. A crystallographic model of the surface may be a useful schematization, but it is not always representative of the working state of the catalyst surface.
2. It is necessary to develop models of the active surface as a function of the reaction conditions (e.g., temperature, feed composition, conversion of hydrocarbon and oxygen, presence of other gas phase components such as water).
3. The idealization of a uniform surface often does not correspond to a realistic situation because apart from the question of the presence of steps and corners on the surface (thus sites with different coordination environments), domains at a microscopic level may form owing to redox surface processes during the catalytic reaction. These domains have different work functions and thus a net electronic flow is induced that changes the surface electronic and reactivity characteristics.

8.2.8. General Conclusions on the Modeling Approach to Selective Oxidation Catalysis

Starting from the analysis of the approach used to model the reaction mechanism of (amm)oxidation of propene over bismuth/molybdate oxides (a stepwise mechanism at a single "ensemble" site), we have shown how breakthroughs in the understanding of the surface mechanisms can derive from a more general view of the catalytic chemistry at oxide surfaces ("living active surface" model) (Scheme 8.7). This view must take into account that the catalytic behavior is determined by the whole surface modified by chemisorbed species and that the schematization of the active centers based on a crystallographic method is an oversimplification. However, since only preliminary information is available, considerable effort will be needed to create the basis for this new approach. In particular, it is useful to summarize the main points discussed in this section in order to highlight some guidelines for future research in this area. It is necessary to:

1. Analyze the factors that determine whether a reaction mechanism will be concerted rather than stepwise, taking into account the effect of the conditions under which the reaction proceeds. It is also necessary to evaluate the possible presence of multiple pathways of reaction and how they can depend on the reaction conditions.
2. Determine whether there is site or active species cooperation, which includes a possible role of adspecies.
3. Determine how the surface reactivity changes as a function of: (i) the chemisorption of reactants or of other species (e.g., water), (ii) reaction temperature, and (iii) degree of reduction. Different crystalline faces may be differently influenced by these effects.

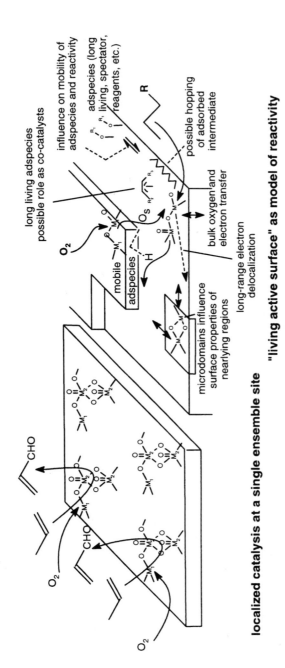

Scheme 8.7. Model of localized catalysis at a single site with respect to a model of a "living active surface."

4. Analyze the rate of surface diffusion of adspecies (simple species such as O and organic intermediates) and determine whether the rate of this surface diffusion is significantly lower with respect to the lifetime of the intermediate species or if a model of surface transport/movement of adspecies should be considered. It also is necessary to examine the possibility of anisotropic transport.

5. Determine the short- or long-range transport of electrons during redox catalytic reactions and see whether the electrons generated are localized at transition metal ion centers or surface defects or delocalized by hopping between surface sites or through adspecies. It is also necessary to analyze the anisotropy in electron transport and the presence of bulk versus surface electron transport.

6. Study the presence of nonuniform characteristics of the working surface both in terms of the presence of adsorbate islands and surface microdomains caused, for instance, by reduction during the catalytic reaction. Aspects to study are: (i) how adsorbate islands interact with each other and influence the migration of adspecies, (ii) the possible differences in the surface reactivity of the surface microdomains and (iii) how the presence of microdomains (generated, e.g., locally by reduction) influence the surface potential and electronic characteristics of the base surface.

8.3. SURFACE OXYGEN SPECIES AND THEIR ROLE IN SELECTIVE OXIDATION

The nature of the surface oxygen species and their reactivity and role in catalysis is obviously a central theme of selective oxidation. Following the literature, we may distinguish between two basic types of oxygen species:

1. Lattice oxygen, which can be either terminal ($M{=}O$) or bridging ($M{-}O{-}M_l$) oxygen. Further differences arise not only from the strength of these metal–oxygen bonds, which depends on the nature and valence state of the transition metal but also from the coordination of the metal–oxygen polyhedra.

2. Adsorbed radical-type oxygen, formed in the process of the reduction of gaseous O_2 to lattice O^{2-} (see Scheme 8.4a) due to electron transfer from the metal to the oxygen. These species are stabilized by coordination with surface metal ions. During this transformation the nature of the metal–oxygen bond and its polarizability changes from an electrophilic character ($M^{\delta-}{-}O^{\delta+}$) for the negatively charged species to a nucleophilic character ($M^{\delta+}{-}O^{\delta-}$) for structural O^{2-}.

There is general agreement, at least for the case of the alkenes conversion, that the electrophilic oxygen species, is associated with nonselective total oxidation of the hydrocarbon, while the nucleophilic oxygen in involved in selective insertion of oxygen into the hydrocarbon molecule to form partial oxygen products (aldehydes, acids, e.g., from the corresponding alkenes) (see Scheme 8.4b) as well as in the mechanism of hydrogen abstraction from the same type of substrate.

The basic supporting evidence is the fact that when hydrocarbons interact with an oxide surface in the absence of gaseous oxygen, a higher selectivity for partial oxidation products with respect to the case of cofed hydrocarbon + O_2 is often observed, although clearly the progressive reduction of the catalyst as a consequence of the consumption of its structural oxygen (incorporated into the hydrocarbon or water molecules, the latter formed as the final product of the mechanism of hydrogen abstraction) leads to a progressive decline in the reactivity.

Although the issue is more complex, as seen, for instance, from the fact that in oxide catalysts having a high bulk oxygen mobility, alkene oxidation is nonselective even in the absence of gaseous oxygen,[89] the observation that in the absence of gaseous oxygen higher selectivities are possible is generally valid. However, gaseous oxygen is needed for catalyst reoxidation and thus to maintain a steady state catalytic activity, and electrophilic oxygen species form in the process of catalyst reoxidation by gaseous O_2. The problem of selectivity can thus be simplified to one of finding a way in which the overall rate from O_2 to O^{2-} is fast enough that the surface concentration of electrophilic oxygen transient labile intermediates is very low. The presence of redox metals that catalyze the reoxidation (e.g., Fe in Sb-based mixed oxides[90]) and a surface restructuring process, which avoids the formation of localized defects (via, e.g., a change in linkage of coordination polyhedra—see Scheme 8.4c) are two effects that are known to limit the surface concentration of electrophilic oxygen intermediates.

There is general agreement on this view of the surface oxidation chemistry, but a closer analysis reveals that there are some fundamental unsolved problems in the way of its acceptance as a general theory for oxidation catalysis:

1. *Extendability:* The validity and applicability of this reaction mechanism when extended to catalysts and substrates other than mixed oxides active for alkene selective oxidation.
2. *Kinetics:* Is the rate of product formation during steady state oxidation (in the presence of both hydrocarbon and oxygen) the same as that estimated under nonsteady state conditions, when hydrocarbon and oxygen react in separate stages? In other words, the question is whether selective oxidation via structural oxygen is the faster process during oxidation in the presence of gaseous O_2 or do other oxygen species participate in determining the overall rate of reaction in the presence of gaseous O_2. It should be noted by

the way that the tests to determine the rate of catalyst reduction by hydrocarbon and reoxidation by O_2 must be done with continuous feeding of the hydrocarbon or oxygen: in the widely employed pulse method the time between pulses allows surface reconstruction through bulk oxygen diffusion and thus may not give representative results.

3. *Uniqueness:* Is the mechanism of oxygen insertion via structural oxygen the only one possible for selective oxidation to partial oxidation products when a hydrocarbon and oxygen are cofed over a solid catalyst?

As was observed in Section 8.2 for analogous problems, these basic questions have been largely ignored in the literature in this case as well and thus there are no specific clear examples in favor or against their validity. Nevertheless, the few examples discussed below (Section 8.3.3) show that more effort should be directed toward reanalyzing the problem of the nature of active species responsible for the mechanism of selective oxidation over mixed oxides. Before discussing these examples, however, it will be useful to discuss the mechanism of interaction of oxygen with oxide surfaces in more detail.

8.3.1. Nature of the Interaction between Molecular Oxygen and Oxide Surfaces and Types of Oxygen Adspecies

According to molecular orbital theory, the free O_2 molecule can be described as the combination of two isolated O atoms. Their $2s$ and $2p$ atomic orbitals combine to form four bonding and four antibonding orbitals. The two $2s$ atomic orbitals [each containing two electrons (e)] combine to form one σ_s bonding orbital and one σ_s^* antibonding orbital, each filled by two electrons (e). The two $2p_z$ orbitals, each containing one e, combine to form a σ_z bonding orbital (containing two e) and an empty σ_z^* antibonding orbital. The two $2p_x$ and $2p_y$ orbitals combine to form π_x and π_y bonding orbitals (occupied by $2e$ each) and π_x^* and π_y^* antibonding orbitals, each of which contains one electron. The ground state of the oxygen molecule is thus a triplet state $(^3\Sigma_g^-)$ with two unpaired electrons. At slightly higher energies there are two low-lying electronically excited states—the singlet $(^1\Delta_g)$ and $^3\Sigma_g^+$ levels. The energy level for singlet oxygen is 22.64 kcal above that of ground-state triplet oxygen.

However, in the presence of a crystal field due to a surface transition metal, this configuration changes, as well as the coordination of dioxygen, which affects the characteristics of the transition metal ion. For an octahedral symmetric transition metal ion,[22] the fivefold degenerate d orbitals $(d_{xy}, d_{xz}, d_{yz}, d_{x^2-y^2},$ and $d_{z^2})$ convert to a triply degenerate t_{2g} set $(d_{xy}, d_{yz},$ and $d_{xz})$ and a double degenerate e_g set at higher energies. If the difference in energy between these two sets of orbitals is small, unpaired electrons will preferably occupy the e_g orbitals (spin-free situation), whereas if the energy difference is greater than the energy required to pair electrons (when the ligand field is strong), the electrons will pair and occupy the t_{2g} orbitals (spin-paired situation).

The orbitals of O_2 molecules also are influenced by the presence of the crystal/ligand field. The partly filled π^* antibonding orbital degenerates to the two π_x^* and π_y^* components. Depending on the strength of the crystal field, the energy difference between these two orbitals may become larger than the energy of pairing the two electrons. The two electrons would then locate in the lower-energy orbital (π_y^*). In this case, the molecule can be viewed as formed from six sp^2-type hybrid orbitals. Two of these hybrid orbitals lie along the internuclear z-axis (Scheme 8.8a), whereas the other four form angles of about 120° with it. The latter four orbitals contain two electrons each ("lone pair" orbitals). The two $2p_x$ orbitals of atomic oxygen form a lower energy π_x orbital, which contains $2e$ and an empty, higher energy π_x^* antibonding orbital (Scheme 8.8a).

There are basically two types of interactions between these orbitals and those of transition metal ions. In one case the metal species is perpendicular to the O–O axis, while in the other the metal is attached to one oxygen atom and the dioxygen molecule bends on one side (Scheme 8.8b). In the former structure the filled π_x orbital of O_2 overlaps an empty e_g atomic orbital ($d_{x^2-y^2}$ in Scheme 8.8b), whereas in the latter complex, the lone pair orbitals of dioxygen overlap an empty e_g orbital of the metal with possible back-donation from the fully occupied t_{2g} metal orbitals to the empty π_x^* orbital of O_2. The back-donation stabilizes the M–O bond, weakening the O–O bond, facilitating, for instance, the possible formation of single O species on the surface of oxidation catalysts.

8.3.1.1. Neutral Dioxygen Species

Different types of neutral dioxygen species may exist on the surface of an oxide, depending on the coordination with transition metal ions. It is also possible to distinguish between species in the ground state and in an electronically excited state (singlet oxygen, in particular). The role of the latter in the mechanism of selective oxidation should also be considered, because at the typical temperatures of catalytic reaction of mixed oxides, the vibrational energy of the catalyst surface is high enough to form singlet oxygen in nonnegligible quantities. Kazansky et al.[91] for instance, have clearly shown by EPR that singlet oxygen forms when oxygen is adsorbed on chromium oxide.

The reactivity of singlet oxygen with different organic substrates is well known,[92,93] and there is also some evidence concerning its role in the selective oxidation of molecules adsorbed on oxide surfaces. Slawson et al.,[94,95] for instance, have shown that linolenic acid on silica gel and 2,5-diphenylfuran on silica or titania undergo autoxidation in air when the sample is heated owing to the formation of singlet oxygen. It can be noted that singlet oxygen may also form from charged (electrophilic) oxygen species. Krylov[96] has demonstrated that singlet oxygen may

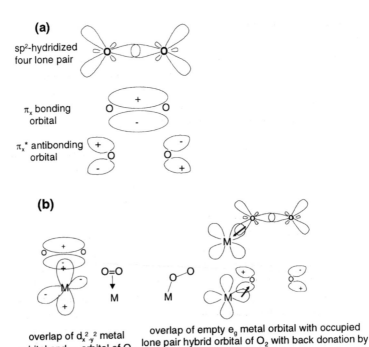

(a)

sp^2-hydridized
four lone pair

π_x bonding
orbital

π_x^* antibonding
orbital

(b)

O=O
↓
M

overlap of $d_{x^2-y^2}$ metal
orbital and π_x orbital of O_2

overlap of empty e_g metal orbital with occupied
lone pair hybrid orbital of O_2 with back donation by
overlap of filled t_{2g} metal orbital with empty π_x^*

Scheme 8.8. (a) sp^2-type hybrid orbitals for O_2 in the presence of a crystal/ligand field. (b) Two possible types of overlap of these hybrid orbitals of O_2 with d-orbitals of a transition metal (M) and consequent different possible structures for adsorption of molecular oxygen on the catalyst surface.[22]

form from O_2^- on transition metal oxides and is the active form of oxygen that interacts with alkenes. It was also shown that O_2^- may convert to singlet oxygen by reaction with water.[97] This observation is important because water is one of the common reaction products in selective oxidation, as well as O_2^- species, which form during catalyst reoxidation by gaseous oxygen (see Scheme 8.4a). Therefore, the formation of singlet oxygen during the catalytic reaction and a role for it in the oxidation mechanism cannot be ruled out, although the limited data and dearth of quantitative studies do not allow more detailed conclusions to be drawn. Further, it should be pointed out here as well that spectroscopic results on a "clean" catalyst surface may be different than those on a "working" catalyst, as indicated by the possible effect of adsorbed water on O_2^- species.

8.3.1.2. Charged Dioxygen Species

Together with neutral dioxygen species, several kinds of charged dioxygen species may exist on oxide surfaces.[98] It is worth noting that although O_2^- (superoxide) is the most commonly reported, other species have been detected including O_2^{2-}, O_2^{3-} and even positively charged species such as O_2^+. In going from O_2^+ to O_2^{2-} the dioxygen bond length becomes progressively longer, and this explains why the latter is usually considered a necessary intermediate for monatomic oxygen species.[99,100]

Although the O_2^+ species is not stable in the gas phase, it can be stabilized on the surface by the additional coulombic stabilization from the lattice. There is no clear evidence for its formation over oxide surfaces, but Davydov[101] studying the adsorption of oxygen on supported molybdena catalysts observed broad IR absorption bands in the 1500–1700 cm^{-1} range, which suggest a partial withdrawal of electrons from the antibonding orbitals of the oxygen molecule to form a partially positively charged adsorbed dioxygen species. When the temperature is increased above about 100°C this species disappears, but similar behavior is also observed for negatively charged oxygen species. It is thus not possible to determine whether similar types of species could be present under the typical conditions in which oxides are active in selective oxidation reactions, but the reactivity of partially positively charged adsorbed dioxygen species should be opposite to that of negatively charged species. It is also probable that adsorbed species that withdraw electrons from the catalyst favor the formation of positively charged dioxygen species.

There has been little investigation of the O_2^{2-} ion, because: (i) it is diamagnetic (thus EPR inactive), (ii) it would be expected to be IR inactive, and (iii) conductivity measurements do not distinguish between the $2O^-$ and a O_2^{2-} ions. It is therefore not possible to obtain direct evidence on the formation of the O_2^{2-} ion and its role in selective oxidation reactions. On the other hand, the O_2^- ion (superoxide) has been extensively characterized and its formation detected over a broad range of oxides and reaction conditions.[98] Its coordination complexes with surface transition metal ions are not significantly different from those described above for the dioxygen molecule and thus will not be discussed further. Polyoxygen species may also form, such as the oxonide ion O_3^-. These species have been clearly characterized by EPR and reflectance spectroscopy,[98] and it should be noted that they play an important role in exchange reactions over oxides through the reaction of gas phase oxygen with O^- on the surface[99]:

$$(^{16}O^-)_{ads} + (^{18}O_2)_{gas} \rightleftharpoons (O_3)^-_{ads} \rightleftharpoons (^{18}O^-)_{ads} + (^{16}O^{18}O)_{gas} \quad (8.1)$$

This observation indicates that there is a continuous equilibrium on the surface among the various adsorbed oxygen species and it is likely that the various species may all form on oxide surfaces, although clearly they have different stabilities and different concentrations on the various oxides. Bielanski and Haber[100] have divided the metal oxides into three main groups according to their interaction with gaseous O_2: (i) p-type semiconducting oxides (NiO, MnO, etc.), which form electron-rich species (O^-, O_2^-); (ii) a group including n-type semiconductors (ZnO, TiO$_2$, V$_2$O$_5$, etc.) and also dilute solutions of transition metal ions in diamagnetic matrices (e.g., Co in MgO), which form O_2^- and O^-; and (iii) binary oxides, where the lattice oxygen is present as O^{2-} in well-defined oxyanions (e.g., bismuth molybdates) and which do not form the adsorbed oxygen species, but only O^{2-} ions. This is an oversimplification because a range of oxygen species can be observed on all these oxides[98,99] given the right conditions, which shows that care must be taken when classifying active oxygen species in selective oxidation reactions.

8.3.1.3. Monoxygen Species

Mononuclear as well as dioxygen species are present on oxide surfaces. The formation of these species requires the dissociation of dioxygen, and since the bond length of O_2^{2-} (0.149 nm) is greater than that of O_2 (0.121 nm) owing to the addition of electrons to the half-filled π_x^* and π_y^* (in the free O_2 molecule) or empty π_x^* orbitals (in the O_2 molecule coordinated to a transition metal ion),[101,102] it can be understood why the process of breaking the O–O bond is usually visualized as involving the prior addition of electrons to form the O_2^{2-} (see Scheme 8.4a).[103] The homolytic breaking of the O_2^{2-} ion produces two O^- species. The rupture of the O–O bond is one of the higher barriers in the conversion of gaseous O_2 into O^{2-}, but the major one is the further addition of an electron to O^- to form O^{2-} ion (844 kJ/mole). Owing to this high endothermic barrier, the rate of this conversion is low, which explains why the O^- species has been detected over a very wide range of oxide catalysts.[99]

It may also be noted that the addition of one electron to an oxygen atom is an exothermic process (–142 kJ/mole in gas phase), and thus oxygen atoms would rapidly be converted to O^- species which, on the surface of a catalyst, would be further stabilized by coulombic interaction with the lattice ions. This suggests that the oxygen atom ($O^\bullet$) may not be present on oxide surfaces, since it would be quickly transformed to O^- locally stabilized at transition metal ions. This indication of the unlikely existence of the oxygen atom ($O^\bullet$) also raises a question concerning the mobility of oxygen monatomic species on the surface (see Scheme 8.6), because O^- species would be much less mobile than $O^\bullet$. It should be noted, however, that when the surface of an oxide becomes negatively charged, as when it is reduced by interaction with a hydrocarbon, the O^- species is destabilized, and thus the further addition of an electron to form an O^{2-} species before it is incorporated into the lattice would become even more difficult. An energetically favorable alternative pathway thus becomes the homolytic dissociation of adsorbed O_2 to form an $O^\bullet$ species, which is then incorporated into a surface oxygen vacancy with a simultaneous two-electron transfer from two metal ions. The overall process of incorporation of oxygen into an oxide lattice would thus be the pathway indicated in Eqs. (8.2a) and (8.2b) rather than the pathway described by Eqs. (8.3a) and (8.3b), which is more usual for oxides:

$$O_2 \rightleftharpoons 2O^{\bullet} \tag{8.2a}$$

$$O^{\bullet} + 2M^{(n-1)+} \rightarrow M^{n+} O^{2-} M^{n+} \tag{8.2b}$$

$$O_2 + 2e^- \rightleftharpoons O_2^- + e^- \rightleftharpoons O_2^{2-} \rightleftharpoons 2O^- \overset{2M^{(n-1)+}}{\rightleftharpoons} 2O^{2-} + 2M^{n+} \tag{8.3a}$$

$$O^{2-} + 2(M^{n+}) \rightarrow M^{n+}O^{2-}M^{n+} \tag{8.3b}$$

It can thus be seen that the process of incorporation of dioxygen into the oxide lattice and the nature of the adsorbed oxygen species involved are less obvious than they might seem on the basis of the energetics of the transformation of the free O_2 molecule without regard for the specific characteristics of the energy of the interaction of oxygen adspecies with the "working" catalyst surface. This is an open question that needs to be answered to understand the nature of surface species and their role in catalysis.

The O^- species on oxides is usually present as a trapped species at surface defects and shows an axial symmetry.[99] Some electron spin is delocalized onto nearby cations and thus the adsorption site has an important effect on the g tensor. The crystal field stabilization energy can be up to 70 kcal/mol for certain geometric arrangements,[104] which indicates that this species is essentially trapped at specific surface sites. The effect is the same for O_2^- ions adsorbed on surfaces,[105] which points to the fact that the specific nature and thus the reactivity of these electrophilic oxygen species depend strongly on the type of the specific interaction with the surface and that these species are characterized by a reduced surface mobility. It also should be noted that in some cases it has been observed that O^- species are located close to but not on the surface.[106,107]

O^- species, besides that from adsorption of O_2 or other reactants such as N_2O, can form by UV irradiation of several supported transition metal oxides, such as Co^{2+}, Ni^{2+}, V^{5+}, or Mo^{6+}:

$$M^{n+}O^{2-} \overset{h\nu}{\rightleftharpoons} [M^{(n-1)+}O^-]^* \tag{8.4}$$

This is a short-lived species, but it has been observed that long-lived O^- species form by heat treatment. Martens *et al.*[108] noted that when $Mg(OH)_2$ is thermally decomposed to form an oxide, O^- is generated according to the following process:

$$OH^- \cdots {}^-HO \rightleftharpoons 2O^- + H_2 \qquad (8.5)$$

The concentration of O^- ions is about 10^4–10^5 times higher than that of O^- ions formed by direct thermal activation of MgO.[109] Tench *et al.*[110,111] observed that O^- on MgO readily reacts with H_2 according to the reaction

$$O^- + O^{2-} + H_2 \rightleftharpoons 2OH^- + e^- \qquad (8.6)$$

which from right to left is an alternative way to generate O^- species. These results point out the role of hydroxyls on the surface chemistry of oxides and the generation of long-lived surface oxygen adspecies.

There is nothing in the literature by way of direct evidence concerning the nature of surface oxygen adspecies in equilibrium with the bulk catalysts at the typical temperatures of catalytic tests (300–500°C) for oxides of more interest in terms of catalytic behavior for selective oxidation reactions. However, it is reasonable to assume that O^- and other oxygen adspecies on mixed oxides are in a dynamic equilibrium with the bulk species. The quantity of these species is a function of the specific characteristics of the bulk oxide, as well as of its surface state (e.g., if it is dehydroxylated or not) and the temperature. It should also be noted that charge-transfer processes involving metal–oxygen double bonds of the kind shown in Eq. (8.7) have been shown to occur easily under a variety of conditions typical for catalytic tests[99]:

$$\qquad (8.7)$$

The charge-transfer process leads to a weakening of the M–O bond and this process becomes energetically more favorable as the coordination of the ion decreases, because the Madelung contribution decreases. An increase in temperature and decrease in coordination of the oxygen in the MO bond (for ions at low coordination sites such as steps, corners, etc.) will thus favor the shift of the equilibrium to the right-hand side of Eq. (8.7).

Thus care must be taken when roughly distinguishing between lattice oxygen (in the absence of gaseous O_2) and electrophilic oxygen adsorbed species (in the presence of gaseous O_2). In a catalyst operating under real conditions, there may be no clear distinction between O^- (or other oxygen adspecies) and O^{2-} ions on the surface at the catalytically active sites.[99] Clearly, this is a major problem in considering the mechanisms of catalytic oxidation reactions, and greater effort is required for a better understanding of this issue.

8.3.2. Reactivity of Adsorbed Oxygen Species

In general, the reactivity of adsorbed oxygen species (briefly called O_{ads}) has been studied for simple once-through reactions, in which the oxygen species are consumed but not renewed. Moreover, as noted in Section 8.3.1, not all O_{ads} species are well renewed, they have not all been well characterized, and they can usually be detected in conditions and for catalysts other than those more relevant for industrial catalytic oxidation reactions.

It must also be noted that it is not correct to think in terms of clearly distinct species (O^-, O_2^-, etc.), because a continuous gradation of species may exist between O_2^+ and O_2^{2-}, depending on the specific environment at the adsorption site. The formal description of an O_{ads} species as O_2^-, for instance, does not reflect its actual charge and bond order when adsorbed on a surface. Owing to the common presence of heterogeneity in site nature over polycrystalline mixed oxides, a gradation of species normally exists at oxide surfaces. Furthermore, at high temperatures (above about 200–250°C) the O_{ads} species may not be readily distinguishable from one another, due to their easy interconversion. It is thus very difficult to obtain conclusive evidence on the catalytic role of O_{ads} species in selective oxidation. Nevertheless, the available indications are useful to further clarify their possible role in selective as well as nonselective oxidation mechanisms, contrary to what is widely accepted.

Oxygen can be involved in oxidation reactions in at least three distinct ways, more than one of which may be operative in any reaction mechanism. The first is the abstraction of a hydrogen or proton from an adsorbed organic molecule to give a radical or carbanion on the surface; the second is the attack on the organic species by a negatively charged oxygen ion whether lattice oxygen or an adsorbed oxygen; and the third is the replenishment of lattice oxygen that has been used in a direct oxidation reaction.

The O^- ion is much more reactive than O_2^-, or O_3^-, or lattice oxygen. The order of reactivity for the reaction with alkanes and alkenes is

$$O^- \gg O_3^- \gg O_2^- \tag{8.8}$$

Iwamoto and Lunsford[112] observed that when C2–C4 alkanes and alkenes are reached over these species generated on a MgO catalyst, in all cases the principal initial reaction appears to be the abstraction of a hydrogen atom from the hydrocarbon by the oxygen, followed by subsequent surface reactions that may involve surface oxide ions. With alkanes the subsequent step is the formation of an alkoxide ion, independently of the oxygen species, whereas with alkenes a carboxylate species forms. However, the type of carboxylate ion formed depends on the type of adsorbed oxygen species present. In the case of O_2^- and O_3^-, there is a scission of the C=C bond following the initial hydrogen abstraction, but not in the case of reaction with O^-.

The reaction of alkanes with O^- has been studied extensively. C1–C4 alkanes react with the O^- ion on the surface of MgO forming an alkyl radical that readily reacts with lattice oxygen to form a surface alkoxide, which at higher temperatures decomposes to form the corresponding alkene selectively.[113] Similar chemistry occurs on various supported transition metal oxides (V_2O_5/SiO_2, MoO_3/SiO_2, WO_3/SiO_2).[114,115] Kazansky,[116] however, noted that a different chemistry is possible in the presence of adsorbed neutral dioxygen:

$$O^- + RH + O_2 \rightarrow OH^- + ROO^\bullet \qquad (8.9)$$

The peroxy radicals may further abstract a hydrogen atom from another alkane molecule or intramolecularly and further transform to selective oxidation products. O^- species may also react with O_2 to form an O_3^- species [see Eq. (8.1)]. Takita and Lunsford[117] found that O_3^-, although less reactive that O^-, abstracts a hydrogen from alkanes generating an O_2 molecule and an alkyl radical over MgO. The latter reacts either with lattice oxygen to give an alkoxide or with adsorbed dioxygen to form an alkyl peroxy radical:

$$C_2H_6 + O_3^- \rightarrow C_2H_5^\bullet + OH^- + O_2 \qquad (8.10)$$

$$C_2H_5^\bullet + O^{2-} \rightarrow C_2H_5O^- + e^- \qquad (8.11)$$

$$C_2H_5^\bullet + O_2 \rightarrow C_2H_5OO^\bullet \qquad (8.12)$$

The ethoxide ion decomposes above 300°C to give ethylene and acetate ions; the peroxy radical was assumed to decompose to form carbonate ions, but the consecutive pathway after formation of the peroxy radical was not investigated in detail. This consecutive pathway is highly dependent on the properties of the oxide catalyst. Furthermore, it is known that peroxy radicals are the selective intermediates in several selective oxidation mechanisms (e.g., autoxidation processes). Thus it is reasonable to expect that the mechanism via the alkyl peroxy radical can lead to the formation of selective oxidation products and not to carbon-chain degradation. It should also be noted that with longer-chain alkanes, the oxygen attacks a methylenic carbon atom to form the alkoxy species, whereas the alkyl peroxy radical is formed from a methylic (terminal) carbon atom.

Reaction of O_3^- with alkenes leads to nonselective oxidation.[117] The first step in this case was also assumed to be the hydrogen abstraction to form a $(CH_2{=}CH)^\bullet$ species, which may react either with gaseous O_2 to form a peroxy radical or with lattice oxygen to give, after further H abstraction, carbonate ions, which then decompose to give $CH_4 + CO_3^{2-}$:

$$H_2C{=}CH_2 + O_3^- \rightarrow (CH_2{=}CH)^{\bullet} + O_2 + OH^- \qquad (8.13)$$

$$(CH_2{=}CH)^{\bullet} + O^{2-} \rightarrow (CH_2{=}C)^- \cdots HO^- \qquad (8.14)$$

$$(CH_2{=}C)^- \cdots HO^- + O^{2-} \rightarrow H_3C{-}COO^- + 3e^- \qquad (8.15)$$

The peroxy radical was suggested as the intermediate in the formation of oxygen-containing organic molecules. The reactivity of ethene with O^- on MgO is similar to that shown in Eqs. (8.13)–(8.15), but an O^- ion formed on various supported oxides (MoO_3/SiO_2, WO_3/SiO_2) adds to the alkene to give a H abstraction.[118–120] The $CH_2CH_2O^-$ intermediate at higher temperatures decomposes to form $(CH_2CH)^{\bullet}$ species, which undergo transformations similar to those described above, indicating that notwithstanding the different initial steps in the reaction mechanism, the final conversion pathway may be similar. This type of chemistry has been observed with higher alkenes (propene, but-1-ene).[119]

The O_2^- ion shows similar chemistry to O^-, but is less reactive. Reaction of O_2^- with propene on MgO gives acetaldehyde and methanol, whereas its reaction with but-1-ene gives 2-butanol, methanol, acetaldehyde, and acrolein.[112] It is suggested that the first step in this case should also be hydrogen abstraction with formation of an allyl radical that reacts with lattice oxygen to form acetate and formate ions, the latter then transforming slowly to carbonate species.

The reaction of O_2^- with propene adsorbed on ZnO gives rise to a π-allyl intermediate as the first step,[121,122] but then this intermediate further reacts with O_2^- to give a hydroperoxide, which then converts to glycidaldehyde and further to acrolein. This indicates that the reactivity of the various intermediates with O_{ads} species depends strongly on the oxide catalyst, as well as on the reaction conditions, thus illustrating the limitations in making generalizations concerning the catalytic behavior of these species.

Akimoto and Echigoya[123,124] suggested that the O_2^- species is involved in the formation of maleic anhydride from butadiene or furan over supported molybdena catalysts. Although their evidence is not conclusive, it would seem that the addition of a dioxygen molecule to a conjugated double bond to form an endoperoxo species is a possible mechanism for the formation of an anhydride. However, working with different oxide catalysts, Fricke et al.,[125] concluded that O^- and O_2^- species are only responsible for the formation of carbon oxides in maleic anhydride synthesis from C4 hydrocarbons. There is thus conflicting evidence, but certainly the chemistry of adsorbed dioxygen species (either charged or neutral) on the surface of oxides in the reaction with organic molecules has not been systematically studied and may hold interesting new synthetic possibilities.

It should be pointed out that no mention has been made here of all the studies regarding the reactivity of adsorbed oxygen species over silver in the epoxidation of ethene, because this is a specific individual case in terms of both the type of catalyst and the substrate and thus does not fit with the general organization of this section of the book. Reference can be made to a major review, where all these aspects are discussed.[126] The problem was also discussed in an earlier paper by Che and Tench on the characteristics and reactivity of molecular oxygen species on solid catalysts.[98]

8.3.3. New Aspects of the Reactivity of Surface Oxygen Species

In the introduction to this section we noted three basic problems (extendability, kinetics, and uniqueness) in generalizing the established ideas on the nature of oxygen species active in selective oxidation reactions. As we observed, in general little effort has been expended to solve these problems, but a few examples can be described that provide interesting perspectives.

The first example relates to the nature of oxygen species over vanadium/phosphorus oxides active in the selective formation of maleic anhydride from n-butane. The general aspects of the nature of the catalyst and surface chemistry were discussed in Chapter 4, so only the items relevant for an understanding of the specific question of the nature of oxygen species active in maleic anhydride synthesis from n-butane will be dealt with here.

The active catalyst for this reaction[127] is formed from a single crystalline phase $[(VO)_2P_2O_7]$ characterized by a specific surface configuration and, in particular, by the presence of pseudosquare pyramidal vanadyl dimers in a *trans* position (see Chapter 4). One of these vanadyl centers is coordinatively unsaturated and thus gives rise to strong Lewis acid sites on the surface.[128] The formal valence state of vanadium in $(VO)_2P_2O_7$ is four and although there is evidence of the formation of V^{3+} ions during the catalytic reaction,[127] there is agreement that the V^{5+}/V^{4+} redox couple is responsible for the oxidation behavior. It has been proposed in a number of studies that various V^{5+} phases (β-VOPO$_4$, α_{II}-VOPO$_4$, γ-VOPO$_4$, and δ-VOPO$_4$) are present on the working catalyst and that the behavior of V/P oxides depends on the synergetic interaction between these surface V^{5+} phases and the $(VO)_2P_2O_7$ matrix.[129–132] This conclusion contradicts results from a study of the change in the structural and reactivity characteristics of industrial V/P oxide catalysts with time-on-stream during n-butane oxidation,[133] indicating that in the starting catalyst ("nonequilibrated" sample) poorly crystallized $(VO)_2P_2O_7$ is present together with an amorphous V^{IV}–P–O and γ-VOPO$_4$, whereas at the end in the "equilibrated" catalyst only well characterized $(VO)_2P_2O_7$ is present. The structural reorganization occurs with an increase in the selectivity in n-butane oxidation to maleic anhydride, thus indicating that the V^{5+}-phase is not necessary for the selectivity. The fact that vanadyl pyrophosphate alone and no other minor V/P oxide phase is responsible for selective oxidation of n-butane to maleic anhydride was confirmed using a combination of physicochemical techniques.[134]

Studies using fast transient catalytic techniques (TAP—temporal analysis of products—reactor) suggested that alkane oxidation involves primarily oxygen species adsorbed at vanadium surface sites,[75,76] and in particular short-lived oxygen species because increasing the time lag from 0.1 to 600 ms between consecutive alternating butane and oxygen pulses leads to a decrease in the formation of maleic anhydride. This seems to contradict the basic evidence indicating that the V/P oxide catalyst is able to convert n-butane to maleic anhydride with high selectivity in the absence of gaseous oxygen (anaerobic conditions).[135,136] Using $^{18}O_2$ in transient catalytic TAP experiments, Kubias et al.[137,138] concluded that a redox mechanism involving only lattice oxygen operates in all reaction steps from n-butane to maleic anhydride, whereas carbon oxide formation involves a mechanism with the participation of adsorbed oxygen species.

Using $^{18}O_2$, but in a closed-recirculation reactor, Abon et al.[139] concluded that both maleic anhydride and carbon oxides form only by reaction with lattice oxygen, because only unlabeled products were observed at the beginning of the reaction. However, they also noted that the active lattice oxygens are only found within a few surface layers, in agreement with the conclusions of Kubias et al.[138] These results reproduce the older data[140] on the quantitative comparison between the number of structural oxygens incorporated into organic molecules in the anaerobic oxidation of hydrocarbons over V/P oxides and Bi/Mo. Unlike the latter, only a number of oxygens corresponding to the first surface layer can be incorporated into the organic molecule in anaerobic oxidation on the V/P oxide catalyst. This evidence does not fit well with the model of lattice oxygen as the only active oxygen species in the V/P oxide catalyst, as it is difficult to justify the idea that only the first surface layer can be removed even in the hypothesis of limited bulk oxygen diffusion. It may also be noted that Schuurman and Gleaves[141] found that the activation energy for aerobic n-butane conversion in transient-response TAP experiments is 12 kcal/mol on a V/P oxide catalyst after oxygen treatment at 450°C for 10 min, but this value increases progressively to 23 kcal/mol with the catalyst reduction. The values of the activation energy in the rate of n-butane depletion, reported by various research groups on the basis of kinetic measurements under steady state conditions also reveal a scattering of values in the 10–30 kcal/mol range.[127] Again, this result does not fit the model of lattice oxygen of vanadyl pyrophosphate as the active species responsible for n-butane selective oxidation, which indicates that the situation is more complex.

In order to clarify these contradictory data on the role of the oxygen species active in the selective oxidation of n-butane to maleic anhydride, it is useful to observe that the cited conclusions were significantly influenced by the rough distinction between structural oxide oxygen and adsorbed oxygen, the latter forming only in the presence of gaseous O_2. This rough discrimination has mainly determined the interpretation of the transient experiments using labeled oxygen. In Section 8.3.1 we criticized this kind of schematization of the active oxygen species on the surface of oxides and thus the validity of the model of nucleophilic structural oxygen as responsible for selectivity and electrophilic adsorbed oxygen as responsible for nonselective conversion (see Scheme 8.4b). Recent results on the V/P oxide catalyst[142,143] further support this conclusion on the limitations of this approach and clarify the above results on the nature of active species in maleic anhydride synthesis.

The starting catalyst in these experiments was an "equilibrated" V/P oxide catalyst that had been used under steady state conditions for approximately 3000 h. The catalyst is monophasic and contains only $(VO)_2P_2O_7$ with a P/V ratio close to 1.0 and an average vanadium oxidation state close to 4.0. Under steady state conditions the catalyst gave selectivities in maleic anhydride of approximately 66% at 78% n-butane conversion. The starting sample is thus representative of an industrial catalyst for this reaction.

This catalyst[142] was pretreated *in situ* before transient catalytic TAP reactor tests at increasing temperatures (410–530°C range) in a flow of pure oxygen before the temperature was decreased to 400°C for the determination of the catalytic behavior while feeding consecutive pulses of a gas blend containing 78% n-butane and 12% argon. The catalytic behavior was followed for up to 16,000 consecutive pulses, feeding about 10^{14} molecules per pulse. Parallel Raman experiments were done to observe the formation of V^{5+}-phosphate species during the oxidation treatment. At temperatures below 500°C in the presence of oxygen no evidence was found for the formation of V^{5+}-phosphate species, even after extended oxidation treatments (24 h). However, experiments indicate a continuous increase in the oxygen uptake when the temperature of V/P oxide pretreatment in oxygen flow increases from 410 to 530°C. The maximum selectivity in maleic anhydride formation was found for an oxygen pretreatment temperature (450°C) significantly lower than the temperature at which V^{5+}-phosphate species appear to be detectable (500°C) together with $(VO)_2P_2O_7$ (Figure 8.4). For the catalyst pretreated at 450°C, the amount of oxygen incorporated into the organic molecule is roughly equivalent to the number of surface V^{4+} sites. Furthermore, the oxygen adsorbed at this pretreatment temperature can all be desorbed at higher temperatures.

Owing to the presence of strong Lewis acid sites,[128] the vanadyl pyrophosphate strongly adsorbs oxygen, forming a rather stable oxygen species coordinated at unsaturated vanadyl surface ions. This oxygen species desorbs from the catalyst at higher temperatures than those of the catalytic activity in n-butane oxidation, although at lower temperatures than lattice oxygen.[143] Thus this species is present on the catalyst surface even during "anaerobic" experiments, because it is not removed by simply flushing the catalyst with an inert gas (N_2, He), but shows a reactivity different than that of real lattice oxygen, as shown from the fact that when

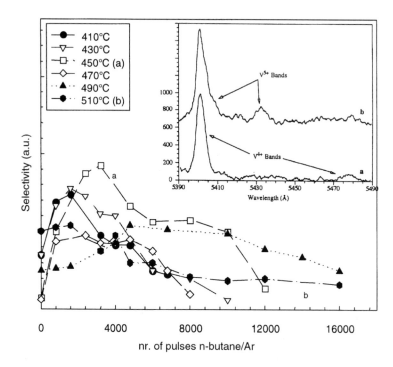

Figure 8.4. Selectivity to maleic anhydride (arbitrary units) at 400°C as a function of the number of *n*-butane pulses in TAP reactor experiments for the same V/P oxide catalyst pretreated at different temperatures in a pure O_2 flow.[142] The inset shows the Raman spectra of the same V/P oxide samples pretreated at 450°C (a) and 510°C (b).[142]

V^{5+}-phosphate species form there is a considerable decline in selectivity. A V^{4+} surface ion with a coordinated oxygen can be viewed as a $V^{5+}-O_n^-$ limiting species, but must be interpreted more correctly as an intermediate situation between lattice oxygen and pure adsorbed oxygen species.

The question, however, is whether or not this suggested oxygen surface species is a chemically reasonable one in terms of both formation and role in the reaction mechanism of maleic anhydride synthesis. A theoretical contribution of Jørgensen *et al.*[144,145] clarifies this point. These authors studied the electronic structure of different types of adsorbed oxygen species on a vanadyl pyrophosphate surface using the frontier orbital approach on a edge-sharing pair VO_5 unit [$(V_2H_8)^{8-}$ cluster], the characteristic unit of the active surface of vanadyl pyrophosphate (see Chapter 4). Calculations were done using a more complex unit instead of a unidimensional bare coupled vanadyl unit [$(V_2H_8)^{8-}]_\infty$. The more complex unit with $(PO_3)^-$ as well as $[P(OH)_3]^{2+}$ units attached as models for pyrophosphate does not show significant differences, however, because the neighboring phosphate groups have little influence on the electronic properties of the vanadium–oxygen bond. Molecular oxygen can be adsorbed onto this vanadyl unit in different ways, and especially as η^1-superoxo and η^2-peroxo (species a and b, respectively, in Figure 8.5). The b type of species is slightly more stable than the a species, although the differences are small. In both cases the V–O overlap population indicates a relatively stable species, although more weakly bound to the surface than the vanadyl oxygens.

The reactivity of these oxygen species with reference to the mechanism of maleic anhydride synthesis was also evaluated using the same approach. The study considered only the final steps from the butadiene intermediate up to maleic anhydride, and thus the first part of the pathway from n-butane to butadiene was not evaluated. The vanadyl ($V^{2+}=O$) was considered to react with butadiene leading to a coordinated 2,5-dihydrofuran type of molecule. This molecule remains adsorbed on the surface while molecular oxygen is adsorbed and activated on adjacent vacant vanadium sites (species c and d in Figure 8.5). The adsorbed molecular oxygen then oxidizes the adjacent coordinated 2,5-dihydrofuran to the corresponding alcohol. The η^2-peroxo species is thought to be the most likely one for activation of the C–H bond in the 2-position of adsorbed 2,5-dihydrofuran (species e in Figure 8.5). The σ_{C-H} bond donates 0.3 electrons to the σ^*_{O-O} orbital, leading to transfer of a hydrogen atom to the peroxo species and the forming of a surface-bound hydroperoxide group (species f in Figure 8.5). The formation of the latter species is energetically favored.[145] With the suggested orientation of the two adsorbed species there is a considerable C^1-O^1 (bond length = 0.16 nm) interaction; an overlap population of 0.635 was calculated.[145] The OH group can therefore be transferred to the neighboring 2,5-dihydrofuran derivate, giving the corresponding 2-hydroxy derivative (species g in Figure 8.5). The asymmetric lactone forms from the latter species by first transferring H^1 to O^2 and then further H^2 with an energy gain of 0.9 eV to give the asymmetric adsorbed lactone and one molecule of water (species h and i in Figure 8.5). It was suggested that the 5-position becomes oxidized in a similar way after desorption of water and adsorption of another molecule. Hence, these results not only indicate that the formation of stable coordinated molecular oxygen species on surface unsaturated vanadyl ions is a reasonable hypothesis, but also confirm their possible role as selective oxidation species.

On the other hand, Gleaves et al.[142] have shown that the V^{5+}-phosphate species also gives rise to maleic anhydride formation from n-butane, although less selectively. Data in Figure 8.4 indicate that on the V/P oxide catalyst in the presence of this V^{5+}-phosphate species maleic anhydride forms for a higher number of pulses than in its absence, which shows that this species can act as a reserve of active oxygen for the vanadyl center during the reaction. This result is confirmed by the

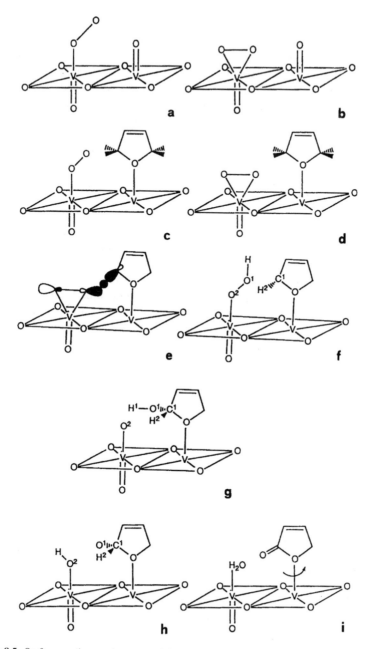

Figure 8.5. Surface reaction species on vanadyl pyrophosphate in relation to the mechanism of maleic anhydride formation.[144]

analysis of the transient response curve for maleic anhydride formation on a "nonequilibrated" V/P oxide catalyst, where a V^{5+}-phosphate species is present together with $(VO)_2P_2O_7$ and an "equilibrated" V/P oxide catalyst consisting only of $(VO)_2P_2O_7$. The results are summarized in Figure 8.6, where the response curve for maleic anhydride production under anaerobic conditions when furan is rapidly pulsed over a V/P oxide catalyst are compared.[78] Furan was used in these tests as a probe molecule because it is a reaction intermediate in n-butane oxidation,[127] and its reaction chemistry is simpler to follow. It is also more easily activated than butane and can react with more of the surface oxygen, which makes it a better probe for following changes in the overall surface properties of the catalyst. With the catalyst showing both V^{5+}-phosphate and $(VO)_2P_2O_7$ (curve a in Figure 8.6), the shape of the response curve for maleic anhydride indicates two distinct reaction pathways: a more rapid pathway that is the same as that for the equilibrated sample (curve b in Figure 8.6) and a slower pathway that is responsible for a larger formation of maleic anhydride. It is thought that the second pathway is determined by the rate of supply of the active oxygen species to the vanadyl centers from the V^{5+}-phosphate reserve, which would explains why from about 2 to 10 s there is a flat region of relatively constant formation of maleic anhydride. Hence, a discussion of selective species does not have much meaning here, because the various active oxygen species act together to determine the overall catalytic behavior, but do not cooperate in determining the selective behavior.

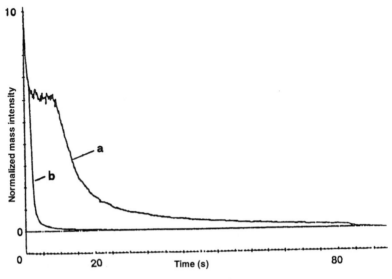

Figure 8.6. Transient response curve for maleic anhydride formation in the interaction of furan under anaerobic conditions with a "nonequilibrated" (a) and an "equilibrated" V/P oxide catalyst.[78]

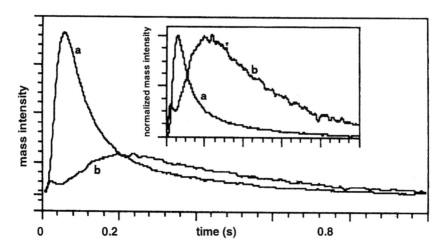

Figure 8.7. Transient response curve for maleic anhydride at 380°C obtained by pulsing an *n*-butane/Ar mixture over V/P oxide samples pretreated in oxygen at 450°C: (a) signal after 1000 pulses, and (b) signal after 8000 pulses.[142] The respective normalized curves are shown in the inset.

Gleaves *et al.*[142] also noted that after pretreatment with oxygen the selectivity to maleic anhydride reaches a maximum after about 2000–4000 pulses, which corresponds to removal of a large fraction of the oxygen taken up (see Figure 8.4). The large number of pulses necessary to reach the maximum in selectivity suggests that the increase in selectivity may not be due to initially present short-lived electrophilic-type adsorbed oxygen species, but rather derives from a change in the concentration of these species which causes the change in selectivity. This observation points out that both the nature of the surface oxygen species and their concentration determine the catalytic behavior. In other words, the problem of selectivity must be analyzed in terms of both the type of oxygen species present and the kinetics of the surface processes. The data in Figure 8.6 illustrate the general problem of how selective behavior is linked to the latter.

In this respect, another interesting observation of Gleaves *et al.*[141,142] is that on the surface of the V/P oxide catalyst after preoxidation the rate of maleic anhydride desorption is significantly more rapid than on the same catalyst after the complete removal of the oxygen taken up during the pretreatment (Figure 8.7). The average surface residence time changes from about 0.1 to 0.3 s and thus it is reasonable to expect that a longer-lived surface species would be more susceptible to consecutive oxidation in the presence of gaseous oxygen than a shorter-lived species. The surface oxidation state (we also include here for the sake of complete accuracy the oxygen species strongly chemisorbed on V^{4+} surface ions as formal V^{5+} species) of the V/P oxide catalyst thus not only determines the nature of the possible oxygen species, but also the rate of desorption of maleic anhydride and the rate of its consecutive oxidation in two opposing ways: (i) increasing the surface oxidation state increases the concentration of oxygen species that can attack the adsorbed maleic anhydride, and (ii) increasing the surface oxidation state decreases the surface lifetime of adsorbed maleic anhydride with a consequent lessened possibility for its consecutive oxidation. This explains why Gleaves et al.[142] observed a maximum in selectivity to maleic anhydride with respect to both the number of pulses for a specific temperature of V/P oxide pretreatment with O_2 and the temperature of O_2 pretreatment, because an optimal surface oxidation state must exist to optimize these two opposing requirements.

This evidence for the case of the V/P oxide catalyst illustrates how care must be taken in analyzing the data in terms of: (i) "anaerobic" behavior and related surface oxygen species, and (ii) factors determining the selective behavior of a catalyst, without detailed information about, e.g., the kinetics of the surface processes, the nature of adsorbed species, and their energy relationships. What also emerges from these data is the fact that the model of nucleophilic vs. electrophilic oxygen species is inadequate to describe the surface catalytic chemistry in selective oxidation reactions at oxide surfaces.

In regard to the three basic problems described at the beginning of this section (extendability, kinetics, and uniqueness), the discussion on the nature of the active species in V/P oxide catalysts clearly indicates that:

1. The model of lattice oxygens as the only type of oxygen responsible for selective catalytic behavior cannot be generalized to explain the catalytic chemistry at oxide surfaces.
2. More complex kinetic models of the surface processes in oxidation reactions that include the concentration and distribution of adspecies, their surface mobility, and lifetime are needed.
3. The various surface properties are linked together, as shown, for instance, by the fact that the rate of desorption of maleic anhydride (thus also its surface lifetime) depends on the redox state of the surface, which also influences the concentrations of oxygen adspecies and their mobility.

It is thus necessary to develop microkinetic models of the surface reactions that include: (i) the lifetime of adsorbed molecules (a function of the rate of all consecutive and competitive surface processes), and (ii) the complete surface reaction network. This will be a challenge for future research, but even a simpler approach following the basic assumptions outlined above can be a good tool for understanding the surface chemistry and designing better catalysts.

An example of this microkinetic approach to surface chemistry is given in the case of the oxidation of n-butane and n-pentane over $(VO)_2P_2O_7$ (V/P oxide) and

V-phosphomolybdo-heteropoly acid ($H_5PV_2Mo_{10}O_{40}$; briefly H/PV/Mo oxide) catalysts.[146,147] V/P oxide is the industrial catalyst for the synthesis of maleic anhydride from n-butane, and optimization of its performance is crucial for its commercial use (Chapter 4). The same catalyst is also active in n-pentane oxidation, but forms maleic and phthalic anhydride.[148,149] Both reactions involve a complex polyfunctional, multistep mechanism which occurs, however, entirely on the adsorbed phase without desorption of any reaction intermediate (Chapter 4). With a surface microkinetic model[146] based on: (i) the surface reaction network indicated by reactivity and spectroscopic studies, (ii) the rate of transformation of the various adsorbed intermediates assumed to be a function of the concentration of surface active oxygens (O^*) divided by the concentration of surface vanadium atoms (V_{sup}), and (iii) the concentration of the various adspecies, it is possible to determine how the selectivity for products depends on this O^*/V_{sup} index. This model thus makes it possible to predict the existence of a maximum in the selectivity (Figure 8.8) as a function of catalyst properties in good agreement with the experimental data. In the case of n-pentane oxidation, the relative selectivity for maleic and phthalic anhydrides is also a function of the O^*/V_{sup} index. This index is an intrinsic property of the catalyst, which depends on the modality of its preparation, the exposition of crystalline planes, and the nature of its defects. It is therefore possible to use it as a tool to optimize the design of V/P oxide catalysts toward a maximum selectivity for phthalic instead of maleic anhydride in n-pentane oxidation.[148]

H/PV/Mo oxide is also a catalyst that is active and selective in maleic anhydride from n-butane, although it is less selective and stable than V/P oxide. In the case of n-pentane, however, it gives only maleic anhydride[150] instead of phthalic and maleic anhydride as V/P oxide.[148,149] It is thus necessary to understand the reasons for this difference. The microkinetic model, whose results are briefly summarized in Figure 8.8, can be used to analyze this problem. The graph shows that if H/PV/Mo oxide has a higher O^*/V_{sup} index with respect to an optimal value to maximize selectivity for maleic anhydride (it is assumed that V/P oxide shows a maximum selectivity for maleic anhydride from n-butane) the selectivity for maleic anhydride is not only lower (Figure 8.8a), but only maleic anhydride forms from n-pentane (Figure 8.8b) rather than a combination of maleic and phthalic anhydride as in the case of the V/P oxide catalyst. The problem then is to demonstrate experimentally that H/PV/Mo oxide shows a higher O^*/V_{sup} index than V/P oxide.

A useful test reaction to evaluate this index is the oxidation of decaline. This alkane is oxidized over both V/P oxide and H/PV/Mo oxide to products of simple oxidative dehydrogenation (tetralin and naphthalene) and products that also contain oxygen (phthalic anhydride and naphthoquinone).[147] With the ratio of the selectivities (S_{ratio}) for these two classes of products it is possible to estimate the O^*/V_{sup} index, because a decrease in this ratio indicates a higher capability of the catalyst

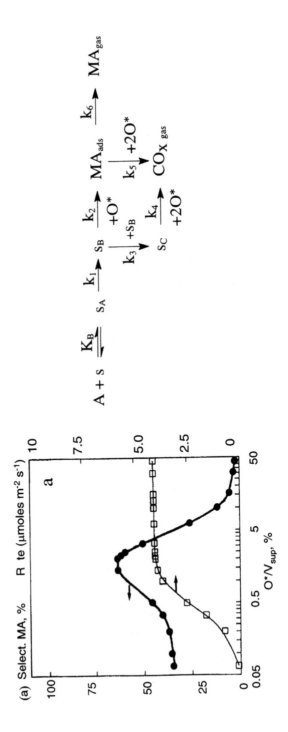

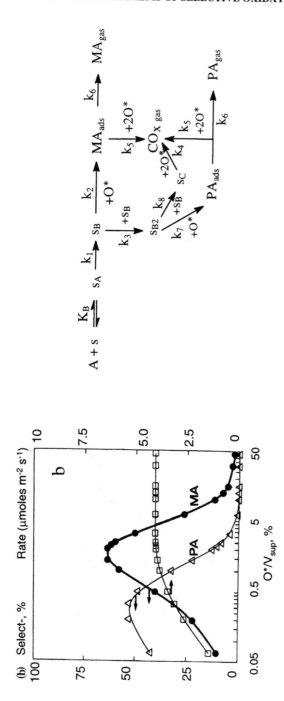

Figure 8.8. (a) Selectivity to maleic anhydride (MA) and rate of butane depletion on $(VO)_2P_2O_7$ as a function of the concentration of surface active oxygens (O^*) divided by the concentration of surface vanadium atoms (V_{sup}); data calculated at 300°C on the basis of a microkinetic model.[146] (b) As in case (a), but for n-pentane oxidation to maleic and phthalic (PA) anhydrides.[146] The relative kinetic reaction networks are also shown.

toward oxygen insertion and thus a higher value of the O^*/V_{sup} index. The results summarized in Figure 8.9 show that H/PV/Mo oxide shows a lower S_{ratio} and thus a higher value of the O^*/V_{sup} index than V/P oxide. When this indication is combined with the results using the microkinetic model shown in Figure 8.8, the differences in the catalytic behavior for alkane oxidation of these two classes of catalysts can be explained.

These results indicate that understanding the kinetics of the surface processes represents a new frontier of catalytic science with practical relevance for the design of new improved catalysts and for the analysis of the possible increase in selectivity by working with the catalyst under nonstationary conditions. Even simple models, such as that discussed above, can be useful for a more quantitative approach to understanding the catalytic chemistry at oxide surfaces.

The last of the three basic issues in this section was the problem of the *uniqueness* of the modality of oxygen activation and interaction with the hydrocarbon to have selective oxidation. We have already pointed out several times that oxygen adspecies may have a role in determining the selective oxidation pathway, but the problem of the uniqueness of the oxygen–hydrocarbon interaction can be addressed by analyzing the results of Frei *et al.*[151-157] These authors observed that in alkaline or alkaline earth exchanged zeolites (NaY and BaY, especially) there is a high electrostatic field inside the zeolite cage. The supercage framework carries

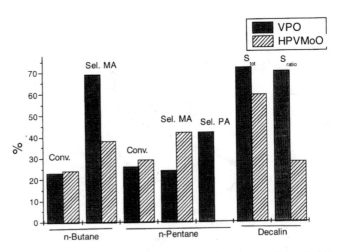

Figure 8.9. Comparison of (i) conversion of *n*-butane and selectivity to maleic anhydride (MA), (ii) conversion of *n*-pentane and selectivity to maleic and phthalic (PA) anhydrides, and (iii) total selectivity (S_{tot}) (all products except carbon oxides) and relative selectivity (S_{rel}) (sum of tetralin + naphthalene to sum of phthalic anhydride + naphthoquinone; this ratio is an index of the H-abstraction properties vs. O-insertion properties) in decaline oxidation. Data refer to a reaction temperature of 300°C.[147]

a formal negative charge of 7 due to the presence of Al in the zeolite lattice, which is counterbalanced by the charges of the alkaline metal ions that are located both inside the hexagonal cages (site S_I) and inside the supercage (sites S_{II} and S_{III}).[158] The cations in the supercage are not well shielded electrically by the zeolite oxygens so very high electrostatic fields are possible. The field is about 30 V/nm at a distance of about 0.1 nm from the Na^+ ion surface.[159]

When a hydrocarbon and oxygen collisional pair forms inside the zeolite cage and is oriented parallel to the field generated by the alkaline metal ion, there is a spontaneous transfer of charge from the hydrocarbon to the dioxygen molecule, forming a hydrocarbon radical cation and a O_2^- complex stabilized by the strong electrostatic field. There is a rapid subsequent proton transfer to the baselike O_2^-, and the hydroperoxy and hydrocarbon radicals so produced undergo cage recombination to yield a hydroperoxide, which then may further convert to selective oxidation products. This reaction pathway is shown in Scheme 8.9 for the case of the cyclohexane and O_2 reaction in NaY zeolite; the final oxidation product is cyclohexanone, which forms very selectively.[155] Irradiation with visible light accelerates the formation of the charge transfer complex, but the reaction also occurs by thermal activation.

In all cases reaction occurs under very mild conditions, around room temperature. Several types of hydrocarbons (alkenes, alkanes, alkylaromatics) have been found to be oxidized selectively under very mild conditions in a similar way[151–157]: (i) toluene to benzaldehyde, (ii) ethylbenzene to acetophenone, (iii) propene to propylene oxide and acrolein, (iv) 2-butene to methyl vinyl ketone, (v) propane to acetone, and (v) ethane to acetaldehyde. These studies were done under conditions quite far removed from those of the possible application, and thus successful upscaling of these experiments with micromolar quantities will require considerable effort to solve problems involving, e.g., effective productivity and operating conditions that allow continuous desorption of the products from the zeolite host. Nevertheless, these studies indicate that effective generation of hydroperoxo species by electrostatic stabilization of the hydrocarbon to oxygen charge-transfer

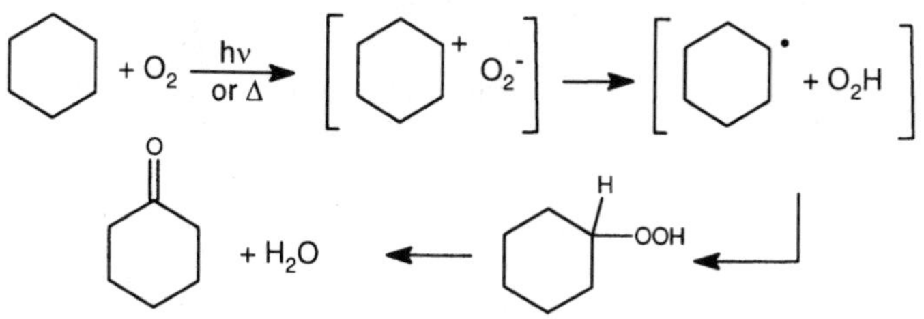

Scheme 8.9. Initial reaction pathway in cyclohexane and O_2 reaction in the presence of a high electrostatic field inside the supercage of a NaY matrix.[155]

complex is possible. The hydroperoxo species can then be converted under mild conditions to various types of selective oxidation products.

In conclusion, selective oxidation products may form over solid surfaces not only by reaction of hydrocarbons with lattice oxygen, but also by direct gas phase hydrocarbon and oxygen interaction, when, due to the presence of a strong local electrostatic field in a confined cleft or cavity of the solid catalyst, the collisional pair further quickly transforms to a charge-transfer encounter complex. The latter is also required to increase the probability of formation of the collisional pair. Three basic types of oxidation reactions at oxide surfaces (we have limited discussion here to gas phase heterogeneous chemistry without analyzing the interesting chemistry in the liquid phase, but with a solid catalyst; an extensive analysis of this topic was published by Sheldon et al.[160–163]) may thus be possible:

1. Reaction of a hydrocarbon with the lattice oxygen.
2. Reaction of a hydrocarbon with adsorbed activated oxygen species on the oxide surface. Included in these oxygen adspecies are: (i) singlet dioxygen formed at the oxide surface, (ii) charged mono- or multinuclear oxygens ($O^{\delta-}$, $O_2^{\delta-}$, $O_2^{\delta+}$, $O_3^{\delta-}$), and (iii) nearly neutral monoxygen formed by homolytic O_2 splitting ($O^{\bullet}$).
3. Reaction of a hydrocarbon and dioxygen in the gas phase, but very near to the surface, which stabilizes the intermediates.

These different types of oxidation chemistries require different catalysts and reaction conditions, although, again, it should not be concluded a priori that more than a single type of surface chemistry determines the catalytic behavior. It should also be noted that while the first type of surface chemistry has been extensively investigated, the other two possibilities have been studied in much less detail and not systematically. More attention should be given to these alternative possibilities for selective oxidation of a hydrocarbon and optimizing the catalysts/conditions. It is also evident that although sometimes the same kinds of products may be obtained, in principle a different chemistry of hydrocarbon and oxygen interactions may lead to a different kind of reaction product, thus opening new synthetic possibilities.

The latter two types of oxidation reactions also point out in general terms the importance of a radical type of surface chemistry in selective oxidation catalysis and the possibility of mixed heterogeneous–homogeneous mechanisms. This subject has been discussed in detail by Garibyan and Margolis[164] and Lunsford et al.,[165] the latter with specific reference to high-temperature catalytic reactions, such as methane oxidative coupling. Some relevant aspects have also been discussed by Sokolovskii[166] in a review on the principles governing oxidative catalysis on solid oxides and by Pyatnitsky and Ilchenko.[167] This subject will not be discussed in detail here, but it must be noted that while a mixed mechanism is reasonable enough for high-temperature reactions, there are data, albeit less precise and convincing, for reactions at temperatures lower than about 500°C. Garibyan and Margolis[164] noted, however, that radicals were detected by mass spectrometry and ESR in propene oxidation over zinc and manganese oxides at temperatures below 500°C.[168–171] Martir and Lunsford,[172] using a matrix isolation ESR method, demonstrated the formation of allyl and allylperoxy radicals in propene oxidation at 450°C over Bi_2O_3, γ-, and α-bismuth molybdate, and MoO_3.

Bismuth oxide was the most effective for generation of both radicals, while MoO_3 was found to be the least effective, although that was probably due in part to the greater rate of further conversion of these radicals on MoO_3. A better estimate of the rate of generation of hydrocarbon radicals over oxide surfaces is thus difficult. Garibyan and Margolis[164] suggested that the propenyl radical forms by interaction of reduced cations (e.g., Bi^{2+}) with the alkene generating the corresponding radical, which may either desorb into the gas phase (and eventually give dimerization reactions in the absence of reactive oxygens) or react with lattice oxygen or O_2 to give the allylperoxy radical. A radical-type surface chemistry is thus one of the aspects to be considered in the analysis of the mechanism of selective oxidation at oxide surfaces.

8.4. MODIFICATION OF THE SURFACE REACTIVITY BY CHEMISORBED SPECIES

In Sections 8.1–8.3 it was shown that the usual mechanistic and kinetic approaches are based on the implicit assumption of an ideal surface with an intrinsic reactivity that is not influenced by the chemisorption of the reactants, intermediates, or products. Some aspects of this influence are often considered such as the presence of competitive chemisorption, i.e., the fact that the number of free active sites decreases when another molecule competes for chemisorption on them, or that the rate of reaction is determined by the adsorption of reactants or desorption of products. However, the basic question of whether or not the "working catalyst surface" shows different reactivity properties from the "clean catalyst surface" has not been systematically addressed, and is an issue related to the general questions discussed in Section 8.2 concerning the differences between the models of "active sites" and "living active surface."

The importance of the role of chemisorbed species in the modification of the surface reactivity can be seen when the behavior of alkanes is compared with that of alkenes on the same catalyst. One example is illustrated in the analysis of the catalytic behavior of V/Sb oxides in the ammoxidation of propane and propene. As described in Chapter 4 V/Sb oxides have been widely patented by BP America and ICI-Katalco, among others, and their interesting catalytic behavior has been confirmed by several research groups. These catalysts are selective in the formation of acrylonitrile from both the C3 alkane and alkene. Propane and propene, owing to their different chemical properties in propene allylic hydrogens and a π-bond system, possess different chemical properties of chemisorption and activation. A comparison of the catalytic behavior of V/Sb oxides in propane and propene ammoxidation is thus useful for understanding how the catalytic properties of the catalyst are influenced by the chemisorption of the reactants on the catalyst surface.

8.4.1. The Role of Alkenes in the Self-Modification of the Surface Reactivity

Propane and propene ammoxidation using the same catalyst and reaction temperature are compared in Figure 8.10, which shows the dependence of the selectivity for acrylonitrile on the conversion of the hydrocarbon at 480°C.[28,173] Acrylonitrile can be further oxidized to give carbon oxides, and thus it is reasonable that at high conversion of the hydrocarbon its selectivity decreases while that for carbon oxides increases. However, the data in Figure 8.10 point out the anomaly that on the same catalyst, and at the same reaction temperature, acrylonitrile converts to carbon oxides faster when formed from propane than when formed from propene, even though propene is the reaction intermediate from propane.[174] In fact, the maximum in acrylonitrile formation from propane is observed at much lower conversions than from propene so the maximum yield is significantly lower. This fact has clear negative implications for the industrial development of a process of acrylonitrile synthesis from propane rather than propene, so it is of critical importance to understand the reasons for it. We shall see that the difference in behavior can be associated with the different modifications of the surface reactivity when feeding in alkane or alkene. This is, however, a more general phenomenon,[175,176] not limited to V/Sb oxides.

Another example of how the reactant induces a change in the surface properties is shown in Figure 8.11, which illustrates the effect of the change in the oxygen to alkane ratio in the feed on the selectivity for propene from propane on a $(VO)_2P_2O_7$ catalyst for two initial concentrations of propane in the feed. As discussed in Chapter 4, $(VO)_2P_2O_7$ is the active phase in the industrial catalysts for the selective oxidation of n-butane to maleic anhydride. It is very selective in n-butane oxidation, but when propane is fed in instead of n-butane, only carbon oxides and traces of other products (propene mainly) are detected. The first question, then, is why does the decrease in the length of the carbon chain produce so drastic a change in selectivity on the same catalyst. The answer might be that a stable product against consecutive oxidation (maleic anhydride) forms from n-butane, but not from propane oxidation.[177,178] However, this answer is a response to only part of the problem. The data shown in Figure 8.10 illustrate the fact that the sensitivity of the reaction product against consecutive oxidation is not the only factor determining the rate of consecutive oxidation to carbon oxides, and Figure 8.11 confirms it. In fact, it is thought that a decrease in the oxygen to alkane ratio may increase the selectivity to the partial oxidation product (propene in the case of propane), because

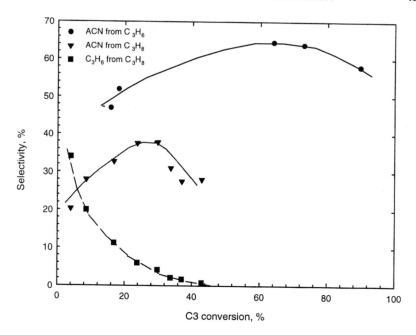

Figure 8.10. Comparison of the selectivity dependence on acrylonitrile in propane and propene ammoxidation at 480°C on a V/Sb oxide catalyst having Sb/V = 2.[28,173]

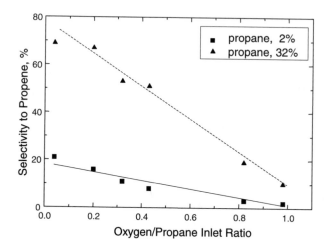

Figure 8.11. Propene formation from propane on V/P oxide as a function of the inlet propane concentration and of the oxygen to propane inlet ratio. Reaction temperature is 322°C.[175]

the formation of carbon oxides requires a larger number of oxygen atoms (propane oxidation to $CO_2 + H_2O$ requires five O_2 molecules, whereas propene formation from propane requires half of an O_2 molecule). However, the data in Figure 8.11 show that this is possible only for high initial concentrations of propane in the feed, whereas for lower initial concentrations the increase in propene selectivity obtained by decreasing the O_2/propane inlet ratio is much less striking. A comparison between the two cases points out that the differences arise from the different formation of alkene product, which is some tenfold greater using the higher inlet concentration of propane in the feed. This indicates that when the formation of propene is higher, the rate of its consecutive oxidation to carbon oxides decreases and thus the selectivity increases. The same type of phenomenon has also been observed by changing the O_2/hydrocarbon ratio in n-butane oxidation on $(VO)_2P_2O_7$, although in this case a wider range of intermediate products was detected.[179]

The formation of alkenes thus induces a self-modification of the surface reactivity. The analysis of the kinetics of but-1-ene oxidation on $(VO)_2P_2O_7$ is useful for understanding kind of effect that is associated with alkene chemisorption. Figure 8.12 shows the effect of but-1-ene and oxygen concentration in the feed on the rate of maleic anhydride and butadiene formation from but-1-ene on the $(VO)_2P_2O_7$ catalyst.[180,181] Unlike alkane oxidation on this catalyst, maleic anhy-

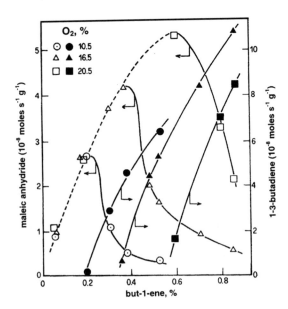

Figure 8.12. Rate of maleic anhydride formation on V/P oxide as a function of the inlet but-1-ene and oxygen concentrations.[181]

dride formation from but-1-ene oxidation requires a higher oxygen concentration, especially when a high alkene concentration is fed in. The presence of a maximum in the rate of maleic anhydride formation, the position of which shifts progressively toward higher values of concentration of but-1-ene in the feed when the concentration of oxygen in the feed increases, clearly indicates the presence of competition between the alkene and oxygen for chemisorption.[180,181]

In Section 8.3 we discussed how recent results indicate that oxygen strongly chemisorbs on surface Lewis acid sites associated with coordinatively unsaturated vanadyl ions. It is quite reasonable that the alkene through its π-bond system can also chemisorb on these sites and thus inhibit oxygen chemisorption. A competitive chemisorption with oxygen on these surface sites is not possible with alkanes, so when the two hydrocarbons are fed in a different population of adspecies is present on the surface of the catalyst, as shown schematically in Figure 8.13. Alkenes, however, are also products of oxidative dehydrogenation of alkanes (propene from propane). Increasing hydrocarbon concentration in the feed also increases the amount of propene formed, which, competing with oxygen for chemisorption, decreases the population of oxygen adspecies. Carbon oxide formation from propane requires a tenfold greater quantity of oxygen than propane to propene transformation, and thus the combined effect of a decrease in the concentration of oxygen in the feed and inhibition of oxygen chemisorption owing to competition from the reaction product (propene) prevents the formation of carbon oxides from the consecutive oxidation of propene. The second effect of self-modification of the surface reactivity by the reaction product is clearly proportional to the total amount of alkene formed, which depends on the concentration of alkane in the feed. For this reason, the considerable increase in selectivity for propene from propane on the $(VO)_2P_2O_7$ catalyst is shown only when the inlet alkane concentration is high (see Figure 8.11).

The same concept can be used to explain why acrylonitrile apparently converts to carbon oxides more rapidly when it forms from propane rather than from propene

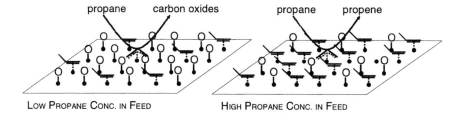

Figure 8.13. Schematic model of the different populations of oxygen and propene adspecies on oxide surfaces as a function of the inlet propane concentration and relevance of this effect on the catalytic behavior.

(see Figure 8.10). Although propene forms from propane as a reaction intermediate, its concentration is clearly higher when it is fed in directly, and thus it limits the concentration of surface oxygen species even for higher conversions of the hydrocarbon. Hence, the maximum in the formation of acrylonitrile is observed at higher hydrocarbon conversions on the same catalyst when the alkene is fed in instead of the alkane (see Figure 8.11). As discussed in the following sections, there are additional effects associated with the modification of the surface reactivity by chemisorbed species that contribute to determining the marked difference in the catalytic behavior in propane and propene ammoxidation, but the competition between alkene and oxygen for chemisorption is one of the critical aspects.

8.4.2. The Role of the Nature of Intermediate Products

In Section 8.4.1 it was observed that on the same catalyst $[(VO)_2P_2O_7]$ active for n-butane selective oxidation to maleic anhydride, propane is oxidized to carbon oxides, apart from special conditions (low O_2/alkane ratio and high alkane concentration in the feed) that allow the selective formation of propene. It was also indicated that the self-modification of the reactivity by the alkene is the key to understanding this phenomenon, but there is no clear explanation of why under the same conditions on $(VO)_2P_2O_7$ butane oxidation is selective and propane is not. One reason is the difference in the sensitivity of the reaction products (maleic anhydride and propene, respectively) toward consecutive oxidation to carbon oxides,[178] but this is only a partial explanation. In fact, it may be noted that in butane oxidation intermediate products are also products with conjugated double bonds (butadiene, furan; see Chapter 4 for further details), which can interact with the surface sites more strongly than propene.

This is confirmed by analyses of the products remaining on the catalyst surface after the feed has been stopped during tests with rapid heating of the catalyst in a nonreactive flow.[182] In these experiments, it was observed that butadiene and furan interact very strongly with the catalyst surface and desorb in large amounts as such (partly also as products of transformation such as crotonaldehyde usually not detected or formed in very small quantities during stationary catalytic tests) when the temperature of the catalyst is raised beyond that of the catalytic run. Much smaller amounts of strongly adsorbed species are observed when propane is fed in. Therefore, owing to the different nature of the intermediate products in butane and propane oxidation on $(VO)_2P_2O_7$ and the formation of species having a stronger interaction with the surface sites in the former case, the self-modification of the surface reactivity in butane oxidation is greater than in propane oxidation. As a consequence, even for low (~ 1%) concentrations of butane in the feed the reaction is selective, unlike propane on the same catalyst.

This suggests the possibility of controlling the surface reactivity in alkane selective oxidation by cofeeding small quantities of alkenes together with alkanes.

However, strong chemisorption of the alkenes may also inhibit the reactivity of the surface sites toward activation of the alkane as well as influence the optimal distribution of active surface oxygen species, which may also affect the selectivity. This is shown in Table 8.1, which summarizes the results of cofeeding experiments in n-pentane oxidation to maleic and phthalic anhydrides on the V/P oxide catalyst.[183] When n-butane is cofed with n-pentane the activity does not change significantly (in Table 8.1 the activity is shown with the index T_{40} for the temperature at which 40% conversion is reached), and only an improvement in the selectivity to maleic anhydride is noted, as it is formed from the oxidation of n-butane. However, the selectivity for phthalic anhydride remains unchanged. On the other hand, when small amounts of conjugated diolefins are added to the feed (butadiene, e.g., see Table 8.1) the activity and selectivity for phthalic anhydride significantly decreases, whereas the selectivity for maleic anhydride does not change. However, as maleic anhydride is also a product of butadiene oxidation, the influence on the formation of maleic anhydride is not indicative.

Similar results were observed when cofeeding small amounts of butadiene together with butane/air or 1-pentene together with pentane/air.[184] In this case as well both the activity and selectivity were depressed by the addition of the alkene, although the added alkene is an intermediate in the reaction network of the corresponding alkanes. Parallel FT-IR experiments[184] showed extensive formation of strongly adsorbed species due to the unsaturated hydrocarbons.

The chemisorption of alkenes on the surface of catalysts active in alkane oxidation can thus be a way to control the surface reactivity, but the formation of overly large quantities of these species can lead to inhibition of the selective catalytic behavior. It may be noted, however, that cofeeding experiments may give negative results when a catalyst that is already highly selective in alkane selective oxidation is used (Table 8.1), but different results can be expected for less selective catalysts. In this case, the cofeeding of alkenes together with alkane makes it possible to control the surface reactivity and improve the catalytic behavior.

The in situ doping of the catalyst by cofeeding small amounts of reaction intermediates can be thus a powerful method for improving the catalytic behavior, but this possibility has never been explored systematically.

TABLE 8.1. Selectivity to Anhydrides at 40% Conversion and Relative Temperature (T_{40}) in the Cofeeding of n-Pentane (2.6%)/Air with Butane (0.9%) or Butadiene (0.2%). $(VO)_2P_2O_7$ Catalyst[183]

Feed	Selectivity, %		T_{40}, °C
	Maleic anhydride	Phthalic anhydride	
n-Pentane/air	17	30	355
n-Pentane/air + n-Butane	29	28	360
n-Pentane/air + Butadiene	29	3	370

8.4.3. Chemisorption and Change in the Surface Pathways of Transformation

The interaction of reactants, intermediates, or products with a particular type of surface site must be viewed not only in terms of steps in the mechanism of surface transformation, because these species can also interact with other types of surface sites. When this interaction is sufficiently strong, the reactivity of the latter surface sites may be inhibited, which can have significant consequences for the surface reactivity.

This concept can be illustrated by an analysis of the effect of the addition of ammonia to a propane/air feed on a V/Sb oxide catalyst.[185] As described earlier in more detail in Chapter 4, these catalysts are selective for the synthesis of acrylonitrile from C3 hydrocarbons, but in the absence of ammonia they give rise mainly to carbon oxides. It may thus be expected that when ammonia is added to the feed, the formation of carbon oxides will decrease and the formation of acrylonitrile will increase. However, for small doses of ammonia in the feed, although the formation of carbon oxides decreases considerably, the parallel increase is in the formation of the propene intermediate.

FT-IR data[185] show that ammonia interacts with the Brönsted acid sites of the catalyst forming ammonium ions, but as their stability is weak, evacuation at temperatures of about 200°C is enough for their nearly complete removal.[186] This suggests that at the typical reaction temperatures for the catalytic tests of propane ammoxidation (400–500°C) the interaction of ammonia with these surface Brönsted acid sites may be considered negligible. Ammonia remains strongly chemisorbed as such on surface Lewis acid sites, up to evacuation temperatures of about 400°C.[186] However, in the presence of water vapor the situation is different, and it was observed[186] that ammonium ions form at high temperatures from ammonia coordinated on Lewis acid sites. Water is one of the main products of reaction in hydrocarbon oxidation, and thus all considerations on the stability of surface species must take its presence into account. Furthermore, desorption of ammonia adspecies is an equilibrium reaction with ammonia in the gas phase. When ammonia is present in the gas phase, Brönsted acid sites are nearly completely converted to ammonium ions even at high temperatures. It is thus reasonable to conclude that during the catalytic reaction of propane conversion on V/Sb oxide in the presence of ammonia, the larger part of the surface Brönsted acid sites is transformed to ammonium ions and consequently the reactivity of these Brönsted acid sites is inhibited.

Brönsted acid sites catalyze various side reactions as discussed in more detail in the following sections, but we will deal here only with one aspect, which, while ignored, is directly related to the effect of chemisorption on the modification of the surface pathways and specifically to the question of the consecutive transformation of the reaction products to carbon oxides.

It has been observed[73] that acrylic acid forms stable and relatively inert acrylate species with V/Sb oxide catalysts up to high temperatures of evacuation. However, in the presence of Brönsted acid sites and water, this species transforms to the corresponding free acid form weakly bound to the surface and easily susceptible to consecutive decarboxylation and further total oxidation (Scheme 8.10). The strong interaction of acrylate with the surface explains why even when formed from propane and propene (as detected by FT-IR spectroscopy[73]), acrylic acid is detected only in traces as a product of C3 hydrocarbon oxidation on V/Sb oxide. The interaction with the surface of the corresponding species formed in the presence of ammonia (acrylonitrile) is weaker, and thus it desorbs more easily (Scheme 8.10b). Acrylonitrile may also be transformed to the corresponding amide by the action of Brønsted acid sites and water (Scheme 8.10b), which easily further transforms to the free acid and then finally to carbon oxides. However, the self-inhibition of ammonia on the activity of Brønsted acid sites converting them to ammonium ions, inhibits this pathway.

The reasons for the much higher selectivity in propane ammoxidation than in propane oxidation on the same V/Sb oxide catalyst thus derive from the two combined effects: (i) less interaction of acrylonitrile with the surface with respect to acrylic acid, and (ii) self-inhibition of ammonia on the degradation of the adsorbed intermediate catalyzed by Brønsted acid sites and water.

When a small amount of ammonia is added to the feed, its primary effect is to block the reactivity of Brønsted acid sites forming ammonium ions and thus to inhibit further transformation of acrylate. The latter species, however, strongly interacts with the surface sites and does not desorb, or desorbs minimally, and thus tends to block the surface reactivity, as demonstrated by IR spectroscopy.[73] As a consequence, propane transformation stops at the first intermediate that can desorb relatively more easily to the gas phase (propene). This explains why the addition

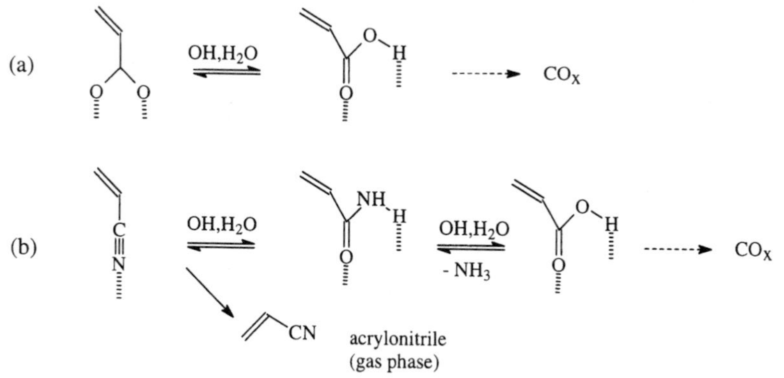

Scheme 8.10. Reaction schemes of acrylate: (a) acrylonitrile and (b) oxidation to carbon oxides.

of small amounts of ammonia to the propane/air feed considerably promotes the selectivity for partial oxidation products, but mainly for propene rather than other products, including acrylonitrile. Weaker coordinated ammonia species also form and react with propene or its conversion products to give acrylonitrile but only for larger quantities of ammonia in the feed. Ammonium ions thus do not play a direct role as reaction intermediates, but they do affect the catalytic behavior by their positive influence on the surface reactivity. This explains why a good correlation was observed in V/Sb oxides[28] between the quantity of ammonium ions detected by IR spectroscopy and selectivity, although the ammonium ions have only an indirect effect. This underscores the fact that care must be taken in assigning a role as a reaction intermediate to surface species, even when their amount correlates nicely with the catalytic behavior. The effect, as shown above, may only be an indirect effect of modification of the surface reactivity.

The influence of ammonia clearly demonstrates the link between selectivity and modification of the surface reactivity by chemisorbed species. It also shows that selective behavior in one reaction and not in another cannot be interpreted simply in terms of different sensitivities of the reaction products toward consecutive oxidation or their different adsorption strengths with the surface sites, but rather that the chemistry of the surface reactions and the effect of chemisorption in the modification of the surface reactivity must be understood in more detail.

Alkenes also have a basic character, although less strong than ammonia, and their chemisorption may induce qualitatively similar effects to those described for ammonia. Their chemisorption, however, like that of ammonia mitigates the effect of Brønsted acid sites in catalyzing the hydrolysis of adsorbed products, the first step in their consecutive oxidative degradation. This is a further reason why acrylonitrile is apparently less sensitive to consecutive oxidation to carbon oxides when it forms from propene rather than from propane (see Figure 8.10).

In the oxidation of C4 and C5 alkanes, anhydrides form that are less susceptible to oxidative attack than the oxidation products of propane but can be hydrolyzed to the corresponding diacid, which can decarboxylate and easily be further degraded. There are thus strong analogies to the case discussed above, despite the fact that the anhydride can be less strongly bound to surface sites than carboxylate forms. This is another reason for the difference in catalytic behavior between C2–C3 alkanes and higher alkanes (C4 and above), as the latter can form cyclic species such as anhydrides, which do not interact with surface sites as strongly as carboxylate species.

8.4.4. Direct Role of Chemisorbed (Spectator) Species in the Reaction Mechanism

Chemisorbed (spectator) species may influence not only the surface reactivity through an electronic long-range effect or a specific interaction with the surface active sites as discussed in the previous sections, but can also participate directly in the reaction mechanism, although there are very few examples of this in the literature, especially regarding selective oxidation. It may be noted, however, that this question of the possible role of coadsorbate in the reaction mechanism has not usually been considered, and thus the absence of published data does not mean that it is not an important aspect of the surface catalytic chemistry at oxide surfaces. Proper attention to the role of coadsorbate on the surface reactivity can be a key factor for tuning surface properties, designing new surface reactions, and explaining catalytic behavior of solid surfaces.

The direct role of carbonaceous species in oxidative dehydrogenation of ethylbenzene or alkanes and, in some cases, of selective oxidation is well documented,[187–190] although the nature of the surface sites on the carbonaceous material responsible for this behavior is not very clear. On the contrary, adsorbed organic species, formed on the surface as products or by-products of the catalytic reaction, have not been thought of as playing a cocatalytic role, although in several cases the bond between organic species and the surface sites is so strong that a significant quantity of these species is present on the surface during the catalytic reaction, as discussed above. However, the complexity of the investigation necessary to distinguish a possible cocatalytic role of coadsorbate species from other effects that are clearly present, such as competition for chemisorption, makes it very difficult to obtain clear information on the direct participation of coadsorbate in the reaction mechanism.

On the other hand, it is known that the activity of homogeneous catalysts depends considerably on the nature of the ligands, owing not only to a modification of the electronic properties of the metal center of the complex, but also to the fact that the ligand "assists" in the catalytic transformation, forming specific bonds with the substrate/intermediate.[191] Furthermore, there are several examples in which an organic molecule "mediates" the selective oxidation. A relatively recent interesting example[192] shows that in the Pd(II)-catalyzed alkene acetoxylation or Wacker-type oxidation, the reoxidation of reduced palladium can be carried out efficiently by a molybdovanadophosphate (HMoVP)/hydroquinone/O_2 system. The primary agent for Pd^0 oxidation is a quinone (e.g., benzoquinone), which forms from the corresponding hydroquinone by reaction with the HMoVP, which is then reoxidized by gaseous oxygen. The hydroquinone–quinone transformation thus mediates the direct oxidation of Pd^0 by HMoVP, which is not a fast reaction.

Solid-supported or unsupported Pd^{2+}-phosphomolybdovanado heteropoly acids are also active in gas phase low-temperature (about 100°C) Wacker-type oxidation,[193,194] in particular for but-1-ene to butan-2-one. The reaction is selective, although the catalytic activity progressively declines owing to both the progressive reduction of the heteropolyacid and the accumulation of organic species on the catalyst surface. However, it was observed[193,194] that the rate of formation of the product initially increases despite the expected progressive deactivation. Although direct evidence is not available, it is not unlikely that the initial induction time is due to the formation of organic species that mediate the reoxidation of reduced palladium. IR data indicate the effective presence of large amounts of organic species on the surface of the catalyst during catalytic reaction.[193]

A different example of selective oxidation mediated by surface-anchored species was described by Tsubokawa et al.[195] Alcohols and diols are barely oxidized by Cu^{2+} salts, but in the presence of nitrosyl radicals immobilized on silica and ferrite they react rapidly to give the corresponding aldehydes, ketones, and lactones. The alcohols are oxidized with nitrosyl moieties (N=O) on the silica and ferrite surfaces, which were formed by the reaction of the nitroxyl radical (N–O$^{\bullet}$) moieties with Cu^{2+}. The nitrosyl moieties are reduced to the corresponding hydroxylamine moieties (N–OH) reoxidized by the Cu^{2+} salt. The surface nitroxyl radical thus mediates the selective oxidation of the alcohol with strong analogies to what is observed in homogeneous catalysis for oxygen transfer reactions from coordinated ligands containing groups such as the nitro group.[160] For example, a cobalt–nitrosyl (Co–NO) complex reacts with O_2 to form a nitro complex (Co–NO$_2$) able to transfer oxygen to various organic species (propene is oxidized to acetone), reforming the starting cobalt–nitrosyl complex. The nitrosyl complex thus mediates the oxygen transfer from gaseous oxygen to the organic molecule.

There have been similar observations on heterogeneous catalysts regarding the rate of hydrocarbon activation. For example, methane reacts very slowly with oxygen over H-zeolites even at very high temperatures (up to 600°C), but in the presence of NO the reaction proceeds quickly at 300°C.[196] Other coadsorbate species, such as sulfate, have also been shown to play an active role in facilitating the dissociative adsorption of saturated hydrocarbons on solid catalysts.[197]

In the ammoxidation of propane to acrylonitrile over Ga/Sb oxide–based catalysts,[198,199] it has been observed that the rate of propane selective oxidation increases with the introduction of ammonia into the reaction mixture, and similarly the rate of acrylonitrile formation also increases. The increase in the product formation rate is not a result of a decline in carbon oxide formation. Hence, ammonia modifies the surface characteristics, enhancing the rate of alkane activation. IR spectroscopic data on the interaction of ammonia with the Ga/Sb oxide catalyst[200] indicate that both ammonia adsorption and catalyst treatment with a reaction mixture produce the NH$_2^-$ amide groups, which can serve as base sites to catalyze the propane activation via a carbanion intermediate further transformed to propene. Adsorbed ammonia species then react with the propene intermediate to form acrylonitrile. Thus in this case ammonia has a double role: (i) it is the reactant necessary to form the product, and (ii) it generates surface-adsorbed species that play a cocatalytic role in the mechanism of alkane activation.

Yoo[201] reported that in the gas phase oxidation with oxygen of alkylaromatics over Fe/Mo/borosilicate, the addition of carbon dioxide to the feed significantly promotes hydrogen abstraction and thus the rate of alkylaromatic conversion, although it adversely affects oxidation of the resulting alkenyl group to oxygenates. Although the mechanism of this effect was not investigated, it is reasonable to assume that the CO_2–acid carbonate cycle is responsible for the increased rate of alkylaromatic hydrogen abstraction. In this case as well a coadsorbate species participates in the reaction mechanism, significantly changing the surface reactivity.

Less clear is the effect of CO_2 in inhibiting the further selective oxidation of the aromatic alkenyl intermediate. In accord with the above hypothesis, the adsorbed CO_2 abstracts hydrogen atoms from the alkylaromatic and reacts with oxygen to form the acid carbonate species, which then dissociates to CO_2 and H_2O. The oxygen to oxidize the aromatic alkenyl intermediate is thus intercepted by the CO_2 to form the acid carbonate, leading on the one hand, to an enhancement in the rate of alkylaromatic activation but, on the other hand, to inhibition of the further oxidation of the intermediate of oxidative dehydrogenation. This example illustrates how it is sometimes possible to significantly change the surface behavior, in terms not only of reactivity, but also of selectivity and the nature of the products formed.

Although the published data are limited, these examples show that various possible effects on the surface reactivity and selectivity in the presence of coadsorbate species together with the reactant or intermediates can be expected: (i) modification of the coordination and electronic properties of a surface metal ion, and (ii) direct participation of the adsorbed species in the reaction mechanism, assisting the metal ions in the catalytic transformation (e.g., helping to hold the intermediate coordinate to the site in a specific mode) or directly mediating oxygen, hydrogen, or electron transfer to or from the reactant/intermediate. A more detailed analysis of this process is thus fundamental for a better understanding of the chemistry of catalytic transformations at oxide surfaces and can also offer valuable indications as to how to influence and control the surface reactivity by gas phase dopants.

8.5. ROLE OF ACIDO-BASE PROPERTIES IN CATALYTIC OXIDATION

In addition to their redox properties, transition metal oxides are characterized by the presence of acido-base sites,[202,203] which also can play a significant role in oxidation reactions. It is thus quite reasonable that in the past many attempts have been made to find correlations, sometimes quantitative,[204–206] between acido-base and redox characteristics and catalytic reactivity. It may be noted, however, that the acido-base properties are not independent of the redox properties. Therefore it is not correct to correlate two factors that cannot be modified independently.

The ideal surface structure of an oxide, as generated by cutting the oxide structure at a specific crystalline plane, is composed by an array of cations and oxygens, some of which are coordinatively unsaturated and accessible to gas reactants, and thus generate Lewis and Brønsted acidic or basic sites. The strength and number of these sites depend on the nature of the cation, the type of metal to oxygen bond, and the packing of the specific crystalline plane. The acido-base characteristics of an oxide may thus be predicted from the oxide structure, but on the oxides of interest for catalytic studies the presence of defects in the oxide and steps and corners on the surface create surface heterogeneities in the acido-base characteristics, which considerably limits the validity of theoretical estimates. Direct determinations, often by FT-IR spectroscopy,[207,208] are thus necessary. It may be noted, however, that these studies usually indicate the acido-base characteristics of the "clean" surface, which may be different from those effectively present during the catalytic reaction, when: (i) chemisorption of the reactants influences the electronic characteristics of the oxide; (ii) some components (water, in particular, but also reactants such as ammonia) alter the acido-base characteristics; and (iii) partial reduction of the metals occurs as a consequence of redox reactions of oxidation, which sometimes also lead to a dynamic reconstruction of the surface.

In spite of these limitations, the study of the acido-base characteristics and their possible influence on catalytic behavior is central for a better understanding of the surface processes and their control to improve reactivity and selectivity. However, it is necessary to closely analyze the reaction mechanism and the influence of acido-base characteristics on the individual steps or possible competitive/consecutive reactions rather than seek general relationships.

The acido-base characteristics of the oxide have three primary roles in relation to the catalytic behavior: (i) influence on the activation of the hydrocarbon molecule, (ii) influence on the rates of competitive pathways of transformation, and (iii) influence on the rate of adsorption and desorption of reactants and products.

Before analyzing these aspects, however, it will be useful to summarize some general concepts regarding the surface acido-base features of metal oxides. Detailed data on the acido-base characteristics of binary and mixed oxides in relation to the role of acidity in catalytic oxidation can be found in Busca et al.[10]

8.5.1. Basic Concepts on the Acido-Base Characteristics of Metal Oxides

Oxygen has a highly electronegative character and thus the bond type with metals is ionic, whereas with nonmetals it is covalent, although there is clearly a gradation between these two extremes, in relation to the nature of the cation as well. The electronegativity of a cation depends on its oxidation state and increases as the oxidation state increases. Oxides of metals in a high oxidation state such as V^{5+} and Mo^{6+} are characterized by a covalent metal to oxygen bond similar to that of some semimetals (Si^{4+} in SiO_2) and behave as acid oxides, whereas the same elements in lower-oxidation-state metal oxides have an ionic character and behave as basic oxides.

On the surface of oxides of low-oxidation-state metals, the metal to oxygen bond has a highly ionic character and thus when the surface is clean, unsaturated metal cations able to act as Lewis acid sites and oxide anions having a basic character are both present. The strength of the surface Lewis sites depends on: (i) the ionicity of the metal to oxygen bond, (ii) the ratio between the charge of the cation and its ionic radius (determines its ability to polarize the oxygen anion), and (iii) cation coordination (as coordination number increases Lewis acid strength decreases). In ionic oxides having a small highly charged cation with a low coordination number (e.g., in some aluminas) Lewis acidity predominates. On the other hand, when the cation has a low charge and larger ionic radius (e.g., BaO), basicity predominates. The presence of water leads to the generation of surface hydroxyl groups, but they are ionically bonded to metal cations and have more a basic character rather than a Brønsted acid behavior.

In the oxides of semimetals (e.g., SiO_2) the metal to oxygen bond is largely covalent and thus there is normally no coordinative unsaturation at the surface, because surface OH groups saturate the coordination vacancy. The oxides show Brønsted acidity but no Lewis acidity or basicity.

The oxides of high-oxidation-state metals also do not show basicity, but due to the possibility of forming terminal metal–oxygen double bonds (e.g., V=O, Mo=O), notwithstanding the mainly covalent character of the metal to oxygen bond, strong Lewis acid sites may be present. The OH groups are covalently bonded to the metal, and the charge generated by dissociation may be delocalized on terminal M=O groups. A medium to strong Brønsted acidity is thus present. The reduction of the metal cations leads to a lower electronegativity and larger cation size, and thus a change from covalent to ionic character. This generally causes a decrease in Brønsted acidity, an increase in Lewis acidity, and an increase in the degree of basicity. A summary of the acido-base characteristics of the different classes of binary oxides discussed above is shown in Table 8.2,[10] which also lists the acido-base characteristics of some examples of mixed oxides.

In multicomponent mixed oxides, the simpler situation is represented by the case in which only two types of cations are present. When the two have similar oxidation states and electronegativities, a true mixed oxide compound forms, whereas the combination of a basic oxide with an acidic oxide results in the formation of an oxysalt. In the latter case, the global acido-base characteristics are often dominated by those of one of the two components.

TABLE 8.2.   Summary of Acido-Base Properties of Some Binary and Mixed Metal
Oxides of Interest as Catalysts in Selective Oxidation Reactions[10,209-218]

Binary oxides	Characteristics of the oxide	M–O bond nature	Acidity	Basicity
SiO_2	Semimetal, small cation size	Covalent	Brφnsted, medium-weak	None
MoO_3, V_2O_5, Nb_2O_5	Semimetal, high oxidation state, medium cation size	Largely covalent	Brφnsted and Lewis, medium to strong	None
γ-Al_2O_3	Metal, small cation size	Ionic	Lewis, strong	Weak
TiO_2	Metal, medium cation size	Ionic	Lewis, medium	Medium-weak
SnO_2, ZrO_2, CeO_2	Metal, large cation size	Ionic	Lewis, medium-weak	Medium-strong
MgO, CaO, CuO	Metal, low oxidation state, large cation size	Ionic	Lewis, medium to weak	Strong to very strong

Mixed metal oxides	Types	M–O bond nature	Acidity	Basicity
Rock-salt solutions	True mixed oxide	Ionic	Lewis, weak	Strong
Spinels (AB_2O_4)			Lewis, medium strong	Medium
Perovskites (ABO_3)			Lewis, weak	Strong
$(VO)_2P_2O_7$	Oxy-salt	Covalent	Brφnsted and Lewis, medium-strong	No
TiO_2–SiO_2		Ionic	Brφnsted (weak) and Lewis (strong)	No
$Mg_3(VO_4)_2$			Lewis, weak	Medium
V_2O_5-TiO_2	Supported (monolayer type)	Covalent on ionic	Brφnsted (medium), Lewis (medium-strong)	Weak

A particularly relevant case for selective oxidation is an oxide supported in the form of more or less a "monolayer" over another oxide. The acido-base characteristics are influenced by both the dispersion of the one oxide on the other and the possibility of stabilizing different surface species with respect to those present in the starting bulk oxides. In general terms, however, the acido-base characteristics are determined by the stronger acid or base properties of the two starting oxides.

8.5.2. Influence of Acido-Base Characteristics on the Activation of Hydrocarbons

The effect of acido-base catalyst properties on the activation of hydrocarbons is complex, because it depends not only on the nature of the oxide, but also on the nature of the hydrocarbon, even within the same class of hydrocarbons.[219–223] Therefore, it is preferable to restrict discussion to a single type of reactant. For different hydrocarbons, the chemistry of transformation is the same, but the pathway of transformation may be different owing to the possibility of having additional sites that can react (e.g., hydrogens in the α-position) and the different stabilities of intermediate species formed (e.g., carbanions).

The discussion is centered on the activation of propane for two main reasons: (i) owing to the absence of reactive sites, alkanes do not chemisorb as such on oxides and thus a better understanding of the role of surface acido-base properties on propane activation and transformation can be obtained, and (ii) propane conversion has been studied on a wide range of mixed oxides with different acido-base characteristics. The term activation here indicates not only the first step of (usually) C–H abstraction, but also the consecutive transformations up to the first relatively stable intermediate (usually propene in propane oxidative conversion).

There is general agreement in the literature that the rate-determining step in propane conversion is the breaking of the first C–H bond, leading to a propyl species.[224–226] However, there are various possibilities for the generation of this species. Breakage of the C–H bond can be homolytic with the formation of a propyl radical.[224,227,228] This radical species transforms to propene either after desorbing in the gas phase and reacting with other gas phase molecules (oxygen or propane) or via a surface reaction.[224,225,229,230] The relative rates of the two competitive routes depend on both the nature of the catalyst (reducibility) and the reaction temperature.[227,228,231] Low temperatures and reducible catalysts promote the surface reaction route, in which the second proton abstraction occurs by the OH$^-$ species created in the first H abstraction step (Scheme 8.11a) with generation of a water molecule or, what is more likely, by a neighboring basic oxygen and subsequent shift of the hydrogen to the OH$^-$ species to generate a H$_2$O molecule. It is clear that the basic characteristic of lattice oxygen and associated OH$^-$ species determines the rates of reaction.

Other authors have suggested a heterolytic splitting of the C–H bond with formation of a carbocation by hydride abstraction[232,233] or of a carbanion by proton abstraction.[166,234] The latter process occurs on strong Lewis acid catalysts, whereas the former takes place on strong basic catalysts. It may be noted, however, that, as expected, redox equilibria coexist at the surface, independently of whether the activation is homolytic or heterolytic, as indicated in Scheme 8.11b. The position of the equilibrium depends on the redox and acido-base properties of the catalyst, as well as on the reaction conditions (temperature and reaction atmosphere).

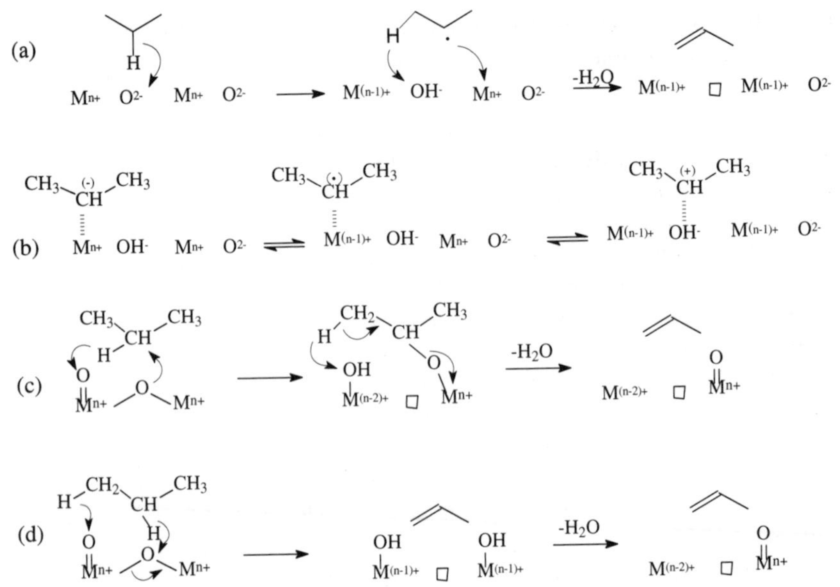

(a)

(b)

(c)

(d)

Scheme 8.11. Alternative routes of propane conversion to propene over oxide surfaces (a, c, and d) and equilibrium existing between the different adspecies after propane hydrogen abstraction (b).

Busca et al.,[10,235,236] using primarily FT-IR spectroscopy, have proposed a different pathway of propane activation and transformation. They suggested the interaction of σ or σ^* C–H orbitals with d-type orbitals of the transition metal cation, with a direct flow of a couple of electrons from the hydrocarbon to the cation, leaving a carbocation that is immediately attacked by lattice oxygens to form a surface alkoxy group (Scheme 8.11c). It may be noted, however, that a quasi-simultaneous two-electron transfer to the same metal, as suggested by Busca et al.,[10,235,236] is quite unlikely from the energy point of view. Secondary alkoxides can evolve in two different ways, either by oxydehydrogenation, giving rise to ketones, or by elimination, giving rise to alkanes and surface hydroxy groups. Surface alkoxides have a carbocation character that increases with increasing catalyst Brønsted acidity. Transformation of the alkoxy species to propene requires the presence of Brønsted sites, and Moro-oka et al.[237] did observe a relationship between propene formation from propane and the number of Brønsted acid sites on Mo/Mg oxide catalysts.

The mechanism of formation of the alkoxy intermediate can be viewed as a two-step concerted mechanism, but a true concerted mechanism is also a reasonable possibility (Scheme 8.11d) and probably results in a lower total energy because it involves simultaneous breaking and forming of chemical bonds. In this case as well both acido-base and redox properties of the catalyst determine the rate of transformation.

Finally it should be noted that when the analysis is extended to other hydrocarbons, the activity trend may be roughly correlated to the stability of carbocations formed by heterolytic dissociation:

$$\text{benzyl} \approx \text{allyl} > \textit{tert}\text{-butyl} > \text{isopropyl} > \text{ethyl} > \text{methyl}$$

This sequence suggests a relationship between the rate of the first C–H breaking (generally the rate-determining step) and Lewis acidity of the catalyst, a relationship that has already been demonstrated for butane oxidation over vanadia-based catalysts.[238] The same mechanism was proposed for butane selective activation over $(VO)_2P_2O_7$, although in concert with the abstraction of the second proton by a neighboring basic bridging oxygen.[128,213]

In conclusion, the mechanism of alkane activation is still being debated and any of the mechanisms proposed may be valid, in one situation or another. It is also clear that depending on both the acido-base and redox characteristics of the oxides one of the proposed mechanisms may predominate over the others, although as noted above both the catalyst and the reaction conditions determine the effective occurrence of the activation mechanism.

Scheme 8.11b shows that whether a carboanion, radical, or carbocation is formed depends on: (i) the stability of these organic intermediates (extending the concept to other substrates); (ii) redox (reducibility) character of the transition metal ion; and (iii) the possibility of effectively delocalizing the electrons transferred from the hydrocarbon to the solid. As discussed earlier, this long-range electron delocalization depends on the characteristics of the solid and the effect of the chemisorbed species on the bulk and surface electronic properties. The same factors also determine the acidity versus basicity characteristics of the OH⁻ type of limiting species formed by the first hydrogen abstraction (Scheme 8.11b). This observation illustrates the interdependence between local and collective properties of the solid, an aspect discussed extensively in Section 8.2, but worth emphasizing again here because it is so often ignored.

8.5.3. Competitive Surface Reactions and Acido-Base Properties

In the discussion of the correlation between acido-base properties and alkane activation it has already been shown that after the first step of C–H breaking, there are various competitive pathways whose relative rates depend on the acido-base properties of the catalyst as do the subsequent transformation steps. Discussion here is limited to some examples that clarify this issue.

An interesting observation is that catalysts based on molybdenum and vanadium[239–243] oxidize propane to propene, acrolein, or acrylic acid, whereas defect perovskite and rutile, Bi/Ba/Te, Fe/Nb/ and Fe/Ta/ oxides,[244] form propanal together with acrolein, but not propene. B/P/ oxides[245] show an intermediate behavior, forming propene, propanal, acrolein, and acrylic acid. Interestingly, however,

B/P/ oxides[245] are active in propane conversion but inactive in propene conversion, suggesting that propene is not an intermediate to acrolein and acrylic acid, unlike catalysts based on molybdenum and vanadium oxides.[239–243]

Following the suggestions of Busca *et al.*[10] and Bettahar *et al.*,[2] we may consider the formation of an alkoxide as the first step in propane transformation (Scheme 8.12a). Depending on the acido-base characteristics of the neighboring surface oxygens, this intermediate species may evolve either to adsorbed acetone (by oxydehydrogenation) or to propene (by elimination of hydrogen). The carbon atom of the carbonyl compound has an electrophilic character and thus is susceptible to nucleophilic attack by surface oxygens, especially when there are hydrogens in the α-position, because both base- and acid-catalyzed reactions easily give rise to enolization. In the latter case, strongly bound enolate anions are formed. Propene may undergo allylic α-hydrogen abstraction followed by nucleophilic oxygen addition to form an allyl alcoholate intermediate, which can transform to acrolein. As in the case of acetone, the carbon atom of the $C=O$ group in acrolein also has an electrophilic character, although weaker due to charge delocalization. Thus on weakly basic oxides such as MoO_3/TiO_2 and V_2O_5/TiO_2[246–248] acrolein is not susceptible to attack by surface oxygens and is adsorbed almost intact (acetone, on the other hand, enolizes immediately), but on more basic oxides such as Co_3O_4 and $MgCr_2O_4$[235,236,249] acrolein also is immediately subject to nucleophilic attack and undergoes oxidation to strongly bonded acrylate. The carboxylate species can desorb in the form of the corresponding organic acids, but in the presence of water and medium-strong Brønsted sites may also decarboxylate and further transform to carbon oxides (see Scheme 8.10).

The same type of surface mechanism may be proposed for propanal formation, but with the formation of 1-propoxy intermediates as the first reaction step (Scheme 8.12b). The attack in the 1-position rather than the 2-position in propane indicates that in this case the attack should be the abstraction of a hydride anion by a Lewis acid site instead of abstraction of a proton by a basic (nucleophilic) oxygen. The 1-propoxy intermediates then easily transform to propanal. The subsequent dehydrogenation of propanal to acrolein is facilitated by the equilibrium formation of its enolate form, in which a hydrogen atom is abstracted from the CH_2 group as a proton on a basic site. The enolate species may then lead to acrolein by losing a hydride anion, probably by a concerted mechanism on a Lewis acid site. This explains the mechanism of formation of propanal and acrolein on perovskite and rutile Bi/Ba/Te, Fe/Nb/ and Fe/Ta oxides.[244] The difference with respect to V- and Mo-based catalysts (Scheme 8.12a) is thus the different nature of the first intermediate (1- vs. 2-propoxy species), which is determined by the different acido-base and redox characteristics of the two classes of catalysts.

It should also be noted that the enolate species formed from propanal is readily susceptible to attack by relatively weak basic (nucleophilic) surface oxygen to form a propionate intermediate (Scheme 8.12b), which can undergo oxidative dehydrogenation similar to that described above for adsorbed propanal giving rise to an acrylate species, the desorption of which forms acrylic acid. The formation of propionate from propane has been observed on V/Sb oxides and its oxidative dehydrogenation to acrylate occurs at lower temperatures than the propane to propene transformation.[73] Thus, the presence of basic oxygen together with strong Lewis acid sites may change the pathway of transformation of the enolate intermediate. This explains the possible reaction mechanism in B/P oxides[245] in which a mixture of propanal, acrolein, and acrylic acid is observed. Other by-products observed in propane oxidation on B/P oxides[245] (acetone and acetic acid, as well as propene) probably derive from the formation of alkoxy species in the 2-position as well as in the 1-position.

Hence, a direct path of oxygenate production from propane, by-passing the propene intermediate, is possible on V/Sb oxide catalysts. It may be recalled, however, that the formation of propanal is not proof of the existence of this direct pathway. Volta et al.,[250] studying the oxidation of propane and propene on α-$Mg_2V_2O_7$, observed the formation of acrolein, propanal, and acetic acid as oxygenated by-products with the same relative selectivities from the alkane and the alkene. They also observed[250] that in propane and propene oxidation on α-$Mg_2V_2O_6$, acetaldehyde is the only oxygenate by-product (apart from carbon oxides) from both hydrocarbons. This indicates that the oxygenate produced in the oxidation of propane comes rather from the propene intermediate (V/Mg oxides are catalysts for propane oxidative dehydrogenation to propene; see Chapter 4).

Propanal may thus also form by hydration of propene (Scheme 8.13). Oxyhydration in the 2-position competes with this reaction, leading to acetone instead of to propanal. The competition between these two reactions depends on the acid strength of the Brønsted sites, which catalyze the reactions, both of which are in equilibrium. The 1-propoxy species is more reactive toward oxidation than the 2-propoxy species. Moreover, on supported V/P oxide catalysts, 1-propanol is oxidized mainly to propanal with propene as a minor product, whereas 2-propanol is oxidized mainly to propene with acetone as a minor product.[2] This can explain why propanal, but not acetone, is observed on V/Mg oxides[250] as the by-product in both propane and propene conversion. The relative rates of these two competitive reactions depend on both the acido-base properties on the catalyst and the reaction conditions.

The nucleophilic oxygen attack on propene (after allylic hydrogen abstraction) to form σ-allyl species (Scheme 8.13) also competes with the formation of the 1- and 2-propoxy species and usually is the dominant reaction, being favored at higher reaction temperatures.[73] The σ-allyl species easily transforms to acrolein, but through a SN$_2$ mechanism[251] may also directly isomerize to propanal bypassing the acrolein stage (Scheme 8.13). The propanal may then undergo oxidative dehydro-

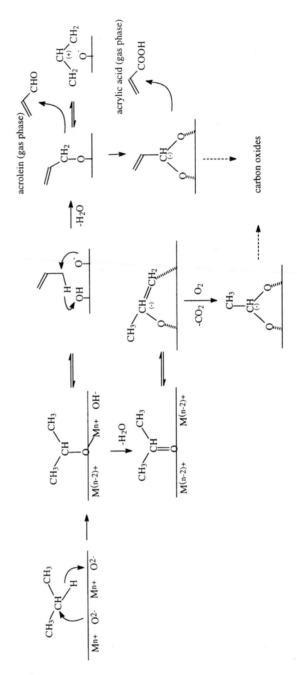

(a)

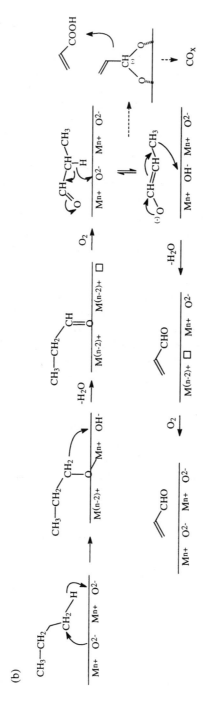

Scheme 8.12. Reaction pathways of propane conversion by oxygen attack in the 2-position (a) or 1 position (b).

genation to acrolein. Formation of carboxylate species, as described above, competes with these reactions.

The relative rates of these competitive pathways clearly depend on the acido-base properties of the catalyst and the reaction conditions, but can also be significantly influenced by the chemisorption of reactants or products. In this respect, an important role is played by water, which even if not specifically added to the feed, is a primary product of the oxidation reactions. The influence of water on the surface reactivity and acido-base properties can be illustrated by recalling the transformation of aldehydes to their corresponding carboxylic acids, discussed in part above. Of course, this reaction is not only part of the overall reaction scheme of hydrocarbon oxidation (Scheme 8.13), but is also an industrially interesting reaction (e.g., oxidation of metacrolein to metacrylic acid).

The catalytic oxidation of an aldehyde to its corresponding carboxylic acid involves its dissociation into a carboxylate species,[252] which then should desorb as carboxylic molecules (Scheme 8.14a). The analogous surface species of gem-diols $[RCH(OM)_2]$ first form by nucleophilic attack by O_2^- or OH species[253] and are then transformed into carboxylates (Scheme 8.14a). In the presence of water, the concentration of surface hydroxyl groups is higher, and thus, the attack of organic molecules probably involves these species (Scheme 8.14b). Moreover, when water is added to the feed in propene oxidation, the selectivity for acrolein decreases and that for acrylic acid increases.[254,255] Similarly, with a high water to propane ratio in the feed, acrolein does not form, but acrylic acid does.[256] On the other hand, water also favors oxyhydration of propene,[257,258] which leads to acetic acid via the

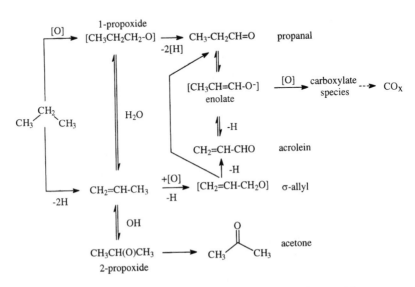

Scheme 8.13. General reaction network of propane conversion on oxides.

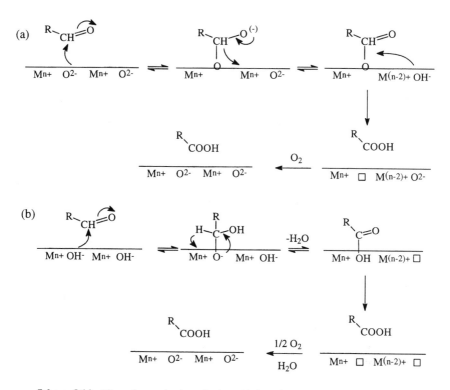

Scheme 8.14. Alternative mechanisms for the oxidation of alkyl aldehydes to alkyl acids.

acetone intermediate and can also enhance the rate of catalyzed decarboxylation of the weakly bound carboxylic acid, as discussed above.

The acido-base properties together with catalyst redox properties thus play a fundamental role in determining the surface reactivity, but there is a complex interdependence between the change in the surface acido-base properties of the catalyst (either by doping or modification by gas phase reactants) and catalytic behavior. A proper analysis of this problem should start with an elucidation of the entire surface reaction network and competitive pathways of reaction and go on to the effect on the various steps in the change of the surface acido-base properties. Thus it is not possible to consider acido-base properties simply as global parameters and try to correlate them with overall catalytic behavior. It is also clear that acido-base properties influence both selective and nonselective routes, although controlled tuning may allow optimization of the catalyst performance, which is the reason that the addition of very small amounts of alkaline or alkaline earth cations to oxidation catalysts often promote a reaction, while larger amounts kill it.

8.5.4. Role of Acido-Base Properties on the Adsorption/Desorption of Reactants and Products

In general, it is quite obvious that to obtain a selective oxidation product it is necessary that it desorb relatively easily from the surface to preclude further transformations. Therefore, a general rule is that the catalyst should not react too strongly with the product. Catalysts used to synthesize acids or anhydrides (e.g., acrylic acid and maleic anhydride) cannot be strongly basic, and thus are usually mixed oxides having a prevailing acid character.[10] In some cases where both the reactant and the final product are strong acids (e.g., isobutyric acid to metacrylic acid) very strong acidic catalysts such as heteropolyacids are required.[259]

Alkenes have a weakly basic character and are certainly more basic than the starting alkane. Therefore, in catalysts for alkene synthesis from alkanes a basic surface is needed to help alkene desorption as well as to prevent its oxidative degradation by Brønsted acid sites. This is the reason that selectivity for propene in supported vanadium and molybdenum oxides is promoted by the addition of alkaline metals as well as why the addition of magnesium oxide to vanadium oxide to form V/Mg oxide species generally leads to promotion of selectivity for propene accompanied by a decrease in the activity.[250,260–265] The effect is certainly more complex, because, for instance, in V/Mg oxides the addition of MgO does not lead only to a change from acidic to basic properties of the catalyst, but also to a change in the reducibility of vanadium, coordination, etc.

Even though, as noted above for V/Mg oxide catalysts, it is nearly impossible to modify the acido-base properties independently of others such as catalyst reducibility, there are no studies in the literature directed toward a systematic analysis of the relationship between acido-base properties and catalytic behavior, apart from the generic indications described above. Similar to conclusions in Section 8.5.3, the simple correlation in terms of acido-base characteristics of the catalyst and desorbing product does not take into account the dynamics of the surface processes. A stronger interaction of a product with the surface does not mean that it is necessarily negative regarding selectivity, but the rates of the consecutive transformations should be analyzed as well as the way in which modification of the surface properties influences the rate of synthesis of the product. Thus, a catalyst modification leading to a stronger interaction of the product with the surface sites may be either positive or negative, depending on the influence on the kinetics of all the surface processes.

A better knowledge of the chemistry of surface transformations, the reaction network, the kinetics of individual surface transformations, and the relationship with surface properties is the key to controlling and tuning surface reactivity and improving catalytic performance. Qualitative overall factors, including general relationships between acido-base properties and global catalytic behavior, are bases for study but do not provide the information necessary for the design of new catalysts.

8.6. REACTIVE INTERMEDIATES IN HETEROGENEOUS OXIDATIVE CATALYSIS

In Section 8.5 it was pointed out how the analysis of the chemistry of surface transformations, which also may be viewed as organic chemistry at oxide surfaces,[1] can be a very useful tool for understanding how to control surface catalytic transformations. Even though various other factors have a part in determining the overall catalytic behavior, as discussed at length in the other sections of this chapter, a detailed study of the chemistry of the possible surface processes is one of the basic components for understanding, controlling, and tuning of surface catalytic processes. We will thus continue the discussion of this topic here using selected examples relevant to the field of selective oxidation. Other interesting aspects have been dealt with by Busca,[1] Lavalley et al.,[2] and Matyshak and Krylov,[266] and have been described in Davydov's book,[267] and the reader is referred to these works for further discussion.

8.6.1. Analysis of the Reactive Intermediates by IR Spectroscopy

The identification of surface reactive intermediates, i.e., a molecular level understanding of surface reactivity, requires the use of physicochemical techniques sensitive to the identification of the nature and coordination of adsorbed species. FT-IR spectroscopy is one of the preferred tools for this purpose,[1,2,266-271] as it provides detailed information on the nature of chemisorbed species and their surface transformations. Using the reactants and the precursors of the reactive intermediates as probe molecules (e.g., isopropyl alcohol to form the isopropylate intermediate species; the selection of the probe molecules should be made on the basis of indications on the general reaction network ascertained by reactivity studies) and studying both their individual surface transformations and their conversion in the presence of other coadsorbate species (e.g., the organic intermediates and ammonia), one can often derive an overall surface reaction network showing the different possible pathways of transformation.

There are basically two ways to obtain mechanistic information using FT-IR: in situ experiments or studies under controlled conditions. In the former, IR spectra are recorded during the catalytic reaction, using reaction conditions close to those of the flow reactor catalytic tests. Usually, the diffuse reflectance technique (DRIFT) and an environmental cell are used for these studies. In the latter, usually done in the transmission mode with self-supported disks of the catalyst, the probe molecules are first adsorbed, often at room temperature, and then their surface transformation as a function of the reaction temperature is monitored in a vacuum or a controlled atmosphere (e.g., in the presence of one or more of the other reactants).

Although the first method appears to be more relevant for catalysis because it provides information on the surface species during the catalytic reaction directly, it has a number of serious drawbacks. In fact, during the catalytic reactions the oxide surfaces are covered by abundant "spectator" species, which except in some special cases or simple reactions, mask the reactive intermediates. Under conditions ideal for flow reactor tests the reactive intermediates are only present in very small amounts and thus are very difficult to detect. At high temperatures due to catalyst lattice vibrations and IR emissions, the bands of adspecies broaden considerably so that the recorded spectra are often useless. Finally, the more intense chemisorption or higher extinction coefficient of bands of a single species can mask the detection of more relevant adsorbed species. One example is found in the study of the ammoxidation of hydrocarbons, where the more intense spectrum of ammonia adspecies requires the use of very small quantities of ammonia together with larger amounts of organics in order to obtain useful information from coadsorption experiments.[28] When the spectra are recorded during *in situ* experiments, they are dominated by the very strong bands of ammonia adspecies, which makes it impossible to obtain useful information on the mechanism of transformation of organic adspecies.

These problems are overcome using the second methodology for the IR studies, which gives information on the surface reactivity under reaction conditions different from those of the catalytic tests. In particular, in order to isolate and detect the intermediate species, it is necessary to operate at much lower reaction temperatures to slow down the rate of the consecutive transformations. It is thus necessary to hypothesize that in going from low to high temperatures the surface reaction network does not change and that only the rate constants of the reaction vary. This hypothesis cannot be proven, but it can be supported by analyzing the agreement of the surface reaction network with the indications obtained from steady state and nonsteady state kinetic studies.

This indirect proof is a necessary step and thus complete analysis of the surface reaction network requires IR spectroscopy, catalytic reactivity data, and complementary techniques [e.g., temperature programmed methods such as TPD (temperature programmed desorption), TPSR (temperature programmed surface reaction), etc.] on the *same* catalyst.

IR spectroscopy thus should be viewed as a tool for analyzing the surface reactivity and identifying the intermediates and their possible modalities of transformation, in order to elucidate the overall reaction network. The further necessary step is to link the overall reaction network to the kinetics of surface transformation and general catalytic behavior, to analyze: (i) the reliability of IR information on the mechanism of the surface transformations, and (ii) whether or not one specific pathway dominates over another. This second step in the study of the surface mechanisms of transformation is clearly critical, but is often not taken or if taken is underestimated.

The combination of the two steps in a single approach using *in situ* studies (Matyshak and Krylov[266] called it the spectrokinetic method) gives better indications in a few very simple cases but often not better than the two-step approach. In this respect, it is worthwhile noting that Matyshak and Krylov[266] classified the reaction intermediates in reactions of partial oxidation of alkenes from a low to high degree of reliability, assigning a low degree of reliability to spectroscopic studies before and after the catalytic reaction; a medium degree for studies under catalytic conditions, but without accompanying kinetic studies; and a high degree of reliability to *in situ* spectral data obtained in parallel with the determination of catalytic reaction data. Most of the studies were rated low or at best medium, suggesting a general lack of reliable information on the surface mechanisms of alkene partial oxidation. On the other hand, it was demonstrated that none of the "true spectrokinetic" studies led to different conclusions for the complex surface transformations of real interest in hydrocarbon selective oxidation than those obtained by a careful two-step approach, which usually enables more complete identification of the chemistry of surface transformations.

The limitations of the two-step approach show up in analyzing catalytic phenomena when: (i) there is a significant influence of coadsorbed species in determining the surface reactivity (even the simple influence of water), (ii) a dynamic restructuring of the catalyst surface driven by chemisorbed species occurs during catalysis, (iii) the mobility of surface species plays a major role in determining the rates of surface transformations, and (iv) coadsorbed species act as cocatalysts. As discussed at length in this chapter, these phenomena are not unlikely in selective oxidation reactions, and *in situ* studies are necessary to clarify these aspects.

IR studies *in situ* or in a controlled atmosphere are complementary rather than competitive techniques, for analyzing the complex chemistry of catalytic transformations at oxide surfaces, a conclusion that is clearly valid not only for IR spectroscopy, but generally for other physicochemical techniques used to elucidate the mechanisms of catalytic reactions.

8.6.2. Chemistry of Oxidation of Methanol

Although oxidation of methanol to formaldehyde over iron/molybdate catalysts is an important process,[272–274] its yields and selectivities are so high (above 95%) that a discussion of the reaction mechanism with a view toward improving it may be superfluous. Nevertheless, it is useful to briefly discuss here the reaction mechanism of the surface transformation of methanol on oxide surfaces from the perspective of the chemistry of formation and transformation of reactive intermediates, because easy activation and reduced modes of activation/transformation with respect to other substrates allow the surface chemistry and mechanism to be studied in detail.

For the same reasons, methanol conversion is a useful test reaction to elucidate surface properties, although in extrapolating the data to other substrates it must be taken into account that the first step of the reaction (methoxy species formation, as discussed below) is so rapid on most oxides that the reaction rate is often determined only by the diffusion of methanol to the surface. This fact was ignored by several authors who focused only on the turnover frequency of methanol oxidation, which is not the right indication and may lead to serious misinterpretations. Methanol oxidation is a good test reaction regarding the redox and acido-base properties of the catalyst, which determine the selectivity behavior not the rate of conversion.

Even though methanol is a simple molecule, a number of different types of adsorbed species (Scheme 8.15) may form through the contact of methanol with an oxide surface. Methoxy groups (species ② in Scheme 8.15) are usually the principal species detectable by IR analysis.[1] The C–O stretching mode is sensitive to the type of coordination and makes it possible to distinguish between the ②a → ②c species.[275] Undissociated adsorbed methanol (species ①) is possible, but usually is rather weakly coordinated and can be easily removed by pumping off. Species ③ → ⑤ represent species already oxidized. Species ③ is adsorbed formaldehyde in molecular form, but observed, only at very low temperatures.[276] On ionic metal oxides and at room temperature or higher, formaldehyde is usually detected as a dioxymethylene species (④a) or polymeric-type species (④b) when higher concentrations are used. The dioxymethylene species can also undergo disproportionation to methanol and formate species (⑤).[276,277]

Methoxy species (often type ②a) are the first that form by contact with methanol, apart from physisorbed species. The consecutive transformation depends on both the reaction conditions and the catalyst properties (Scheme 8.16). If the concentration of methanol is high enough and the rate of consecutive oxidation of

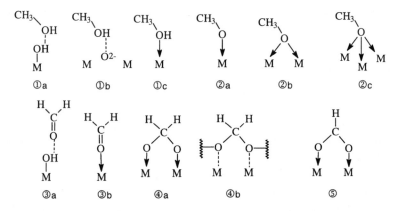

Scheme 8.15. Different forms of adsorbed methanol or of its products of oxidation over oxide surfaces.[1]

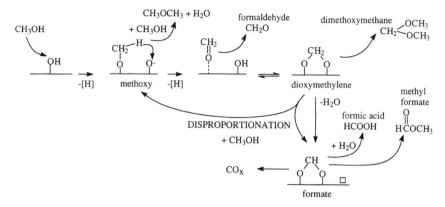

Scheme 8.16. Reaction network in methanol oxidation on oxides.

the methoxy species low (i.e., high concentration of methoxy species on the surface) a condensation reaction can occur leading to dimethylether. This reaction occurs more often over acid oxides of nonreducible cations such as alumina.[278–281] The reaction mechanism over alumina[282] probably involves the reaction of a bridging methoxy species (species ②b in Scheme 8.15) having a nucleophilic character with an undissociated methanol species coordinated to strong Lewis acid sites (species ①c in Scheme 8.15). The oxygen of the bridging methoxy species attacks the carbon atom (having a strong electrophilic character) of the coordinated methanol species to give the dimethylether. However, when strong Brønsted acid sites are present, protonation of the coordinated alcohol can occur followed by nucleophilic substitution.

Oxidation (H abstraction and electron transfer) of the methoxy groups leads to the formation of coordinated formaldehyde, which is in equilibrium with the dioxymethylene species (Scheme 8.16). The reaction occurs, as is reasonable, by a nucleophilic attack by surface oxygen giving rise to the dioxymethylene species. If the metal to oxygen bond has a mainly covalent character, the equilibrium between the dioxymethylene and coordinated formaldehyde is shifted toward the latter, which is the precursor for gas phase formaldehyde. In fact, catalysts such as bulk or supported vanadium oxide[278,283–285] or molybdenum oxide[286,287] are selective in the oxidative dehydrogenation of methanol to formaldehyde. However, over ionic oxides dioxymethylene is the more stable species, and is the intermediate for further transformations. It can undergo a Cannizzaro-type disproportionation to a methoxy and a formate species (Scheme 8.16).[276] The formate species, alternatively, may also form by direct oxidative dehydrogenation of dioxymethylene. By reaction with methanol, the dioxymethylene forms dimethoxymethane, which desorbs to the gas phase. Moreover, this species is usually detected at low conversions of methanol, where the concentration of the reactant is higher.

The formate species may also react with methanol to give methyl formate, which forms with relatively high selectivities over catalysts such as vanadium oxide supported on TiO_2[288–290] or Sn/Mo oxides.[291] The formation of methyl formate over these catalysts[292] requires that the desorption of formaldehyde is slow (the equilibrium between coordinated formaldehyde and dioxymethylene shifted toward the latter product—see Scheme 8.16) and the rate of oxidation/disproportionation is relatively high. Both pathways contribute to formate formation; the relative rates between these two reactions probably depend on the surface coverage of dioxymethylene species and the reaction conditions. However, as the formate species can undergo further transformation (Scheme 8.16), the selective formation of methyl formate requires a relatively inert formate species (with respect to competitive further transformations) and a relatively high concentration of methanol.

The formate ion, in fact, can either form formic acid by reaction with (coordinated) water and desorption or be decomposed to carbon oxides[293–297]:

$$H\text{–}COO^- \rightleftharpoons CO + OH^- \qquad (8.16)$$

Clearly the rates of the two pathways of reaction of formate ions with methanol to give methyl formate or formate decomposition are a function of the concentration and conversion of methanol. Moreover, methyl formate forms over Zn-chromites[295] at low conversion, whereas at higher conversion the decomposition reaction predominates.

These data thus further demonstrate several aspects of the surface reactivity noted earlier:

1. The role of the stability/reactivity of surface species, i.e., more correctly the role of the surface lifetime of adspecies, which depends not only on the intrinsic catalyst characteristics but also on other factors such as the concentration of adspecies (e.g., in the disproportionation reaction) and the concentration of gas phase species (methanol concentration and influence of its conversion on the formation of methyl, formate, etc.).
2. The presence of surface equilibria between adspecies, the equilibrium of which determines the catalytic behavior even if desorption occurs after the rate-determining step (e.g., equilibrium between coordinated formaldehyde and dioxymethylene). The equilibrium depends on various factors and not only on the intrinsic reactivity properties of the catalyst.
3. The dependence of the macroscopic catalytic behavior on the rate of desorption of products, even if they form after the rate-determining step (e.g., the desorption of coordinated formaldehyde).
4. The influence on the surface reactivity of other gas phase components (besides the reactant) or of reaction products (water, in particular). Water, e.g., influences the possibility of desorption of formate as formic acid.

The above observations regarding the mechanism of methanol conversion also indicate that the intrinsic kinetics of reaction clearly change as a function of the coverage by adspecies and concentration of reactants in the gas phase, and thus should vary along the axial profile of an integral fixed-bed reactor. This is not a case restricted to this specific example, but is more general, as also discussed later, and again demonstrates the limitations of the traditional kinetic approach based on the rate-determining step and the Langmuir–Hinshelwood type of reaction rates.

The kinetics of surface transformation should thus be linked to the surface chemistry of active intermediates. IR spectroscopy can be a very useful method for reaching this objective, but must be employed in parallel with other time-resolved spectroscopic methods and transient reactivity tests. In all cases, attention must be given to reproducing the relevant features of the working catalyst in terms of surface coverage by adspecies and of gas phase species in equilibrium with the surface; otherwise indications can be obtained on the surface chemistry but not on its link with the surface reactivity (transformation kinetics). This concept, while often not well understood, is of central importance as well as being a key to avoiding the improper use of certain characterization techniques. For example, thermodesorption in a nonreactive or reactive environment is useful for characterizing the surface chemistry, similar to IR spectroscopy, but extrapolation of the data to describe the kinetics of the surface processes must be done with caution when the differences in the surface properties in terms of coverage by adspecies are not taken into account.

It should be also noted that the kinetics of the surface processes may also depend on the presence of "spectator" species, which may be very similar to active intermediates. An interesting example in this regard was described by Schlögl et al.[298] in a study of the reaction pathways and kinetic oscillations during methanol oxidation over copper/oxygen catalysts. They observed in agreement with Bowker et al.[299] that stable surface methoxy species are present on the catalyst surface during catalytic reaction and do not play a role as direct reaction intermediates. However, these species do have a part in controlling the supply of gaseous oxygen to the surface, one of the factors that determines the surface reactivity and oscillatory phenomena. However, Schlögl et al.[298] also agree that methoxy-type species should be reactive intermediates in the process, but are more weakly coordinated to surface sites than the above stable, strongly bound methoxy species. This problem will be discussed in Section 8.8 but this example leads to two further important observations:

1. It is important to study the energies involved in the surface interactions of active intermediates, which depend on the surface lifetime of these species and the type of coordination. It is reasonable to assume that only labile species are really determining, although their identification is more critical. The "frozen" approach used, for instance, in IR spectroscopy, to study the

surface species at lower temperatures than those of the catalytic tests so that their further transformation is slowed (Section 8.6.1), may eliminate the differences between more or less strongly bound adspecies. Microcalorimetric studies can be useful here.

2. Strongly bound "spectator" species may also influence the surface reactivity, although in an indirect way, besides having the other possible effects discussed above (e.g., blocking surface sites, giving or withdrawing electrons from the surface, changing mobility of adspecies).

8.6.3. Oxidation of Linear C4 Hydrocarbons

The oxidation of linear C4 hydrocarbons is considerably more complex than that of methanol, owing to the possibility of multiple types of activation modalities as well as a wide range of possible products. The principal products observed include: butenes (but-1-ene, but-2-ene in *cis* and *trans* conformation) and butadiene,[223,224,263] acetic acid,[300] tetrahydrofuran,[301] furan,[302] but-1-ene-3-oxy (methyl vinyl ketone), butan-2-one (methyl ethyl ketone), crotonaldehyde and maleic anhydride,[25,127,180,303] depending on the hydrocarbon, the catalyst and the reaction conditions. Various other minor by-products have also been found.

While the initial formation of methoxy species in methanol conversion is quite obvious, what can be considered the first stage of the reaction in linear C4 hydrocarbon conversion is much less defined. Let us consider, for instance, *n*-butane. It is possible to have abstraction of a hydrogen in the 1- or 2-position (the former energetically less favorable), which may be a concerted process with the abstraction of a second hydrogen or the insertion of an oxygen on the same or on an adjacent position.

Hydrogen abstraction in 1,4-positions may also be possible, with a concerted oxygen insertion to form a tetrahydrofuran-like species.[301] The formation of the latter is the driving force to overcome the higher energy required to abstract two hydrogen atoms from methylic positions, although tetrahydrofuran formation from butane has been observed only by a single team[301] and no other researchers have been able to reproduce the data. Furthermore, kinetic isotopic experiments on a similar catalyst showed that the two abstracted hydrogen atoms come from the methylenic groups and not from the methylic groups.[140]

The problem with linear C4 hydrocarbons is thus to identify whether the different products mentioned above depend on different modalities of hydrocarbon activation (on different catalysts or on the same catalyst due to a different type of active site) or whether there is a common modality of activation, with the products depending on the various pathways possible after this common initial intermediate. Busca[1] has suggested the second hypothesis. Independently of the catalyst and type of final product, there is a first common step of formation of the 2-butoxide (sec-butoxy) species either from butane or butene. This 2-butoxide species, together with various others (formate, butan-2-one, acetate) is detected by contacting n-butane at room temperature with magnesium/chromite,[304] which is a combustion catalyst.[305] However, on the same catalyst but-1-ene forms but-1-ene-3-oxide (vinyl methyl ketone) instead of 2-butoxide, a species that has *not* been observed in butane or but-1-ene interaction with magnesium/vanadate[304,306] or vanadium/phosphorus oxides,[87,304–310] selective catalysts for butane oxidative dehydrogenation or maleic anhydride synthesis, respectively. However, the 2-butoxide is detected by but-1-ene adsorption over vanadium oxide supported on TiO_2,[311] which is again a catalyst for oxidative degradation of n-butane giving rise to acetic acid formation.

Therefore, although the hypothesis of formation of 2-butoxide as the first intermediate common to different products and catalysts is reasonable, it is not sufficiently well supported by experimental evidence. Clearly, the problem may be that when this species is formed over selective catalysts it is transformed to other products so quickly as to be undetectable, whereas on catalysts for butane oxidative degradation the milder conditions of formation allow its detection. However, this hypothesis does not account for its not being found with but-1-ene, with the rates of butane depletion over catalysts of oxidative degradation and selective oxidation, or for the type of products of selective oxidation as discussed below. This example shows the limitations in extrapolation of spectroscopic data, when it is not adequately supported by reactivity tests on the *same* catalysts but is based solely on generic published data on the reactivity.[1]

After formation of 2-butoxide, the hydrogen linked to carbon of the C–O bond is easily susceptible to abstraction by basic neighboring lattice oxygen with rapid formation of 2-butanone, as indicated by IR data.[1] Over strong basic or acidic oxides having little oxidizing power, the alkoxide decomposes forming the corresponding alkene. There are abundant published data showing that on catalysts that do not give oxidative degradation of alkanes and are highly selective (e.g., vanadium/phosphorus oxides) the oxidative dehydrogenation of alkanes to alkenes is the first step of the reaction.[127,176] The mechanism of formation of an alkene from an alkane based on an initial step of alkane reaction with the surface to form an alkoxide that is not stable and thus quickly decomposes to give alkene does not seem to be a highly probably one. A mechanism of two hydrogen abstractions on the C4 alkane, through a concerted or stepwise process is more reasonable. Owing to lower C–H bond energy, the abstraction should occur on the two methylenic groups and as no products of skeletal isomerization are found the mechanism probably involves a concerted 2H abstraction. In n-butane selective oxidation over $(VO)_2P_2O_7$, a mechanism has been suggested that involves the contemporaneous hydratelike abstraction (i.e., a proton and two electrons) from a Lewis acid site (unsaturated surface vanadyl—see Chapter 4) and proton abstraction from the neighboring methylenic group by a basic surface oxygen bridging two vanadium atoms[127,128,176]:

$$(8.17)$$

Alternatively, the simultaneous two-hydrogen abstraction from an η^2-peroxo species formed by O_2 chemisorption on the V Lewis sites (see Section 8.3.3) has also been proposed.[176] Both mechanisms seem to be consistent with experimental evidence, but further studies are needed. On the other hand, based on the above discussion, it is possible to conclude that the mode of activation determines the pathway of transformation and the catalytic behavior, limiting the observation to the case of a common intermediate (2-butoxide).[1] This is valid at least to distinguish between catalysts for which oxidative degradation is the main pathway (such as Mg/chromite, V/TiO_2) and catalysts for which it is carbon-chain rupture (such as Mg/vanadate and V/P oxide). In fact, 2-butanone is very susceptible to C(2)–C(3) bond breaking, as discussed earlier, giving acetate species and finally carbon oxides. The C4 alkene can undergo fast allylic hydrogen abstraction and oxygen insertion to give allyl–alkoxide species, either but-1-ene-3-oxide or but-3-ene-1-oxide, depending on when the attack occurs on but-1-ene or but-2-ene. It must be noted, however, that catalysts able to provide allylic oxidation also have enough acid properties to catalyze a fast isomerization of the alkenes (in 1- or 2-positions and the latter in *cis* or *trans* configurations). At the temperatures at which these catalysts are active in allylic oxidation, the quantities of the various alkenes depend on the thermodynamic equilibrium.

But-1-ene-3-oxide is the precursor of but-1-ene-3-one (methyl vinyl ketone), which can also evolve to acetate, and but-3-ene-1-oxide is the precursor of 2-butenal (crotonaldehyde). Both these products can form in significant amounts starting from but-1-ene over catalysts such as V/P oxide,[127] although their formation depends on the modality of catalyst preparation. Both but-1-ene-3-oxide and but-3-1-oxide can transform to butadiene although, a direct second hydrogen abstraction on the allyl intermediate to form butadiene directly instead of passing through the allyl-alkoxide intermediate is more reasonable.

Butadiene can undergo 1–4 oxygen insertion (a Diels–Alder type of reaction) to give 2,5-dihydrofuran more easily than oxidative dehydrogenation to furan. The latter can be oxidized to maleic anhydride,[302] but the direct alternative oxidation of dihydrofuran to maleic anhydride is also possible. In fact, the hydrogens in 2-

positions on 2,5-dihydrofuran are allylic hydrogen and can be easily removed. By consecutive oxygen attacks the same carbon atom butyrrolact-2-one forms and then attacks the symmetric carbon atom (in position 5) forming maleic anhydride (2,5-furandione).

It is also worth noting that crotonaldehyde is detected as an adsorbed species on V/P oxide in large amounts (main adsorbed product) when butadiene rather than but-1-ene is fed in. Moreover, during catalytic tests this species forms in larger amounts from butadiene than from but-1-ene. Thus an additional, more efficient, pathway of crotonaldehyde formation that does not involve the but-3-ene-1-oxide intermediate must exist. Although specific studies are not available, the formation of crotonaldehyde from butadiene probably involves the ring-opening and rearrangement of the 2,5-dihydrofuran intermediate, which in all likelihood is catalyzed by Brønsted acid sites when further oxidation of dihydrofuran is not fast enough. The reaction network in butane oxidation is shown in Scheme 8.17.

Alternative pathways are also possible, including the direct oxidation and oxygen insertion on butane at 1,4-positions (not very probable as already discussed) and maleic anhydride formation by condensation of a 1,4-dialdehyde or by formation first of maleic acid, but no experimental evidence clearly supports them. However, several of the steps in the surface reaction network shown in Scheme 8.17 have not been fully demonstrated, nor has it been clearly shown how their relative importance depends on the catalyst characteristics and reaction conditions. It may thus be concluded that the mechanism of C4 linear hydrocarbon oxidation, unlike that of methanol oxidation, must still be clarified through further studies.

8.6.4. Oxidation of Alkylaromatics

Although apparently even more complex than the previous case, the oxidation of alkylaromatics may be considered simpler because the reactivity of the carbon and hydrogen atoms of the aromatic ring is significantly different than that of the carbon and hydrogen atoms in the lateral alkyl groups, and this translates into fewer possible pathways of activation and surface transformation. Furthermore, owing to the aromatic ring, the hydrogen atoms of the methyl group in toluene can be considered allylic hydrogen and thus are easily removed because of the stabilization effect of the aromatic ring on the reaction intermediate. Owing to an inductive effect, the situation for xylenes or other substituted toluenes is more complex, but the reactivity remains simpler than in the case of C4 hydrocarbons and similar to that of C3 alkenes. Unlike the latter, however, a simple π-bond interaction and formation of an allyl coordinated species (Section 8.2.1) is not possible. The aromatic ring can interact with the surface sites in either a planar or a side-on form (nearly perpendicular to the surface). These modalities of interaction probably influence the reactivity, although IR data generally indicate a weak perturbation of the aromatic ring by adsorption on a range of oxides,[1] suggesting that the main

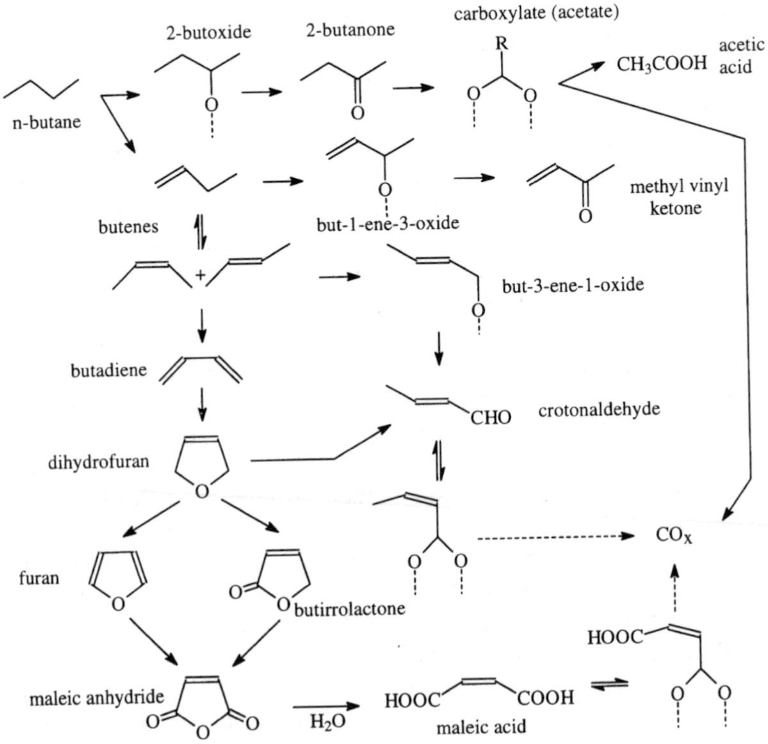

Scheme 8.17. Reaction network in *n*-butane oxidation on oxides.

interaction with the surface sites involves only lateral alkyl groups, at least in the selective oxide catalysts.

In alkylaromatic oxidation (here we shall discuss only toluene oxidation, but all the observations are valid for other alkylaromatics) the first step is usually assumed to be hydrogen abstraction from the methyl group to form a benzyl-type species (charged or radical type). There are contradictory indications in the literature concerning the modalities of activation,[312-324] but the more chemically reasonable suggestion is the formation of a radical-type species because this is the energetically more favorable pathway owing to stabilization of electrons on the aromatic ring. Spectroscopic IR data[312-324] are in general agreement with this suggestion, although often the same spectra are interpreted in difficult ways. A discussion of the various IR spectra and relevant interpretations can be found in the review by Busca.[1] It is worth noting here, however, that IR data indicate that the activation of toluene leads to a rather stable species that does not contain oxygen, has a nearly intact aromatic ring, and is thus assigned to a benzyl radical.[312] The extensive delocalization of the radical center over the aromatic ring creates a relatively stable intermediate species, which can be detected over various oxides, including combustion catalysts such as α-Fe$_2$O$_3$,[325] whereas the corresponding allyl species (e.g., in propene oxidation) cannot be detected.[235] This indicates that a rather stable species with a long surface lifetime forms by activation of toluene. Sanati and Andersson[318] indicated that the rate-determining step is the interaction of the benzyl species with the surface oxide. Therefore, the benzyl species cannot be interpreted as a reaction intermediate, but is simply an adsorbed form of toluene in equilibrium with the gas phase species and with the H$^\bullet$ species delocalized on the surface around the adsorption site.

One question is why this benzyl radical species, having a long surface lifetime and being in a relatively high surface concentration, does not give rise to a significant formation of dimerization products, which are detected, but in small quantities.[326,327]

The first true intermediate is the one that forms from the benzyl species. Similar to what was discussed earlier, the benzyl species can be attacked by surface oxygen to give a benzyloxide species,[313] but one which has never been detected because it quickly further transforms to the corresponding aldehyde. The adsorbed aromatic aldehyde desorbs in the gas phase or is further oxidized to the corresponding carboxylate species, being quite reactive with the hydrogen atoms of the benzylic carbon. Thus, to avoid this step it is necessary to have fast desorption of the coordinated benzaldehyde and avoid the formation of dioxyphenyl methylene (the equivalent species from benzaldehyde of di-oxymethylene from formaldehyde; see Section 8.6.2), which is favored over oxides with ionic metal to oxygen bonds. The carboxylate species can decar-boxylate rather easily and be further transformed to carbon oxide, whereas the desorption of benzoic acid requires an acid-catalyzed hydrolysis of the carboxy-late species. In o-xylene, the mechanism is analogous, but the presence of two adjacent methyl groups allows the synthesis of an anhydride (phthalic anhy-dride), which, interacting less strongly with the surface, desorbs faster in com-parison with acid species (benzoic acid). The other aspects of the reaction mechanism are quite similar to those discussed for toluene and do not add further relevant information. Apart from the already cited review by Busca,[1] further discussion can be found in several other reviews and papers.[326-333]

The comparison between toluene and o-xylene oxidation again shows the importance of the type of interaction between the product and the surface sites and underscores the role of carboxylate species in the mechanism of oxidative degra-dation. Details of this mechanism have not yet been studied, but the comparison provides indications regarding the factors that determine the rate of reaction and thus possible ways to limit this nonselective transformation.

It should also be noted that although in catalytic oxidation of alkylaromatics small quantities of products deriving from aromatic ring rupture (e.g., maleic anhydride) are observed, they have not been detected in IR tests in the absence of gas phase oxygen. However, they can be seen in IR tests in the presence of gaseous oxygen, which suggests that they can form in a parallel pathway involving a different form, possibly adsorbed, of oxygen than that responsible for the main, nondegradative pathway (tolualdehyde from toluene or phthalic anhydride from o-xylene). The dioxygen addition to benzene to form an endo-peroxide species that rearranges to form 1,4-diphenol or benzoquinone, leading to further conversion to maleic anhydride was proposed some time ago[334,335] and subsequently supported by quantum-chemical calculations.[336-338] It is thus reasonable that the oxidative degradation pathway from alkylaromatics to products such as maleic anhydride can involve the formation of peroxide adducts due to electrophilic attack of oxygen on the benzene molecule.

An interesting observation from the quantum-chemical calculations[336-338] is that different products can be formed depending on the direction of the approach and mutual interaction of the reacting molecules. If the oxygen molecule with its axis lying in the plane of the benzene ring, parallel to the side of the ring, approaches this side a peroxide intermediate can be formed that then rearranges into o-hydroquinone. On the other hand, if the oxygen molecule with its axis perpendicular to benzene ring approaches the side of the benzene molecule there is cleavage of C–C bonds in the ring and two carbon atoms are pulled away by the O_2 molecule to form a $C_2O_2H_2$ fragment, which further transforms to carbon oxides. When the oxygen molecule with its axis parallel to the diagonal of the ring approaches the benzene molecule along the reaction pathway perpendicular to the ring and passing through its center, an O–O bridge is formed between 1,4-carbon atoms. The rearrangement by insertion of oxygen into C–H bonds leads to p-hydroquinone, which is the intermediate to maleic anhydride. The mutual direction of approach is thus an important factor and an interaction between the aromatic ring and the surface that can force a preferential configuration may be a key to determining whether one pathway occurs or another. On the other hand, it should be noted that these quantum-chemical calculations do not take into account the thermal vibrational energy of molecules at the typical temperatures of catalytic tests (300–400°C) nor the energy associated with translational motion of colliding molecules. When these elements and the sticking factor are included the above conclusions lose some validity under real catalytic conditions, but certainly remain indicative of the limiting situations and of the various factors that determine catalytic behavior.

An interesting point is whether it may be possible to influence this mutual interaction of hydrocarbon and oxygen molecules and thus determine the selectivity. Naked metal ions at the surface may not only have Lewis acido-base properties but may also generate a local electrostatic field that probably influences the alignment of adsorbing molecules (often having a dipole character and being polarizable, as, e.g., in the case of toluene and oxygen), when they approach the surface. Similarly, when active sites are located in clefts or surface cavities of molecular dimensions, the possibility of a modification in the energy associated with the approach of gas phase molecules to active centers with respect to an "ideal" flat surface is a reasonable one.

Frei *et al.*[151-157] (see also Chapter 7) have shown, for instance, for room temperature toluene oxidation to benzaldehyde on BaY zeolite, that key aspects of the reactivity are: (i) the alignment of toluene and oxygen molecules due to the strong electrostatic field generated by naked Ba^{2+} ions inside the supercages of Y zeolite, and (ii) the confinement action of zeolite walls to form the hydrocarbon–oxygen encounter complex. Under these conditions there is a spontaneous electron transfer from hydrocarbon to oxygen and the charge-transfer complex (stabilized by the electrostatic field) undergoes consecutive transformation to an organic peroxide species. The model of the toluene–oxygen aligned encounter complex is shown in Figure 8.14.

These results indicate the possibility of having *noncontact* catalysis (i.e., catalysis that is close *to*, but not on the surface) as a result of the influence of surface sites on the near environment. This also may mean that active intermediates in

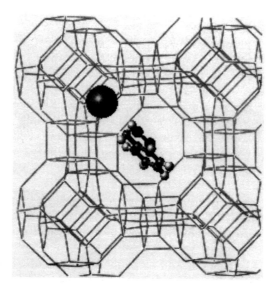

Figure 8.14. Model of toluene–oxygen encounter complex aligned by an electrostatic field of naked Ba^{2+} ions in BaY zeolite.

heterogeneous catalysis are not necessarily adsorbed species. Although at this stage these observations are only speculative, it is clear that they can open new perspectives both in the understanding of catalytic quasi-surface phenomena and the design of new types of catalysts.

8.6.5. The Case of Ammonia

There are several industrially relevant cases of selective oxidation of hydrocarbons in the presence of ammonia (ammoxidation reactions) over mixed oxides, such as the propene and propane conversion to acrylonitrile[8,11–13,28,174,339–341] and the synthesis of aromatic nitriles from alkylaromatics.[342–353] A key aspect of these reactions is the nature of the ammonia intermediate species, both those responsible for the generation of the nitrile species and those responsible for the side reaction of direct ammonia oxidation to N_2.

Ammonia coordination modes to oxide surfaces are well known because ammonia is a widely used probe molecule for IR studies.[354] Tsyganenko *et al.*,[355] in particular, have discussed specifically the surface species arising from ammonia adsorption on oxide surfaces. Different types of species can be distinguished:

1. Hydrogen-bonded species via one of the ammonia hydrogen atoms to a surface oxygen atom (or to the oxygen of a surface hydroxyl group).
2. Hydrogen-bonded species via ammonia nitrogen atoms to the hydrogen of a surface hydroxyl group (Brønsted acid site) to form an ammonium ion. Different types of the latter species may exist owing to the interaction of one or more hydrogens of the ammonium ion with the neighboring surface oxygen atoms to form distorted ammonium ions bent toward the surface instead of an end-on symmetrical species. The different species are clearly detectable by IR because the deformation mode (triply degenerate) splits into the various components.
3. Ammonia coordinated to an electron-deficient metal atom (Lewis acid sites).

In addition to these species, chemisorbed molecules formed by homo- or heterolytic dissociative chemisorption (NH_2 or NH type species) can also be present, but generally have been poorly defined with respect to the first three.

Ammonia may also react with the oxide to form metal–imido ($M=NH$) or amino ($M-NH-M$) species. These species appear to be the only ones responsible for the mechanism of nitrogen insertion in the propene substrate to form acrylonitrile over different types of oxides such as Bi-molybdate ($Mo=NH$ species)[12,13] and Fe-antimonate ($Sb-NH-Sb$ species).[356]

The concept of a two-step mechanism of nitrogen insertion into hydrocarbons (formation of metal–imido or amino species followed by addition of these species to the substrate—see Scheme 8.2) has been widely used, accepted, and extended to other catalytic systems and the ammoxidation of other hydrocarbons. For example, Andersson et al.[357–359] have indicated that the formation of V=NH or Cu=NH species is responsible for the ammoxidation activity in the toluene to benzonitrile transformation in supported and unsupported vanadium oxide and barium-cuprate catalysts, respectively. The same team also suggested that Sb–NH– Sb species are responsible for the ammoxidation of propane on Sb/V oxides.[360,361]

The NH group is electronically equivalent to lattice oxygen, and since the latter's involvement is the basis of one of the mechanisms of selective oxidation (Mars–van Krevelen mechanism, see the discussion in Section 8.2), the extrapolation to the case of ammonia was quite obvious. However, the presence of the proton on the NH group makes it less stable with respect to oxygen and thus in the presence of oxygen and water (both being present in the reaction medium during ammoxidation reactions), the presence of M=NH or M–NH–M (M = metal) groups is less obvious. Moreover, Wise and Markel,[362,363] studying the formation of molybdenum nitride by the reaction of ammonia with MoO_3, indicated that the substitution of lattice oxygen by nitrogen is inhibited by the presence of oxygen and water and occurs at much higher temperatures than those of the catalytic tests of hydrocarbon ammoxidation.

Furthermore, it must be noted that none of the metal–imido or amino species has spectroscopically proved to be really present; their participation in the reaction mechanism during hydrocarbon ammoxidation has never been clearly demonstrated. Theoretical studies have shown the possibility of formation of metal–imido species in the case of molybdate catalysts[364] and vanadium oxide.[365] However, these studies do not prove either that the metal–imido species are stable in the presence of ammonia *and* oxygen/water, or that they form at the temperatures of ammoxidation catalytic tests.

It also should be noted that direct oxidation of ammonia to N_2 competes with the nitrogen insertion into the organic substrate. Zanthoff et al.[366] observed that the decline of acrylonitrile formation from propane is almost identical to the decline in N_2 formation, when the flow of ammonia over a V/Sb oxide catalyst is stopped. This indicates the involvement of a common type of ammonia intermediate species in both reactions. Although not studied in detail, it is difficult to accept the idea that the mechanism of NH_3 oxidation to N_2 involves the first step of formation of metallo–imido or amino species, which suggests that these species do not play a role in the mechanism of nitrogen insertion into the organic substrate. In fact, none of the several studies on the mechanism of ammonia conversion over oxide or mixed oxides analogous to those also active in the ammoxidation of hydrocarbons (V/TiO$_2$ catalysts, e.g., are active in both alkylaromatics ammoxidation and ammonia oxidation or conversion to N_2 in the presence of NO) produced evidence for or even suggested the possibility of metal–imido or amino species formation.[367,368] In *all* mechanism of ammonia conversion to N_2 ammonia adspecies have been indicated as the reaction intermediates, although different types of intermediate species have been suggested.

In the case of ammoxidation of hydrocarbons, more recent studies have pointed out the role of adsorbed ammonia species in the mechanism of nitrogen insertion on the substrate. Buchholz and Zanthoff[369] found that the formation of acrylonitrile in transient TAP reactor catalytic experiments is associated with the presence of short-lived NH_x species that cannot be either metal–imido or amino species. Martin *et al.*[350] made similar observations for the mechanism of toluene ammoxidation to benzonitrile over vanadium/phosphorus oxides. However, although there is increasing evidence regarding the role of ammonia adspecies in the reaction mechanism of hydrocarbon ammoxidation,[370] there are not sufficient systematic data to justify general conclusions. It is also worth noting that it is more reasonable to consider that activated ammonia species formed by hetero- or homolytic splitting of ammonia should be more reactive than metallo-imido or metallo-amino species. Further, the fact that IR spectroscopy can demonstrate their reaction with hydrocarbons to give selective ammoxidation products (NB, however, that IR spectroscopy indicates that this is a possible pathway, not that it is the dominant pathway during reaction) suggests that these ammonia adspecies should also be the intermediates in the primary pathway of hydrocarbon ammoxidation.

In conclusion, contrary to what is usually assumed, there is no clear evidence in the literature about the role of metallo-imido or metallo-amino species as the reactive intermediates in hydrocarbon oxidation. Rather, recent evidence points to a role for ammonia adspecies in the reaction mechanism, but limited data do not allow generalization of the conclusions, and a specific research effort is needed to clarify this aspect of the surface reactivity in a major class of industrial selective oxidation reactions.

8.7. PRESENCE OF COMPETITIVE PATHWAYS OF CONVERSION AND FACTORS GOVERNING THEIR RELATIVE RATES

In Sections 8.5 and 8.6 it was shown that multiple pathways of surface transformation are possible, and that they sometimes lead to the same final product. It was also demonstrated that the nature and quantity of both chemisorbed and gas phase species determine the relative rates of the competitive pathways and thus that it is not enough to discuss just a reaction mechanism, but rather that the dominant pathway under a given reaction condition must be indicated. Owing to the importance of this concept for a good understanding of the surface reactivity of mixed oxides in selective oxidation reactions, some further examples will be discussed here.

8.7.1. Propane Ammoxidation on (VO)$_2$P$_2$O$_7$

(VO)$_2$P$_2$O$_7$ is the active phase in catalysts for the oxidation of n-butane to maleic anhydride (see Chapter 4). The same catalyst also shows a selective behavior in propane ammoxidation, but only when there is an optimal surface concentration of ammonia adspecies. In fact, studying the transient catalytic reactivity of (VO)$_2$P$_2$O$_7$ as a function of the ammonia coverage,[371] it was observed that three different regimes of catalytic behavior as a function of the amount of chemisorbed ammonia can be demonstrated (Figure 8.15):

1. On the surface of (VO)$_2$P$_2$O$_7$ completely covered with chemisorbed ammonia (right-hand side in Figure 8.15) propane can be selectively activated and oxidatively dehydrogenated to propene C3$^=$), but a limited amount of acrylonitrile forms. Under these conditions, however, the formation of carbon oxides is very limited and thus the selectivity for the sum of the products of selective oxidation (propene, acrylonitrile, and acetonitrile) is very high (about 98%).

2. When the surface coverage with adsorbed ammonia decreases, the rate of formation of acrylonitrile from the intermediate propylene increases and at the same time the rate of propane conversion also increases. The formation

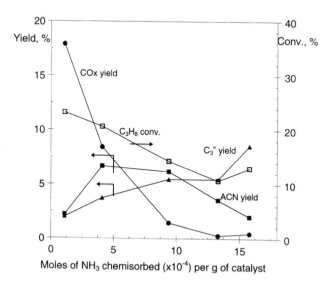

Figure 8.15. Relationship between the amount of ammonia chemisorbed on (VO)$_2$P$_2$O$_7$ and the change in the conversion of propane (C$_3$) and yield of propene (C$_3^=$), acrylonitrile (ACN), and carbon oxides (CO$_x$) at 406°C during transient catalytic tests.[370,371]

of carbon oxides remains very low and the selectivity to acetonitrile is nearly constant.

3. With a further decrease in the amount of chemisorbed ammonia, the selectivity for acrylonitrile decreases with a parallel considerable increase in carbon monoxide formation and propane conversion. It should also be noted that the formation of N_2, due to the side reaction of oxidation of NH_3, becomes apparent after the maximum in the selectivity for acrylonitrile, and its formation parallels that of CO formation.

Therefore, the optimal behavior (in terms of selectivity for acrylonitrile) does not correspond to the maximum concentration of adsorbed ammonia on the surface (reactant for the synthesis of acrylonitrile), nor to the optimal behavior in terms of minimal selectivity for carbon oxides.

The analysis of the reaction mechanism based on IR spectroscopy[372] and in particular the results of ammonia and hydrocarbon coadsorption experiments provide further insight on this particular catalytic behavior of the surface of $(VO)_2P_2O_7$. Propane is oxidatively dehydrogenated to propene through a mechanism that was described in Section 8.6.3 [see, in particular, Eq. (8.17)]. After this first step, the propene intermediate coordinates strongly over vanadium Lewis acid sites and, in the absence of gaseous oxygen, slowly transforms to acrylonitrile by reaction with amido-like species (NH_2^-) formed by dissociative chemisorption of ammonia. The intermediate propylenamine then further transforms to acrylonitrile by consecutive H abstractions by surface lattice oxygens. In the presence of gaseous oxygen, however, coordinated propene reacts with oxygen coordinated over Lewis acid sites to form an adsorbed acrolein or acrylate species that reacts with the neighboring ammonium ions to give acrylonitrile and water via an imine intermediate.

Hence, two pathways of formation of acrylonitrile from propane are present on $(VO)_2P_2O_7$ (Scheme 8.18), the first under anaerobic conditions involving the intermediate formation of propylenamine and the second, in the presence of oxygen, involving the reaction between surface acrylate species and ammonium ion species. The second pathway is much more rapid than the first, but coordinated ammonia inhibits oxygen chemisorption and thus the formation of the acrylate species. When all surface sites for oxygen chemisorption are occupied by chemisorbed ammonia, propene transformation to acrylonitrile occurs mainly through the first, less efficient pathway of reaction, whereas when some of the surface coordination sites are free, the rate of the second pathway increases with an increase in the selectivity for acrylonitrile as well as in the rate of propane depletion. However, IR data[372] show that acrylate species are also intermediate for the formation of carbon oxides even when neighboring ammonium ions are absent (see Scheme 8.10). For this reason, the further depletion of chemisorbed ammonia leads to a decrease in the maximum in acrylonitrile formation and an increase in the formation of carbon oxides (Figure 8.15).

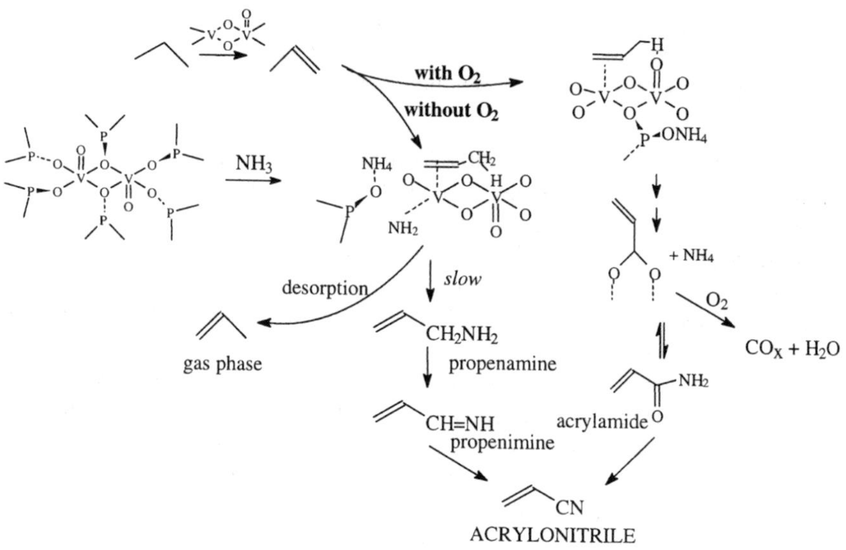

Scheme 8.18. Reaction network in propane ammoxidation on $(VO)_2P_2O_7$.

This example shows that both the rate of hydrocarbon depletion and the selectivity of its transformation depend on the active state of the catalyst surface in terms of the amount of surface adsorbed species. Furthermore, it also points out that the pathway of transformation may change as a function of the nature and amount of chemisorbed species.

8.7.2. Toluene Ammoxidation on $(VO)_2P_2O_7$ and V/TiO_2 Catalysts

In the ammoxidation of alkylaromatics both on $(VO)_2P_2O_7$[373–378] and other types of catalysts (e.g., V/TiO_2)[313,344,346] multiple pathways and surface intermediates similar to the case discussed in the previous section were found. Lücke *et al.*[373–377] suggested the following sequential reaction mechanism on the basis of combined spectroscopic and transient reactivity data:

1. Toluene is adsorbed coordinatively with the ring π-system on Lewis acid sites.
2. An oxidative H abstraction from the methyl group leads to a benzyl radical and the methylene group will interact with the V–OH group formed by H abstraction.
3. The generated aldehyde reacts with neighboring ammonium ions (on phosphate groups) to form a benzylimine species.

4. A further dehydration step leads to the generation of benzonitrile, which desorbs from the surface and leaves a vacancy.
5. The reduced site is reoxidized though a Mars–van Krevelen mechanism.

This reaction mechanism is similar to that discussed in the previous section for propane ammoxidation (Scheme 8.18), but the possible route via benzylamine is excluded. Lücke *et al.*,[373–375] in fact, found that benzylamine is rapidly hydrolyzed to benzoate and ammonium ions, and thus discarded this route. The hydrolysis of benzylamine is probably a reversible reaction with an equilibrium that should depend on both the amount of adsorbed water and the presence of free Brønsted sites. It is also probable that under the conditions of the tests done by Lücke *et al.*[373–375] the hydrolysis of benzylamine is favored, but this does not exclude the possibility that during catalytic reaction and in the presence of a higher ammonia surface coverage benzylamine is more stable and is an active reaction intermediate. This comment demonstrates the general problem of extrapolating data on the reaction mechanism to different surface coverages by adspecies.

There is no conclusive evidence as to whether there are or are not competitive pathways in toluene ammoxidation. In fact, in IR and kinetic experiments Cavalli *et al.*[346,379,380] and Busca *et al.*[313,348,381] pointed out that toluene ammoxidation to benzonitrile on V/TiO_2 catalysts occurs by two parallel pathways from the common intermediate benzyl radical: (i) partial oxidation of the first benzylamine generated via a surface imine to benzonitrile, and (ii) reaction of benzaldehyde formed as an intermediate with ammonia to give benzonitrile via an imine intermediate. The latter mechanism has also been suggested by Forni *et al.*[382] for methylpyrazine ammoxidation over Sb/V/M oxides and by Niwa *et al.*[353,383] for toluene ammoxidation on vanadium oxide supported on alumina.

The ratio of the reaction rates of these two pathways depends on: (i) the rate of oxygen insertion and reaction with ammonia (i.e., formation first of a C–O bond and then substitution of O with NH or direct formation of C–N bonds), and (ii) stability of the intermediate species. The first aspect depends on the amount of adsorbed ammonia and the intrinsic oxidation characteristics of the catalyst. The stability of the benzylamine and benzaldehyde types of species depends on similar factors. In fact, benzylamine can be hydrolized to benzoate and ammonium ions when there is a small quantity of adsorbed ammonia species and in the presence of free Brønsted acid sites. The benzoate species is also in equilibrium with the benzylamide species (see Scheme 8.10), which also can further transform to benzonitrile. The rate of oxidation of benzaldehyde with activated ammonia to form the benzylimine species or benzaldehyde oxidation to the benzoate species depends on (i) the amount of ammonia adsorbed and (ii) the rate of oxygen versus nitrogen insertion on the intermediate. These competitive rates depend on both adsorbate geometry and catalyst characteristics. By changing the nature of the catalyst as well as the experimental conditions (temperature, concentrations of reactants in the feed, etc.), it is possible to alter the relative rates of these competitive pathways with a consequent possible change from one dominant pathway to another.

This discussion of the chemistry of transformation of toluene to benzonitrile shows that several equilibria and competitive pathways exist on the surface and that the dominant reaction mechanism can be significantly altered by the reaction conditions. Care must be taken in excluding any possible pathway of reaction before proper account has been taken of how the experimental evidence may be affected by the different reaction conditions and populations of surface adspecies with respect to conditions during the catalytic tests.

8.7.3. Propane Ammoxidation on V/Sb Oxides

The surface reaction network for propane ammoxidation as derived from combined IR and transient reactivity studies[28,73,185,186,384–386] is reported in Scheme 8.19. Some of the reaction steps in the scheme are numbered because they will be used to indicate specify reaction rates in the discussion of the catalytic behavior in terms of the overall surface network.

The primary route of propane conversion is through the formation of propene as an intermediate, but a side reaction of acrylate formation via a propionate intermediate is possible. The relative rate of this side reaction is higher in oxidation than in ammoxidation because chemisorption of ammonia inhibits the activity of the Lewis acid sites responsible for the alkane activation by this side route. Direct conversion of propane to carbon oxides is also indicated, but it should be noted that the supported vanadium oxide species primarily responsible for this route transforms during the first minutes of catalytic reaction, so that this pathway is less important under steady state conditions especially in the presence of ammonia.

Propene may be oxidized via two routes, the first reversible and favored at low temperatures with the formation of an isopropylate species by reaction with Brønsted sites and the second with the formation of an allyl alcoholate by abstraction of allylic hydrogen and nucleophilic lattice oxygen attack. The first adspecies easily transforms to acetone, which may undergo oxidative cleavage to an acetate species and a C1 fragment. The reaction of Brønsted acid sites with ammonia to give ammonium ions slows down the rate of the first nonselective side reaction. The second adspecies (allyl alcoholate) easily transforms to acrolein, which further converts to an acrylate species. The latter interacts strongly with the surface and is thermally stable up to relatively high temperatures, but reacts faster with ammonium ions to give an acrylamide intermediate. At higher temperatures the amide transforms to acrylonitrile, but when Brønsted sites are present, the amide is hydrolized to reform ammonia and free, weakly bonded acrylic acid. The latter easily decarboxylates forming carbon oxides. In the presence of strong Brønsted

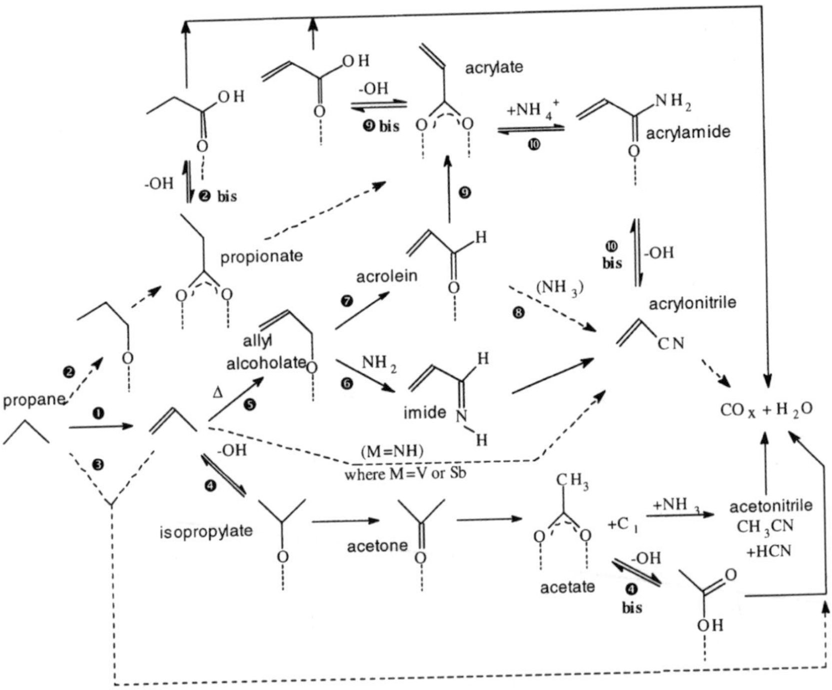

Scheme 8.19. Surface reaction network for propane ammoxidation on V/Sb oxides.[28]

acid sites acrylonitrile also can be hydrolized back to free acrylic acid and ammonium ions, and then transformed to carbon oxides. A different activated ammonia adspecies (NH_2-type species formed by heterolytic dissociation of ammonia coordinated to Lewis acid sites) reacts with the allyl alcoholate or acrolein intermediate to form the acrylimine intermediate, which transforms to acrylonitrile more rapidly than the acrylamide intermediate.

The reaction illustrated in Scheme 8.19 thus shows that there are different pathways on the surface of V/Sb oxides to form acrylonitrile and that the intrinsic selectivity of each is different owing to the various possible side reactions. The overall selectivity for acrylonitrile depends on the relative rates of these competitive pathways, which, in turn, depend on the reaction conditions and the nature of the catalyst.

There are several key features in the catalytic behavior of V/Sb oxides in propane ammoxidation that can be rationalized using the surface reaction network described above. A first general question comes out of a comparison of propane and propene ammoxidation on the same catalyst and at the same reaction temperature.[28] The rate constant of propene depletion is one order of magnitude higher than that of propane depletion and thus when propene forms from propane its rapid further transformation is expected, but flow reactor reactivity data show that relatively large amounts of propene are present at the maximum in the selectivity for acrylonitrile.[174] This characteristic feature has been observed either as a function of the reaction temperature or of the hydrocarbon conversion and illustrates the apparent contradiction that propene alone is converted faster than propane to acrylonitrile, whereas the propene to acrylonitrile step (in propane conversion) is slower than the propane to propene step.

There is another characteristic feature of the ammoxidation of C3 hydrocarbons on V/Sb oxides. While selectivity to acrylonitrile in propene conversion is maximal at high conversion, selectivity to acrylonitrile from propane is maximal at much lower conversions with a considerable decrease as conversion increases further, indicating that on the same catalyst and at the same reaction temperature acrylonitrile converts faster to carbon oxides when formed from propane than when formed from propene, even though propene is the reaction intermediate from propane.

Another feature of the reactivity of V/Sb oxides is the effect of the NH_3/C_3H_8 ratio on the product distribution in propane ammoxidation. When very small amounts of ammonia are added to a propane–oxygen feed[185] the formation of carbon dioxide drops rapidly from over 70% to 20–30%, with a parallel increase in the selectivity not for acrylonitrile but for propene, which increases from about 20 to 60%. Acrylonitrile starts to form only after a further increase in the ammonia concentration, notwithstanding that: (i) acrylonitrile formation requires ammonia, but propene formation does not, and (ii) propene forms in larger amounts than acrylonitrile under the conditions of the cited tests[185] and thus the effect cannot be due simply to the necessity to form enough propene before it can be converted to acrylonitrile.

The effect of the reaction temperature on the acetonitrile to acrylonitrile ratio in propane ammoxidation is also interesting.[28,386] Acetonitrile is a product of carbon-chain degradation and thus it is expected that carbon-chain rupture is favored by acid-catalyzed reactions at high temperatures. However, much to the contrary, reactivity data[386] indicate that the acetonitrile to acrylonitrile ratio decreases with increasing reaction temperature. It also should be noted that this ratio decreases with increasing Sb:V ratio in the catalyst and increasing ammonia to propane ratio.[386]

This last characteristic of the reactivity of V/Sb oxide in propane ammoxidation can be easily rationalized on the basis of the surface reaction network summarized in Scheme 8.19. IR data show that the rate of reaction 5 vs. reaction 4 (see Scheme 8.19 for the numbers of the relative reaction steps) increases with increasing reaction temperature, which promotes lattice oxygen attack in the allylic position, explaining why the acetonitrile (formed along pathway 4) to acrylonitrile (formed along pathway 5) ratio decreases with increasing reaction temperature. The action of Brφnsted acid sites in catalyzing reaction 4 is inhibited when they react with ammonia to form ammonium ions, which explains the change in the acetonitrile to acrylonitrile ratio with the increase in NH3/C3H8 ratio. Finally, allylic oxygen insertion (pathway 5) is favored at higher Sb:V ratios[174] and thus the acetonitrile to acrylonitrile ratio decreases with increasing Sb:V ratio.

The effect of ammonia in promoting selectivity, but only for small quantities of propene, can also be rationalized on the basis of the surface reaction network (Scheme 8.19). Ammonia has various effects on the surface reactivity of V/Sb oxides,[186] apart from being a reactant in the synthesis of acrylonitrile. Chemisorption on the surface Lewis acid sites due to coordinatively unsaturated vanadium sites inhibits the rate of activation of the alkane with a decrease in its depletion rate,[185] but especially hinders the nonselective pathway 2 → 2bis vs. pathway 1. Ammonia, interacting with Brφnsted acid sites, also inhibits their effect in acid-catalyzed reactions and, in particular, inhibits reactions 4 and 2bis, as well as 4bis and 9bis which lead to carbon-chain degradation (4) and free, weakly coordinated acids (2bis, 4bis, 9bis), which easily decarboxylate and further rapidly convert to carbon oxides. It is thus possible to rationalize the fact that ammonia at low concentrations acts mainly as a modifier of the surface reactivity (inhibits pathways 2, 4, and others associated with Brφnsted acidity), whereas at higher concentrations it also acts as a reactant for acrylonitrile synthesis.

Olefins interacting with Brφnsted acid sites have an effect comparable to that previously discussed for ammonia. It is useful to discuss this concept on the basis of the surface reaction network (Scheme 8.19) to determine (i) why there is an apparent different stability of acrylonitrile when it forms from propane or propene (different rate of consecutive oxidation under the same conditions and on the same catalyst), and (ii) why a high selectivity for acrylonitrile requires operation with a high propane concentration in the feed.[339,341]

The chemisorption on the Brφnsted acid sites of propene as a reactant combines with ammonia chemisorption to moderate the activity of the sites toward catalyzing side reactions. In particular, in propane ammoxidation at high conversion, the rates of steps 9bis (degradation of acrylate species) and 10bis → 10 → 9bis (degradation of acrylonitrile) are quite important. It should be noted that due both to the higher rate of transformation and to stronger chemisorption of the alkene with respect to alkane, the ratio of the rates of O vs. N addition and thus the ratio of rates of pathway 7 → 9 vs. 6 and 8 is higher for alkane than for alkene ammoxidation. The formation of acrylate is thus faster for propane transformation than using propene directly as a reactant, even though propene is the main intermediate in propane to acrylonitrile, because the surface population of adspecies is different and thus the surface reactivity during catalytic reaction is different (see Section 8.4). Although the acrylate species is rather stable, it can be transformed to weakly coordinated acrylic acid in the presence of Brφnsted acid sites and water. This weakly coordinated acid is rather unstable and quickly decarboxylates and further oxidizes to carbon oxides.

Owing to the presence of this side reaction and the more rapid rate of pathway 5 → 7 → 9 with respect to pathway 5 → 6 in propane ammoxidation with respect to propene ammoxidation, the selectivity for acrylonitrile from propane is lower than from propene (on the same catalyst and under the same reaction conditions). Furthermore, the propene concentration is lower when it forms from propane with respect to the case of directly feeding propene, and thus the "self-protection" effect on the nonselective catalysis by Brønsted acid sites is also reduced. At high propane conversion, where the concentrations of ammonia and propene intermediates are lower (thus their inhibiting effect on Brønsted acid activity is smaller) and the concentration of water higher (water is a main reaction product, needed to shift the equilibrium toward the acids), the rates of reaction steps 4 and 9*bis* and of the pathway 10*bis* → 10 → 9 (acrylonitrile degradation) become dominant and thus selectivity for acrylonitrile becomes very low. In propene conversion, on the other hand, the residual propene is still enough to "shield" acrylonitrile toward its degradative pathway (10*bis* → 10 → 9) and thus selectivity remains high up to nearly complete conversion.

Indirect proof that this analytical approach is correct can be found in the observation that in light of the above discussion in regard to propane ammoxidation increasing the concentration of propane in the feed increases selectivity[339,341] because it increases the formation of propene intermediate, which in turn increases its beneficial effect on the surface reactivity. In fact, the increase in selectivity for acrylonitrile with increasing propane concentration in the feed is a common observation on V/Sb oxide for propane ammoxidation[174] and one of the aspects of the catalytic behavior mentioned before as requiring explanation.

The lower concentration of surface adspecies in propane ammoxidation with respect to propene ammoxidation not only favors the side reactions on organic adspecies, but also favors the side oxidation of ammonia to N_2, which reduces the effective number of surface species available for N insertion on the intermediates and thus acrylonitrile formation. This is the primary reason that the rate of propene → acrylonitrile transformation is apparently slower with respect to the same reaction when propene is fed in directly.

Use of the surface reaction network and an analysis of the effect of chemisorption of reactants and intermediate species on the surface properties[28] make it possible to rationalize a series of apparent contradictions when propane and propene ammoxidation are compared on the same catalyst and under the same reaction conditions or, in more general terms, to understand the key features of the reactivity on the basis of the reaction mechanism.

The analysis of the surface reaction network in propane ammoxidation also points out that optimal catalytic behavior depends on a balance between catalyst properties and the effect of chemisorption of the reactants and intermediate species on the surface reactivity. A consequence of this observation is that the optimal catalytic behavior in a homogeneous series of samples depends on the feed composition, and different catalyst compositions for V/Sb oxides can be optimal depending on the reaction conditions. Moreover, in flow reactor catalytic tests[387] it was observed that the selectivity and productivity for acrylonitrile from propane for a series of V/Sb oxide catalysts with increasing Sb:V ratio depends effectively on the feed composition, because a change, for instance, in the rates of the $5 \rightarrow 6$ pathway vs. $5 \rightarrow 7 \rightarrow 9$ pathway (Scheme 8.19) should be reflected in different optimal reaction conditions. This shows how comparing the behavior of catalysts for only a single set of reaction conditions can be misleading.

The discussion on the relationship between surface reaction network and catalytic behavior also demonstrates that the addition to the feed of small doses of gas dopants can be a useful way to tune the surface reactivity in a manner analogous to that previously discussed regarding the effect of ammonia and alkenes on the surface selectivity. This is a very open substantially untouched field of research, but one that can be very fruitful in terms of improving catalytic behavior. Study of the surface reaction network is thus not only a good approach for understanding the surface catalytic chemistry in complex selective oxidation reactions but can also be a useful tool for analyzing the modifications in catalyst properties or reaction conditions that optimize catalytic behavior.

8.7.4. *o*-Xylene Ammoxidation on V/TiO$_2$ Catalysts

In the surface reaction networks of both propane and toluene ammoxidation it has already been shown that there are alternative pathways as a result of the competition between oxygen and nitrogen insertion on a common intermediate but that both eventually lead to the same type of final product. However, in the case of *o*-xylene ammoxidation on V/TiO$_2$ catalysts, the competition between the alternative routes leads to different reaction products, and it may be possible to demonstrate more clearly that the distribution of products can be altered by influencing the relative rates of these two competitive pathways.

o-Xylene is selectively converted to tolunitrile (TN) in the presence of oxygen and ammonia, but TN can be converted to phthalimide (PI) or phthalonitrile (PN).[388,389] While the latter product derives from the conversion of the second methyl group to a nitrile group, the formation of the former involves first the oxidation (O insertion) of the second methyl group and then consecutive rearrangement to form the PI. The relative selectivity for these two products, therefore, depends roughly on the relative rates of nitrogen versus oxygen insertion on the TN intermediate.

Vanadium oxide on titania catalysts may be thought of as being composed of a first vanadium species interacting directly with the titania surface, an overlayer of amorphous vanadium oxide weakly interacting with titania, and, for higher loadings, crystalline V$_2$O$_5$.[86] The last species, however, is inactive and does not play a role in the behavior of V/TiO$_2$ catalysts. While the first type of vanadium species (strongly interacting with the surface) has stronger Lewis and Brønsted acid properties than the amorphous vanadium oxide overlayer (both properties are important in the activation of ammonia), the latter species has higher redox properties and more labile oxygen species, thus showing a higher rate of oxygen insertion into the substrate. Therefore, changing the relative rates of formation of these two vanadium species using *ad hoc* preparation methods should make it possible to change the relative selectivity for PN (favored by the vanadium species strongly interacting with the titania surface) and for PI (favored by the presence of the amorphous V oxide overlayer). Figure 8.16 shows that there is a good linear relationship between relative selectivity to PI and PN species and the ratio of amorphous overlayer V oxide (V_{over}) and strongly interacting vanadium species (V_{SI}).[389]

This example illustrates the fact that starting from the analysis of the surface reaction pathway and the possible dependence of the competitive pathways on catalyst properties it is possible to tune and optimize catalytic behavior toward the desired product.

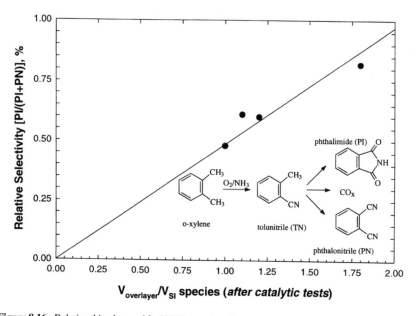

Figure 8.16. Relationship observed for V/TiO$_2$ catalysts between the relative selectivity in phthalimide (PI) formation and that in phthalonitrile (PN) formation during *o*-xylene ammoxidation as a function of the ratio between amorphous vanadium oxide overlayer species ($V_{over.}$) and strongly interacting vanadium species (V_{SI}).[389] Data for the quantity of vanadium species in the catalysts refer to the value observed after the catalytic tests.

8.7.5. Oxidation of *n*-Pentane on (VO)₂P₂O₇

On the same $(VO)_2P_2O_7$ catalyst that is active and selective for maleic anhydride formation from *n*-butane, two anhydrides are formed from *n*-pentane: maleic and phthalic anhydride.[87,148,390–392] The latter clearly involves a dimerization of two hydrocarbon molecules (or of products of partial transformation) at some stage of the reaction network. The problem, however, is to understand why the increase of one carbon atom in the alkane chain can change the reactivity to such a degree. In fact, in *n*-butane oxidation phthalic anhydride is detected in the reaction products, but only in traces (see Chapter 4).

The mechanism of *n*-pentane oxidation, unlike that of *n*-butane (Chapter 4) has not yet been elucidated in detail.[148] However, it is reasonable to assume that the basic reaction network is the same, but that in one of the stages the difference in the nature of the active intermediates results in a different pathway. In *n*-butane oxidation, the main initial intermediates are butenes and butadiene (all products of oxidative dehydrogenation) after which come the products of oxygen insertion (dihydrofuran, butyrrolactone, furan, maleic anhydride (see Scheme 8.17). In the case of *n*-pentane, a similar sequence of transformations should lead to the formation of pentadiene. However, at this stage there is a major difference between butadiene (from *n*-butane) and pentadiene (from *n*-pentane) (Scheme 8.20). In fact, the latter has allylic hydrogens not present in butadiene and thus further hydrogen

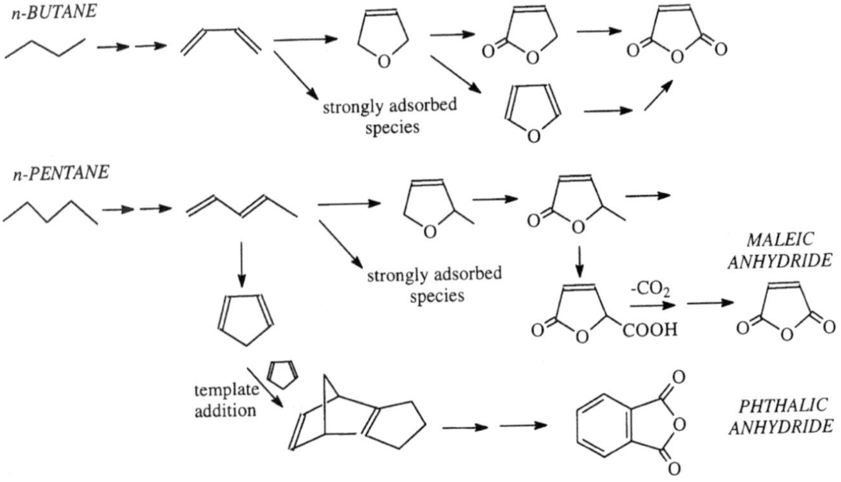

Scheme 8.20. Comparison of the surface reaction networks in *n*-butane and *n*-pentane oxidation on a vanadyl pyrophosphate catalyst.

abstraction is possible, which leads to the formation of cyclopentadiene as an intermediate. Cyclopentadiene is a rather stable species that easily interacts strongly with the surface Lewis acid centers on the surface of vanadyl pyrophosphate (these centers are due to coordinatively unsaturated vanadyl ions—see Chapter 4). These strongly coordinated cyclopentadiene species are relatively inert toward consecutive oxidation, but susceptible to Diels–Alder type of reactions with other adsorbed intermediates. For example, by reaction with another cyclopentadiene species a C10 intermediate forms, which then evolves to phthalic anhydride (Scheme 8.20). However, other Diels–Alder type of reactions are possible.

Although the details of the mechanism of further transformation are not clear and should be studied more in depth, this example shows that the pathways of surface transformation are also a function of the possibility of having a correct sequence of consecutive transformation steps. A small change in the nature of the substrate may lead to a change in the rate of one of these surface steps and thus promote alternative transformation pathways.

8.8. DYNAMICS OF CATALYTIC OXIDATION PROCESSES

Over 35 years ago, Tamaru[393,394] wrote that: "Catalysis is a dynamic process. Without studying the dynamic behavior of catalyst surfaces, no real nature of catalysts can be elucidated." While this statement is certainly true, unfortunately the knowledge and experimental techniques necessary for a precise description of the surface chemical processes on real catalysts in dynamic terms are still too sparse. On the other hand, surface science is progressively, albeit slowly, moving from the study of single-crystal metal surfaces to single-crystal metal oxide surfaces and extending the characterization to conditions to come closer to those typical of catalytic processes. Surface science researchers are thus gradually bridging the gap between theory and the catalysis that actually occurs in practice, but that gap will continue to exist for a while. In the meantime the evidence concerning surface processes derived from surface science studies must be combined with macroscopic models of surface reactivity to determine how and when surface catalytic reactivity depends on the microscopic events suggested by the surface science approach.

In this chapter we have given several illustrations that point to the need for a more advanced model of surface reactivity in selective oxidation reactions, which should include several of the recent results of surface science studies, but on a more generalized and less-defined basis. In this section some further interesting examples and concepts relevant to the development of macroscopic models of the surface reactivity in selective oxidation reactions will be discussed, although it does not cover all the interesting aspects emerging from recent evidence systematically.

8.8.1. Relevant Evidence from Surface Science Studies

When a single crystal of a metal is oxidized in a controlled way, it is known that progressively ordered surfaces form. For example, when the Cu (110) surface is oxidized at room temperature the $(-Cu-O-)_n$ chain grows along the <001> direction.[395] For higher coverages with oxygen, the chain coalesces with a $p(2 \times 1)$ arrangement to form islands.[395] The surface structure depends on the type of metal, as well as on the crystal plane. On the Ag (110) surface one-dimensional $(-Ag-O)_n$ chains grow along the <001> direction, but for higher coverages they form an ($n \times$ 1) arrangement (n varies from 7 to 2 depending on the coverage). It is also known that these surface structures influence both the reactivity and the mobility of oxygen adspecies, but it is difficult to correlate this information with the properties of oxide surfaces and their reactivity. Although it is difficult, this information can be correlated with the catalytic reactivity of the metal itself. Metallic silver is a unique catalyst for ethylene epoxidation, but its catalytic reactivity depends considerably on various factors such as the presence of dopants and reaction conditions.[126] Furthermore, the ordered surface structure detected at room temperature and the related reactivity phenomena (e.g., anisotropy in surface diffusion of adspecies) would be expected to be neither present nor important at the temperatures of catalytic tests, or at least there is a significant gap between surface science data and applied catalysis.

There are, on the other hand, other types of information from surface science studies that focus attention on surface phenomena that could be directly relevant for understanding catalytic phenomena in applied cases. We discuss here only one topic related to the modification of surface properties by chemisorption of gas phase molecules, because this argument has already been addressed from different perspectives in this chapter and thus can be a good example to show what applied catalysis and surface science related to catalysis have in common.

Interesting information results from the analysis of CO oxidation on the Ni (100) surface. Takagi et al.,[396] studying the dynamic equilibrium between preadsorbed $^{13}C^{18}O$ and gas phase $^{12}C^{16}O$ by time-resolved IR reflection absorption spectroscopy, have shown that $^{13}C^{18}O$ desorption is considerably enhanced by the presence of $^{12}C^{16}O$ in the gas phase. This effect is due to a repulsive interaction between the preadsorbed CO and incident CO molecules from the gas phase arriving at nearby sites. Consequently, the activation energy is reduced from 127 kJ/mol (in vacuum) to 25 kJ/mol for surface coverages of 0.1–0.3 monolayers. An even more marked effect may be expected for higher coverages, although that was not determined. The effect is not specific to the single crystal studied, as it was also observed for the Pt (111) surface.

In the classical Langmuir–Hinshelwood approach, the basis of almost all kinetic models for selective oxidation reactions, desorption is a function of the coverage of the adsorbed species, but the possibility of adsorption-assisted desorption is usually not considered. This is one of the differences that could arise when the kinetics on single crystals under high-vacuum conditions are extrapolated to atmospheric or higher pressures. However, more generally, the critical question to be answered is when effects of adsorption-assisted desorption may be relevant for the catalytic behavior. At high temperatures, such as those during catalytic tests, the surface mobility of adspecies, owing to their weaker bond with the surface (largely as a consequence of thermal vibrational modes of the catalyst surface), may prevent strong repulsive effects by adsorption of near-colliding molecules, but, on the other hand, the energy of colliding molecules is also higher and the energy released when they impact on the surface can give rise to local transient overheating, which enhances desorption. It is thus also expected that the sticking coefficient changes at high temperatures.

While this concept should be demonstrated by specific studies, it is supported by recent data on the modifications induced in surface characteristics by adsorption. Recent time-resolved EXAFS experiments[397] indicate that the colliding molecule induces a change in the local distortion in the neighboring atoms of the catalyst surface and thus in the local structure. In fact, time-resolved EXAFS experiments[397] show that a finite time is necessary for the relaxation of the structure to return to the starting situation after collision and adsorption of a gas phase molecule. This result clearly indicates that the adsorbing molecule induces a local modification in the surface structure.

It is also possible to speculate on what occurs when the number of colliding and adsorbing molecules is higher than in the case of these EXAFS experiments. When the number and energy of colliding molecules is sufficiently great (as probably occurs in several catalytic reactions of industrial interest), it is reasonable to assume the creation of surface wave modes owing to the creation of self-organized modes of surface relaxation to minimize the energy of the process. It is reasonable to expect that these waves may also create local distortion in the metal packing mode to allow, for instance, fast subsurface oxygen diffusion, which would otherwise not be possible based on the crystallographic metal packing. This is probably the mechanism that allows subsurface oxygen diffusion in Cu and Ag metal surfaces during ethylene epoxidation, a fundamental phenomenon that determines the catalytic properties of these samples.[126,398] This is also demonstrated by recent studies of Schlögl et al.,[298] who investigated the kinetic oscillations in methanol oxidation in the copper/oxygen system. They observed the following sequence of events. Initially (at zero time of oscillations) the surface consists of metallic copper and is inactive due to the low sticking coefficients of all gas phase species. Then, subsurface oxygen diffuses to the interlayer and leads to chemisorption and transformation of methanol, creating formate species which rapidly cover the whole surface. The subsurface reservoir rapidly decreases and thus also the efficiency of the reaction. At this stage, oxygen is able to coadsorb with methanol on the surface and the energy released from this reaction causes an increase in the temperature, which destabilizes the copper/metal adsorbate system. There is thus a continuous restructuring of the surface during the catalytic reaction and consequently of the surface reactivity. The formation of subsurface oxygen plays a special role in this mechanism.

The adsorbate-induced restructuring of the surface is a well-known phenomenon on metals. Somorjai[399] has reported several examples on metals, but not on oxide surfaces, probably due to the greater difficulty in characterizing well-defined oxide catalysts. Haller and Coulston[400] pointed out that one of the main differences can be that between short- and long-range restructuring. For example, the Pd(110)(1 × 1) surface restructures to a Pd(110)(1 × 2) surface when CO is chemisorbed, because this increases CO binding, although it returns to the Pd(110)(1 × 1) structure at full coverage. On oxide-type materials the examples are often short range. Van Santen et al.[401] showed that a change in the Si–O and Al–O bond length of the zeolite four-ring cluster occurs upon protonation or generally upon adsorption of the reactants, although the IR evidence of a modification of lattice zeolite vibrations indicates that this is not a local phenomenon but that the deformation probably extends over the entire structure. Haller and Coulston[400] noted that presumably a similar alteration of the bond length of Pd in the (110) surface occurs upon CO adsorption, but that when the coverage is high enough this propagates, resulting in restructuring of the Pd(110) (1 × 2) surface. Thus, the restructuring of a site is probably general; how local or long range it will be depends on the surface, and it is surely more likely to be long range for crystalline surfaces, e.g., zeolites or single-crystal metals than for amorphous or disordered surfaces (as often is the case with selective oxidation catalysts) because the ordered surface will provide a mechanism for propagating the restructuring away from the adsorption (reaction) site.

Iwasawa[38] described various examples showing that adsorption of the reactants may also have a direct role not only in changing the adsorption/desorption energetics or the (local) structure of the active sites but in changing the path of transformation. On a clean MgO surface formic acid decomposes to CO and H_2, but when a water molecule is adsorbed near the formate ion (also intermediate species for decomposition to CO/H_2) CO_2/H_2 forms instead. On Nb monomer on a SiO_2 surface a switchover of the reaction path from dehydration (γ-hydrogen abstraction) in ethanol oxidation to dehydrogenation (β-hydrogen abstraction) was observed when a second molecule of ethanol coordinated on the same site. Thus the first reactant molecule is converted to an intermediate, which is converted to the product with the assistance of the second reactant molecule due to an intermediate–reactant interaction. The first and second reactant molecules have different roles and display nonlinear behavior. In general terms, the study of these nonlinear phenomena in catalysis is one of the research frontiers.[402]

Iwasawa[35] reported on a different example of the role of adspecies in modifying the intrinsic surface reactivity. On Mo metal surface the major products of methanol conversion are H_2 and CO, but on the oxidized surface formaldehyde becomes the major product. However, this reaction occurs selectively only on the fully oxidized surface, whereas CH_4, H_2, C_{ads}, and O_{ads} are the main products at lower coverages with preadsorbed oxygen. The characterization of this effect indicates the peculiar role of oxygen adatoms (not incorporated into the surface structure—see Figure 8.17) in determining the reactivity and selective oxidation pathways. On the Mo(112)–$p(1 \times 2)$–O surface [molybdenum (112) crystal plane oxidized to form a $p(1 \times 2)$ arrangement of oxygen atoms—see Figure 8.17] methoxy species are stabilized by the oxygen forming the $p(1 \times 2)$–O phase, which avoids the presence of molybdenum atoms with high coordination and thus does not rapidly decompose, but gives rise to the formation of formaldehyde, although with only 50% selectivity. However, when coadsorbed oxygen adatoms are present the selectivity for formaldehyde is enhanced to over 90%, and oxygen adatoms decrease the activation energy for the transformation. Thus, the catalytic behavior is determined by two types of coadsorbed oxygens, modifier atoms and adatoms.

The difference between the two forms of oxygen on the molybdenum surface is related to the different stabilities of these species, the less stable (oxygen adatoms) also being the more reactive. A similar observation also may be made for reactive intermediates. In the case of methanol (see also Section 8.6), methoxy species are often indicated as reaction intermediates, but the analysis of the stability of these

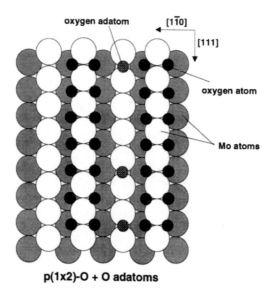

p(1x2)-O + O adatoms

Figure 8.17. Models for an oxygen-modified Mo(112) surface. Adapted from Iwasawa.[35]

species indicates that they are long-lived species[299,403] and cannot be considered directly as reaction intermediates, although they should be involved in the reaction mechanism. Vibrational spectroscopic data of methoxy[404] and analogous ethoxy[405] species further support this idea, but do not clarify the details of the structural differences between reactive and nonreactive adsorbed forms, but recent theoretical calculations,[406] although for a different type of catalytic material (zeolites), provides useful indications in this regard. The determination of the pathway of methanol transformation and the structure of intermediates by quantum-mechanical methods suggest that the end-on methoxy species is the first and more stable one, but before the transformation products develop a second higher-energy intermediate species is formed that represents the true rate-determining reactive intermediate. The calculated structure of this second intermediate is a side-on species (Figure 8.18)[406] and it is usually not detected by spectroscopic methods, which evidence only the first stable species, the end-on species, which however, is not the one that determines the catalytic reactivity.

These few examples, although not exhaustive of the wide range of direct suggestions on the catalytic chemistry of oxide-type materials for selective oxidation reactions, clearly illustrate the necessity of extending the concept of Langmuir-

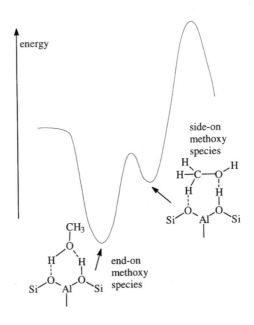

Figure 8.18. Schematic representation of the first two intermediate species in methanol transformation over zeolites and the corresponding energy diagram of the transformation. Adapted from van Santen.[406]

type chemisorption and reaction mechanisms at (ideal) surfaces to a non-Langmuir type of approach and nonideal surfaces. This concept is well presented in studies of reactions over metal surfaces,[399] but is usually not much considered in catalysis at oxide-type surfaces, as discussed extensively here. Thus a major possible contribution of surface science studies in the coming years is in this direction, because the development of basic knowledge on the surface structure and reactivity of well-characterized (from the surface point of view) oxide surfaces is a slow process and there is still a considerable gap to be bridged to arrive at a correct description of catalytic phenomena at real oxide surfaces.

8.8.2. Dynamics of Oxide Phase Transformation in the Active Form of the Catalysts

Although several authors have pointed out the role of the dynamic transformations of phase during catalytic oxidation reactions,[167,407–409] few experimental results have provided real proof that as redox changes at a single site and phase cooperation to enhance selectivity or activity, a dynamic behavior in phase interchanges is also a key factor in catalytic behavior.

An interesting example is found in the study of heteropoly acids (HPA) active in selective oxidation reactions. Heteropoly compounds are a wide class of oxoanions formed by condensation of more than two different mononuclear oxoanions.[410–413] The structure of a heteropolyanion is called a "primary structure." There are several different possible types of heteropolyoxoanions, but a major class of compounds relevant for selective catalytic oxidation applications is that formed by phosphomolybdo oxoanions (Keggin structure), where some of the molybdenum atoms are substituted by vanadium atoms. In the acid form, the general composition is $H_{3+n}PV_nMo_{12-n}O_{40}$. The H^+ can be substituted (totally or partially) by other cations such as Cs^+ (useful to improve the thermal stability of the sample) or transition metal cations with redox behavior (such as Cu^{2+}), which have a more direct influence on the catalyst reactivity.

The modality of connection of the oxoanions constitutes the secondary structure, where the cations compensating for the charge of the anions, water, and eventually other adsorbed molecules are localized (Figure 8.19). Finally, a higher-order "tertiary" can be also distinguished, depending on the modality of assembling the macro secondary structures. All three types of structures influence the reactivity, although the main effect is associated with the first two.

Schlögl et al.,[414–416] using in situ UV/VIS, ^{31}P-NMR, and IR spectroscopy to study the mechanism of $H_3PMo_{12}O_{40}$ and $H_4PVMo_{11}O_{40}$ modification during the catalytic reaction of methanol oxidation, observed the formation of defective HPA at the temperature of catalytic tests and reversible restructuring into the intact Keggin structure triggered by water, even that formed during the catalytic reaction

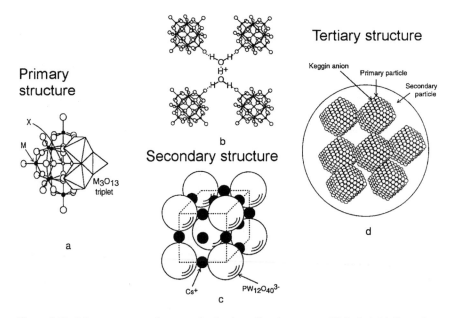

Figure 8.19. Primary structure of an oxoanion having a Keggin structure ($XM_{12}O_{40}$) (a). Secondary structure [(b) for $H_3PW_{12}O_{40}\cdot6H_2O$, and (c) for $Cs_3PW_{12}O_{40}$]. Tertiary structure [$Cs_{2.5}H_{0.5}PW_{12}O_{40}$, cubic structure as in (c)] (d). Adapted from Misono et al.[412]

itself. They thus suggest a continuous interchange during the catalytic reaction between formation of a defective (partially decomposed or lacunar) oxoanion and its reconstruction in the presence of oxygen and water. The catalytic activity of the HPA samples is associated with a continuous interchange between these two forms. They also observed that the partial reduction of the oxoanion leads to enhanced storage of hydrogen atoms produced during the dehydrogenation of methanol. These hydrogen atoms may then react with gaseous oxygen to form H_2O. A similar observation was made by Jalowiecki-Duhamel et al.[417] studying the oxidation of isobutane to methacrolein on $C_{1.6}H_{2.4}P_{1.7}Mo_{11}V_{1.1}O_{40}$ samples. When this catalyst becomes active in isobutane oxidation, anionic vacancies form, which store reactive hydrogen species, thus promoting dehydrogenation activity. However, when the lacunar structure is not rapidly converted to the intact Keggin oxoanion form, further decomposition may occur, leading to irreversible degradation of constituent oxides.

Albonetti et al.[418] found similar results in studying 12-molybdophosphoric acid catalysts for isobutyric acid oxidehydrogenation to methacrylic acid. Using several combined physicochemical techniques (IR, ESR, XRD), they also concluded that at high temperatures a partial reduction of molybdenum to Mo^{5+} occurs and a fraction of the molybdenum ions located in the Keggin unit move to the cationic position of neighboring undecomposed anions, stabilizing them. By comparing the catalytic activity of these compounds, they also observed that the formation of heteropoly compound with molybdenum ions in catatonic positions leads to greater activity in isobutyric acid oxidehydrogenation with respect to the original heteropoly compound.

Similarly, in studying the reactivity of vanadophosphomolybdic acids in C4 and C5 hydrocarbon oxidation,[150,419] the structural changes during catalytic reaction and the effect of having vanadium in cationic and anionic positions as well as their interconversion during the catalytic tests, it was concluded that: (i) the migration of vanadium from inside to outside of the Keggin ion is easy and occurs rapidly in the conditions of maximum activity, and (ii) the heteropolyacid works in a partially decomposed state, but close to irreversible deactivation. The model of the active phase suggested was the formation of defective oxoanions with vanadium bridging two oxoanions and stabilizing them. The role of water in reconstruction of the intact Keggin anion was also shown.

In conclusion, several authors studying different selective oxidation reactions have concluded that the activity of heteropoly acids is a maximum in a partially decomposed state for which the irreversible transformation to oxide constituents is prevented by the continuous *in situ* water-mediated reconstruction of the intact Keggin structure. There is thus a continuous interchange between lacunar and intact Keggin oxoanions, and this is one of the key features explaining the behavior of these catalysts in selective oxidation reactions. On the other hand, this also underscores the stability problems of these catalytic materials.

8.8.3. Dynamics of Surface Species and Their Effect on Catalyst Surface Properties

There is a continuous modification of surface population during the catalytic reactions of oxidation as well as often a continuous rapid interchange between different phases on a timescale that makes it difficult to detect. Schlögl et al.[43] observed, for instance, that on polycrystalline silver during methanol oxidation there is a continuous very rapid interchange between the different microstructures, that can only be detected when the phenomenon is frozen using specific reaction conditions. Sometimes, this phenomenon is responsible for the presence of macroscopic oscillations in the catalytic behavior, as detected, e.g., in propene oxidation by N_2O over dispersed platinum catalysts.[420] The oscillating behavior is caused by the reconstruction of platinum surface atoms, which is associated with surface fluctuations of weakly adsorbed propene. Schlögl et al.[298] studying the oscillations of methanol oxidation on copper also concluded that phase restructuring during the catalytic reaction is associated with fluctuations in oxygen adsorption. More generally, the analysis of oscillatory phenomena in oxidation reactions indicates a relationship between the nature and mobility of adsorbed species and continuous restructuring of the catalyst during the catalytic reaction.[421,422]

Although relevant for an understanding of catalyst surface chemistry, the analysis of these oscillatory phenomena does not add specific concepts useful for understanding the catalytic chemistry in the selective oxidation at oxide surfaces, other than pointing out the mobility of adspecies as a part of the mechanism leading to dynamic surface restructuring during the catalytic reaction. However, it should be noted that analogous concepts are found in the model of the dynamic modification of oxide catalysts by remote control[30–32,423–426] although the basis of the model differs from the phenomena in oscillating reactions.

The remote control theory (see Scheme 8.6) is based on the idea that one phase (called the "donor") is able to activate molecular oxygen into a highly active mobile species, which migrates onto the surface of the other phase (the "acceptor") with which it reacts. This reaction brings about the creation of new selective sites and/or the regeneration of sites deactivated during the redox cycles of the oxidation mechanism. Initially, the remote control model developed out of observations of synergetic effects when various phases are mixed gently, but more recently supporting kinetic data[32,423,425] as well as results of the migration of ^{18}O-labeled oxygen have been published.[427]

Recent evidence has also indicated the presence of *in operandi* reconstruction of active sites and in particular the change in form of MoO_3 crystallites.[424,426] When MoO_3 was catalytically reacted in the presence of a spillover oxygen donor phase, the (010) faces of the crystallites got reconstructed to (100) steps. This does not happen when MoO_3 is tested alone. The spillover oxygen thus maintains (or restores) the activity of MoO_3 by keeping its surface covered by more selective corner-sharing pairs of octahedra at the expense of nonselective sharing ones.[428,429] This indicates that there is a continuous reconstruction of active sites, triggered by mobile oxygen adspecies. Note should be taken of the analogy between the reconstruction phenomena related to oscillating phenomena as discussed above and the effect of *in operandi* reconstruction of MoO_3 crystallites.

8.9. CONCLUSIONS

Analysis of the reaction mechanism was central to the study of selective oxidation reactions, because maximization of selectivity generally requires not only an investigation of the structure–activity and selectivity relationships, but also an analysis of the details of the reaction mechanism to derive guidelines for tuning the surface reactivity. There have been several recent advances in connection with reaction mechanisms, especially regarding new areas of study such as the functionalization of alkanes, but the discussion in this chapter has focused on open questions and novel research directions to establish a basis for new models of reactivity in selective oxidation and outline possible new trends.

In particular, throughout this chapter, it has been emphasized that the reaction mechanism should not be discussed in terms of the nature of the active sites (the "common" approach), but rather that the preferred approach should be in terms of collective catalyst surface and bulk properties and organic chemistry of transformation at the oxide surface. The discussion also centered on the necessity of using new models of the surface reactivity that include more recently acquired information on aspects such as: (i) the role of chemisorbed species on the surface reactivity, (ii) the presence of multiple pathways of reaction, (iii) the dynamics of catalyst reconstruction, and (iv) the mobility of surface adspecies. Consideration of all these possible effects in analyzing the surface reactivity will make the design of new catalysts possible as well as improve understanding of surface reactivity at oxide surfaces.

Extensively in past years considerable effort was devoted to correlating catalytic reactivity with one or more parameters, and only recently, Masel and Lee[430] suggested that the intrinsic activation barrier (assumed to be the same in the gas phase and over solid catalysts) is the parameter that can be correlated with the surface reactivity. Hodnett et al.[431,432] correlated the selectivity with the ability of the catalyst to activate a C–H bond in a reactant rather than a similar C–H or C–C bond in a selective oxidation product. The discussion in this chapter points out the limitations in this type of relationship, demonstrating, for instance, how the intrinsic selectivity of a surface depends on the surface coverage by adspecies and how different selectivity patterns for a homologous series of catalysts can be obtained depending on the feed composition. These effects illustrate that the specific nature and dynamics of surface catalytic phenomena must be taken into account—and this will be the challenge for future research on surface selective oxidation transformations.

Recent advances using in situ techniques and in particular high-resolution electron microscopy[88] and EXAFS now also offer new possibilities for direct monitoring of catalyst surface reconstruction and dynamics during a reaction with catalysts such as mixed oxides, which, owing to the fact that they are difficult to study, have not been investigated in as much detail (in terms of surface science) as, e.g., metals. These new possibilities will enable further and more precise understanding of the surface catalytic chemistry of oxidation reactions at metal oxide surfaces and lead to the development of better catalysts using the analysis of the reaction network and mechanism as a tool to unravel the complex multistep chemical characteristics of selective oxidation reactions.

The possibility of selectively carrying out in a single step complex reactions that would otherwise require several steps, and at the same time meeting the two demands of "green chemistry" (to reduce environmental impact and the complexity of processes) is a particular advantage of selective oxidation reactions at solid surfaces. It was for this reason that in a survey of U.S. industrial concerns selective oxidation catalysis was the first choice among the catalytic technologies that are the most promising for a sustainable chemistry.

Better understanding of the reaction mechanism is the key to developing catalysts that are more selective or capable of new complex catalytic reactions. This chapter has illustrated that to reach this objective, selective oxidation catalysts must be seen not only in terms of their solid state chemistry and surface characteristics, but also in terms of how these properties depend on interaction with feed and reaction products, including electron and heat transfer during the catalytic reaction. Furthermore, properties must be related to a microkinetic model of both dominant and side reactions. The integration of all these aspects will be central to the development of the next generation selective oxidation catalysts.

REFERENCES

1. G. Busca, *Catal. Today* **27**, 457 (1996).
2. M.M. Bettahar, G. Constentin, L. Savary, and J.C. Lavalley, *Appl. Catal. A: General* **145**, 1 (1996).
3. P. Mars and D.W. van Krevelen, *Chem. Eng. Sci.* **3**, 41 (1954).
4. J. Haber, in: *3rd World Congress on Oxidation Catalysis*, (R.K. Grasselli, S.T. Oyama, A.M. Gaffney, and J.E. Lyons, eds.), Studies in Surface Science and Catalysis Vol. 110, Elsevier Science: Amsterdam (1997), p. 1.
5. J.C. Vedrine, G. Coudurier, and J.-M.M. Millet, *Catal. Today* **32**, 115 (1996).
6. J.C. Vedrine, G. Coudurier, and J.-M.M. Millet, *Catal. Today* **33**, 3 (1997).
7. J. Haber, in: *Proceedings 8th International Congress on Catalysis, Berlin 1984*, Vol. 1 - Plenary Lectures, Verlag Chemie-Dechema (1984), p. 85.
8. R.K. Grasselli, in: *Surface Properties and Catalysis by Non-Metals* (J.P. Bonelle, B. Delmon, and E. Derouane, eds.), Reidel: Dordrecht (1983), p. 273, 289.
9. R.K. Grasselli, *J. Chem. Educ.* **63**, 216 (1986).
10. G. Busca, E. Finocchio, G. Ramis, and G. Ricchiardi, *Catal. Today* **32**, 133 (1996).
11. J.D. Burrington and R.K. Grasselli, *J. Catal.* **59**, 79 (1979).
12. R.K. Grasselli, and J.D. Burrington, *Adv. Catal.* **30**, 133 (1981).
13. R.K. Grasselli, J.D. Burrington, and J.F. Brazdil, *Faraday Soc. Discuss.* **72**, 204 (1982).
14. J.D. Burrington, C.T. Kartisek, and R.K. Grasselli, *J. Catal.* **63**, 235 (1980).
15. J.D. Burrington, C.T. Kartisek, and R.K. Grasselli, *J. Catal.* **75**, 225 (1982).
16. J.D. Burrington, C.T. Kartisek, and R.K. Grasselli, *J. Catal.* **81**, 489 (1983).
17. J.D. Burrington, C.T. Kartisek, and R.K. Grasselli, *J. Catal.* **87**, 363 (1984).
18. L.C. Glaeser, J.F. Brazdil, M.A. Hazel, M. Menieic, and R.K. Grasselli, *J. Chem. Soc. Faraday Trans. 1* **81**, 2903 (1995).

19. J.F. Brazdil, D.D. Suresh, and R.K. Grasselli, *J. Catal.* **66**, 347 (1980).
20. J.N. Allison and W.A. Goddard, in: *Solid State Chemistry in Catalysis* (R.K. Grasselli and J.F. Brazdil, eds.), ACS Symposium Series 279, American Chemical Society: Washington D.C. (1985), p. 23.
21. T.P. Snyder and C.G. Hill, *Cata. Rev.-Sci. Eng.* **31**, 43 (1989).
22. D.B. Dadyburjor, S.S. Jewur, and E. Ruckenstein, *Cata. Rev.-Sci. Eng.* **19**, 293 (1979).
23. J. Belgacem, J. Kress, and J.A. Osborn, *J. Mol. Catal.* **86**, 267 (1994).
24. G.W. Keulks, L.D. Krenzke, and T.M. Notermann, *Adv. Catal.* **27**, 183 (1978).
25. A. Bielanski, and J. Haber, *Oxygen in Catalysis*, Marcel Dekker: New York (1991).
26. L.D. Krenzke and G.W. Keulks, *J. Catal.* **64**, 295 (1980).
27. J. Haber, in: *Surface Properties and Catalysis by Non-Metals* (J.P. Bonelle, B. Delmon, and E. Derouane, eds.), Reidel: Dordrecht (1983), p. 1.
28. G. Centi and S. Perathoner, *CHEMTECH* **2**, 13 (1998).
29. Y. Moro-oka and W. Ueda, *Adv. Catalysis, Vol. 40* (D.D. Eley, H. Pines, and W.O. Haag, eds.), Academic Press: New York (1994), p. 233.
30. B. Delmon and P. Ruiz, *Catal. Today* **1**, 1 (1987).
31. B. Delmon and P. Ruiz, *Catal. Today* **3**, 199 (1988).
32. B. Delmon and G.F. Froment, *Catal. Rev.-Sci Eng.* **38**, 69 (1996).
33. G.M. Pajonk, S.J. Teichner, and J.E. Germain, *Spill-over of Adsorbed Species*, Elsevier Science: Amsterdam (1983).
34. R.A. van Santen and M.N. Neurock, *Catal. Rev.-Sci. Eng.* **37**, 557 (1995).
35. Y. Iwasawa, *11th International Congress on Catalysis - 40th Anniversary* (J.W. Hightower, W.N. Delgass, E. Iglesias, and A.T. Bell, eds.), Studies in Surface Science and Catalysis Vol. 101, Elsevier Science: Amsterdam (1996), p. 21.
36. Y. Iwasawa, *Advances in Catal.* Vol. 35, Academic Press: Orlando (1987), p. 187.
37. Y. Iwasawa, *Catal. Today* **18**, 21 (1993).
38. Y. Iwasawa, in: *Elementary Reaction Steps in Heterogeneous Catalysis* (R.W. Joyner and R.A. van Santen, eds.), NATO ASI Series C Vol. 398, Kluwer Academic: Dordrecht (1993), p. 287.
39. H. Onishi, T. Aruga, and Y. Iwasawa, *J. Catal.* **146**, 557 (1994).
40. X. Bao, M. Muhler, B. Pettinger, Y. Uchida, G. Lehnpfuhl, R. Shlögl, and G. Ertl, *Catal. Lett.* **32**, 171 (1995).
41. C. Rehren, G. Isaaac, R. Shlögl, and G. Ertl, *Catal. Lett.* **11**, 253 (1991).
42. X. Bao, M. Muhler, B. Pettinger, R. Shlögl, and G. Ertl, *Catal. Lett.* **22**, 215 (1993).
43. H. Schubert, U. Tegtmeyer, D. Herein, X. Bao, M. Muhler, and R. Shlögl, *Catal. Lett.* **33**, 305 (1995).
44. J.L. Callahan and R.K. Grasselli, *AIChe J.* **9**, 755 (1963).
45. J.C. Volta and B. Morawek, *J. Chem. Soc. Chem. Comm.* **338** (1980).
46. J.C. Volta, W. Desquesnes, B. Moraweck and G. Coudurier, *React. Kinet. Catal. Lett.* **12**, 241 (1979).
47. J.M. Tatibouet and J.E. Germain, *C.R. Acad. Sci.* **290**, 321 (1980).

48. J.M. Tatibouet, J.E. Germain, and J.C. Volta, *J. Catal.* **82**, 240 (1983).

49. M. Abo, B. Mingot, J. Massadier, and J.C. Volta, in: *Structure–Activity and Selectivity Relationships in Heterogeneous Catalysis* (R.K. Grasselli and A.W. Sleight, eds.), Studies in Surface Science and Catalysis Vol. 67, Elsevier Science: Amsterdam (1991), p. 67.

50. J.C. Volta and J.L. Portefaix, *Appl. Catal.* **18**, 1 (1985).

51. A. Guerrero-Ruiz, A. Massardier, D. Duprez, M. Abon, and J.C. Volta, in: *Proceedings 9th International Congress on Catalysis* (M.J. Phillips and M. Ternan, eds.), Chemical Institute of Canada: Ottawa (1988), p. 1601.

52. M. Abon, M. Roullet, J. Massardier, and A. Guerreo-Ruiz, in: *New Developments in Selective Oxidation II* (V. Corberan and S. Vic Bellon, eds.), Studies in Surface Science and Catalysis Vol. 82, Elsevier Science: Amsterdam (1994), p. 67.

53. J. Klaas, K. Kulawik, G. Schulz-Ekloff, and M.I. Jaeger, in: *Zeolites and Related Materials: State of the Art 1994* (J. Weitkamp, H.G. Karge, H. Pfeifer, and W. Hölderich, eds.), Studies in Surface Science and Catalysis Vol. 84C, Elsevier Science: Amsterdam (1994), p. 2261.

54. G. Centi, S. Perathoner, F. Trifirò, A. Aboukais, C.F. Aïssi, and M. Guelton, *J. Phys. Chem.* **96**, 2617 (1992).

55. A. Guerrero-Ruiz, I. Rodriguez-Ramos, P. Ferreira-Aparicio, M. Abon, and J.C. Volta, *Catal. Today* **32**, 223 (1996).

56. A. Guerrero-Ruiz, J.M. Blanco, M. Aguilar, I. Rodriguez-Ramos, and J.L.G. Fierro, *J. Catal.* **137**, 429 (1992).

57. J. Ziolkowski, E. Bordes, and P. Courtine, *J. Catal.* **122**, 126 (1990).

58. J. Ziolkowski, E. Bordes, and P. Courtine, *J. Catal.* **84**, 307 (1993).

59. J. Ziolkowski, *J. Catal.* **100**, 45 (1986).

60. E. Bordes, in: *Structure–Activity and Selectivity Relationships in Heterogeneous Catalysis* (R.K. Grasselli and A.W. Sleight, eds.), Studies in Surface Science and Catalysis Vol. 67, Elsevier Science: Amsterdam (1991), p. 21.

61. J.C. Volta, M. Forissier, F. Theobald, and T.P. Pham, *Faraday Discuss. Chem. Soc.* **72**, 225 (1981).

62. J. Ziolkowski, *J. Catal.* **80**, 263 (1993).

63. J. Ziolkowski, *J. Catal.* **84**, 317 (1983).

64. J. Ziolkowski and J. Jamas, *J. Catal.* **81**, 298 (1983).

65. J.C. Vedrine, in: *Catalytic Oxidation—Principles and Applications* (R.A. Sheldon and R.A. van Santen, eds.), World Scientific: Singapore (1995), p. 53.

66. A. Kapoor, R.T. Yang, and C. Wong, *Catal. Rev.-Sci. Eng.* **31**, 129 (1984).

67. P.A. Sermon and G.C. Bond, *Catal. Rev.-Sci. Eng.* **8**, 211 (1973).

68. F. Solymosi, A. Erdohelyi, and M. Kocsis, *J. Catal.* **65**, 428 (1980).

69. F. Solymosi, I. Tombacz, and M. Kocsis, *J. Catal.* **75**, 78 (1982).

70. G. Centi, S. Militerno, S. Perathoner, A. Riva, and G. Brambilla, *J. Chem. Soc. Chem. Comm.*, 88 (1991).

71. G. Centi and S. Perathoner, *Ind. Eng. Chem. Res.* **36**, 2945 (1997).
72. G. Centi, N. Passarini, S. Perathoner, and A. Riva, *Ind. Eng. Chem. Res.* **31**, 1947 (1992).
73. G. Centi, F. Marchi, and S. Perathoner, *J. Chem. Soc. Faraday Trans.* **92**, 5141 (1996).
74. G. Gleaves, J.R. Ebner, and F.C. Kuechner, *Catal. Rev.-Sci. Eng.* **30**, 49 (1988).
75. J. Ebner and J. Gleaves, in: *Oxygen Complexes and Oxygen Activation by Transition Metals* (A. Martell and D. Sawyer, eds.), Plenum Press: New York (1988), p. 273.
76. Y. Schuurman, J.T. Gleaves, J. Ebner, and M. Mummey, in: *New Developments in Selective Oxidation II* (V. Corberan and S. Vic Bellon, eds.), Studies in Surface Science and Catalysis Vol. 82, Elsevier Science: Amsterdam (1994), p. 203.
77. Y. Schuurman and J.T. Gleaves, *Ind. Eng. Chem. Res.* **33**, 2935 (1994).
78. G. Centi and J.T. Gleaves, *Catal. Today* **16**, 69 (1993).
79. S. Sundaresan and K.R. Kaza, *Chem. Eng. Comm.* **35**, 1 (1983).
80. R. Gomez, in: *Aspects of the Kinetics and Dynamics of Surface Reactions* (U. Landman, ed.), *AIP Conference Proceedings No. 61*, Springer Verlag: New York (1980).
81. R. Gomez, R. Wartman, and R. Lundy, *J. Chem. Phys.* **26**, 1147 (1957)
82. R. Gomez, R. Wartman, and R. Lundy, *J. Chem. Phys.* **26**, 1363 (1957)
83. R. Gomez, R. Wartman, and R. Lundy, *J. Chem. Phys.* **26**, 2376 (1957).
84. J. Haber and E. Lalik, *Catal. Lett.* **33**, 119 (1997).
85. J.M. Hermann, *Catal. Today* **20**, 135 (1994).
86. G. Centi, *Appl. Catal. A: General* **147**, 267 (1996).
87. G. Busca and G. Centi, *J. Am. Chem. Soc.* **111**, 46 (1989).
88. P.L. Gai-Boyes, *Catal. Rev.-Sci. Eng.* **34**, 1 (1992).
89. G. Centi and F. Trifirò, *Appl. Catal.* **12**, 1 (1984).
90. G. Centi and F. Trifirò, *Catal. Rev.-Sci. Eng.* **28**, 165 (1986).
91. N.I. Lipatkina, L.K. Przheral'skaya, V.A. Shvets, and V.B. Kazansky, *Dokl., Akad., Nauk SSSR* **242**, 1114 (1978).
92. A. Frimer, *Chem. Rev.* **79**, 359 (1979).
93. A.A. Goman and M.A.J. Rodgers, *Chem Soc. Rev.* **10**, 205 (1981).
94. V. Slawson and A.W. Adamson, *Lipids* **11**, 472 (1976).
95. V. Slawson, A.W. Adamson, and R.A. Stein, *Lipids* **13**, 128 (1978).
96. O.V. Krylov, *Kinet. Katal.* **14**, 24 (1973).
97. A.U. Khan, *J. Am. Chem. Soc.* **103**, 6516 (1981).
98. M. Che and A.J. Tench, *Adv. Catal.* **32**, 1 (1983).
99. M. Che and A.J. Tench, *Adv. Catal.* **31**, 77 (1982).
100. A. Bielanski and J. Haber, *Catal. Rev.* **19**, 1 (1979).
101. Yu.A. Likhov, Z. Musil, and A.A. Davydov, *Kinet. Katal.* **20**, 256 (1979).
102. J.S. Valentine, *Chem. Rev.* **73**, 235 (1973); G.K. Borevskov, *Kinet. Katal.* **14**, 2 (1973).
103. V.J. Choy and C.J. O'Connor, *Coord. Chem. Rev.* **9**, 145 (1972).
104. R.J. Kollrack, *J. Catal.* **12**, 321 (1968).

105. J.H. Lunsford, *Catal. Rev.-Sci. Eng.* **8**, 135 (1973).

106. N.B. Wong, Y. Ben Taarit, and J.H. Lunsford, *J. Chem. Phys.* **60**, 2148 (1974).

107. S. Abdo, R. Howe, and W.K. Hall, *J. Phys. Chem.* **82**, 969 (1978).

108. R. Martens, H. Gentsch, and F. Freund, *J. Catal.* **44**, 366 (1976).

109. M. Boudart, A.J. Delbouille, E.G. Derouane, V. Indovina, and A.B. Walters, *J. Am. Chem. Soc.* **94**, 6622 (1972).

110. A.J. Tench and T. Lawson, *Chem. Phys. Lett.* **7**, 459 (1970).

111. A.J. Tench, T. Lawson, J.F. Kibblewhite, *J. Chem. Soc. Faraday Trans. 1*, **68**, 1169 (1972).

112. M. Iwamoto, J.H. Lunsford, *J. Phys. Chem.* **84**, 3079 (1980).

113. L. Aika and J.H. Lunsford, *J. Phys. Chem.* **81**, 1393 (1977).

114. N.I. Lipatkina, V.A. Shvets, and V.B. Kazansky, *Kinet. Katal.* **19**, 979 (1978).

115. F.S. Yang and J.H. Lunsford, *J. Catal.* **63**, 505 (1980).

116. V.B. Kazansky, *Kinet. Katal.* **19**, 279 (1978).

117. Y. Takita and J.H. Lunsford, *J. Phys. Chem.* **83**, 683 (1979).

118. V.B. Sapozhnikov, V.A. Shvets, N.D. Chuvylkin, and V.B. Kazansky, *Kinet. Katal.* **17**, 1251 (1976).

119. J.F. Hemidy and A.J. Tench, *J. Catal.* **68**, 17 (1981).

120. V.A. Shvets, V.B. Sapozhnikov, N.D. Chuvylkin, and V.B. Kazansky, *J. Catal.* **52**, 459 (1978).

121. B.L. Kugler and R.J. Kokes, *J. Catal.* **32**, 170 (1974).

122. B.L. Krugler and J.W. Gryder, *J. Catal.* **44**, 126 (1976).

123. M. Akimoto and E. Echigoya, *J. Catal.* **29**, 191 (1973).

124. M. Akimoto and E. Echigoya, *J. Chem. Soc. Faraday Trans. 1* **73**, 193 (1977).

125. R. Fricke, H.G. Jerschkewitz, G. Lischke, and G. Ohlmann, *Z. Anorg. Allg. Chem.* **448**, 23 (1979).

126. R.A. van Santen and H.P.C.E. Kuipers, *Adv. Catal.* **35**, 265 (1987).

127. G. Centi, F. Trifirò, J.R. Ebner, and V.M. Franchetti, *Chem. Rev.* **88**, 55 (1988).

128. G. Busca, G. Centi, and F. Trifirò, *J. Am. Chem. Soc.* **107**, 7757 (1985).

129. Z. Zhang, R. Sneeded, and J. Volta, *Catal. Today* **16**, 39 (1993).

130. Y. Zhang, M. Forissier, R. Sneeden, and J. Vedrine, *J. Catal.* **145**, 267 (1994).

131. J.C. Volta, *Catal. Today* **32**, 29 (1996).

132. M.T. Sananes, S. Sajip, C. Kiely, G.J. Hutching, A. Tuel, and J.C. Volta, *J. Catal.* **166**, 388 (1997).

133. S. Albonetti, F. Cavani, F. Trifirò, P. Venturoli, G. Calestani, M. Lopez Granados, and J.L.G. Fierro, *J. Catal.* **160**, 52 (1996).

134. V.V. Guliants, J.B. Benziger, S. Sundaresan, I.E. Wachs, J.-M. Jehng, and J.E. Roberts, *Catal. Today* **28**, 275 (1996).

135. R. Contractor, J. Ebner, and M.J. Mummey, in: *New Developments in Selective Oxidation* (G. Centi and F. Trifirò, eds.), Studies in Surface Science and Catalysis Vol. 55, Elsevier Science: Amsterdam (1990), p. 553.

136. G. Emig, K. Uihlein, and C.J. Hächer, in: *New Developments in Selective Oxidation II* (V. Corberan and S. Vic Bellon, eds.), Studies in Surface Science and Catalysis Vol. 82, Elsevier Science: Amsterdam (1994), p. 243.

137. U. Rodemerck, B. Kubias, H.W. Zanthoff, and M. Baerns, *Appl. Catal. A: General* **153**, 203 (1997).

138. U. Rodemerck, B. Kubias, H.W. Zanthoff, G.U. Wolf, and M. Baerns, *Appl. Catal. A: General* **153**, 217 (1997).

139. M. Abon, K.E. Bére, and P. Delichère, *Catal. Today* **33**, 15 (1997).

140. M.A. Pepera, J.L. Callahan, M.J. Desmond, E.C. Milberger, P.R. Blum, and N.J. Bremer, *J. Am. Chem. Soc.* **107**, 4883 (1985).

141. Y. Schuurman and J.T. Gleaves, *Catal. Today* **33**, 25 (1997).

142. D. Dowell, J.T. Gleaves, and Y. Schuurman, in: *3rd World Congress on Oxidation Catalysis* (R.K. Grasselli, S.T. Oyama, A.M. Gaffney, and J.E. Lyons, eds.), Studies in Surface Science and Catalysis Vol. 110, Elsevier Science: Amsterdam (1997), p. 199.

143. J.T. Gleaves, Lecture given at 3rd World Congress on Oxidation Catalysis, San Diego, September 1997.

144. B. Schiott and K.A. Jørgensen, *Catal. Today* **16**, 79 (1993).

145. B. Schiott, K.A. Jørgensen, and R. Hoffmann, *J. Phys. Chem.* **95**, 2298 (1991).

146. G. Centi and F. Trifirò, *Chem. Eng. Science* **45**, 2589 (1990).

147. G. Centi and F. Trifirò, in: *Catalysis Science and Technology, Vol. 1*, (S. Yoshida, N. Takezawa, and Y. Ono, eds.), VCH Verlag: Basel (1991), p. 225.

148. G. Centi, J.T. Gleaves, G. Golinelli, and F. Trifirò, in: *New Developments in Selective Oxidation by Heterogeneous Catalysis* (P. Ruiz and B. Delmon, eds.), Studies in Surface Science and Catalysis Vol. 72, Elsevier Science: Amsterdam (1992), p. 231.

149. M. Burattini, G. Centi, and F. Trifirò, *Appl. Catal.* **32**, 353 (1987).

150. G. Centi, J. Lopez Nieto, C. Iapalucci, K. Bruckman, and E.M. Serwicka, *Appl. Catal.* **46**, 197 (1989).

151. F. Blatter and H. Frei, *J. Am. Chem. Soc.* **115**, 7501 (1993).

152. F. Blatter and H. Frei, *J. Am. Chem. Soc.* **116**, 1812 (1994).

153. H. Sun, F. Blatter, and H. Frei, *J. Am. Chem. Soc.* **116**, 7951 (1994).

154. F. Blatter, F. Moreau, and H. Frei, *J. Phys. Chem.* **98**, 13403 (1994).

155. H. Sun, F. Blatter, and H. Frei, *J. Am. Chem. Soc.* **118**, 6873 (1996).

156. H. Frei, F. Blatter, and H. Sun, *CHEMTECH* **26**, 24 (1996).

157. F. Blatter, H. Sun, and H. Frei, *Catal. Lett.* **35**, 1 (1995).

158. H. van Bekkum, E.M. Flanigen, and J.C. Jansen (eds.), *Introduction to Zeolite Science and Practice*, Studies in Surface Science and Catalysis Vol. 58, Elsevier Science: Amsterdam 1991.

159. E. Preuss, G. Linden, and M. Peuckert, *J. Phys. Chem.* **89**, 2955 (1985).

160. R.A. Sheldon and J.K. Kochi, *Metal-Catalyzed Oxidation of Organic Compounds*, Academic Press: New York (1981).

161. R.A. Sheldon, *CHEMTECH* **9**, 566 (1991).

162. R.A. Sheldon, in: *New Developments in Selective Oxidation* (G. Centi and F. Trifirò, eds.), Studies in Surface Science and Catalysis Vol. 55, Elsevier Science: Amsterdam (1990), p. 1.

163. R.A. Sheldon and J. Dakka, *Catal. Today* **19**, 215 (1994).

164. T.A. Garibyan and L. Ya. Margolis, *Catal. Rev.-Sci Eng.* **31**, 355 (1990).

165. D.J. Driscoll, K.D. Cambell, and J.H. Lunsford, *Adv. Catal.* **35**, 139 (1987).

166. V.D. Sokolovskii, *Catal. Rev.-Sci. Eng.* **32**, 1 (1990).

167. Y.I. Pyatnitsky and N.I. Ilchenko, *Catal. Today* **32**, 21 (1996).

168. P.G. Hart and H.R. Friedly, *J. Chem. Soc. Chem. Comm.* **11**, 621 (1976).

169. J. Novakova and Z. Dolejek, *J. Catal.* **37**, 540 (1975).

170. Z.R. Ismailov and V.P. Anufrienko, *React. Kinet. Catal. Lett.* **3**, 301 (1975).

171. Z. Dolejek and J. Novakova, *Coll. Czech. Chem. Comm.* **44**, 673 (1979).

172. W. Martir and J.H. Lunsford, *J. Am. Chem. Soc.* **103**, 3728 (1981).

173. R. Nilsson, Ph.D. Thesis, University of Lund, Lund, Sweden, (1997).

174. G. Centi, S. Perathoner, and F. Trifirò, *Appl. Catal. A: General* **157**, 143 (1997).

175. G. Centi, *Catal. Lett.* **22**, 53 (1993).

176. G. Centi, in: *Elementary Reaction Steps in Heterogeneous Catalysis*, (R.W. Joyner and R.A. van Santen, eds.), NATO ASI Series C Vol. 398, Kluwer Academic: Dordrecht (1993), p. 93.

177. F. Cavani and F. Trifirò, *CHEMTECH* **4**, 18 (1994).

178. F. Cavani, S. Albonetti, and F. Trifirò, *Catal. Rev.-Sci. Eng.* **38** (1996) 413.

179. G. Centi, G. Fornasari, and F. Trifirò, *J. Catal.* **89**, 44 (1984).

180. F. Cavani, G. Centi, I. Manenti, A. Riva, and F. Trifirò, *Ind. Eng. Chem. Prod. Res. Dev.* **22**, 565 (1983).

181. F.Cavani, G.Centi, and F.Trifirò, *Ind. Eng. Chem. Prod. Res. Dev.* **22**, 570 (1983).

182. G.Busca, G.Centi, and F.Trifirò, in: *Catalyst Deactivation 1987*, (B. Delmon and G.F. Froment, eds.), Elsevier Science: Amsterdam (1987), p. 427.

183. G. Centi, J. Lopez Nieto, D. Pinelli, F. Trifirò, and F. Ungarelli, in: *New Developments in Selective Oxidation* (G. Centi and F. Trifirò, eds.), Elsevier Science: Amsterdam (1990), p. 635.

184. G. Centi, J. Gleaves, G. Golinelli, S. Perathoner, and F. Trifirò, in: *Catalyst Deactivation 1991* (J.B. Butt and C.H. Bartholomew, eds.), Elsevier Science: Amsterdam (1991), p. 449.

185. G. Centi, F. Marchi, and S. Perathoner, *Appl. Catal. A: General* **149**, 225 (1997).

186. G. Centi , F. Marchi, and S. Perathoner, *J. Chem. Soc. Faraday* **92**, 5151 (1996).

187. R.S. Drago and K. Jurczyk, *Appl. Catal. A: General* **112**, 117 (1994).

188. G.E. Vrieland and P.G. Menon, *Appl. Catal.* **77**, 1 (1991).

189. D. Pinelli, F. Trifirò, A. Vaccari, E. Giamello, and G. Pedulli, *Catal. Lett.* **164**, 21 (1996).

190. F.J. Maldonado-Hodar, L.M. Palma Madeira, and M. Farinha Portela, *J. Catal.* **164**, 399 (1996).

191. D.J. Berrisford, C. Bolm, and K.B. Sharpless, *Angew. Chem. Int. Ed.* **34**, 1059 (1995).
192. T. Yokota, S. Fujibayashi, Y. Nisiyama, S. Sakaguchi, and Y. Ishii, *J. Mol. Catal. A: Chem.* **114**, 113 (1996).
193. G. Centi, S. Perathoner, and G. Stella, in: *Catalyst Deactivation 1994* (B. Delmon and G.F. Fromant, eds.), Elsevier Science: Amsterdam (1994), p. 393.
194. A.W. Stobbe-Kreemers, G. van der Lans, M. Makkee, and J.J.F. Scholten, *J. Catal.* **154**, 187 (1995).
195. N. Tsubokawa, T. Kimoto, and T. Endo, *J. Mol. Catal. A: Chem.* **101**, 45 (1995).
196. K. Yogo, M. Umeno, H. Watanabe, and E. Kikuchi, *Chem. Lett.* **229** (1993).
197. R. Burch and M.J. Hayes, *J. Mol. Catal. A: Chem.* **100**, 13 (1995).
198. V.D. Sokolovskii, A.A. Davydov, and O. Yu. Ovsitser, *Catal. Rev.-Sci. Eng.* **37**, 425 (1995).
199. Z.G. Osipova and V.D. Sokolovskii, *React. Kinet. Catal. Lett.* **9**, 193 (1978).
200. A.A. Davydov, A.A. Budneva, and V.D. Sokolovskii, *Kinet. Katal.* **22**, 213 (1981).
201. J.S. Yoo, *Appl. Catal. A: General* **142**, 19 (1996).
202. K. Tanabe, M. Misono, Y. Ono, and H. Hattori, *New Solid Acids and Bases: Their Catalytic Properties*, Elsevier Science: Amsterdam (1989).
203. H.H. Kung, *Transition Metal Oxides: Surface Chemistry and Catalysis*, Studies in Surface Chem. and Catalysis Vol. 45, Elsevier Science: Amsterdam (1990).
204. W.-P. Han, and M. Ai, *J. Catal.* **78**, 281 (1982).
205. M. Ai, *J. Catal.* **54**, 426 (1978).
206. K. Tanaka and A. Ozaki, *J. Catal.* **8**, 1 (1967).
207. J.A. Lercher, C. Gründling, and G. Eder-Mirth, *Catal. Today* **27**, 353 (1996).
208. H. Knözinger, in: *Elementary Reaction Steps in Heterogeneous Catalysis* (R.W. Joyner and R.A. van Santen, eds.), NATO ASI Series C Vol. 398, Kluwer Academic: Dordrecht (1993), p. 267.
209. G. Busca, V. Lorenzelli, G. Ramis, and R.J. Willey, *Langmuir* **9**, 1492 (1993).
210. G. Busca, V. Buscaglia, M. Leoni, and P. Nanni, *Chem. Mat.* **6**, 955 (1994).
211. M. Daturi, G. Busca, and R.J. Willey, *Chem. Mat.* **7**, 2115 (1995).
212. G. Ramis, G. Busca, V. Lorenzelli, A. la Ginestra, P. Galli, and M.A. Massucci, *J. Chem. Soc. Dalton*, 881 (1988).
213. G. Busca, G. Centi, F. Trifirò, and V. Lorenzelli, *J. Phys. Chem.* **90**, 1337 (1986).
214. E. Astorino, J. Peri, R.J. Willey, and G. Busca, *J. Catal.* **157**, 482 (195).
215. C. Odenbrand, J. Brandin, and G. Busca, *J. Catal.* **135**, 505 (1992).
216. G. Ramis, G. Busca, and V. Lorenzelli, *J. Chem. Soc. Faraday Trans.* **90**, 1293 (1994).
217. G. Busca, *Langmuir* **2**, 577 (1986).
218. G. Busca and A. Zecchina, *Catal. Today* **3**, 61 (1994).
219. M.A. Chaar, D. Patel, and M.C. Kung, *J. Catal.* **109**, 463 (1988).
220. A. Galli, J.M. Lopez-Nieto, A. Dejoz, and M.I. Vazquez, *Catal. Lett.* **35**, 51 (1995).
221. P. Conception, J.M. Lopez-Nieto, and N. Paredes, *Catal. Lett.* **28**, 9 (1994).
222. C.Y. Kao, H.T. Huang, and B.Z. Wang, *Ind. Eng. Chem. Res.* **33**, 2066 (1994).

223. J.G. Eon, T. Olier, and J.C. Volta, *J. Catal.* **145**, 318 (1994).

224. P.M. Michalakos, M.C. Kung, J. Jahan, and N.H. Kung, *J. Catal.* **140**, 226 (1993).

225. B. Sulikowski, J. Krysciak, R.X. Valenzuela, V. Cortes-Corberan, in: *New Developments in Selective Oxidation II* (V. Corberan and S. Vic Bellon, eds.), Studies in Surface Science and Catalysis Vol. 82, Elsevier Science: Amsterdam (1994), p. 133.

226. D. Patel and H.N. Kung, in: *Proceedings 9th International Congress on Catalysis* (M.J. Phillips and M. Ternan, eds.), Chemical Institute of Canada: Ottawa (1988), p. 1553.

227. K.T. Nguyen and H.H. Kung, *J. Catal.* **122**, 415 (1990).

228. R. Burch and S. Swarnakar, *Appl. Catal.* **70**, 129 (1991).

229. Y. Chang, G.A. Somorjai, and H. Heinemann, *Appl. Catal. A: General* **96**, 305 (1993).

230. E. Morales and J.H. Lunsford, *J. Catal.* **118**, 255 (1989).

231. R. Burch and E.M. Crabb, *Appl. Catal. A: General* **100**, 111 (1993).

232. G. Busca, G. Centi, and F. Trifirò, *Appl. Catal.* **25**, 265 (1986).

233. S. Matsuura and N. Kimura, in: *New Developments in Selective Oxidation II* (V. Corberan and S. Vic Bellon, eds.), Studies in Surface Science and Catalysis Vol. 82, Elsevier Science: Amsterdam (1994), p. 271.

234. R. Wang, M. Xie, P. Li, and C. Nguyen, *Catal. Lett.* **24**, 67 (1994).

235. E. Finocchio, G. Busca, V. Lorenzelli, and R.J. Willey, *J. Chem. Soc. Faraday Trans. 1* **90**, 3347 (1994).

236. E. Finocchio, G. Busca, V. Lorenzelli, and R.J. Willey, *J. Catal.* **51**, 204 (1995).

237. Y.S. Yoo, W. Ueda, and Y. Moro-oka, *Catal. Lett.* **35**, 57 (1995).

238. G. Busca, V. Lorenzelli, G. Olivieri, and G. Ramis, in: *New Developments in Selective Oxidation II* (V. Corberan and S. Vic Bellon, eds.), Studies in Surface Science and Catalysis. Vol. 82, Elsevier Science: Amsterdam (1994), p. 253.

239. Y.C. Kim, W. Ueda, and Y. Moro-oka, *Appl. Catal.* **70**, 175 (1991).

240. Y.C. Kim, W. Ueda, and Y. Moro-oka, *Catal. Today* **13**, 673 (1992).

241. M. Ai, *J. Catal.* **101**, 389 (1986).

242. M. Ai, *Catal. Today* **13**, 679 (1992).

243. N. Mizuno, M. Tateishi, and M. Iwamoto, *Appl. Catal. A: General* **128**, L165 (1995).

244. W.C. Conner, S. Soled, and A. Signorelli, in: *New Horizons in Catalysis* (T. Seiyama and K. Tanabe, eds.), Studies in Surface Science and Catalysis Vol. 7, Elsevier Science: Amsterdam (1981), p. 1224.

245. K. Komatsu, Y. Uragami, and K. Otsuka, *Chem. Lett.* 1903 (1988).

246. G. Ramis, G. Busca, and V. Lorenzelli, *Appl. Catal.* **32**, 305 (1987).

247. E. Sanchez Escribano, G. Busca, and V. Lorenzelli, *J. Phys. Chem.* **94**, 8939 (1990).

248. E. Sanchez Escribano, G. Busca, and V. Lorenzelli, *J. Phys. Chem.* **94**, 8945 (1990).

249. E. Finocchio, G. Busca, and V. Lorenzelli, *J. Chem. Soc. Faraday Trans. 1*, **92**, 1587 (1996).

250. D. Siew Hew Sam, V. Soenen, and J.C. Volta, *J. Catal.* **123**, 417 (1990).

251. R. Hubaut, J.P. Bonnelle, and M. Daage, *J. Mol. Catal.* **55**, 170 (1989).

252. E. Serwicka, J.B. Black, and J. Goudenough, *J. Catal.* **106**, 23 (1987).
253. Y. Konishi, K. Sakata, M. Misono, and Y. Yoneda, *J. Catal.* **77**, 169 (1982).
254. P. Jaeger and J.E. Germain, *Bull. Soc. Chim. Fr.* **1**, 407 (1982).
255. M. Ai, *J. Catal.* **101**, 473 (1986).
256. Y. Takita, H. Yamashita, and K. Moritaka, *Chem. Lett.* 1733 (1989).
257. J. Buiten, *J. Catal.* **10**, 188 (1968).
258. Y. Moro-oka, S. Tan, and A. Osaki, *J. Catal.* **12**, 241 (1968).
259. T. Okuhara, N. Mizuno, and M. Misono, *Adv. Catal.* **41**, 113 (1996).
260. D. Patel, P.J. Anderson, and H.H. Kung, *J. Catal.* **125**, 132 (1990).
261. O.S. Owen, M.C. Kung, and H.H. Kung, *Catal. Lett.* **12**, 45 (1992).
262. A. Corma, J.M. Lopez,-Nieto, and N. Paredos, *Appl. Catal. A: General* **144**, 425 (1993).
263. D. Bhattacharya, S.K. Bej, and M.S. Rao, *Appl. Catal. A: General* **87**, 29 (1992).
264. M.A. Chaar, D. Patel, M.C. Kung, and H.H. Kung, *J. Catal.* **105**, 483 (1987).
265. K.T. Nguyen and H.H. Kung, *Ind. Eng. Chem. Res.* **30**, 352 (1991).
266. A. Matyshak and O.V. Krylov, *Catal. Today* **25**, 455 (1995).
267. A.A. Davydov, *Infrared Spectroscopy of Adsorbed Species on the Surface of Transition Metal Oxides,* John Wiley & Sons: New York (1990).
268. J.B. Peri, in: *Catalysis: Science and Technology, Vol. 9,* J.R. Anderson and M. Boudart, eds.), Springer-Verlag: Berlin (1984), p. 1.
269. C.M. Quinn (ed.), Identification of Intermediate Species in Catalytic Reactions, *Catal. Today* **12** (special issue) (1992).
270. R. Burch (ed.), *In situ* Methods in Catalysis, *Catal. Today* **9** (special issue) (1991).
271. A.T. Bell, in: *Vibrational Spectroscopy of Molecules on Surfaces* (J.T. Yates and T.E. Madey, eds.), Plenum Press: New York (1987), p. 105.
272. K. Weissermel and H.J. Arpe, *Industrial Organic Chemistry, 2nd Ed.*, Verlag Chemie: Weinheim (1993).
273. M. Carbucicchio and F. Trifirò, *J. Catal.* **45**, 77 (1976).
274. M. Carbucicchio and F. Trifirò, *J. Catal.* **62**, 13 (1980).
275. J. Lamotte, V. Moravek, M. Bensitel, and J.C. Lavalley, *React. Kinet. Katal. Lett.* **36**, 113 (1988).
276. G. Busca, J. Lamotte, J.C. Lavalley, and V. Lorenzelli, *J. Am. Chem. Soc.* **109**, 5197 (1987).
277. J.C. Lavalley, J. Lamotte, G. Busca, and V. Lorenzelli, *J. Chem. Soc. Chem. Comm.*, 1006 (1985).
278. F. Roozenboom, P.D. Cordingley, and P.J. Gellings, *J. Catal.* **68**, 464 (1981).
279. F. Abbatista, S. Del Mastro, G. Gozzelino, D. Mazza, M. Vallino, G. Busca, V. Lorenzelli, and G. Ramis, *J. Catal.* **117**, 42 (1989).
280. R.S. Schiffino and R.P. Merrill, *J. Phys. Chem.* **97**, 6425 (1993).
281. G. Busca, P.F. Rossi, V. Lorenzelli, M. Benaissa, and J.C. Lavalley, *J. Phys. Chem.* **89**, 5433 (1985).

282. P.F. Rossi, G. Busca, and V. Lorenzelli, *Z. Phys. Chem.* **149**, 99 (1986).

283. M. Ai, *J. Catal.* **77**, 279 (1982).

284. C.M. Sorensen and R.S. Weber, *J. Catal.* **142**, 1 (1993).

285. G. Deo, I.E. Wachs, *J. Catal.* **146**, 323 (1994).

286. J.S. Chung, R. Miranda, and C.O. Bennet, *J. Chem. Soc. Faraday Trans. 1*, **81**, 19 (1985).

287. R.P. Groff, *J. Catal.* **86**, 215 (1984).

288. P. Forzatti, E. Tronconi, G. Busca, and P. Tittarelli, *Catal. Today* **1**, 209 (1987).

289. G. Busca, A.S. Elmi, and P. Forzatti, *J. Phys. Chem.* **91**, 5263 (1987).

290. E. Tronconi, A.S. Elmi, N. Ferlazzo, P. Forzatti, G. Busca, and P. Tittarelli, *Ind. Eng. Chem. Res.* **26**, 1269 (1987).

291. M. Ai, *J. Catal.* **77**, 279 (1982).

292. F.S. Feil, J.G. van Ommen, and J.R.H. Ross, *Langmuir* **3**, 668 (1987).

293. R.G. Greenler, *J. Chem. Phys.* **37**, 2094 (1962).

294. D.L. Roberts and G.L. Griffin, *J. Catal.* **101**, 201 (1986).

295. A. Riva, F. Trifirò, A. Vaccari, L. Minthev, and G. Busca, *J. Chem. Soc. Faraday Trans. 1* **84**, 1422 (1988).

296. C. Chauvin, J. Saussey, J.C. Lavalley, H. Idriss, A. Kiennemann, P. Chaumette, and P. Courty, *J. Catal.* **121**, 56 (1990).

297. M. He and J.G. Heckerdt, *J. Catal.* **90**, 17 (1988).

298. H. Werner, D. Herein, G. Schulz, U. Wild, and R. Schlögl, *Catal. Lett.* **49**, 109 (1997).

299. M. Bowker, S. Poulston, R.A. Bennett, and A.H. Jones, *Catal. Lett.* **43**, 267 (1997).

300. W.E. Slinkadt and P.B. Degroot, *J. Catal.* **68**, 423 (1981).

301. V.A. Zazhigalov, J. Haber, J. Stoch, G.A. Komashko, A.I. Pyatniskaya, and I.V. Bacherikova, in: *New Developments in Selective Oxidation II* (V. Corberan and S. Vic Bellon, eds.), Studies in Surface Science and Catalysis Vol. 82, Elsevier Science: Amsterdam (1994), p. 265.

302. G. Centi and F. Trifirò, *J. Mol. Catal.* **35**, 255 (1986).

303. G. Centi, I. Manenti, A. Riva, and F. Trifirò, *Appl. Catal.* **9**, 177 (1984).

304. E. Finocchio, G. Ramis, G. Busca, V. Lorenzelli, and R.J. Willey, *Catal. Today* **28**, 381 (1996).

305. G. Busca, M. Daturi, E. Finocchio, V. Lorenzelli, G. Ramis, and R.J. Willey, *Catal. Today* **33**, 239 (1997).

306. G. Busca, V. Lorenzelli, G. Oliveri, and G. Ramis, in: *New Developments in Selective Oxidation II* (V. Corberan and S. Vic Bellon, eds.), Studies in Surface Science and Catalysis Vol. 82, Elsevier Science: Amsterdam (1994), p. 253.

307. S.V. Gerei, E.V. Rozhkova, and Ya.B. Gorokhovatsky, *J. Catal.* **28**, 341 (1973).

308. S.J. Puttock and C.H. Rochester, *J. Chem. Soc. Faraday Trans. 1* **82**, 3033 (1986).

309. A. Ramstetter and M. Baerns, *J. Catal.* **109**, 303 (1988).

310. R.W. Wenig and G.L. Schrader, *J. Phys. Chem.* **91**, 1911 (1987).

311. E. Sanchez Escribano, G. Busca, and V. Lorenzelli, *J. Phys. Chem.* **95**, 5541 (1991).

312. G. Busca, *J. Chem. Soc. Faraday Trans. 1* **89**, 753 (1993).
313. G. Busca, F. Cavani, and F. Trifirò, *J. Catal.* **106**, 471 (1987).
314. G. Busca, in: *Selective Catalytic Oxidation* (T. Oyama and J. Hightower, eds.), ACS Symposium Series, American Chemical Society: Washington D.C. (1993), p. 168.
315. M.D. Lee, W.S. Chen, and H.P. Chiang, *Appl. Catal. A: General* **101**, 269 (1993).
316. A.A. Davydov, *Mat. Chem. Phys.* **19**, 97 (1988).
317. A.B. Azimov, V.P. Vislovskii, E. Mamedov, and R.G. Rizayev, *J. Catal.* **127**, 354 (1991).
318. M. Sanati and A. Andersson, *J. Mol. Catal.* **81**, 51 (1993).
319. H. Miyata, T. Mukai, T. Ono, T. Ohno, and F. Hatayama, *J. Chem. Soc. Faraday Trans. 1* **84**, 2465 (1988).
320. F. Hatayama, T. Ohno, T. Yoshida, T. Ono, and H. Miyata, *React. Kinet. Catal. Lett.* **44**, 451 (1991).
321. A.J. van Hengstum, J. Pranger, S.M. van Hengstum-Nijhuis, J.G. van Ommen, and P. Gellings, *J. Catal.* **101**, 323 (1986).
322. M.C. Nobbenhuis, A. Baiker, P. Baernickel, and A. Wokaum, *Appl. Catal. A: General* **85**, 157 (1992).
323. E.A. Mamedov, V.P. Vislovskii, R.M. Talyshinskii, and R.G. Rizayev, in: *New Developments in Selective Oxidation by Heterogeneous Catalysis* (P. Ruiz and B. Delmon, eds.), Studies in Surface Science and Catalysis Vol. 72, Elsevier Science: Amsterdam (1992), p. 279.
324. S. Meijers, E. Finocchio, G. Busca, and V. Ponec, *J. Chem. Soc. Faraday Trans. 1* **91**, 1861 (1995).
325. G. Busca, T. Zerlia, V. Lorenzelli, and A. Girelli, *React. Kinet. Catal. Lett.* **27**, 429 (1985).
326. C.R. Dias, M. Farinha Portela, and G.C. Bond, *J. Catal.* **162**, 284 (1996).
327. S.L.T. Andersson, *J. Catal.* **98**, 138 (1996).
328. C.R. Dias, M. Farinha Portela, and G.C. Bond, *Catal. Rev.-Sci. Eng.* **39**, 169 (1997).
329. V. Nikolov, D. Klissurski, and A. Anastasov, *Catal. Rev.-Sci. Eng.* **33**, 319 (1991).
330. M.S. Wainwright and N.R. Foster, *Catal. Rev.-Sci. Eng.* **19**, 211 (1979).
331. I.E. Wachs, R.Y. Saleh, S.S. Chan, and C. Chersich, *CHEMTECH* **12**, 756 (1985).
332. G.C. Bond and K. Brukman, *Disc. Faraday Soc.* **72**, 235 (1981).
333. J.N. Papageorgiou, M.C. Abello, and G.F. Froment, *Appl. Catal. A: General* **120**, 17 (1994).
334. R.W. Petts and K.C. Waugh, *J. Chem. Soc. Faraday Trans.* **78**, 803 (1982).
335. B. Dmuchovsky, M. Freerks, E. Pierron, R. Munch, and F. Zienty, *J. Catal.* **4**, 291 (1965).
336. E. Broclawik, J. Haber, and M. Witko, *J. Mol. Catal.* **26**, 249 (1984).
337. M. Witko, E. Broclawik, and J. Haber, *J. Molec. Catal.* **35**, 179 (1986).
338. J. Haber and M. Witko, *Catal. Today* **23**, 311 (1995).
339. G. Centi, F. Trifirò, and R.K. Grasselli, *Chim. Ind. (Milan)* **72**, 617 (1990).
340. R.K. Grasselli, G. Centi, and F. Trifirò, *Appl. Catal.* **57**, 149 (1990).

341. G. Centi, R.K. Grasselli, and F. Trifirò, *Catal. Today* **13**, 661 (1992).

342. R.G. Rizayev, E.A. Mamedov, V.P. Vislovskii, and V.E. Sheinin, *Appl. Catal. A* **83**, 103 (1992).

343. Y. Moro-oka and W. Ueda, *Catalysis, Vol. 11*, Royal Society of Chemistry, London (1994), p. 223.

344. F. Cavani and F. Trifirò, *Chim. Ind. (Milan)* **70**, 48 (1980).

345. A. Andersson, S.L.T. Andersson, G. Centi, R.K. Grasselli, M. Sanati, and F. Trifirò, *Appl. Catal. A*, **113**, 43 (1994).

346. P. Cavalli, F. Cavani, I. Manenti, F. Trifirò, and M. El-Sawi, *Ind. Eng. Chem. Res.* **26**, 804 (1987).

347. M. Sanati and A. Andersson, *J. Mol. Catal.* **59**, 233 (1990).

348. G. Busca, F. Cavani, and F. Trifirò, *J. Catal.* **106**, 25 (1987).

349. M. Sanati, L.R. Wallenberg, A. Andersson, S. Jansen, and Y. Tu, *J. Catal.* **132**, 128 (1991).

350. A. Martin, H. Berndt, B. Lücke, and M. Meisel, *Top. Catal.* **3**, 377 (1996).

351. J. Fu, I. Ferino, R. Monaci, E. Rombi, V. Solinas, and L. Forni, *Appl. Catal. A* **154**, 241 (1997).

352. L. Forni, *Appl. Catal.* **20**, 219 (1986).

353. M. Niwa, H. Ando, and Y. Murakami, *J. Catal.* **49**, 92 (1977).

354. M.C. Kung and H.H. Kung, *Catal. Rev.-Sci. Eng.* **27**, 425 (1985).

355. A.A. Tsyganenko, D.V. Pordnyakov, and V.M. Filimov, *J. Mol. Struct.* **29**, 299 (1975).

356. R.K. Grasselli and J.F. Brazdil, in: *Proceedings 8th International Congress on Catalysis*, Vol. 5, Dechema: Frankfort AM (1984), p. 369.

357. J.C. Otamiri and A. Andersson, *Catal. Today* **3**, 211 (1988).

358. J.C. Otamiri, S.L.T. Andersson, and A. Andersson, *Appl. Catal.* **65**, 159 (1990).

359. M. Sanati and A. Andersson, *Ind. Eng. Chem. Res.* **30**, 312 (1991).

360. R. Nilsson, T. Lindblad, and A. Andersson, *J. Catal.* **148**, 501 (1994).

361. R. Nilsson, T. Lindblad, A. Andersson, C. Song, and S. Hansen, in: *New Developments in Selective Oxidation II* (V. Cortes Corberan and S. Vic, eds.) Studies in Surface Science and Catalysis Vol. 82, Elsevier Science: Amsterdam (1994), p. 293.

362. R.S. Wise and E.J. Markel, *J. Catal.* **145**, 335 (1994).

363. R.S. Wise and E.J. Markel, *J. Catal.* **145**, 344 (1994).

364. J.N. Allison and W.A. Goddard, in: *Solid State Chemistry in Catalysis* (R.K. Grasselli and J.F. Brazdil, eds.), ACS Symposium Series, American Chemical Society: Washington D.C. (1985), p. 279.

365. J. Otamiri, A. Andersson, and S.A. Jansen, *Langmuir* **61**, 365 (1990).

366. H.W. Zanthoff, S.A. Buchholz, and O.Y. Ovsitzer, *Catal. Today* **32**, 291 (1996).

367. H. Bosh and F. Janssen, *Catal. Today* **2**, 369 (1988).

368. G. Centi and S. Perathoner, *Appl. Catal. A* **132**, 179 (1995).

369. S.A. Buchholz and H.W. Zanthoff, in: *Heterogeneous Hydrocarbon Oxidation* (B. Warren and T. Oyama, eds.), ACS Symposium Series 638, American Chemical Society: Washington, D.C. (1996), p. 259.
370. G. Centi and S. Perathoner, *Catal. Rev.-Sci. Eng.* **40**, 175 (1998).
371. G. Centi, T. Tosarelli, and F. Trifirò, *J. Catal.* **142**, 70 (1993).
372. G. Centi and S. Perathoner, *J. Catal.* **142**, 84 (1993).
373. B. Lücke and A. Martin, in: *Catalysis of Organic Reactions* (M.G. Scaros and M.L. Prunier, eds.), Marcel Dekker: New York (1995), p. 479.
374. A. Martin, B. Lücke, H. Seeboth, and G. Ladwig, *Appl. Catal.* **49**, 205 (1989).
375. Y. Zhang, B. Lücke, A. Martin, A. Brückner, and M. Meisel, in: *Proceedings 11th International Congress on Catalysis—40th Anniversary* (J.W. Higtower, ed.), Rice University Printing Center: Baltimore (1996), Po-229.
376. A. Martin, B. Lücke, G.-U. Wolf, and M. Meisel, *Catal. Lett.* **33**, 349 (995).
377. A. Martin and B. Lücke, *Catal. Today* **32**, 279 (1996).
378. H. Berndt, K. Büker, A. Martin, S. Rabe, Y. Zhang, and M. Meisel, *Catal. Today* **32**, 285 (1996).
379. P. Cavalli, F. Cavani, I. Manenti, and F. Trifirò, *Catal. Today* **1**, 246 (1987).
380. P. Cavalli, F. Cavani, I. Manenti, and F. Trifirò, *Ind. Eng. Chem. Res.* **26**, 639 (1987).
381. G. Busca, *ACS Symp. Series* **523**, 168 (1992).
382. L. Forni, M. Toscano, and P. Pollesel, *J. Catal.* **130**, 392 (1991).
383. Y. Murakami, M. Niwa, T. Hattori, S. Osawa, I. Igushi, and H. Ando, *J. Catal.* **49**, 83 (1977).
384. G. Centi and S. Perathoner, *J. Chem. Soc. Faraday Trans.* **93**, 1147 (1997).
385. G. Centi and S. Perathoner, *Appl. Catal. A*, **124**, 317 (1995).
386. G. Centi and F. Marchi, in: *11th International Congress on Catalysis—40th Anniversary* (J.W. Hightower, W.N. Delgass, E. Iglesia, and A.T. Bell, eds.), Studies in Surface Science and Catalysis Vol. 101, Elsevier Science: Amsterdam (1996), p. 277.
387. G. Centi and S. Perathoner, *J. Chem. Soc. Faraday Trans.* **93** (1997) 3391.
388. G. Centi, D. Pinelli, and F. Trifirò, *Gazz. Chim. It.* **119**, 133 (1989).
389. F. Cavani, G. Centi, J. Lopez Nieto, D. Pinelli, and F. Trifirò, *Fine Chemistry and Catalysis*, Studies in Surface Science and Catalysis Vol. 41, Elsevier Science: Amsterdam (1988), p. 345.
390. M. Burattini, G. Centi, and F. Trifirò, *Appl. Catal.* **32**, 353 (1987).
391. G. Centi, J. Nieto Lopez, D. Pinelli, and F. Trifirò, *Ind. Eng. Chem. Res.* **28**, 400 (1989).
392. G. Centi, G. Golinelli, and G. Busca, *J. Phys. Chem* **94**, 6813 (1990).
393. K. Tamanu, *Bull. Chem. Soc. Jpn.* **31**, 666 (1958).
394. K. Tamaru, *Adv. Catal.* **15**, 65 (1965).
395. M. Taniguchi, K. Tanaka, T. Hashizume, and T. Sakurai, *Chem. Phys. Lett.* **192**, 117 (1992).
396. N. Takagi, J. Yoshinobu, and M. Kawai, *Phys. Rev. Lett.* **73**, 292 (1994).

397. R. Schlögl, private communication.
398. D.J. Sajkowski and M. Boudart, *Catal. Rev.-Sci. Eng.* **29**, 325 (1987).
399. G.A. Somorjai, in: *Elementary Reaction Steps in Heterogeneous Catalysis* (R.W. Joyner and R.A. van Santen, eds.), NATO ASI Series C Vol. 398, Kluwer Academic: Dordrecht (1993), p. 3.
400. G.L. Haller and G.W Coulston, in: *Elementary Reaction Steps in Heterogeneous Catalysis* (R.W. Joyner and R.A. van Santen, eds.), NATO ASI Series C Vol. 398, Kluwer Academic: Dordrecht (1993), p. 473.
401. R.A. van Santen, G.J. Kramer, and W.P.J. Jacobs, in: *Elementary Reaction Steps in Heterogeneous Catalysis*, (R.W. Joyner and R.A. van Santen, eds.), NATO ASI Series C Vol. 398, Kluwer Academic: Dordrecht (1993), p. 113.
402. T. Inui, *Res. Chem. Intermed.* **24**, 373 (1998).
403. D. Herein, W. Werner, Th. Schedel-Niedrig, Th. Neisius, A. Nagy, S. Bernd, and R. Schlögl, in: *3rd World Congress on Oxidation Catalysis* (R.K. Grasselli, S.T. Oyama, A.M. Gaffney, and J.E. Lyons, eds.), Studies in Surface Science and Catalysis Vol. 110, Elsevier Science: Amsterdam (1997), p. 103.
404. R. Miranda, C.O. Bennet, and J.S. Chung, *J. Chem. Soc. Faraday Trans. 1* **81**, 19 (1985).
405. W. Zhang and S.T. Oyama, *J. Am. Chem. Soc.* **118**, 7173 (1996).
406. S.R. Blaszkowski and R.A. van Santen, *J. Phys. Chem. B* **101**, 2292 (1997).
407. J.-M. Jehng, H. Hu, X. Gao, and I.E. Wachs, *Catal. Today* **28**, 335 (1996).
408. J. Volta, *Catal. Today* **32**, 1 (1996).
409. M. Caldararu, A. Ovenston, D. Sprinceana, J.R. Walls, and N.I. Ionescu, *Appl. Catal. A* **141**, 31 (1996).
410. G.A. Tsigdinos, *Top. Curr. Chem.* **76**, 1 (1978).
411. M.T. Pope, *Heteropoly and Isopoly Oxometallates*, Springer-Verlag: Berlin (1983).
412. M. Misono, *Catal. Rev.-Sci. Eng.* **30**, 339 (1988).
413. T. Okuhara, N. Mizuno, and M. Misono, *Adv. Catal.* **41**, 113 (1996).
414. B. Herzog, M. Wohlers, and R. Schlögl, *Microchim. Acta*, **14**, 703 (1997).
415. T. Ilkenhans, H. Siegert, and R. Schlögl, *Catal. Today* **32**, 337 (1996).
416. Th. Ilkenhans, B. Herzog, Th. Braun, and R. Schlögl, *J. Catal.* **153**, 275 (1995).
417. L. Jalowiecki-Duhamel, A. Monnier, Y. Barbaux, and C. Hecquet, *Catal. Today* **32**, 237 (1996).
418. S. Albonetti, F. Cavani, F. Trifirò, M. Gazzano, M. Koutyrev, F.C. Aissi, A. Aboukais, and M. Guelton, *J. Catal.* **146**, 491 (1994).
419. G. Centi, V. Lena, F. Trifirò, D. Ghoussoub, C.F. Aissi, M. Guelton, and J.P. Bonnelle, *J. Chem. Soc. Faraday Trans.* **86**, 2775 (1990).
420. M. Kobayashi, T. Kanno, H. Takeda, and S. Fujisaki, *11th International Congress on Catalysis—40th Anniversary* (J.W. Hightower, W.N. Delgass, E. Iglesias, and A.T. Bell, eds.), Studies in Surface Science and Catalysis Vol. 101, Elsevier Science: Amsterdam (1996), p. 1059.
421. F. Schüth, B. Henry, and L.D. Schmidt, *Adv. Catal.* **39**, 51 (1993).

422. G. Ertl, *Adv. Catal.* **37**, 213 (1991).
423. S. Hoornaerts, D. Vande Putte, F.C. Thyrion, P. Ruiz, and B. Delmon, *Catal. Today* **33**, 139 (1997).
424. E.M. Gaigneaux, P. Ruiz, and B. Delmon, *Catal. Today* **32**, 37 (1996).
425. D. Vande Putte, S. Hoornaerts, F.C. Thyrion, P. Ruiz, and B. Delmon, *Catal. Today* **32**, 255 (1996).
426. E. M. Gaigneaux, J. Naud, P. Ruiz, and B. Delmon, in: *3rd World Congress on Oxidation Catalysis* (R.K. Grasselli, S.T. Oyama, A.M. Gaffney, and J.E. Lyons, eds.), Studies in Surface Science and Catalysis Vol. 110, Elsevier Science: Amsterdam (1997), p. 185.
427. G. Mestl, P. Ruiz, B. Delmon, and H. Knözinger, *J. Phys. Chem.* **98**, 11269, 11276, 11283 (1994).
428. B. Delmon, *Surf. Rev. Lett.* **2**, 25 (1995).
429. B. Delmon, *Heter. Catal. Lett.* **1**, 219 (1994).
430. R.I. Masel and W.T. Lee, *J. Catal.* **165**, 80 (1997).
431. C. Batiot and B.K. Hodnett, *Appl. Catal. A* **137**, 179 (1996).
432. C. Batiot, F.E. Cassidy, A.M. Doyle, and B.K. Hodnett, in: *3rd World Congress on Oxidation Catalysis* (R.K. Grasselli, S.T. Oyama, A.M. Gaffney, and J.E. Lyons, eds.), Studies in Surface Science and Catalysis Vol. 110, Elsevier Science: Amsterdam (1997), p. 1097.

GENERAL CONCLUSIONS

Selective oxidation using solid catalysts plays a central role in the chemical industry and, as the production cycle of many materials and commodities includes a selective oxidation process, it has also contributed significantly to modern manufacturing. More than half the products made using catalytic processes and nearly all the monomers used for the production of fibers and plastics are obtained through selective oxidation. The *1998 White Paper on Catalysis and Biocatalysis Technologies*[1] ranked catalytic selective oxidation processes or oxidative dehydrogenation first in its classification of the reactions and process innovations of greatest industrial importance for 21st-century catalysis. Selective oxidation processes thus constitute one of the central building blocks of the chemical industry, but there is still great need for innovation both in catalyst development (Chapter 3) and in the engineering of the processes, in particular in reactor design (Chapter 2).[2]

Technological challenges include increasing product selectivity, stereoselectivity, reducing undesirable by-products, minimizing energy consumption, utilizing and controlling exothermicities, designing catalysts for aqueous environments, and reducing process steps with multifunctional catalysts (Chapter 1).[2] These goals require not only a better understanding of the chemistry of the catalytic phenomena associated with selective oxidation catalysts (Chapter 8), but more integration between catalysts and reactor design (Chapter 2).[3] Recent major developments in this area [e.g., a new process for the oxidation of *n*-butane to maleic anhydride in a riser reactor (Chapter 4), selective oxidation at very short contact times (Chapter 5), a new process of formaldehyde synthesis in a combined fixed bed–monolith reactor (Chapter 1), a new process for the oxidation of *o*-xylene to phthalic anhydride in a composite bed reactor (Chapter 2)] derive from an integrated view of the catalyst, the reactor, and their relationship.

The economic potential of any improvements in this area is enormous. Dautzenberg,[3] in analyzing the challenges for R&D in catalysis, noted that improvement of selective oxidation processes has the greatest potential economic impact and that any expected selectivity improvement in the catalysts currently used for the seven large-scale petrochemical oxidation processes would lead to an annual saving in feedstock materials of approximately US\$ 1.4 billion worldwide.

The search for new catalysts and selective oxidation reactions has progressed slowly in recent years.[2] Much of the research activity has been focused on obtaining a further understanding of known catalytic materials more than toward innovative directions such as:

- Design of new materials based on advances in understanding the working state of solid catalysts (Chapter 8).
- Search for new classes of catalytic materials that can open a field of new applications rather than be just a solution for a single, specific problem (Chapter 3).
- Development of advanced catalysts that take effective advantage of new reactor options developed recently and are thus based on a combined solid state and reaction engineering approach (Chapter 2).

Nevertheless, the general thrust in this volume is that selective oxidation with solid catalysts is not a mature field of research, but rather quite an innovative area that requires new approaches to develop breakthrough technologies and reactions. More intensive study of novel materials, types of oxidants, and reactions as well as efforts to integrate reactor engineering and catalyst design are necessary.

The environmental factor has also been a driving force behind innovation over the last two decades (Chapter 1). The main characteristic of the technological progress of oxidation processes during these years has been the increasing complexity of the catalysts and the concomitant change in reactor technology to improve performance. Process simplification and flexibility have been the other two key words for innovation in selective oxidation processes.

Recent trends and developments in industrial selective oxidation technology can be summarized as follows:

1. The use of new raw materials and alternative oxidizing agents: Alkanes are increasingly replacing aromatics and alkenes as raw materials (Chapters 4 and 5). Phenol can be directly synthesized from benzene using N_2O and zeolite-type catalysts (Chapter 7). In liquid-phase oxidation, hydrogen peroxide is being used more frequently as an oxidizing agent in place of traditional oxygen transfer agents (Chapter 6). In some cases, the hydrogen peroxide is generated *in situ* (Chapter 7).
2. The development of new catalytic systems and processes: Heterogeneous rather than homogeneous catalysts are being employed (Chapter 1), and oxidative dehydrogenation is increasingly being used in place of simple dehydrogenation (Chapter 5). New processes are being developed that generate fewer or no undesired coproducts.

3. The conversion of air-based to oxygen-based processes in vapor phase oxidation to reduce polluting emissions (Chapter 2): Examples are: (i) the synthesis of formaldehyde from methanol, (ii) the ethene epoxidation to ethene oxide, and (iii) the oxychlorination of ethene to 1,2-dichloroethane.
4. Existing processes are being fine-tuned to improve each stage of the overall process (Chapter 1). Profit margins are being increased through changes in process engineering rather than through economies of scale (Chapter 2).

New developments are being investigated in other areas as well. The most significant examples are: (i) the selective oxidation of hydrogen to hydrogen peroxide using Pd-based catalysts, a reaction discovered at DuPont,[4] but now improved in terms of lifetime and safety, which permits the synthesis of hydrogen peroxide at greatly reduced cost with respect to the commercial anthraquinone route (Chapters 3 and 7), and (ii) the catalytic partial oxidation of methane to syngas, in place of the thermal noncatalytic process.

Catalytic science has been the driving force for these innovations. In addition to the development of new catalytic materials (Chapter 2), new fields of application of solid catalysts (Chapter 6), and new ways of activating oxygen or hydrocarbons (Chapters 7 and 8), the understanding of complex surface catalytic phenomena during selective oxidation reactions has provided new opportunities for tuning the surface reactivity (Chapter 8) thus increasing the selectivity or devising single-step reactions to replace others that would require several and more polluting steps. Both these aspects are central in the development of a sustainable chemistry for the 21st century.

Thus several current challenges are driving intense research effort in the field of selective oxidation and basic and applied research can lead to important breakthroughs, but innovation requires that the study be addressed from a range of different perspectives. We hope and trust that this book has provided ideas and guidelines in this direction.

REFERENCES

1. R. Bloksberg-Fireovid and J. Hewes, *1998 White Paper on Catalysis and Biocatalysis Technologies*, NIST (US National Institute of Standards and Technology) Advanced Technology Program Report.
2. G. Centi and S. Perathoner, *Current Opinion in Solid State & Materials Science* **4**, 74–79 (1999).
3. F.M. Dautzenberg, *CATTECH* **3**, 54–63 (1999).
4. L.W. Gosser and M.A. Paoli, *US Patent* 5,135,731 (1992).

INDEX

Two-phase processes. *See also* Phase transfer
     catalysts
  propene oxide, 108
  using heteropolyacids, 307

UOP, propane dehydrogenation, 206

Vanadium
  antimony oxides, 173
  on MgO, 232
  oxidative dehydrogenation, 222
  phosphorus oxides, 143
  role in solid Wacker catalysts, 312
  role on species in *o*-xylene ammoxidation,
     467
  on silica, 221
  silicalite, 130
  on tin oxide, 223
  V/Nb/Mo oxides, 225
Vanadium–Antimony oxides
  comparison with Nb/Sb oxides, 251
  mechanism in propane ammoxidation, 186,
     462
  microstructure, 184
  nature of phases, 176
  non-stoichiometry, 173
  properties in propane ammoxidation, 174
  structure, 175
Vanadium phosphorus oxides
  activation, 153
  alkene vs. alkane oxidation, 170
  catalysts for maleic anhydride, 147
  dopants, 148, 157
  "equilibrated" catalyst, 155
  ethane oxidation, 245
  in situ restructuring, 166

Vanadium phosphorus oxides (*continued*)
  microkinetic, 169, 406
  microstructure, 161
  modeling surface reaction, 402
  n-pentane oxidation, 266
  properties, 162
  properties in relation to reactor, 168
  reactivity of surface oxygen species, 398
  role of P/V ratio, 156
  structure, 159
  surface topology, 164
  synthesis, 151
Vanadyl pyrophosphate. *See also* Vanadium
     phosphorus oxides
  n-butane oxidation, 150
  comparison with V_4O_9, 245
  propane ammoxidation, 457
  properties, 162
  structure, 158
  synthesis, 151
Vinyl acetate
  by ethene acetoxylation, 352
  model of reaction, 354

Wacker
  in electrochemical reactor, 331
  heteropolyacid catalysts, 303
  mechanism, 305
  model reaction in thin liquid films, 353
  role of coadsorbed species, 423
  solid catalysts, 310
  in supported thin liquid films, 350

Zeolites
  framework substitution, 289
  host materials for oxide, 129, 131
  isomorphic substitution, 126, 287